Organic Reactivity: Physical and Biological Aspects

Organic Reactivity: Physical and Biological Aspects

Edited by

Bernard T. Golding, Roger J. Griffin, and Howard Maskill
Department of Chemistry, University of Newcastle upon Tyne, UK

The Proceedings of the 4th European Symposium on Organic Reactivity and the 2nd Newcastle Meeting on Molecular Mechanisms in Bioorganic Processes, organized by the Royal Society of Chemistry Perkin Division and held in Newcastle upon Tyne, UK, on 11–16 July 1993

The cover diagram shows the active site of horse liver alcohol dehydrogenase complexed with NAD^+ and a substituted benzyl alcohol (courtesy of Brian Bahnson, Dan Peisach, and Judith Klinman). The three-dimensional coordinates used to generate this picture were taken from S. Ramaswamy, H. Eklund, and B.V. Plapp, *Biochemistry*, 1994, **33**, 5230–5237. See also Figure 2 on page 44.

Special Publication No. 148

ISBN 0-85404-710-7

A catalogue record for this book is available from the British Library

© The Royal Society of Chemistry 1995

All Rights Reserved
No part of this book may be reproduced or transmitted in any form or by any means – graphic, electronic, including photocopying, recording, taping, or information storage and retrieval systems – without written permission from The Royal Society of Chemistry

Published by The Royal Society of Chemistry,
Thomas Graham House, Science Park, Milton Road,
Cambridge CB4 4WF, UK

Printed in Great Britain by Bookcraft (Bath) Ltd

PREFACE

This book contains plenary and section plenary lectures from an International Symposium ('Organic Reactivity: Physical and Biological Aspects') held at the University of Newcastle upon Tyne, 11 - 16 July 1993. In addition, two oral contributions and one poster, selected for their excellence, are included. The aim of the conference was to bring together physical organic and bioorganic chemists in order to introduce new bioorganic problems to physical organic chemists and to demonstrate the relevance of new thinking in physical organic chemistry to bioorganic chemists. Hence, this book is a timely statement at the interface between physical organic and bioorganic chemistry. It should be a useful teaching aid and provide pointers to future research.

The sessions on Thursday 15 July were designated the 'Ingold Symposium', to commemorate the 100th anniversary of the birth of Sir Christopher Ingold. The speakers in this 'Mini-symposium' were chosen to reflect a variety of current mechanistic themes and included Professor Alwyn Davies, the Royal Society of Chemistry's Ingold Lecturer for 1993.

We thank speakers for their lectures and manuscripts, and also the poster presenters. All contributions have been converted into a common format (Microsoft Word), although authors' original diagrams have mainly been used. We are grateful to Dr Christine Bleasdale for translating some documents into Word files, and to Lindsey Brown for assistance with the re-typing and formatting of documents. The book has been organised into two sections: Section A (Biological Aspects of Organic Reactivity) contains lectures concerned with enzyme mechanisms, whilst Section B (Physical Aspects of Organic Reactivity) comprises lectures on other aspects. In both sections the plenary lectures come first.

We wish to thank the many people who assisted with the organisation of the Symposium, especially Ms Paula Elliot and Dr John Gibson (Royal Society of Chemistry) and our postgraduate students.

Finally, we very much regret to have to record that Elena Peña, one of the section plenary speakers, died on the 17th December, 1993.

Bernard Golding, Roger Griffin, and Howard Maskill
July 1994

CONTENTS

Section A: Biological Aspects of Organic Reactivity   1

Studies on Bioorganic Reactions and Mechanisms   3

Ronald Breslow

1. Introduction
2. RNA cleavage catalyzed by imidazole buffer
3. RNA cleavage catalyzed by morpholine buffer
4. Further discussion of the mechanistic conclusions
5. Bifunctional catalysis by cyclodextrin bis-imidazoles
6. Properties of isomeric DNA with a 2',5" linkage

Mechanisms Selected by Enzymes   25

Sir John Cornforth

Hydrogen Tunneling in Enzyme Reactions   38

Judith P Klinman

1. Introduction
2. Results and discussion
3. Summation and future directions

Cobalamin-Dependent Methionine Synthase: Dissection of a Large   58
Protein into Functional and Structural Domains

James T Drummond, Sha Huang, and *Rowena G Matthews*

1. Introduction
2. Model studies of the methionine synthase reaction
3. Structure of the enzyme
4. Catalysis of the physiological reaction
5. Inactivation of methionine synthase by nitrous oxide

Biomimetic Reactions with Hydrophobic Vitamin B_{12} 73

Yukito Murakami, Yoshio Hisaeda, and Teruhisa Ohno

    1   Introduction
    2   Hydrophobic vitamin B_{12}
    3   Electrochemical reactions mediated by hydrophobic vitamin B_{12}
    4   Artificial enzyme composed of hydrophobic vitamin B_{12} and synthetic bilayer membrane

Chasing the Enzymes of Alkaloid Biosynthesis   89

Meinhart H Zenk

    1   Function and activity of benzylisoquinoline alkaloids
    2   The biosynthesis of isoquinoline alkaloids
    3   Future aspects of research in the alkaloid field

Mechanistic and Structural Features of the Picornaviral 3C Protease   110

Racheli Kreisberg, Michael Shocken, Dietmar Schomburg, and *Dorit Arad*

    1   Introduction
    2   Results and discussion

Neighbouring Group Participation: A Model for Enzymic Catalysis?   123

Keith Bowden

    1   Introduction
    2   Results and discussion
    3   Conclusions

Enzyme-Catalysed Hydroxylations of Aromatic Substrates: Stereochemical and Mechanistic Aspects   130

Derek R Boyd, Narain D Sharma, and Howard Dalton

    1   Introduction
    2   Discussion
    3   Conclusions

Iron-Sulfur and Flavin-Dependent Dehydrations in Anaerobic Bacteria   140

Klaus Bendrat, Ulrich Eikmanns, Antje E M Hofmeister, Anne-Grit Klees, Uta Müller, Uwe Scherf, and *Wolfgang Buckel*

    1   Introduction
    2   (R)-2-hydroxyacyl-CoA dehydratases
    3   4-Hydroxybutyryl-CoA dehydratase
    4   5-Hydroxyvaleryl-CoA dehydratase
    5   Conclusions

Enzyme Engineering by *In Vitro* Selection Using Genetic and Organic Tools   149

Patrice Soumillion, Pascale Sartiaux, Michèle Bouchet, Jacqueline Marchand-Brynaert, and *Jacques Fastrez*

    1   Introduction
    2   Results
    3   Discussion

Isotope-Aided NMR Studies of Protein-Ligand Interactions   161

James Feeney

    1   Introduction
    2   Dihydrofolate reductase
    3   Conclusion

Sequential Electron Transfer and Base Catalysis in the N-Dealkylation of Amines by Cytochrome P450 Enzymes

185

F Peter Guengerich

1. Introduction
2. Sequential electron transfer in P450 reactions
3. Comparison of P450s and peroxidases
4. Formation and decomposition of N,N-dialkylaniline N-oxidases by P450s

Structure and Mechanism of Porphobilinogen Deaminase

196

Peter M Jordan

1. Introduction
2. The discovery of preuroporphyrinogen
3. Cloning and over-expression of *Escherichia coli* porphobilinogen deaminase and the discovery of the dipyrromethane cofactor
4. Enzyme intermediate complexes with one, two, three and four molecules of substrate covalently linked to the enzyme
5. X-ray structure of *E coli* porphobilinogen deaminase
6. Key amino acids in the catalytic cleft
7. The catalytic reaction
8. The role of the enzyme in catalysis
9. The chemistry of porphobilinogen and the influence of the enzyme
10. Summary

Organometallic B_{12}-Chemistry

209

Bernhard Kräutler

1. Introduction
2. Effects of the nucleotide function
3. Conformational effects of the organic ligand

Hydrolytic Catalysts as Metalloenzyme Models 223

Paolo Scrimin, Paolo Tecilla, and *Umberto Tonellato*

1. Introduction
2. Functionalized ligand surfactants as enzyme models
3. The leaving group effect in the cleavage of picolinate esters employing ligands 2
4. Enantioselectivity in the cleavage of α-amino acid esters

Section B: Physical Aspects of Organic Reactivity 233

Gas Phase Ion Chemistry: Kinetic and Mechanistic Aspects 235

Fulvio Cacace

1. Introduction
2. Proton migration in aromatic systems
3. Aromatic nitration
4. Aromatic silylation
5. The cyclohexyl cation
6. Conclusion

Hydrogen-ene and Metallo-ene Reactions 263

Alwyn G Davies

1. Metals as hydrogen equivalents
2. The mechanism of the H-ene reaction
3. The Schenk and Smith rearrangements
4. The reaction of metalloenes with enophiles
5. Conclusion

Chemistry by Computer: A Theoretical Approach to Structure and Mechanism   278

Leo Radom

    1   Introduction
    2   Methods
    3   Applications
    4   Recent case studies
    5   Concluding remarks

Direct Studies of the Reactivities of Short-Lived Carbocations   301

Robert A McClelland

    1   Introduction
    2   Flash photolysis studies of carbocations - requirements
    3   Indentification of transients as carbocations
    4   The azide 'clock' method
    5   Effect of solvent on lifetimes of carbocations
    6   Direct observation of cationic intermediates in Friedel-Crafts alkylation
    7   Photoprotonation of aromatic compounds by 1,1,1,3,3,3-hexafluoroisopropyl alcohol
    8   Lifetime of the phenyl cation in 1,1,1,3,3,3-hexafluoropropyl alcohol
    9   Observation of the initial step in the cationic polymerization of styrene
    10   Fragmentation of bicumene
    11   C-C Fragmentation of a cyclohexadienyl cation - a reverse Friedel-Crafts reaction
    12   Silylsubstituted cyclohexadienyl cations - a photochemical protodesilylation
    13   Summary

Nitrosation and Nitric Oxide Chemistry 320

D Lyn H Williams

1. Introduction
2. Reagents for effecting nitrosation
3. Aliphatic C-nitrosation
4. Nitric oxide chemistry
5. Nitrovasodilators

Formation and Stability of Reactive Intermediates of Organic Reactions in Aqueous Solution 334

Tina L Amyes, John P Prichard, and Vandanapu Jagannadham

1. Formation and stability of simple oxygen ester and simple thiol ester enolates
2. Kinetic and thermodynamic stability of α-oxygen and α-sulfur stabilized carbocations

Structure Reactivity Effects in the Aqueous Chemistry of Alkane Diazoates 351

Jian Ho, Jari I Finneman, and James C Fishbein

1. Introduction
2. Primary anti-alkane diazoates
3. *Syn* versus *anti*: reactivity and chemistry
4. Summary

Recent Advances in the Application of Cross-Interaction Constants 361

Ickchoon Lee

1. Introduction
2. Non-interactive (or isoparametric) phenomenon
3. Sign of ρ_{XZ} in S_N2 TS
4. Magnitude of ρ_{ij}
5. Non-interactive and TS imbalance phenomena
6. Kinetic solvent isotope effect (KSIE)
7. Temperature and solvent effects on ρ_{ij}

Ambident Reactivity in Reactions Involving the Nitroso Group   374

*M Elena Peña**

    1   Nitrosation of ambident nucleophiles
    2   Reactions with ambident electrophiles
    3   Reaction of ambident nucleophiles with ambident electrophiles

Self-Assembly of [n]Rotaxanes   387

Martin Bělohradský, Douglas Philp, Françisco M Raymo, and *J Fraser Stoddart*

    1   Preamble
    2   Introduction
    3   The threading approach
    4   The slipping approach
    5   Conclusions

The Superelectrophilic Character of the 4,6-Dinitrobenzofuroxan Structure   399

François Terrier

    1   Introduction
    2   σ-Complexation of DNBF by very weak carbon bases
    3   Concluding remarks

Ion Pairs and Ion-Molecule Pairs in Solvolytic Substitution, Elimination and Rearrangement Reactions   415

Alf Thibblin

    1   Introduction
    2   Elimination *via* an irreversibly formed ion pair or ion-molecule pair $(D_N^{\#*}A_{xh}D_H)^1$
    3   Elimination and substitution *via* a reversibly formed ion pair $(D_N^*A_{xh}D_H^{\ddagger})^3$
    4   Rearrangement and substitution reactions *via* ion-molecule pairs

---

* Deceased, 17th December 1993

The Course of Oxidation Processes of Organic Substrates Mediated by Mo(VI) and W(VI) Polyoxoperoxo Complexes 429

Francesco P Ballistreri, *Gaetano A Tomaselli*, and Rosa Maria Toscano

- 1 Introduction
- 2 Results and discussion
- 3 Conclusions

Theoretical Modelling of Mechanisms for Glycoside Hydrolysis 437

John A Barnes and *Ian H Williams*

- 1 Introduction
- 2 Modelling of AMP hydrolysis
- 3 Hydrolysis of glucopyranosides
- 4 Concluding remarks

Selected Poster

Rotational Isomerism of Disubstituted Benzenes in the Alkylphenyldi(1-adamantyl)methanol Series 444

John S Lomas and V Bru Capdeville

Subject Index 445

SECTION A

Biological Aspects
of Organic Reactivity

STUDIES ON BIOORGANIC REACTIONS AND MECHANISMS

Ronald Breslow

*Department of Chemistry, Columbia University
New York, New York 10027, USA*

ABSTRACT

In this chapter, several related topics are discussed. First is a study on the cleavage of RNA by imidazole buffers. In this work a novel mechanism was deduced that led to the proposal of a novel mechanism for the enzyme ribonuclease A. The conclusions from this work are further supported by studies on RNA cleavage catalyzed by morpholine buffer. In both these cases a critical piece of evidence was the finding that these buffers catalyze the isomerization of 3',5" linked RNA to its 2',5" isomer at the same time that they catalyze cleavage. The isomerization reaction shows negative catalysis by one buffer component, an important aspect of the mechanistic evidence.

The mechanistic conclusions have also guided the synthesis of a mimic of ribonuclease whose geometry results in improved properties. The set of bifunctional acid-base catalysts prepared in this work also is used to study the enolization of a ketone; only one of the geometric isomers of the catalyst set is an effective bifunctional catalyst of this process.

Finally, a consideration of the properties of the 2',5" isomers produced in the mechanistic study has led to the synthesis of some DNA isomers with the unnatural 2',5" linkage. They differ strikingly from natural DNA, with its 3',5" linkage, and in ways that are explicable in terms of the predictions from computer modelling of the molecules.

1   INTRODUCTION

Enzymes differ in several important ways from simple chemical catalysts. For one, they selectively bind their substrates so as to induce high rates because of proximity and favorable entropy. For another, enzymes normally perform bifunctional or multifunctional catalysis, in which two or more catalyst groups operate *simultaneously*. The advantage of such a process is that unstable intermediates can be avoided. For instance, in the conversion of a ketone to its enol, simultaneous catalysis would use a base to remove a proton from the α-carbon while an acid group is putting a proton on the carbonyl oxygen atom. In this way the ketone is directly converted to the enol, avoiding a less stable enolate ion that would have been an intermediate if catalysis had been *sequential*, with the base operating first.

One of the well-known enzymes in which an acid and a base group cooperate in catalysis is ribonuclease A. This enzyme cleaves RNA by a two-stage process, in which an imidazole ring and an imidazolium ring of histidine residues 12 and 119 convert the RNA to a 2',3'-cyclic phosphate ester in one step, and then again cooperate to hydrolyze this cyclic phosphate in the second step. There is also some electrostatic help from a protonated lysine residue.

The classic mechanism for this process - that appears in most textbooks - is shown in Figure 1. Here the cyclic phosphate is formed by an ester exchange in which the basic imidazole ring deprotonates the attacking 2' hydroxyl group, while the acidic imidazolium ion protonates the leaving 5" oxygen of the next residue. Some years ago we initiated studies of the bioorganic chemistry related to this process.[1-11] We addressed several questions.

1   Can simple imidazole buffer, containing imidazole and imidazolium ion, catalyze the cleavage of RNA?
2   If so, can we use modern kinetic methods to learn the mechanism by which this occurs?
3   Will this mechanism give us any insight into the mechanism used by the enzyme ribonuclease A and related enzymes?
4   Will our mechanistic results help us design a mimic of this enzyme?

The answer to all these questions has proven to be yes. Furthermore, the bifunctional catalyst systems we have devised are able to perform other reactions as well, including in particular the simultaneous bifunctional conversion of a ketone to its enol. Finally, our work on the reactions of RNA with such catalysis has stimulated us to investigate an isomer of DNA in which the linkage is 2',5" instead of the normal 3',5". The properties of this DNA isomer give us considerable insight into the special character of natural DNA.

2   RNA CLEAVAGE CATALYZED BY IMIDAZOLE BUFFER

In our first study, we devised a technique to determine the number of cuts that had been made in an RNA strand,[1] and used this technique to follow the cleavage of RNA catalyzed by imidazole buffer.[2] Interestingly, we saw a bell-shaped curve (Figure 2) in which the maximum rate occurred when both the basic imidazole and acidic imidazolium ion were present. This indicates that the best mechanism - with the fastest rate - uses *both* catalysts. We also saw that this bifunctional catalysis was not mandatory, since there was an appreciable rate at each end of the plot of Figure 2 when only one of the catalysts is present.

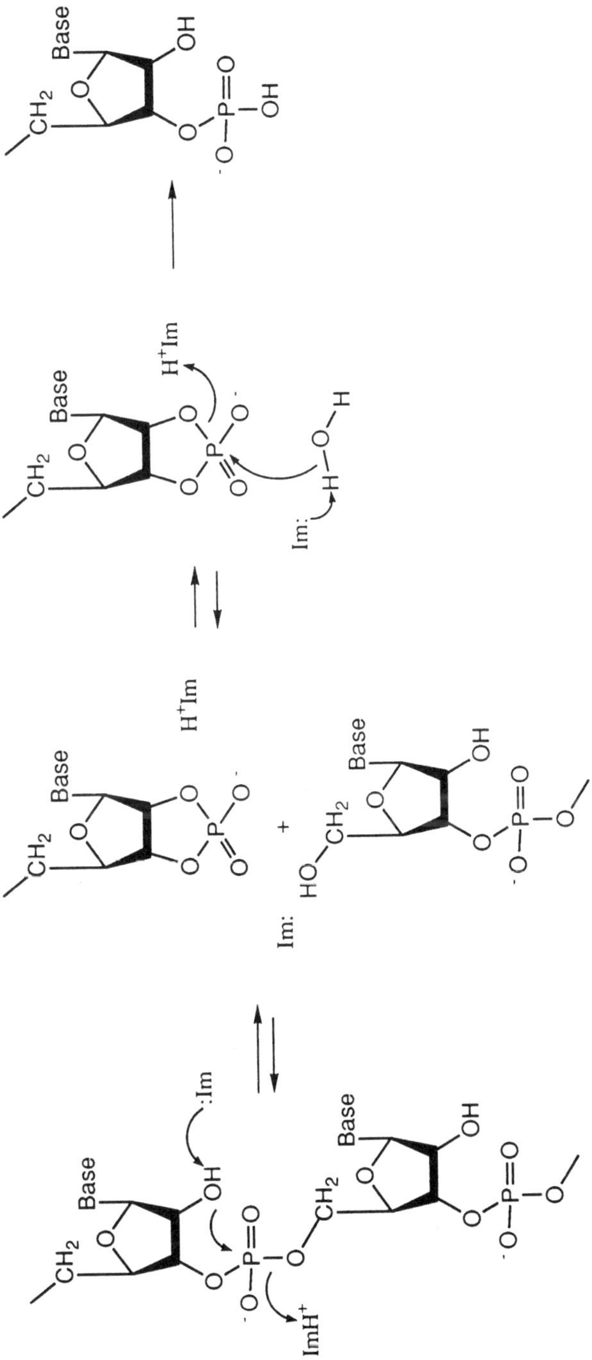

Figure 1: The classical mechanism for ribonuclease action.

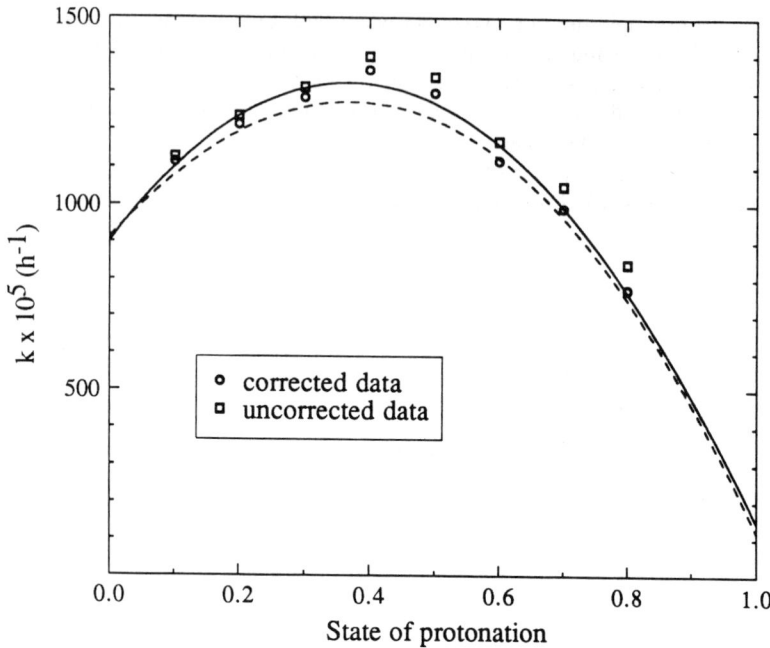

Figure 2: The first order rate constant for cleavage of polyU by imidazole buffers at 1.3 M as a function of the protonation state ($BH^+/[BH^+ + B]$). Both the uncorrected data and the corrected data, obtained by subtracting the observed rate when [buffer] is extrapolated to zero, are shown. The solid curve is the theoretical curve from fitting Equation 1, modified with the addition of terms for catalysis as well by ImH^+ or by Im, to the corrected points. The data are from reference 2. The dashed curve uses parameters common to those from Figure 3, as described there.

In work of this kind it is essential to distinguish the reactions catalyzed by the buffer species from those catalyzed by H^+, OH^-, and water; as the buffer is changed, the pH will change so non-buffer catalysis will also be affected. We solved this, as is always done, by extrapolating the buffer concentration to zero at each buffer ratio, to obtain the rate correction from reaction at that pH in the absence of buffer. As Figure 2 shows, with the rather high 1.3 M buffer concentration used, the non-buffer correction was only 8% or less so a bell-shaped curve was seen whether one performed this subtraction or not.[2,9] As we will describe later, with the mechanism probably used by this system a simple subtraction of the buffer = zero rate at each pH is not precisely valid. However, the correction is so small that this nuance is not important.

In principle the rate maximum when both buffer components are present could indicate a simultaneous bifunctional mechanism, as it does for the enzyme. However, such a termolecular reaction of unattached species is quite unlikely. Instead, we are dealing with a *sequential* bifunctional mechanism in which first one catalyst, then the other, acts to cleave the RNA. The evidence for this is shown in Figure 3, a plot of the observed rate of cleavage vs the total buffer concentration at two different buffer ratios.

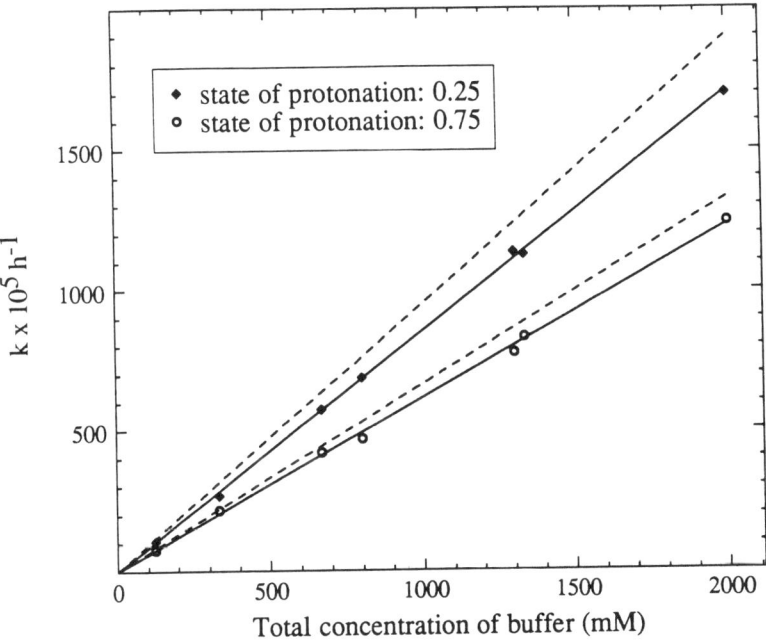

Figure 3: The corrected rate constant for cleavage of polyU by imidazole buffers as a function of the total buffer concentration. The solid points are for 3/1 Im/ImH$^+$ while the open circles are for 3/1 ImH$^+$/Im. The solid curves are calculated by fitting Equation 1, modified as in Figure 2, to the data from reference 2. These data were collected later than those of Figure 2, and seem to differ in a regular way by ca. 10%, possibly reflecting some small difference in conditions. I also show in Figures 2 and 3 some dashed line curves that are the best fit to the equation if I ignore the apparent systematic small differences in data and set all parameters the same for the two plots.

The plot is linear, and various control reactions showed that there was no complexing that could have distorted the picture. Thus only one buffer species is present in the transition state of the reaction. There is a two-step mechanism in which the substrate is converted by the first catalyst to an *intermediate*, and this is then converted to the product by the second catalyst. Equation 1 describes this two-step process.[2,10] Of course a more complex equation is needed to handle the real situation, in which monofunctional catalysis was also observed.

$$S \underset{k_{-1}}{\overset{\substack{k_1 \\ \text{cat 1}}}{\rightleftarrows}} I \xrightarrow[k_2]{\text{cat 2}} P$$

$$\frac{dP/dt}{[S]} = \frac{k_1 k_2 [1][2]}{k_2[2] + k_{-1}[1]} \qquad (1)$$

The starting material is a linear phosphate diester, and the product is a cyclic phosphate diester whose formation accompanies cleavage of the chain. Any later hydrolysis of this cyclic phosphate ester would play no role in the rate observed, which simply reflects chain cuts from the ester interchange process. The intermediate must be a phosphorane with five-coordinated phosphorus. Detailed kinetic arguments[4] show that it must be a phosphorane *mono*anion. Thus the reaction sequence is as shown in Figure 4.

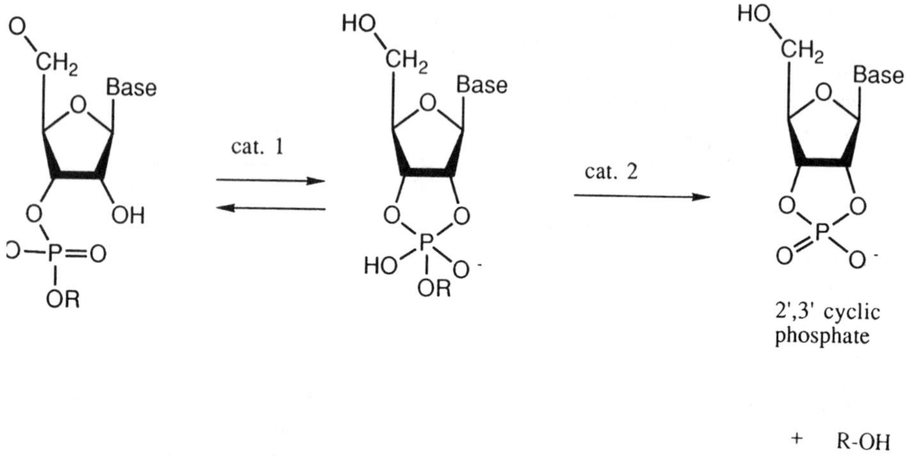

Figure 4: The general mechanism involved in buffer-catalyzed cleavage of RNA.

From our simple kinetic studies we could not tell which was the first catalyst; equation 1 has significant symmetry. A theoretical treatment by Taira[12] favored one alternative, with the base as the first catalyst and the acid as the second. However, we developed a method to establish the sequence, and it was the reverse: first the acid catalyst converted the substrate to a phosphorane anion, then the base converted this to the cleaved product. This mechanism was established by studies[4,5] on simple dinucleotides - 3',5"-uridyluridine (3',5"-UpU) and 3',5"-adenosyladenosine (3',5"-ApA). These also cleaved with imidazole buffer catalysis to the simple cyclic phosphate 2',3'-cyclic uridylic acid and uridine - and the corresponding compounds in the adenosine series. However, with these simple dinucleotides we could also detect an additional process in our hplc analysis. The buffer catalyzed the conversion of 3',5" linked dinucleotide to the 2',5" linked isomer at the same time that it catalyzed cleavage. This isomerization process let us resolve the kinetic ambiguity in the cleavage mechanism.

In contrast to the cleavage reaction, with its sequential bifunctional catalysis, the isomerization was catalyzed only by the *acid* component of the buffer.[4,5] Such an isomerization is expected to involve a phosphorane intermediate, because of the stereochemistry of nucleophilic substitution reactions at phosphorus.[13] The attacking and departing groups must occupy *apical* positions in the phosphorus trigonal bipyramid. Thus there cannot be a simple direct migration of phosphorus from one oxygen to the other, because the first phosphorane formed will have the C-2 oxygen apical but the C-3 oxygen equatorial. A pseudorotation of a phosphorane reaction intermediate is required to interconvert their character, so the C-3 oxygen can depart as an apical group. Since we knew that the cleavage also involves a real phosphorane intermediate, it was attractive to postulate[4] that the two processes went through the *same intermediate*, branching from that intermediate either by base-catalyzed cleavage or by migration that did not require base catalysis.

Other studies established that indeed the two reactions did involve a common intermediate. The evidence was the finding of 'negative catalysis' of the isomerization process.

$$\frac{k_{isomerization}}{\text{of UpU}} = \frac{k_1 k_3 [BH^+] + k''_w k_3}{k_{-1}[BH^+] + k_2[B] + k_3 + k_w} \qquad (2)$$

B = imidazole or morpholine

$$\frac{k_{cleavage}}{\text{of UpU}} = \frac{k_1 k_2 [BH^+][B] + k'_w}{k_{-1}[BH^+] + k_2[B] + k_3 + k_w} + \frac{k'[B] + k''[BH^+]}{k_{-1}[BH^+] + k_2[B] + k_3 + k_w} \qquad (3)$$

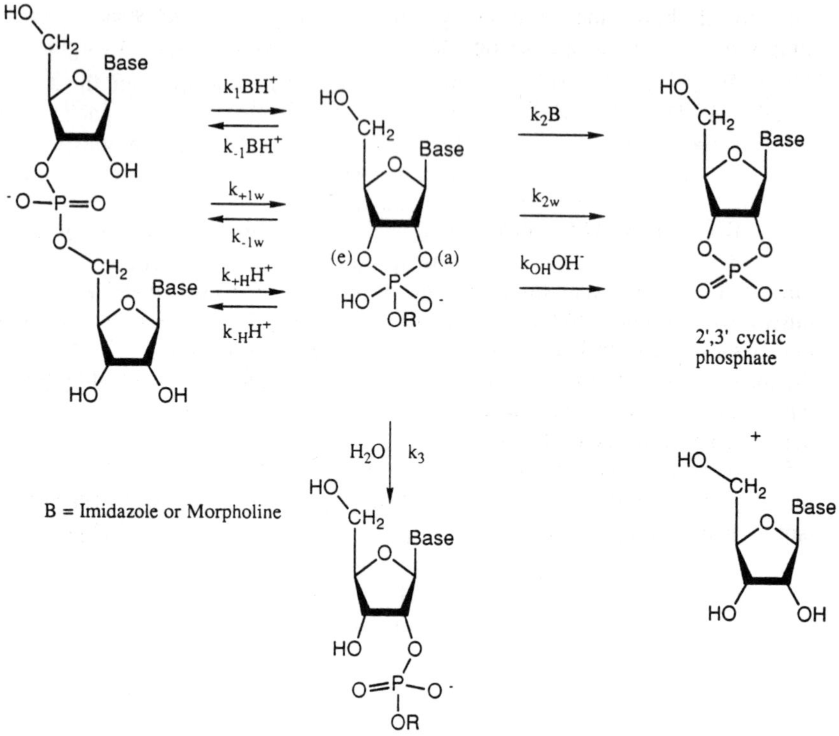

Figure 5: The detailed kinetic scheme for cleavage and isomerization of ribodinucleotides catalyzed by buffers and by solvent species.

Consider the complete proposed mechanism (Figure 5) of the cleavage and isomerization reactions, formulated as involving a common intermediate phosphorane anion with branching along either a cleavage path or an isomerization path. The kinetic equations for this mechanism are Equation 2 and Equation 3. As Equation 2 shows, the isomerization path is catalyzed by the buffer acid BH^+ but the buffer base B appears only in the denominator of the rate expression. Thus it is predicted that increasing the concentration of BH^+ will speed up the isomerization but increasing the concentration of B will slow it. The physical reason is that B promotes the cleavage leg of the mechanism, catalyzing conversion of the phosphorane intermediate to the cleavage products. Since the intermediate is at a low steady state concentration, this catalysis by B will *decrease* the concentration of the intermediate. Such a decrease slows the observed rate of the isomerization process.

This was observed.[5] When we kept $[BH^+]$ constant, but increased [B], the isomerization reaction slowed; the cleavage reaction accelerated, since in Equation 3 the [B] appears as a multiplier in the numerator, not just as one term in the denominator. Of course holding $[BH^+]$ constant, while increasing [B], leads to a pH increase. We corrected for the rate effect of this pH change by extrapolating [buffer] to zero, but there is a question about such a procedure. As

Equation 2 shows, the catalyses by buffer and by non-buffer species are not simply additive, so simple extrapolation to zero [buffer] and subtraction is not necessarily a way to isolate the buffer-only process. For this and other reasons, we decided to look for related evidence in which no such pH correction was needed.

3 RNA CLEAVAGE CATALYZED BY MORPHOLINE BUFFER

Since the negative catalytic effect was seen with the basic component of the buffer, we examined[9,11] a buffer - morpholine and morpholinium ion - in which that basic component had an even higher pK than did imidazole. We also kept the buffer ratio constant but changed [buffer], keeping the ionic strength constant with added NaCl. Under these conditions the pH did not vary, so no pH correction was needed. Finally, we looked at high B/BH^+ ratios, so that the effect of the basic buffer component would be maximized. As Figure 6 shows, under these conditions we saw a negative catalytic effect on the rate of isomerization of 3',5"-UpU to 2',5"-UpU simply as a result of increasing [buffer].

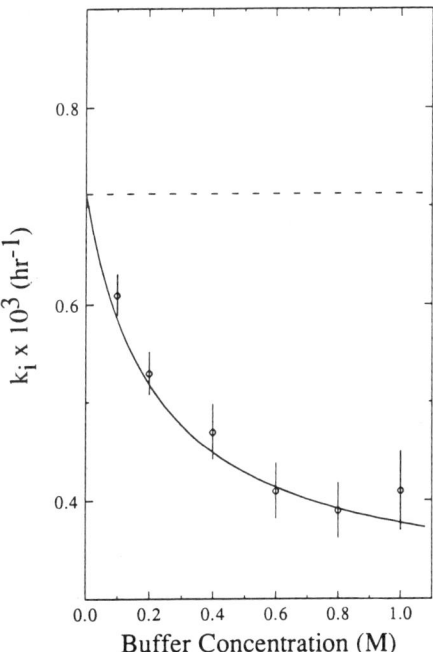

Figure 6: The pseudo-first-order rate constant for isomerization of 3',5"-UpU to 2',5"-UpU at 80 °C and pH 9.3, catalyzed by 9/1 morpholine/morpholinium ion, varying the buffer concentration but maintaining constant ionic strength at 0.2 M with NaCl. The curve was generated by fitting Equation 2 to these points.

In the early part of the curve the rate dropped, then leveled off. This is the behavior expected from Equation 2, and the curve in Figure 6 is simply that calculated from Equation 2. Increasing [buffer] at first leads to an increase in the denominator of Equation 2. The numerator also increases, but with less effect since the B/BH$^+$ ratio is 9/1 and the high pK of morpholine makes it a good base catalyst and BH$^+$ a relatively weak acid catalyst. As [buffer] nears 1.0 M, it dominates the non-buffer terms in the denominator so the curve levels. At concentrations higher than 1.0 M buffer, there was a rate rise as a new process set in; this new process is discussed in detail elsewhere.[11]

We saw the rate decrease with increasing [buffer] at the 9/1 B/BH$^+$ buffer ratio, and also at a 95/5 ratio, but not at a 4/1 ratio.[11] Thus the negative effect is indeed being contributed by the basic buffer component, as we had concluded for the earlier imidazole study.

4 FURTHER DISCUSSION OF THE MECHANISTIC CONCLUSIONS

Some criticism of the early work on this problem merits discussion. One critic[14] formulated a mechanism that he ascribed to us, but this mechanism appears nowhere in our papers nor is it implied by us. As has been discussed elsewhere,[11] we wrote equations for the buffer-only catalysis that had terms like the k_w of Equation 2 in the denominator, but no corresponding water term in the numerator. These equations resulted simply from subtracting an approximate correction for the non-buffer catalysis, as our paper[4] clearly indicates. With such a subtraction, terms like k'_w disappear from the numerator, but of course no terms disappear from a common denominator such as that of Equation 2.

This critic also objected to our characterization of the observation that increasing the concentration of the basic buffer component led to a decreased isomerization rate. We called the observed coefficient of this buffer component an experimental rate constant, and pointed out that it was negative in this case.[5] Of course idealized rate constants of theoretical rate treatments can never be negative, and as Equation 2 shows the term $k_B[B]$ has a negative effect on the rate because it appears in the denominator of the equation, not because k_B is negative. However, we took the position that one should distinguish between an experimental observation and a theoretical treatment; experimentally, [B] had a negative coefficient in its contribution to the rate of isomerization. Many chemists do not like the term 'experimental rate constant' to describe such an observed negative coefficient.

A second critic[15] has asserted that the cleavage reaction and the isomerization reaction do not proceed through a common intermediate, although he did not offer an alternative mechanism for either process. As we have shown elsewhere,[11] all our data - including our early data - fit the predictions of our

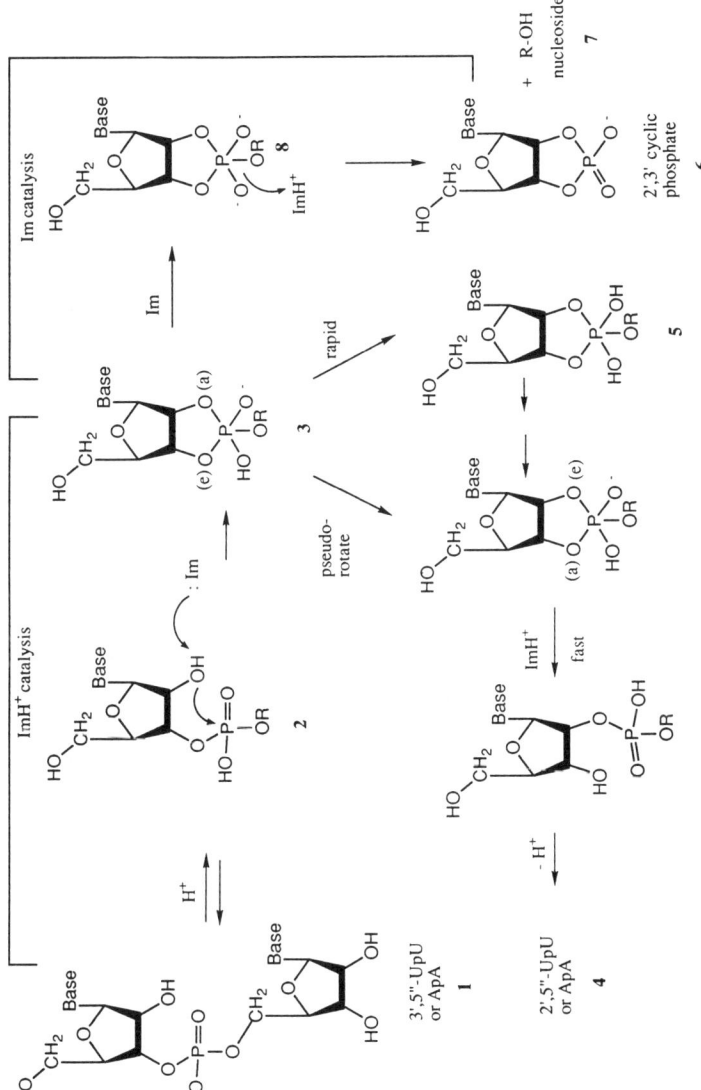

Figure 7: The detailed mechanism by which sequential bifunctional buffer catalysis of the cleavage and isomerization of ribodinucleotides occurs.

proposed mechanism, including an equation proposed by the critic as a test of a common intermediate. With the new data on morpholine catalysis that confirms the earlier picture with imidazole catalysts, there is really no doubt that the mechanism of Figure 5 is correct in its essentials. It has been a very useful guide in other work.

First of all, we must consider the details of the catalytic steps that this kinetic scheme implies. In the first step, the substrate **1** is converted to the phosphorane anion **3** with catalysis by BH^+, which is ImH^+ in our earlier work. This conversion requires that a proton be placed on the phosphate oxygen, and another one removed from the attacking OH group. This conversion is catalyzed by BH^+, but not necessarily in one step. The most reasonable scheme, shown in Figure 7, is reversible protonation of the phosphate anion to form a neutral phosphoric acid derivative **2**, then attack on that by the 2'-OH group with catalyzed removal of a proton by the buffer base. Such a scheme - reversible protonation and then catalysis by a general base - is the specific acid/general base alternative mechanism for BH^+ catalysis.

The fate of the phosphorane anion intermediate is interesting. There are three paths open to it, and each shows a different mode of catalysis. For one, it can reverse its formation, going back to starting material. According to the mechanism of Figure 7, this would of course have to be an exact reversal of the formation process. That is, BH^+ would have to catalyze the release of the C-2 oxygen by general acid protonation to form an OH group, and then the resulting phosphoric acid derivative **2** would have to lose a proton. Both the forward and the reverse steps were catalyzed by BH^+, but in a different fashion. The forward reaction used the specific acid/general base mechanism, but the reverse reaction uses the general acid mechanism. This distinction in mechanism for BH^+ catalysis in the forward and reverse directions is quite common and reasonable.

A second process open to the phosphorane **3** is conversion to the 2',5" isomer **4**. This requires a pseudorotation, and we have argued that this will probably be faster if **3** is fully protonated to **5**. Under most conditions this will be a fast process, not rate limiting, and after pseudorotation the conversion to **4** will also be fast. Thus under most conditions no catalysis by buffer is actually observed for the isomerization, whose slow step is the pseudorotation.

The third path available is cleavage of **3** to the cyclic phosphate **6** and the second fragment **7**. This is catalyzed by the buffer base B. The only sensible function of such a base is the deprotonation of **3** to the phosphorane dianion **8**, which then fragments. This fragmentation could be assisted if BH^+ protonates the leaving oxygen, but such a process will be invisible kinetically. As we have pointed out elsewhere,[10,16] if a B is converted to BH^+ in the first step of a catalytic sequence, and the BH^+ is then used in a later step, the BH^+ will not appear in the kinetics.

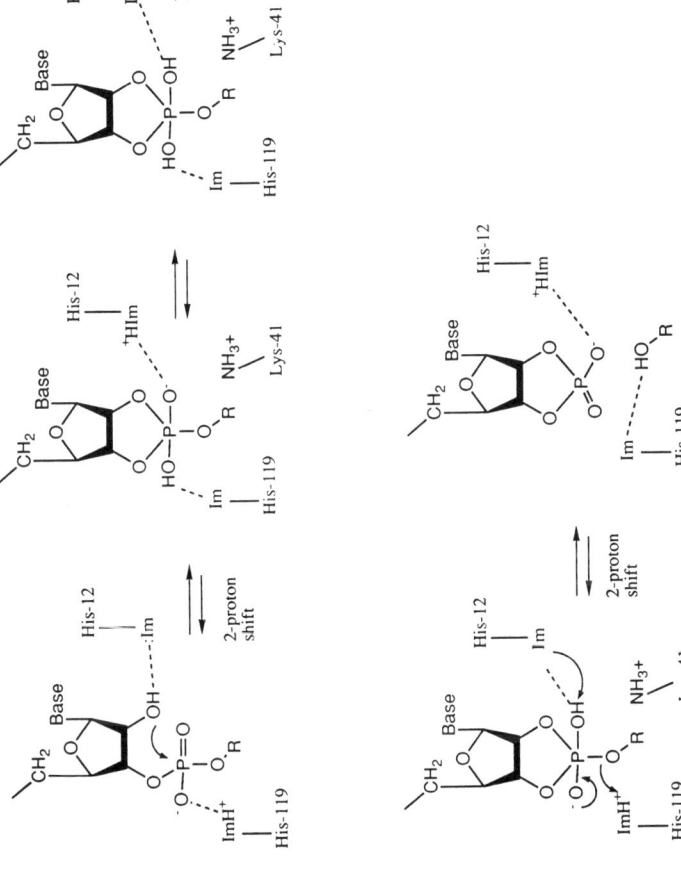

Figure 8: Our proposed mechanism for the enzyme ribonuclease A. After the cyclic phosphate is formed, hydrolysis occurs by reversing the steps but with H-OH substituted for R-OH.

It is interesting that ring opening of the phosphorane anion **3** by departure of the 2' oxygen can occur with BH^+ catalysis, but that fragmentation by loss of the 5" oxygen of the next residue requires prior conversion of **3** to the phosphorane dianion **8**. We have discussed this point earlier.[4] The cyclic phosphate **6** is sufficiently strained that it needs to be formed as its stable *anion*, but unstrained **2** can be formed as the protonated species and lose its proton later. Since such mechanisms need to make sense when considered in either direction, another description involves the reverse reactions. The substrate phosphate group needs to be protonated before it is attacked by the C-2 OH group. However, cyclic phosphate **6** is sufficiently reactive that it can be directly attacked by a nucleophilic OH group - probably with assistance by B - without being protonated first.

This sequential acid then base mechanism cannot be directly extended to the enzyme. There the catalysis is surely *simultaneous*, not sequential. This is probably generally true for bifunctional catalysis in enzymes, and for ribonuclease there is proton inventory evidence[17] that two protons are moving in the rate-determining step, as required for simultaneous bifunctional acid/base catalysis. However, one aspect of our mechanism may well carry over - the finding that the first function of the acid catalyst is to protonate the phosphate oxyanion, not the leaving oxygen atom. We believe that this is probably true for the enzyme as well, and have proposed such a mechanism (Figure 8).[3,4,8,9]

The mechanism in Figure 8 shows simultaneous bifunctional catalysis in several steps. First of all, a phosphorane monoanion is formed by the simultaneous two-proton transfer catalyzed by Im and ImH^+; later, it decomposes to the cyclic phosphate and departing nucleoside by another simultaneous bifunctional two-proton transfer. As we have pointed out, this mechanism is consistent with physical data on the enzyme, and it has also been examined by the Karplus group by molecular mechanics simulation.[18] They find that it is a lower-energy path than is the classical textbook mechanism. Thus our model studies may have given us insight into the biological process of catalysis. The studies have also been helpful in guiding the design of enzyme mimics, as discussed in the next section.

5 BIFUNCTIONAL CATALYSIS BY CYCLODEXTRIN BIS-IMIDAZOLES

Many years ago we started a study on potential ribonuclease mimics consisting of a β-cyclodextrin ring carrying two imidazole rings. β-Cyclodextrin is a cyclic heptaglucoside with a cavity into which hydrophobic substrates will bind in water solution. We and others have based a number of enzyme mimics on this building block, since hydrophobic binding of substrates mimics an important feature of enzyme catalysis. We set out to attach the two imidazole rings to the cyclodextrin with a well defined geometry, so the two could cooperate in a simultaneous bifunctional catalytic reaction related to that performed by ribonuclease A. This was done using some chemistry developed by Tabushi, in which rigid disulfonyl chlorides are allowed to bridge across the primary face of the cyclodextrin, reacting with the C-6 hydroxyl groups. Depending on the reagent used, it is possible to functionalize neighboring glucose units (positions 6A and 6B of the cyclodextrin), glucoses one unit apart (6A, 6C) or glucoses as far apart as possible across the seven-unit ring (6A, 6D).

To give the maximum possible definition to the substrate geometry, we decided to examine the catalyzed hydrolysis of a cyclic phosphate (Figure 9). t-Butylphenyl groups are strongly bound into the β-cyclodextrin cavity, so we selected the cyclic phosphate of 4-t-butylcatechol **9** as substrate. At the time we believed in the textbook enzyme mechanism - a water molecule delivered to the phosphorus atom by an imidazole general base while the leaving oxygen was assisted to depart by the imidazolium ion acid group. In this mechanism, the attacking and leaving oxygen atoms should be 180° apart, in line with the phosphorus. Thus it seemed that the imidazole rings of an effective catalyst should be on opposite sides of the cavity, so we prepared the 6A,6D cyclodextrin bis-imidazole **10**. It was indeed a bifunctional catalyst, with a pH/rate profile showing a rate maximum when the imidazoles were half titrated.[19] We also found a similar catalytic activity for the A,C isomer **11**.

With the mechanistic information from our imidazole buffer studies, it became apparent that we had ignored an important geometric isomer. Our mechanism for the enzyme involved the protonation of the phosphate oxyanion, not the leaving group, in the first step of hydrolysis. Looking at the substrate from the top (Figure 10), the projection angle of this phosphate oxygen is 90° from the direction of attack of the water molecule, not 180°. Thus with this mechanism the A,B isomer **12** of the catalyst should be able to work. The angle between the two attachment points in this isomer is only 52°, but with the angles made by protons at the oxygen atoms of the attacking water and the phosphate oxyanion it was clear from molecular models that the A,B isomer could easily fit the geometric requirements of the new mechanism. It had not been examined previously since it was hopelessly unfit for the previous classical mechanism.

To our delight, the 6A, 6B cyclodextrin-bis-imidazole **12** was not only able to catalyze the hydrolysis of substrate **9**, and with excellent regioselectivity, it was actually much better than the other isomers (*cf* Figure 11).[20] Studies using the proton inventory method established[21] that **12** was indeed performing a simultaneous bifunctional hydrolysis reaction, just as is proposed for the enzyme. In these studies we were also able to obtain additional information supporting the validity of the proton inventory method as a mechanistic tool.

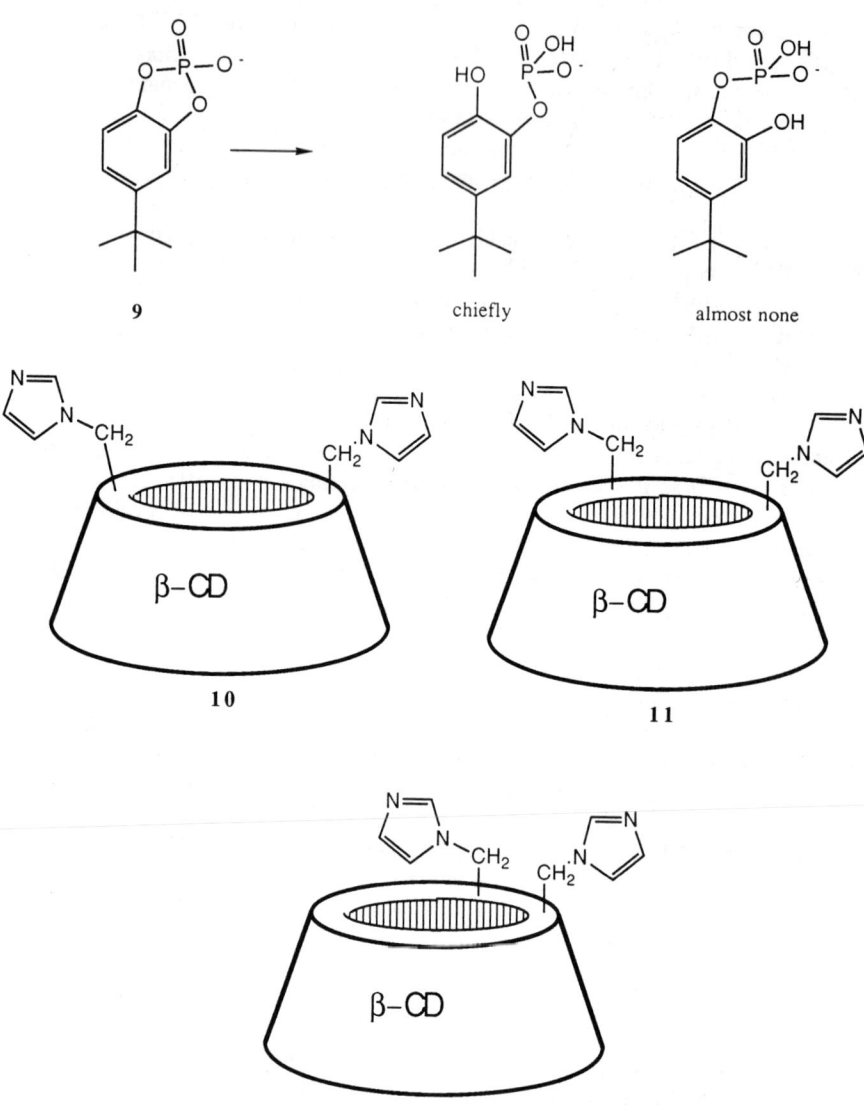

Figure 9: The hydrolytic cleavage of substrate 9, and the three cyclodextrin bis-imidazole catalysts with imidazoles mounted on β-cyclodextrin on the primary C-6 positions of neighboring glucose units 12, of next nearest neighbors 11, and of maximally separated units 10.

Studies on Bioorganic Reactions and Mechanisms

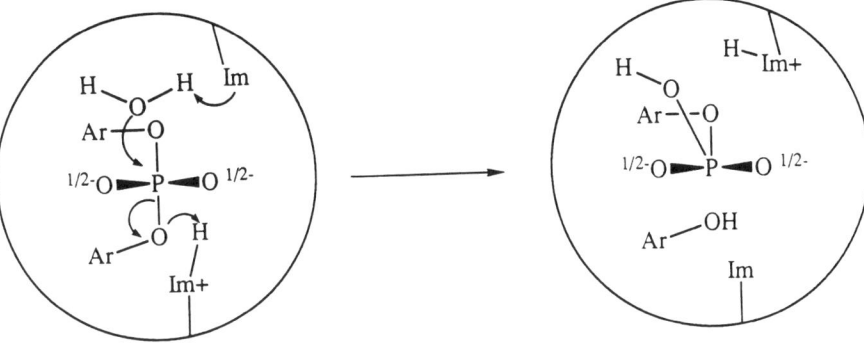

Protonation of the leaving group.

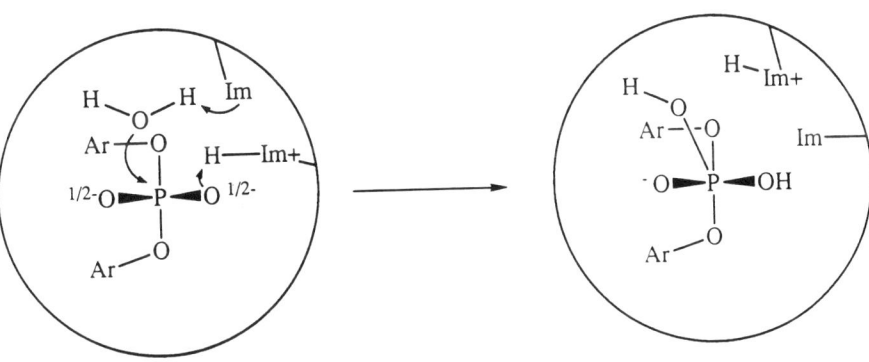

Protonation of the phosphate anion, forming a phosphorane that cleaves in a later step.

Figure 10: A top view of the bifunctionally catalyzed addition of water to substrate 9, catalyzed by a cyclodextrin bis-imidazole. The top sequence shows the classical scheme, while the bottom sequence shows our favored process in which the first function of the ImH+ is to protonate the phosphate oxyanion, forming a phosphorane intermediate that cleaves in a later step.

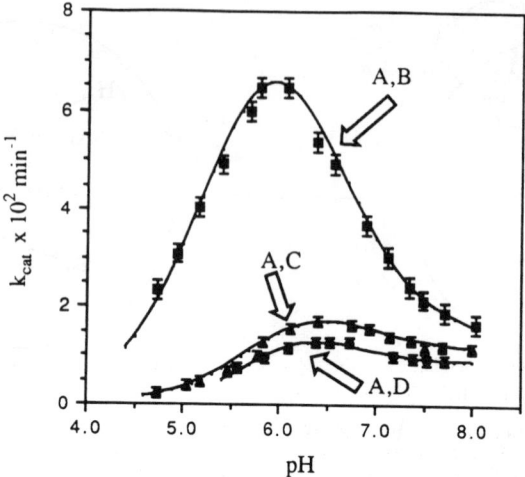

Figure 11: The observed pseudo-first-order rate constant as a function of pH for cleavage of substrate **9** at kinetic saturation with catalyst **10** (A,D), **11** (A,C), or **12** (A,B). Taken from reference 20.

Compound **12** is actually a reasonably good catalyst for the hydrolysis of substrate **9**. Comparing it with the hydrolysis of 2',3'-cyclic cytidylic acid catalyzed by ribonuclease A, the k_{cat} is 4000-fold less but the K_a is 17-fold greater, so k_{cat}/K_m is only 230-fold smaller than for the enzyme case. Of course our substrate, a derivative of a phenol instead of ribose, is intrinsically more reactive. Thus we still have some to go to rival the enzyme with its substrate or another one of comparable reactivity.

With the availability of the three cyclodextrin bis-imidazole isomers, we have the opportunity to look for bifunctional catalysis of other interesting reactions, and to examine their geometric preference as well. We have just started this program, but have already seen striking results in the bifunctional conversion of a ketone to its enol (Figure 12).[22] We examined ketone **13**, with a t-butylphenyl group that binds well into the β-cyclodextrin cavity. Enolization was followed by examining deuterium incorporation into the methyl group of **13**, running the reaction in D_2O with controlled pH.

In the absence of any cyclodextrin catalyst the reaction was unobservably slow under our conditions, but it was accelerated by the A,B (**12**) and A,C (**11**) bis-imidazole cyclodextrins. However, their rates were essentially the same as

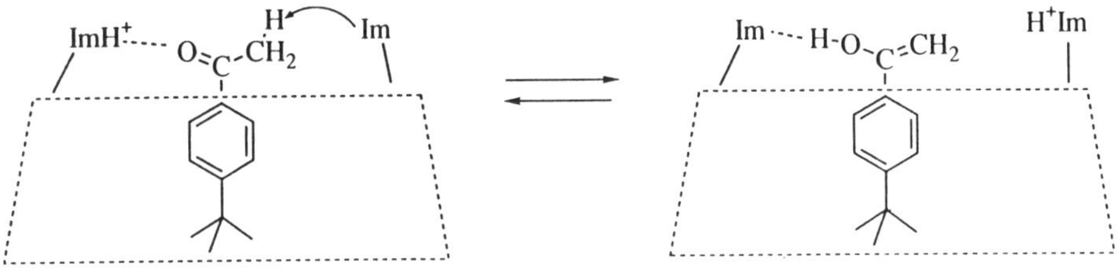

Studies on Bioorganic Reactions and Mechanisms

Figure 12: Reversible conversion of ketone **13** to its enol, monitored by the rate of deuterium incorporation, with bifunctional catalysis by **10**.

those with cyclodextrin 6-monoimidazole, so these bis-imidazoles are apparently not bifunctional catalysts. By contrast, the A,D isomer **10** is significantly better. Computer modelling shows that this isomer can indeed use its imidazolium ion to put a proton on the carbonyl oxygen while the imidazole removes the α-proton of the methyl group. With the availability of the three isomers **10, 11,** and **12** we hope to learn about the geometric preferences in this and related cases, an important clue to the preferred structures of the transition states.

6 PROPERTIES OF ISOMERIC DNA WITH A 2',5" LINKAGE

A key element in our mechanistic studies was the interconversion of the 3',5"- and 2',5"-isomers of UpU and ApA. This and other considerations led us to investigate the properties of 2',5"-linked DNA. Nature has of course chosen to remove the 2'-hydroxyl group of ribonucleosides, providing us with DNA that has a 3',5" link. However, the biochemistry involved in this oxygen removal could also have occurred at carbon 3, except for the geometric preferences of the enzymes involved. There is a real possibility that evolutionary *selection* occurred, and that the isomeric DNA with a 2',5" link is inferior in some way to the normal isomer. It seemed to us important to learn what we could about this alternative to one of the most important molecules ever made.[23] Thus we started such a study.

We have described elsewhere[24] the syntheses of building blocks for the machine synthesis of DNA containing A* (**14**) and T* (**15**) (Figure 13). The * indicates a 3-deoxynucleoside, which forms 2',5" links. These were used[25] in three ways: 1) they were incorporated into normal DNA sequences, replacing one or two of the normal nucleosides; 2) they were used to synthesize the homopolymers $(A*)_{16}$ and $(T*)_{16}$; and 3) they were used to synthesize a mixed sequence A*T* 16-mer and its complement. The results from all three investigations are of interest.

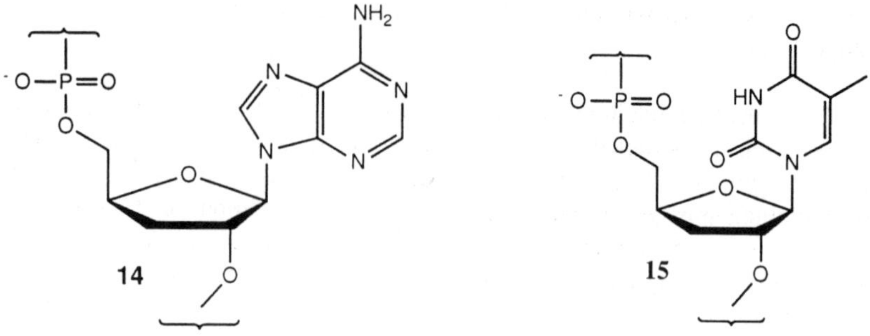

Figure 13: The isomeric 3-deoxyadenosine **14** and the corresponding thymidine derivative **15** that must form 2',5" links in DNA, not the normal 3',5" links.

First of all, the incorporation into one strand of an A* instead of an A, or a T* instead of a T, led to a considerable weakening of the double helix formed with the normal DNA complement. However, the weakening - judged by a lowering of the T_m - was not as great as that from a total mismatch. Thus the A* and T* can apparently bind to at least some extent to their normal T or A partners, but with a distortion in geometry that leads to weakening of the helix.

The results with polymers containing *only* the isomeric A* and T* were striking. We saw that $(A^*)_{16}$ could bind to $(T^*)_{16}$ in media with high salt concentrations, but not under the normal salt concentrations in which $(A)_{16}$ binds to $(T)_{16}$. The evidence came from T_m studies, from circular dichroism studies, and from non-denaturing gel electrophoresis. As we will describe later, further studies reveal a very unusual situation with respect to this binding.

In contrast to the situation with the homopolymers, the 16-mers with mixed sequence of A* and T* showed *no* binding to their complement under any conditions, including high salt. This striking contrast with the behavior of the homopolymers suggested that we were not seeing simple double helix formation in any case. It also indicated that these isomeric nucleosides could not be used as the basis of a genetic coding system; no significant code could be based on molecules that recognized only *homo*polymers.

More recently, in collaboration with Ken Breslauer at Rutgers we have further examined the homopolymer binding.[26] It was found that the *only* association that occurs is the formation of a *triple* helix, with two $(T^*)_{16}$ and one $(A^*)_{16}$, and that only at high salt. No double helix formation could be detected. In contrast, normal $(T)_{16}$ and $(A)_{16}$ form a double helix at moderate salt concentrations and then a triple helix at high salt. This result explains why the heteropolymers containing only A* and T* did not associate, since there is no way to make a two-T*/one-A* triplex when the chains are of mixed A*/T* composition.

Molecular modelling helps account for all this. We had seen that a double helix could be constructed in the computer between the A* 16-mer and the T* 16-mer, and with normal hydrogen bonds, but that the stacking was poor. There was considerable slippage, so that the hydrophobic surface on one purine-pyrimidine pair was poorly covered by the next one.[27] Hydrophobic effects are a principal cause of the stacking that stabilizes a DNA helix. In a triple helix each 'step' of the helix is now longer, with the third base added, so we might expect that the stacking interaction would be better in spite of the slippage of the steps relative to each other.

Wilma Olson at Rutgers has examined this situation further with computer modelling.[26] She finds that the twist angle for the polyA*/polyT* double helix is smaller than that normal for a double helix, but is better for a triple helix. Her studies indicate that the double helix is unlikely to form, but the triple helix seems fine. These calculations are in good agreement with our findings.

In summary, one or two 2',5" links can be tolerated within a normal DNA sequence without preventing association as a double helix, but we see no evidence for double helix formation with 16-mers composed exclusively of A* and T* units. However, a triple helix is formed with the homopolymers. These properties help us appreciate the special chemistry present in natural DNA. They also suggest that if some early organism tried to base a genetic system on the 2',5" isomers of DNA, that organism would not have succeeded.

REFERENCES

1. R Corcoran, M Labelle, A W Czarnik, and R Breslow, *Anal Biochem*, 1985, **144**, 563.
2. R Breslow and M Labelle, *J Am Chem Soc*, 1986, **108**, 2655.
3. R Breslow, Deeng-Lih Huang, and E Anslyn, *Proc Natl Acad Sci USA*, 1989, **86**, 1746.
4. E Anslyn and R Breslow, *J Am Chem Soc*, 1989, **111**, 4473.
5. R Breslow and D-L Huang, *J Am Chem Soc*, 1990, **112**, 9621.
6. R Breslow, E Anslyn, and D-L Huang, *Tetrahedron*, 1991, **47**, 2365.
7. R Breslow and D-L Huang, *Proc Natl Acad Sci USA*, 1991, **88**, 4080.
8. R Breslow, *Acc Chem Res*, 1991, **24**, 317.
9. R Breslow, and R Xu, *Proc Natl Acad Sci USA*, 1993, **90**, 1201.
10. R Breslow, *Proc Natl Acad Sci USA*, 1993, **90**, 1208.
11. R Breslow and R Xu, *J Am Chem Soc*, 1993, **115**, 10705.
12. K Taira, *Bull Chem Soc Japan*, 1987, **60**, 1903.
13. F H Westheimer, *Acc Chem Res*, 1968, **1**, 70.

14 F Menger, *J Org Chem*, 1991, **56**, 6251.
15 A Haim, *J Am Chem Soc*, 1992, **114**, 8384.
16 R Breslow, *J Chem Ed*, 1990, **67**, 228.
17 M S Matta and D T Vo, *J Am Chem Soc*, 1986, **108**, 5316.
18 K Haydock, C Lim, A T Brünger, and M Karplus, *J Am Chem Soc*, 1990, **112**, 3826.
19 R Breslow, J Doherty, G Guillot, and C Lipsey, *J Am Chem Soc*, 1978, **100**, 3227.
20 E Anslyn and R Breslow, *J Am Chem Soc*, 1989, **111**, 5972.
21 E Anslyn and R Breslow, *J Am Chem Soc*, 1989, **111**, 8931.
22 R Breslow and A Graff, *J Am Chem Soc*, 1993, **115**, 10988.
23 For a prior study of nucleic acid analogs with glucose substituted for ribose, see A Eschenmoser and M Dobler, *Helv Chim Acta*, 1992, **75**, 218.
24 C J Rizzo, J P Dougherty, and R Breslow, *Tetrahedron Lett*, 1992, **33**, 4129.
25 J P Dougherty, C J Rizzo, and R Breslow, *J Am Chem Soc*, 1992, **114**, 6254.
26 R Jin, W H Chapman, Jr, A R Srinivasan, W K Olson, R Breslow, and K J Breslauer, *Proc Natl Acad Sci USA*, 1993, **90**, 10568.
27 R Breslow, *Supramol Chem*, 1993, **1**, 111.

MECHANISMS SELECTED BY ENZYMES

Sir John Cornforth

*School of Chemistry and Molecular Sciences,
University of Sussex, Falmer, Brighton BN1 9QJ, UK.*

For some centuries now, chemists have been making and maltreating mixtures of the substances available to them, in the hope that something new would happen; and if something did they recorded, as chemists still do, what they imagined they had done. A collection of recipes began to grow, and some of these even included numbers: Carl Scheele for example knew quite well in 1784 that when you have made solid calcium citrate by dissolving chalk in hot lemon juice, you have to decompose the solid with a solution of half its weight of oil of vitriol to get the best yield of citric acid. That type of regularity promoted the atomic theory and led in due course to theories of molecular structure. Considering that the new structures often had to be deduced by studying the products of chemical reactions that were little understood, it is surprising but highly creditable that satisfactory structural theories did arise. The logic underpinning organic structures, before X-ray crystallography permitted more direct inspection, is in some ways like the roots of a large tree: tortuous, tangled and very strong.

I first learned molecular structure from a 1930 edition of Bernthsen's 'Organic Chemistry'. There, condensations were indicated by enclosing the extruded atoms in a little lasso of dotted lines, and it was not considered important if one lassoed the wrong oxygen in, *eg*, an esterification: the printer's convenience mattered more. But even then, the electrical theory of matter had prompted theories of bonding and reactivity; and in this country Robinson and Ingold were laying some of the foundations of modern mechanistic theory.

Neither of these two great chemists worked with enzymes. Perhaps Robinson came closer: quite early in his life he became interested in what he called the structural relations of natural products. Even before the chemical nature of enzymes and their ubiquitous role in constructing all molecules of life were recognized, he sensed that the chemistry generating these structures was not essentially different from the organic chemistry that he learned and extended and taught; and that because the conditions of construction were mild, the reactions must be easy. His spectacular feat[1] of preparing tropinone by mixing three chemicals in aqueous solution at room temperature was not, as it turned out, the replication of a natural synthesis; but he was certainly using reactions that have close analogues in enzyme chemistry.

Ingold, whose centenary is celebrated during this meeting, became progressively more interested in the physical aspects of chemical change. If he had ever worked with enzymes we can feel sure that he would have studied the kinetics, the replacement of aqueous media by deuterium oxide and by other compatible solvents of different dielectric constants, and perhaps low-temperature enzymology. He was far more interested in stereochemistry and its connexion with reactivity than Robinson ever was, but enzyme stereochemistry came late for him. We still can applaud the contribution that his sharp and perspicacious intellect must have made to the Cahn-Ingold-Prelog[2] system of naming stereoisomers - a very fine piece of legislation, founded on reality, that I for one adopted with joy as soon as it appeared.

As it happened, the decisive advances in the knowledge of organic reactions preceded by a decade or two a similar advance with enzymic reactions. In the result, we have impressive inventories of reactions of both kinds, and our respect for Nature as an organic chemist has grown considerably. Among many other examples, we did not invent the aldol or Claisen condensations, the Wagner-Meerwein rearrangement, the alkylation and cyclopropanation of alkenes, the formation and chemistry of imines and enamines, catalytic hydrogenation, or the Knorr pyrrole synthesis. Indeed, our useful Stetter reaction between aldehydes and Michael acceptors owes its origin to lessons taught to us, with Professor Breslow as interpreter,[3] by enzymes expert in ylide chemistry. In organometallic chemistry, in the chemistry of free radicals and the associated field of antioxidants, the enzymes were there before us. All this chemistry had been practised for millions of years before there were anthropoid apes, let alone organic chemists.

I know well enough that in lecturing about mechanisms selected by enzymes I am attributing to these catalysts a sense of purpose that they are not supposed to possess. An enzyme molecule in its natural setting spends its short life as the creature and slave of an organism, and its success is measurable by its master's survival and increase. We still have the haziest ideas of what primitive enzymes might have been like, but if they were less efficient than their modern counterparts they surely catalysed easy reactions (Scheme 1). Aldol condensations, cyanohydrin and Strecker syntheses are among the simplest reactions; they all occur in mild conditions, without chemical activation and with minimal catalysis, and they all involve addition to sp^2 centres. Further, they almost all generate chiral centres. Any surface that can bind sp^2 centres so as to shield one of two non-equivalent sides is potentially a catalyst of enantiospecific addition, and the binding is easily associated with an activating acid or base. Such a catalyst does not need to be highly active in comparison with spontaneous synthesis in order to provide a useful excess of chiral product. It is interesting and may be significant that Inoue's diketopiperazine **1**, which catalyses the addition of hydrogen cyanide to some aromatic aldehydes with almost complete enantiospecificity, exerts this effect not in solution but as a solid phase.[4]

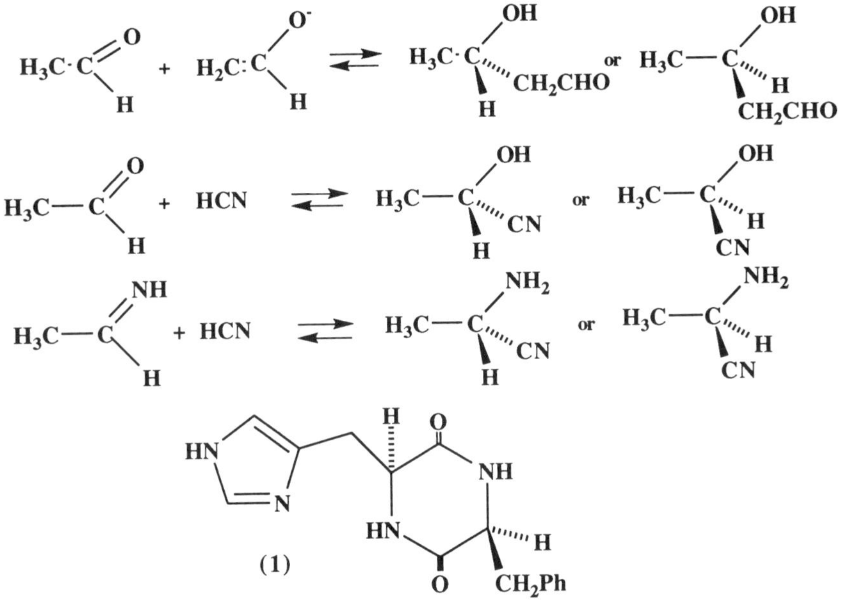

Mechanisms Selected by Enzymes   27

[Scheme 1 structures]

Scheme 1: Chiral products from planar precursors.

Whatever the beginnings, there was scope enough for development. Where two substrates, neither of them water, reacted with each other, primitive enzymes with some binding affinity could help simply by increasing the effective concentrations. When redox systems developed, the same effect applied: the systems were incorporated as coenzymes. When activated groups like thiolesters were available, aldol-catalysing enzymes were well placed to become catalysts of Claisen condensations. Catalysis of more difficult reactions could develop so long as catalysts of easier but mechanistically similar processes were available and adaptable. Development of substrate specificity in an enzyme would not always be an evolutionary advantage, unlike efficiency of catalysis or of stereochemical control.

Enzymes produced by a slightly erratic control mechanism are likely to show the same kind of versatility that human chemists are famed for - that is, they may make an unexpected product that turns out to be useful, or they may aimlessly manipulate a novel raw material and discover a new process. The master organism would find this behaviour of its slaves most useful when a new source of food could thereby be connected speedily to the normal sequences fuelling and supplying repair and reproduction. Some of the adaptations by bacterial enzymes to bring unfamiliar nutrients into the conventional pathways look like the work of a brilliant organic chemist. How, for example, do you convert a molecule of camphor **2** (see Scheme 2) into three molecules of eminently edible acetate and one of nutritious isobutyrate, wasting no carbon and very little energy? In the hope that some of you will be led to read Gunsalus' beautiful work[5] on this, I won't reveal the plot; but here is a less well-known example. Some industrial detergents contain straight chain alkane-1-sulfonates

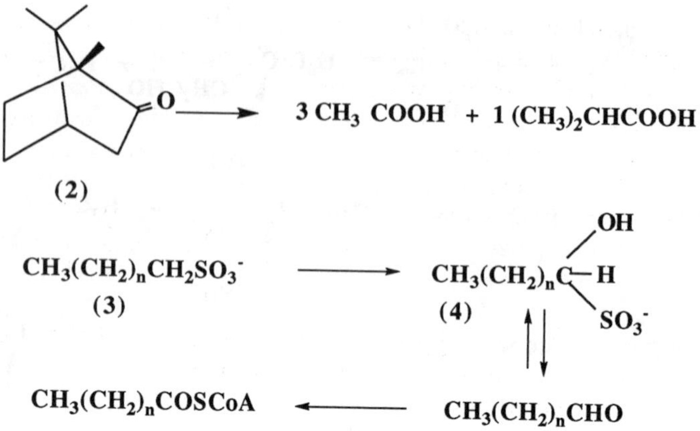

Scheme 2: Microbial digestion.

3, and some strains of *Pseudomonas* have learned to use these as their sole sources of carbon and energy. They do this, as Thysse and Wanders[6] showed, by using an oxygenase to hydroxylate the substrate at the 1-position. The older chemists among you will recognize the product **4** as an aldehyde bisulfite, in equilibrium with the aldehyde at cellular pH; and in fact, formation of aldehyde and sulfite anion was demonstrated, as was the now straightforward conversion to fatty acids in a form suitable for normal assimilation. What could be more elegant? The organism, which is tuned to reinforce any enzymic activity having survival value, effectively puts pressure on to the enzymes to find the shortest and cheapest route, preferably using existing equipment. And, as with many industrial syntheses, the route often acquires a purely incidental beauty.

Since a full comparison of enzymic with chemical mechanisms is not possible in the available time, I shall concentrate on one type only: chemical and enzymic carboxylation and decarboxylation. I make no apology for this concentration; I have been interested in these reactions for more than half my life, and you are a captive audience; so I shall try to interest you too. But the subject is by no means unimportant, especially on the enzymic side. We could not live without the carboxylations, essentially of enols, by which green plants assimilate carbon dioxide; and decarboxylations are what we use to return the carbon, drained now of energy, for recycling.

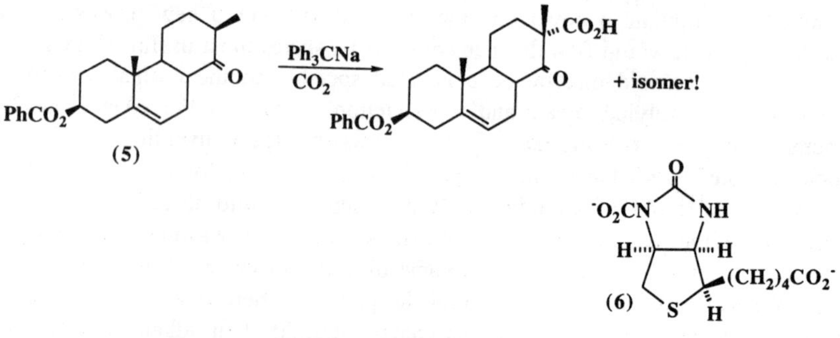

Scheme 3: Carboxylation reagents, human and enzymic.

Mechanisms Selected by Enzymes

On the chemical side, the carboxylation of enolate anions goes back to Kolbe's salicylic acid synthesis in 1860. It is not much otherwise used, but I remember with affection that in our synthesis of the non-aromatic steroids we applied, in 1950, Charles Hauser's pioneering carbanion chemistry[7] by using tritylsodium to form an enolate anion from a tricyclic ketone **5** (see Scheme 3) and, by pouring the solution on solid carbon dioxide, added a carbon atom vital for construction of the fourth ring.[8] But even here the enzymes do better; they neatly put the strong base and the high CO_2 concentration into a single package: *N*-carboxybiotin **6**.

Carboxylation of an enol and decarboxylation of a 3-oxo acid are balanced sensitively enough to be swayed in either direction and both reactions occur in mild conditions; so it is not surprising that enzyme chemistry abounds in examples, even at the level of primary metabolism. In the overall process (Scheme 4) ketonization of the enol is the largest free energy change, and the protonation at carbon is stereochemically ambiguous. This feature is used with delightful economy by the enzyme system that removes the *gem*-dimethyl group from ring A in the biosynthesis of sterols. The 4α methyl group is first oxidized to carboxyl; dehydrogenation at the 3 position generates a 3-oxo acid **7**; this is decarboxylated and the intermediate enol is protonated on the β face, presenting to the oxygenase another 4α methyl for similar treatment. Effectively, the same methyl group is removed twice.[9] A 3-oxo acid decarboxylates easily because the heterolytic detachment of carbon dioxide leaves a stabilized carbanion. The process could also be described as a 1,2-elimination in which the leaving groups are carbon dioxide and a pair of electrons from the carbonyl double bond. There are plenty of chemical examples of such eliminations. The simplest is the formation of alkenes, carbon dioxide, and halide ion from 3-halocarboxylic acids in mildly basic media (Scheme 5).

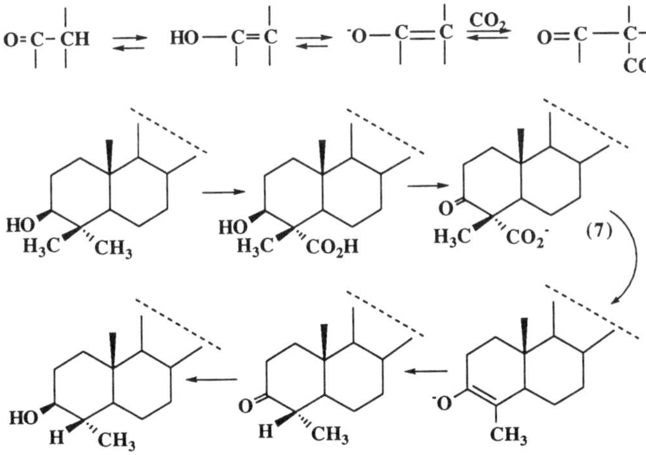

Scheme 4: A subtlety of enzymic 3-oxo acid decarboxylation.

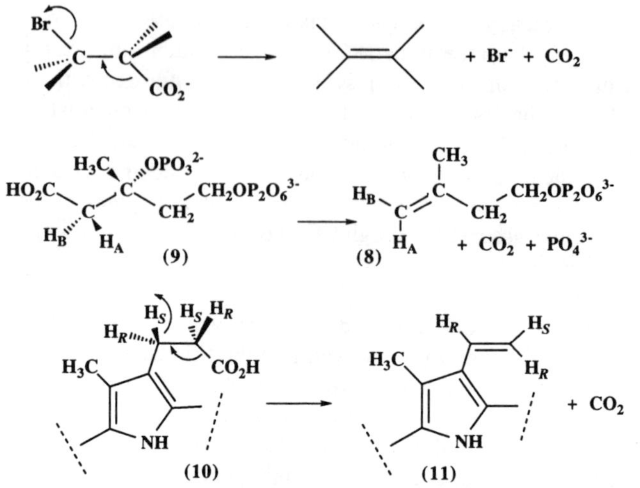

Scheme 5: Decarboxylative 1,2-eliminations.

Here, too, the enzymes have anticipated us. A key step in terpenoid biosynthesis produces carbon dioxide and 3-methyl-3-butenyl diphosphate **8** from mevalonic acid 5-diphosphate **9**, the tertiary hydroxyl group being prepared for departure by phosphorylation with ATP. In both examples *anti* elimination is the preferred stereochemical course.[10] The oxidative process in porphyrin biosynthesis that converts propionic acid side chains **10** to vinyl groups **11** removes hydride and carbon dioxide, also by an *anti* elimination.[11]

The well-known thermal decarboxylation of βγ-unsaturated acids[12] with migration of the double bond, has its counterpart in the enzymic formation (Scheme 6) of itaconic **12** from aconitic acid **13**; this can also be done thermally. Bentley and Thiessen[13] showed, by making use of the known dissymmetry of citric acid biosynthesis, that the double bond does migrate. Again in porphyrin biosynthesis,[11] the decarboxylation of acetic acid side chains in uroporphyrinogen (as **14**) is thought to proceed by a similar mechanism, with later rearrangement of the double bonds back into the pyrrole rings.

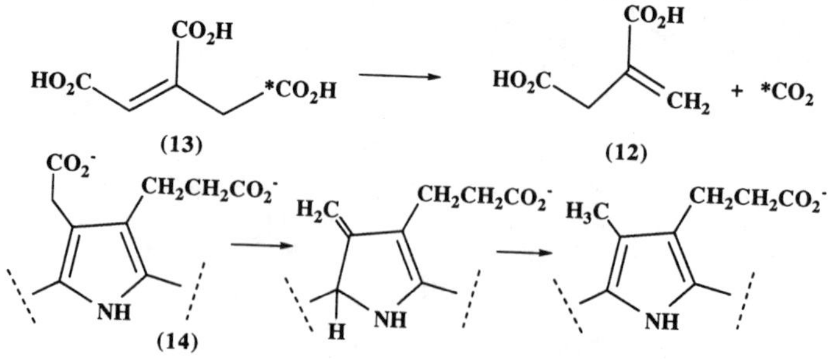

Scheme 6: Decarboxylation and prototropy.

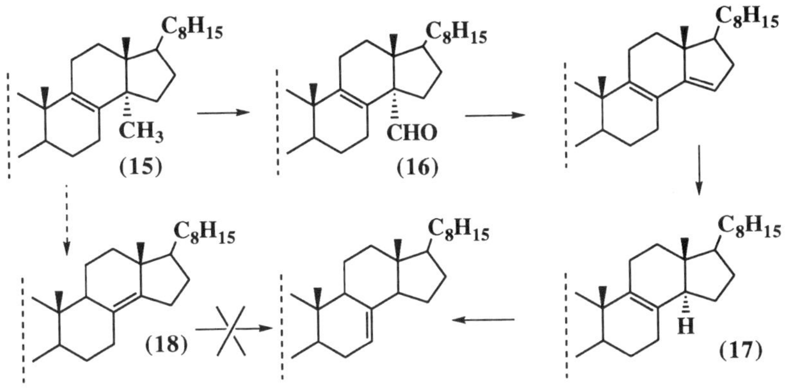

Scheme 7: A finesse in cholesterol biosynthesis.

Reverting to sterol biosynthesis (Scheme 7) it is worth remarking that enzymes have *not* selected a similar reaction for removal of the superfluous 14α methyl group in lanosterol (as **15**). Here, oxidation could have generated a βγ-unsaturated carboxylic acid for decarboxylation; instead, we have a longer and metabolically more expensive route in which the intermediate aldehyde **16** undergoes an oxidative elimination of formic acid, leaving a 14(15) double bond that then has to be reduced.

The subtle point may be that this reduction can be arranged to give a 14-norlanosterol **17** with unchanged location of the double bond, whereas the direct decarboxylation is bound by its mechanism to deliver a $\Delta^{8(14)}$ isomer **18**; and although an 8(9) double bond can be rearranged enzymically at a later stage to the 7(8) position, this does not seem to be possible for an 8(14) double bond, though 8(14) unsaturated sterols can certainly function as precursors of cholesterol.[9] There is one organism, *Methylococcus capsulatus*, in which the sterol end-products have 8(14) double bonds.[14] The method used by this organism to remove 14α methyl groups has yet to be investigated.

Decarboxylations featuring the departure of carbon dioxide from a carboxylate, leaving a negative charge to be absorbed by an electrophilic system, account for a considerable fraction of known decarboxylases: the enzymes harnessed to thiamine, which decarboxylate 2-oxo acids, and the pyridoxal phosphate enzymes which act on amino acids and can effect either decarboxylation or oxidative deamination. There are chemical counterparts to all these processes and they can occur in quite mild conditions even without specific catalysis. But they have been well reviewed[15] and I shall not dwell on them here.

A decarboxylation that - so far - has no parallel in enzyme chemistry generates alkenes from αβ-unsaturated acids, particularly those that have

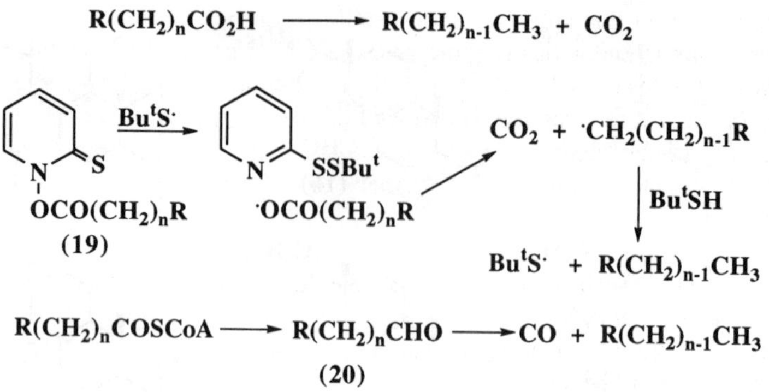

$$R(CH_2)_nCO_2H \longrightarrow R(CH_2)_{n-1}CH_3 + CO_2$$

Scheme 8: Evasions of catalytic decarboxylations of fatty acids.

(19) N-acyloxy-2-thiopyridone: pyridine with S and OCO(CH$_2$)$_n$R, reacts with Bu^{t}S· to give pyridyl-SSBu^t and ·OCO(CH$_2$)$_n$R, which gives CO$_2$ + ·CH$_2$(CH$_2$)$_{n-1}$R; then with Bu^{t}SH gives Bu^{t}S· + R(CH$_2$)$_{n-1}$CH$_3$.

$$R(CH_2)_nCOSCoA \longrightarrow R(CH_2)_nCHO \longrightarrow CO + R(CH_2)_{n-1}CH_3$$
(20)

additional unsaturation. Heat alone is sufficient but it is often supplemented by the use of quinoline as a solvent and of copper in various forms as a catalyst. The catalysed process has been studied and is thought to proceed by way of a complexed copper(I) carboxylate that yields carbon dioxide and a vinylcopper. This does not happen with saturated acids and the driving force may come from the relative stability of vinyl carbanions. These also have some configurational stability and, mostly, hydrogen replaces carboxyl with retention of geometry.

I come now to decarboxylations of a type that has apparently given grave problems both to chemists and to enzymes: that is, the smooth cleavage of a saturated acid to carbon dioxide and the complementary hydrocarbon (Scheme 8). Every schoolboy knows, or used to know, that you can make methane by heating a mixture of sodium acetate and soda-lime, but fewer people know that this trick works badly if at all with any homologous acid. When chemists want to decarboxylate a saturated acid to examine isotopic labelling of the carboxyl group, they use the Hofmann, Curtius or Schmidt rearrangements, or a sequence that depends on the spontaneous cleavage of a carboxylate free radical. None of these processes delivers the saturated hydrocarbon except Barton's most ingenious use of N-acyloxy-2-thiopyridones 19 in a radical-chain reaction in which the closing link is acceptance, by the decarboxylated alkyl radical, of hydrogen from a tin hydride or a thiol.[16] That is magnificent, but it is not catalysis. I should like very much to find an efficient catalyst for decarboxylation of saturated acids. The problem is not the free energy - such decarboxylations are slightly exergonic - but the activation barrier; and it has to be said that enzymes seem to have the problem too. This is not because no organism wants to decarboxylate a fatty acid. The leaf waxes of most higher plants contain long-chain paraffins derived formally from fatty acids by loss of carbon dioxide; and the mould *Botryococcus braunii* includes similar paraffins in the mixture of hydrocarbons that it accumulates. But Kolattukudy,[17] who has studied for many years the biosynthesis of these paraffins, has obtained good evidence that not decarboxylation, but prior reduction of the fatty acid *via* its coenzyme A thiolester to the aldehyde 20, and catalytic cleavage of carbon

Mechanisms Selected by Enzymes 33

monoxide from this aldehyde on a metalloenzyme, is the route adopted in both cases. This is energetically quite an expensive process, ATP being required to form the thiolester and NADPH to reduce it (though when *B braunii* is working anaerobically it recoups some energy by oxidizing part of the carbon monoxide), so that although there was some incentive for the organisms to find the more direct route, they did not find it. Perhaps this augurs ill for chemists seeking the same goal.

I will finish this survey by discussing two decarboxylations - one chemical, one enzymic - that seem to me incongruous and unclarified. The chemical example comes from sugar chemistry (Scheme 9). It has long been known that the hexuronic acids, glucuronic and galacturonic, decarboxylate slowly in hot water, even when no additional acid is present. The decomposition is complex, ending with 2-furaldehyde in the more acid treatments, and the actual decarboxylation in such conditions has been plausibly explained[18] as occurring with intermediate 3-oxo acids or their vinylogues. But in 1956 Zweifel and Deuel[19] showed that decarboxylation of galacturonic acid **21** is catalysed by some metallic ions, notably lead; and that although many products are formed one of them is the formal decarboxylation product, arabinose **22**. Further, catalysed decarboxylation occurs more readily and completely in pyridine; here, nickel ions were the best catalyst found and the neutral reaction product was simpler, consisting largely of four sugars. One of these was certainly arabinose, which was isolated in the crystalline form; ribose was also indicated and there were two unidentified sugars. Because pyridine is a known epimerizing agent for sugars and could thus account for the presence of ribose, the two unknowns were thought to be the ketoses arabulose and ribulose, which could be formed from the same enediol intermediate. Unfortunately, this explanation will not do: arabulose *is* ribulose. It was also established that galacturonic acid α-methylglycoside did not decarboxylate in pyridine with nickel acetate, but that the other three hydroxyls could be methylated without stopping decarboxylation.

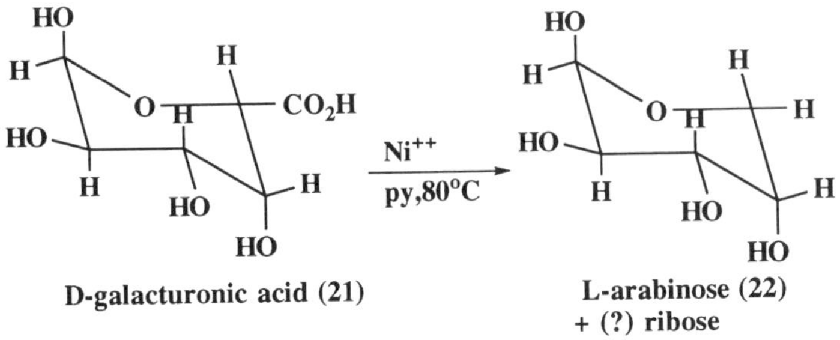

D-galacturonic acid (**21**) $\xrightarrow{Ni^{++}, py, 80°C}$ L-arabinose (**22**) + (?) ribose + 2 neutral sugars

Scheme 9: An unexplained catalytic decarboxylation.

Now this is a very remarkable decarboxylation. 2-Alkoxy and 2-hydroxy acids just do not decarboxylate in conditions like this, and one also has the evidence that the remote hemiacetal group is essential. Zweifel and Deuel postulated an inverting electrophilic displacement of the carboxylate anion by the hemiacetal proton made more acidic by complexation with the metal ion. If this unprecedented mechanism were true one would expect to find other examples of suitably placed acidic hydrogens promoting other easy decarboxylations; but so far as I know there is none.

Is there an enzymic counterpart to this reaction? Yes and no; there are two decarboxylases that act on UDP-glucuronate and UDP-galacturonate, transforming them into UDP-xylose and UDP-arabinose.[21] Unfortunately they are both of the malic enzyme type: they require NAD^+ and there is evidence that they operate by a transient dehydrogenation at C-4 generating a 3-oxo acid. This is not a mechanism that could explain the chemical catalysis.

So what is the mechanism? Recently, Grigg and his co-workers[21] have published an interesting series of papers in which they show that Schiff bases of α-amino acids decarboxylate by generating azomethine ylides, which can be trapped with 1,3-dipolarophiles. An analogous decarboxylation to an oxonium ylide (23, Scheme 10) seems possible with galacturonic acid, and this ylide could be hydrated to yield L-arabinose and/or L-lyxose. The mysterious congeners of arabinose in the catalytic decarboxylations immediately become of great interest, and this unexplored territory cries out, in my opinion, for more investigation, especially now that modern analytical methods have made it easier to separate and identify sugars. The chemistry of simple analogues also awaits exploration because, I think, the singularity of this catalytic decarboxylation has not been appreciated.

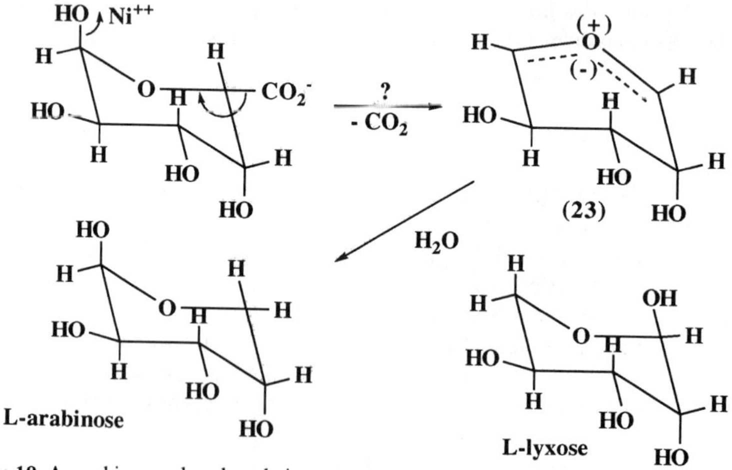

Scheme 10: An ambiguous decarboxylation.

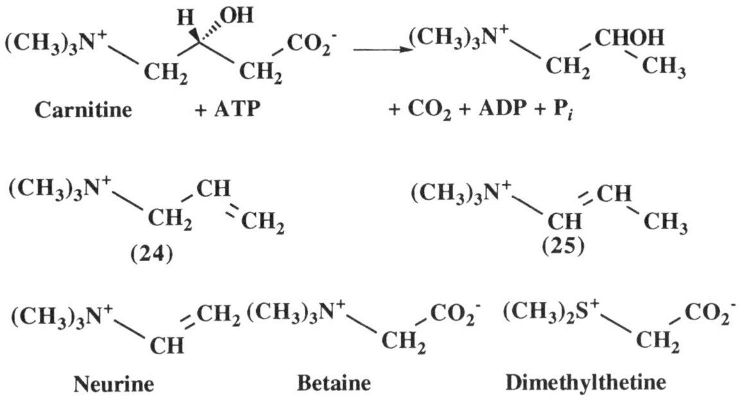

Scheme 11

$(CH_3)_3N^+-CH_2-CH(H)(OH)-CH_2-CO_2^-$ → $(CH_3)_3N^+-CH_2-CHOH-CH_3$
Carnitine + ATP   + CO_2 + ADP + P_i

$(CH_3)_3N^+-CH_2-CH=CH_2$ **(24)** $(CH_3)_3N^+-CH=CH-CH_3$ **(25)**

$(CH_3)_3N^+-CH=CH_2$ Neurine $(CH_3)_3N^+-CH_2-CO_2^-$ Betaine $(CH_3)_2S^+-CH_2-CO_2^-$ Dimethylthetine

Scheme 11: A singular carboxy-lyase.

I come now to the singular decarboxylase (Scheme 11). Carnitine, or Vitamin B_T as it was called until the discovery that it is a vitamin for mealworms but not for us, is concerned with the transport of lipids; and there is an enzyme, carnitine decarboxylase, that transforms it to carbon dioxide and a homologue of choline. The enzyme seems to be quite widely disseminated. The preparation first explored came from the mitochondria of rat heart,[22] but another preparation has been made from the gut of blowfly larvae,[23] and this preparation was considered homogeneous enough to justify amino acid analysis (one unidentified component was found). The singularity is in the co-factor: the enzyme requires ATP and it was satisfactorily shown that one equivalent of orthophosphate and one of ADP are formed along with each equivalent of carbon dioxide.

So what is going on? The only other enzyme of this class requiring ATP is diphosphomevalonate decarboxylase and it does not effect a formal decarboxylation, as here, but a 1,2-elimination forming a double bond. If that happens here the product allyltrimethylammonium **24** must be a substrate, hydrated in a separate step. Enzymic hydration of unconjugated double bonds is rare, but it is not quite unknown. Assuming that 1-propenyltrimethylammonium ion **25** is a possible intermediate, for example by prototropic isomerization of the allyl isomer, would this be hydrated more readily? It is certainly true that vinylsulfonium and vinylphosphonium ions add nucleophiles very easily, as shown by Doering and Schreiber;[24] but they also found that vinyltrimethylammonium (neurine) is much more resistant. This difference is mirrored in the decarboxylation of dimethylthetine and betaine: the sulfonium zwitterion decarboxylates on gentle heating[25] but betaine rearranges to methyl dimethylaminoacetate at 290 °C (a demonstration, by the way, that mere creation of even a full positive charge at a position β to a carboxyl group is not in itself sufficient to promote easy decarboxylation). But although Doering and Schreiber could not isolate a neurine adduct, they did find that neurine reacted slowly with 2-hydroxyethanethiol at 78 °C in the presence of a catalytic amount of the sodium mercaptide. One product was trimethylamine, suggesting that addition here was followed by a Hofmann elimination. Thus the possibility remains open that hydration of the propenyl isomer could be easy enough to encourage enzymic catalysis.

But this whole subject is full of unanswered questions. It has not even been established whether the decarboxylation product has the same chirality as carnitine; whether the hydroxyl oxygen of carnitine is retained in the product or, if not, what happens to it and what is the origin of the oxygen that replaces it; whether any exchange of carbon-bound hydrogen occurs and, if so, where and with what chirality; and whether hydrogen replaces carboxyl with overall retention or inversion of configuration. All these questions could be answered by a suitable mixture of known chemical and biochemical techniques, and it has been a part of my purpose to persuade someone who hears or reads this lecture that this unique decarboxylation, having at present no parallel either in chemistry or in enzymology, is worth that inquiry. And although decarboxylation may have been a suitable topic for comparison of methods selected by chemists and by enzymes, it was not chosen because I thought it complete.

REFERENCES

1 R Robinson, *J Chem Soc*, 1917, **111**, 762.
2 R S Cahn, C K Ingold, and V Prelog, *Angew Chem Int Ed Engl*, 1966, **5**, 385.
3 R Breslow, *Chem and Ind*, 1957, 893.
4 K Tanaka, A Mori, and S Inoue, *J Org Chem*, 1990, **55**, 181.
5 I C Gunsalus, C A Tyson, and J D Lipscomb, in *Oxidases and Related Systems*, Univ Park Press, 1973. Vol 2, p 580.
6 G J E Thysse and T H Wanders, *Antonie van Leeuwenhoek*, 1974, **40**, 25.
7 E Baumgarten, R Levine, and C R Hauser, *J Am Chem Soc*, 1944, **66**, 862.
8 H M E Cardwell, J W Cornforth, S R Duff, H Holtermann, and Sir R Robinson, *Chem and Ind*, 1951, 389.
9 G J Schroepfer, jr, *Ann Rev Biochem*, 1982, **51**, 555 (review).
10 J W Cornforth, R H Cornforth, G Popjak and L Yengoyan, *J Biol Chem*, 1966, **241**, 3970; and references cited therein.
11 M Akhtar, in *Biosynthesis of Tetrapyrroles*, (ed P M Jordan), Elsevier, 1991, p 67.
12 R T Arnold, O C Elmer, and R M Dodson, *J Am Chem Soc*, 1950, **72**, 4359.
13 R T Bentley and C P Thiessen, *J Biol Chem*, 1957, **226**, 703.
14 P Bouvier, M Rohmer, P Benveniste, and G Ourisson, *Biochem J*, 1976, **159**, 267.
15 M H O'Leary, *Bio-organic Chem*, 1977, **1**, 259 (review).
16 D H R Barton, D Crich, and W B Motherwell, *J Chem Soc, Chem Comm*, 1983, 939.

17 M E Dennis and P E Kolattukudy, *Arch Biochem Biophys,* 1991, **287**, 268; and references cited therein.
18 K B Hicks and M S Feather, *Carbohydrate Res,* 1977, **54**, 209; and references cited therein.
19 G Zweifel and H Deuel, *Helv Chim Acta,* 1956, **39**, 662.
20 D-F Fan and D S Feingold, *Arch Biochem Biophys,* 1972, **148**, 576.
21 M F Aly, R Grigg, S Thianpatanagul, and V Sridharan, *J Chem Soc, Perkin Trans I,* 1988, 949, and earlier papers cited therein.
22 E I Khairallah and G Wolf, *J Biol Chem,* 1967, **242**, 32.
23 M Habibulla and R W Newburgh, *J Insect Physiol,* 1972, **18**, 1929.
24 W v E Doering and K C Schreiber, *J Am Chem Soc,* 1955, **77**, 514.
25 E H Letts, *Jahresbericht,* 1878, 684.

HYDROGEN TUNNELING IN ENZYME REACTIONS

Judith P Klinman

Department of Chemistry, University of California, Berkeley, CA 94720, USA.

1 INTRODUCTION*

Hydrogen isotope effects have played a major role in our understanding of enzyme catalysis over the past two decades. In the course of extending this valuable probe from physical organic chemistry to enzymology, it has been necessary to describe a new set of guide lines. A major difference between reactions in solution vs those on an enzyme surface is the formation in the latter case of complexes between enzyme and substrate (E•S) and product (E•P). As we now know, the rate of formation or loss of such complexes is often the kinetically limiting step in an enzyme catalyzed conversion of free substrate to product. For this reason, kinetic hydrogen isotope effects have most frequently been used in enzymology to determine the extent to which the chemical step is rate limiting. Under favorable circumstances, the magnitude of the observed isotope effect can be compared to that for an 'intrinsic value' on the chemical step itself. This can be expressed by a simple equation originally described by Northrop:[1]

$$k_H/k_D(obs) = k_H/k_D(int) + C_f + K_H/K_D\, C_r\, /\, 1 + C_f + K_H/K_D\, C_r \quad (1)$$

---

* I want to thank the organizers, in particular Bernard Golding, for the invitation to speak at this excellent meeting. I first met Bernard 17 years ago, and it is a pleasure to see and *recognize* each other after almost two decades. It is also an honor to speak in the Ingold Symposium.

A fascination for most people is the tracing of their lineage. The late biochemist Alan Wilson from Berkeley had analyzed mitochondrial DNA, which is maternally transmitted, and concluded that all individuals are descended from a single woman in Africa (not that long ago - *ca.* 200,000 years). In this light, it is not surprising that practicing scientists can trace themselves back to a common lineage. In the US, Frank Westheimer has kept a family tree, showing the evolution of a large 'family' of practicing bioorganic and biochemists. In the United Kingdom, Sir Christopher Ingold has played the role of the father of physical organic, and, more recently, bioorganic chemistry. Although I am clearly one of Westheimer's granddaughters, I had not realized until this week that I can also make, albeit a weak, claim to Ingold's lineage. In fact, in the late 1960's I spent eight months making my first foray into biochemistry in the laboratory of Charles Vernon at University College. During the past week I learned that Charles Vernon had been a graduate student of C K Ingold. So....the world of science is indeed a small one.

where K_H/K_D is an equilibrium isotope effects and C_f and C_r, referred to as commitments, are ratios of rate constants for steps which precede or follow the isotope sensitive step.

Determination of the value for k_H/k_D(int) is often the limiting factor in the application of eqn (1). In straight forward cases, an estimate for k_H/k_D(int) can result from rapid mixing experiments which allow direct detection of the formation of E•P from E•S (eg ref 2). With regard to steady state kinetics, Northrop devised an approach for the estimation of k_H/k_D(int) using measured values for k_H/k_D(obs) and k_H/k_T(obs).[1] Although this method has been criticized for its relative insensitivity, ie, the requirement for precision in measurements which can exceed the capability of the experimental system under study, it has been used successfully in some instances (eg ref 3). The basis for the calculation of k_H/k_D(int) from experimental deuterium and tritium isotope effects is the Swain-Schaad relationship,[4] which breaks down in a predictable manner when chemistry is not fully rate limiting:

Chemistry Rate Limiting: $k_H/k_D(obs)^{1.44} = k_H/k_T(obs)$ (2)

Kinetic Complexity: $k_H/k_D(obs)^{1.44} < k_H/k_T(obs)$ (3)

Inherent in the application of eqns (2) and (3) is the validity of the Swain-Schaad relationship. Early investigators had shown through experiment and calculation that eqn (2) was likely to hold, even under conditions of significant hydrogen tunneling (cf ref 5). Given the three isotopes for hydrogen, H, D and T, it is possible to measure and express isotope effects in a variety of forms, for example as k_D/k_T and k_H/k_T. Although, in principle, all isotope effects should be interconvertible (related by the reduced masses for H, D and T) Saunders has demonstrated that comparison of k_D/k_T to k_H/k_T may be especially sensitive to tunneling,[6] ie,

No Tunneling: $k_D/k_T(obs)^{3.26} = k_H/k_T(obs)$ (4)

Tunneling: $k_D/k_T(obs)^{3.26} < k_H/k_T(obs)$ (5)

A qualitative way of appreciating the subtle but important difference between comparisons of H/D to H/T isotope effects vs D/T to H/T isotope effects is the common isotope in each pair. In the former case, the common isotope effect is H, which is the nucleus undergoing the greatest degree of tunneling; whereas in the latter case it is tritium, which is expected to undergo relatively little tunneling and provides a frame of reference for the relative amounts of H and D tunneling. Grant and Klinman have performed a detailed comparison of the relationship between the H/D and H/T pair of isotope effects vs. the D/T and H/T pair, using the full Bell correction to estimate tunneling.[7] These calculations confirm the significantly greater sensitivity of the D/T and H/T pair toward the detection of tunneling. Although the advantage to a comparative study of D/T and H/T isotope effects is offset, to some extent, by a greater propagation of error,[8] the approach advanced by Saunders for detection of H-tunneling remains an important addition to our arsenal of tools for the investigation of enzymatic reaction mechanism.

2 RESULTS AND DISCUSSION

2.1 Initial Detection of H-tunneling in an Enzymatic System from a Comparative Study of k_D/k_T and k_H/k_T

Among redox enzymes, alcohol dehydrogenases (ADH's) are one of the best studied systems. Numerous X-ray structures are available for the enzyme from horse liver, including a structure for a product complex consisting of the cofactor NAD^+ and a p-bromobenzyl alcohol substrate.[9] Although an X-ray structure is not available for the enzyme from yeast, an overlay of the sequence for yeast ADH upon the solved structure for horse liver ADH has been performed, showing considerable conservation at the enzyme active site.[10]

During the early 1970's, a detailed study of both structure reactivity correlations and isotope effects was pursued with yeast ADH. It was found that replacement of the normal substrate ethanol by larger aromatic alcohols changed the nature of the rate limiting step from product release to H-transfer.[11] This fortuitous observation allowed for a straight forward application of physical organic chemistry toward the elucidation of transition state structure. Quite unexpectedly, the conclusion from structure reactivity correlations[12] - of an early transition state resembling alcohol - was found to conflict with the conclusion from secondary isotope effects[13] - of a late transition state resembling aldehyde. Subsequent studies of horse liver ADH presented an even more dramatic instance of inconsistencies,[14] with the kinetic secondary isotope effect for reduction of [4-^{2}H]-NAD^+ exceeding the equilibrium limit (k_H/k_D = 1.22 vs K_H/K_D = 0.89). Prior to this observation, it had been assumed that secondary kinetic isotope effects would lie between unity and the equilibrium value, with the magnitude of the secondary kinetic isotope effect indicating the position of the transition state.

Initial efforts at an explanation for the observed anomalies in the ADH reaction invoked a two step mechanism for the H-transfer process, leading to the formation of free radical intermediates.[13] Although radical intermediates in NAD^+-dependent reactions were 'fashionable' for a time, it is now generally accepted that these reactions occur by direct hydride transfer. In a major step forward toward a resolution of the experimental inconsistencies, Huskey and Schowen performed a theoretical analysis of isotope effects in the ADH reaction.[15] It was found that invoking coupled motion between the primary and secondary hydrogens at the reaction center was not sufficient to cause inflation of secondary isotope effects to the magnitude seen experimentally. Rather, *a combination of coupled motion and tunneling* was found necessary to reproduce the experimental numbers.

With the underpinning of Huskey and Schowen's study,[15] together with the methodology advanced by Saunders,[16] a system was sought which would permit an unambiguous demonstration of H-tunneling. Yeast alcohol dehydrogenase appeared to be an ideal system for such a study, given the previously demonstrated role for a single rate limiting chemical step with aromatic alcohols as substrate. The final results of a comparative study of k_D/k_T and k_H/k_T with yeast alcohol dehydrogenase are summarized in Table I. As can be seen, values for k_H/k_T(obs) are significantly larger than k_D/k_T(obs)$^{3.26}$. The magnitude of the exponent required to 'match' k_D/k_T(obs) to k_H/k_T(obs) has been found to be especially elevated in the case of secondary isotope effects (exponent = 10.2 at 25 °C!). This provides direct evidence for (i) significant H-tunneling, (ii) coupled motion between the primary and secondary hydrogens and as a consequence, (iii) a 'propagation' of the tunneling phenomenon from the primary into the secondary hydrogen position.[16]

Table I: Magnitude of tritium isotope effects in the yeast alcohol dehydrogenase reaction.

t(°C)	k_D/k_T(obs)	k_H/k_T(calc)^b	k_H/k_T(obs)	Exp^c
		1° KIE		
0	1.852±0.047	7.456±0.617	8.611±0.193	3.49
25	1.73±0.02	5.91±0.20	7.13±0.07	3.58^a
		2° KIE		
0	1.059±0.017	1.205±0.063	1.401±0.027	5.89
25	1.03±0.006	1.11±0.02	1.35±0.015	10.2 ^{a}

^{a} From ref 16.
^{b} k_D/k_T(obs)$^{3.26}$ = k_H/k_T(calc).
^{c} Exponent relating k_D/k_T(obs) to k_H/k_T(obs).

2.2 Kinetic Complexity Obscures H-Tunneling

As originally introduced by Northrop in the context of H/D and H/T isotope effects,[1] kinetic complexity has a direct and predictable influence on the exponential relationship between multiple isotope effects [*cf* eqns (1) and (2)]. In the case of D/T and H/T measurements it can be shown[16] that the exponential relationship between k_D/k_T and k_H/k_T will become deflated when chemistry is only partially rate limiting, *ie,*

Kinetic Complexity: k_D/k_T(obs)$^{3.26}$ > k_H/k_T (6)

To a first approximation, the relationship expressed in eqn (6) is quite fortuitous, since it is *opposite* to that expected in the instance of tunneling, eqn (5). Thus, with rare exception (*cf* ref 7), the observation of exponents larger than 3.26 relating measured D/T isotope effects to H/T values can be attributed in an unambiguous manner to H-tunneling.

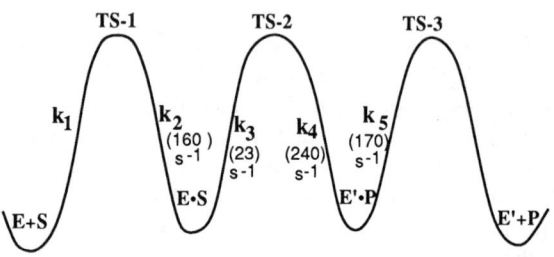

Figure 1: Rate constants for the conversion of benzyl alcohol to benzaldehyde, catalyzed by horse liver alcohol dehydrogenase. E refers to E•NAD$^+$ and E' to E•NADH. k_2 and k_5 are the loss of alcohol and aldehyde from the ternary complex; k_3 and k_4 refer to the hydride transfer step (from ref 17).

Many enzymes are, of course, characterized by reaction profiles in which multiple steps contribute in a fairly equivalent manner to the turnover rate. Figure 1 shows such a free energy profile for the horse liver ADH catalyzed interconversion of benzyl alcohol to benzaldehyde, obtained from a combination of stopped flow kinetic experiments and curve fitting routines.[17] It can be seen that the rate constant for alcohol loss from ternary complex is approximately 7-fold greater than NAD$^+$ reduction by alcohol. In the reverse direction, however, benzaldehyde dissociates at only 70 % the rate of the reverse hydride transfer step. Under this circumstance, it is expected that tunneling and kinetic complexity will work in opposite directions and that our ability to detect H-tunneling will be compromised. This possibility is, in fact, borne out by the experimental results. As summarized in Table II, exponents relating k_D/k_T to k_H/k_T in the horse liver ADH reaction are found to be close to the semi-classical limit of 3.26 (av = 3.25 for primary measurements and 3.92 for secondary isotope effects). Although there is every reason to expect that tunneling contributes to the horse liver system in a manner analogous to yeast alcohol dehydrogenase, a simple comparison of D/T to H/T isotope effects is not sufficient in this case.

Table II: Magnitude of tritium isotope effects in the horse liver alcohol dehydrogenase reaction.[a]

t(°C)	k_D/k_T(obs)	k_H/k_T(calc)[b]	k_H/k_T(obs)	Exp[c]
		1° KIE		
0	1.840±0.024	7.300±0.310	7.365±0.149	3.28
25	1.726±0.021	5.926±0.235	6.355±0.116	3.39
45	1.726±0.032	5.926±0.358	5.415±0.124	3.09
				(Av = 3.25)
		2° KIE		
0	1.074±0.010	1.262±0.038	1.337±0.018	4.07
25	1.070±0.009	1.247±0.034	1.297±0.012	3.84
45	1.064±0.006	1.224±0.023	1.269±0.018	3.84
				(Av = 3.92)

[a] From ref 18.
[b] k_D/k_T(obs)$^{3.26}$ = k_H/k_T(calc).
[c] Exponent relating k_D/k_T(obs) to k_H/k_T(obs).

2.3 Unmasking of H-Tunneling through Site Specific Mutagenesis

The above analysis suggests that a failure to detect tunneling may result from the nature of rate limiting steps, rather than an inherent property of the H-transfer step. Given the power of molecular biological recombinant technology, we set about modifying side chains in the active site of horse liver ADH with the goal of speeding up the release of aldehyde from the ternary complex and hence, unmasking tunneling. A computer drawn ribbon structure of the horse liver ADH active site, generated from the available X-ray data, is shown in Figure 2.

The zinc at the center of the drawing is the active site metal, which is illustrated complexed to the oxygen of bound p-bromobenzyl alcohol in close proximity to the nicotinamide ring of the bound cofactor, NAD^+. Using this structure as a guide, hydrophobic residues within 4 Å (C-C distance) of the bound alcohol were selected as targets for mutagenesis. The liver ADH mutants examined include Leu^{57}→Val, Leu^{57}→Phe, Phe^{93}→Trp and a chimeric combination of E- and S-isozymes (ESE, ref 19) which differs from the E-isozyme by Thr^{94}→Ile, Arg^{101}→Ser, Phe^{110}→Ser and the deletion of Asp^{115}. This deletion is believed to make the binding pocket larger by shifting Leu^{116} away from the alcohol binding pocket, generating an enzyme form which is active on steroidal substrates. The positions of these residues are indicated in Figure 2, in relation to bound alcohol.

Steady state parameters (V_{max} and V_{max}/K_M), isotope effects [k_D/k_T(obs) and k_H/k_T(obs)], and their exponential relationship have been determined for each of the available mutant proteins, Table III.

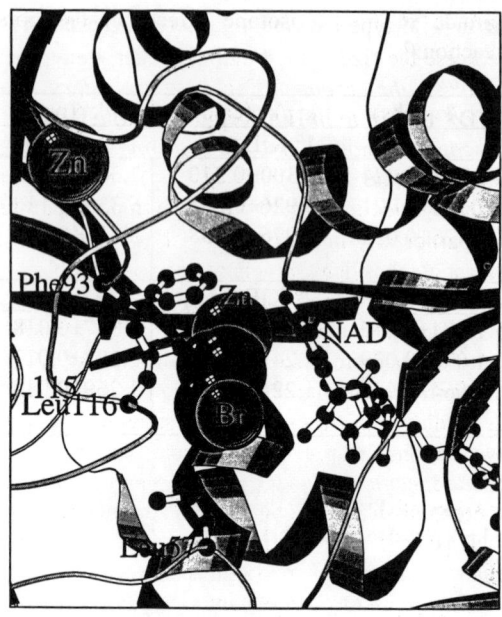

Figure 2: Computer drawn ribbon structure of the active site of horse liver alcohol dehydrogenase (from ref 20).

Table III: Kinetic parameters and exponential relationship between tritium isotope effects for wild-type and mutant forms of horse liver ADH.[a]

	wild-type	larger site ESE	larger site leu^{57}→val	smaller site leu^{57}→phe	smaller site phe^{93}→trp
V_{max} (s^{-1})	0.32	1.7	0.095	0.24	0.13
V_{max}/K_m (mM^{-1}s^{-1})	8.8	3.3	3.5	8.6	4.7
Exp(1°KIE)[b]	3.08	3.23	3.14	3.30	3.31
Exp(2°KIE)[b]	4.10	3.96	4.55	8.50	6.13

[a] T = 25 °C. From ref 20.
[b] $[k_D/k_T(obs)]^{Exp} = k_H/k_T(obs)$.

It can be seen that mutants can be divided into two groups, involving those which increase the size of the substrate bonding pocket and those which diminish this pocket. The two mutants leading to an enlarged pocket, ESE and Leu^{57}→Val, indicate kinetic isotope effects (data not shown) and exponential relationships which are similar to wild type enzyme. A more interesting result is seen with mutants which decrease the size of the binding pocket, Leu^{57}→Phe and Phe^{93}→Trp. In both instances, the exponents relating secondary isotope effects are significantly elevated relative to the semi-classical value of 3.26, providing clear cut evidence for coupled motion and H-tunneling. This occurs without a significant alteration in either V_{max} or V_{max}/K_m, suggesting subtle changes in rate constants for the hydride transfer step relative to product release. The Leu^{57}→Phe mutant is a particularly interesting one, showing very similar exponential relationships to yeast alcohol dehydrogenase (compare Table I to Table III). In the case of the yeast enzyme, Trp is found at both positions 57 and 93. It appears that enlarging the residues at these positions in horse liver ADH converts the enzyme to a form which more closely resembles yeast ADH, enhancing the contribution of chemistry to the rate limiting step and hence, our ability to detect tunneling. These studies confirm our initial suspicion that H-tunneling may be obscured in many reactions due to a lack of rate limitation by the H-transfer step. *As demonstrated, site specific mutagenesis can be employed to unmask tunneling.* Alternatively, in situations where mutagenesis is premature, due to the lack of a DNA encoding the protein of interest or the failure to achieve expression of active enzyme, changes in reaction conditions using wild type enzyme can lead to the same result (*cf* the discussion of monoamine oxidase below).

2.4 Demonstration of H-Tunneling in a Proton Abstraction Reaction

Given the early stage of investigations of H-tunneling in enzyme reactions, it is important to establish the scope of this phenomenon - in particular whether it can be generalized beyond that for NAD$^+$-dependent hydride transfer reactions. The reaction catalyzed by bovine plasma amine oxidase appeared a perfect candidate for such an investigation, in light of the observation of an unusually large primary isotope effect, $k_H/k_T = 35$ at 25°C (in excess of a semi-classical limit for k_H/k_T of 23).[21] Recent structure reactivity correlations with a set of ring substituted benzylamine substrates have established a proton activation mechanism for this enzyme.[22]

Table IV: Comparison of stopped flow and steady state deuterium isotope effects for the oxidation of benzylamine, catalyzed by bovine serum amine oxidase.[a]

t(°C)	k_H/k_D(stopped flow) (E•S → E•P)	k_H/k_D (steady state) (E + S → E + P)
0	14.8±0.7	19.0±0.8
5	15.7±1.2	18.7±0.2
15	16.6±1.4	16.5±0.4
25	16.1±1.8	13.5±0.4
35	13.8±1.1	13.0±0.4
45	10.4±0.9	10.4±0.2

[a] From ref 21.

Initially, non-competitive stopped flow isotope effect experiments were performed as a function of temperature, in an effort to determine the size of the deuterium isotope effects for the direct conversion of E·S to E·P. The results of this study are compared to steady state isotope effects on V_{max}/K_m, which reflect the somewhat more complicated reaction involving the conversion of E + S to E + P. It can be seen (Table IV) that the magnitude of effects determined by these two techniques agree almost exactly between 10 and 45 °C, allowing us to conclude that V_{max}/K_m is controlled by the chemical step under these conditions.[21] Somewhat unexpectedly, the stopped flow isotope effects decrease relative to steady state values below 10 °C, which has been interpreted as an onset of substrate inhibition which is greater with the protio- than deuterio-substrate. This substrate inhibition is not expected to influence V_{max}/K_m tritium isotope effects measurements, and k_D/k_T and k_H/k_T were determined as a function of temperature.

The temperature dependence of isotope effects has been a fairly standard tool for the demonstration of H-tunneling for reactions in solution. Since tunneling is, to a first approximation, temperature independent, reaction rates are expected to level off as the temperature is lowered, producing curvature in Arrhenius plots. Given the mass dependence of tunneling, the degree of curvature in Arrhenius plots is expected to be isotope dependent in the order: H > D > T. A consequence of the above ordering is that extrapolation of Arrhenius plots to infinite temperature leads to a crossing of lines with the resulting ordering of Arrhenius pre-factors: $A_H < A_D < A_T$.[23,24] As Koch and Dahlberg have described, it is important to restrict this approach to reactions which are kinetically limited by the H-transfer step, since changes in rate limiting step can give artifactual curvature in Arrhenius plots.[25] When the criterion of limiting H-transfer is met and tunneling is significant, A_L/A_T is expected to lie below semi-classical limits of 0.6 and 0.9 (for L = H and D, respectively).

The results of the temperature dependence of k_H/k_T and k_D/k_T in the bovine serum amine oxidase reaction are shown in Figure 3.

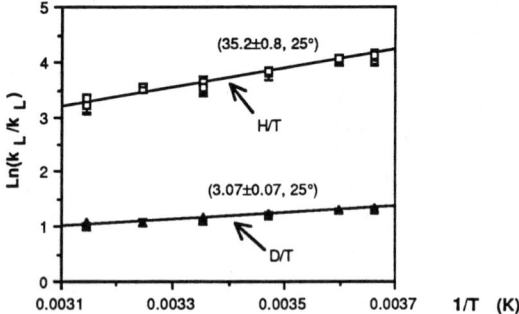

Figure 3: Temperature dependence of primary tritium isotope effects in the oxidation of benzylamine catalyzed by bovine serum amine oxidase (from ref 21).

Hydrogen Tunneling in Enzyme Reactions

The data appear linear in the experimental temperature range and can be extrapolated to infinite temperature, yielding the Arrhenius pre-factors of Table V.

Table V: Arrhenius parameters for 1° KIE's with plasma amine oxidase.[a]

	A_L/A_T(obs)	A_L/A_T(sc)
k_H/k_T	0.12±0.04	> 0.6 [b]
k_D/k_T	0.51±0.10	> 0.9 [b]

[a] From ref 21.
[b] Semi-classical limits, from ref 26.

The value for A_H/A_T = 0.12±0.04 is well below the semi-classical limit, providing clear evidence for protio-tunneling. Additionally, the intercept value from k_D/k_T measurements is reduced from a semi-classical limit of 0.9 to 0.51, showing that significant deuterio-tunneling is occurring as well. One of the advantages of an analysis of D/T isotope effects is that they can be assumed to be completely free of kinetic complexity under conditions where H/T measurements may be slightly reduced due to small commitments. This can be understood in terms of eqn (1), such that any commitment terms (C_f and C_r) in D/T measurements are reduced by the magnitude of the intrinsic H/D isotope effect.[21]

In light of the evidence from temperature dependencies for tunneling in bovine serum amine oxidase, the exponential relationship between experimental k_D/k_T and k_H/k_T values was examined. Values for exponents are summarized in Table VI, showing an average value of 3.17 for primary effects and 4.04 for secondary effects.

Table VI: Temperature dependence of the exponential relationship between tritium isotope effects in the bovine serum amine oxidase reaction.[a]

t(°C)	Exp(1°KIE)	Exp(2°KIE)
0	3.11	4.15
5	3.14	4.80
15	3.13	4.17
25	3.17	3.56
35	3.31	4.00
45	3.16	3.13
	(Av = 3.17±0.05)	(Av = 4.04±0.46)

[a] From ref 21.

It is of considerable interest that both primary and secondary k_D/k_T values are related to k_H/k_T by an exponent close to semi-classical exponential value of 3.26. The failure to see a greatly inflated value for the secondary exponent argues against any significant coupling of motion between the primary and secondary hydrogens of substrate in the transition state.[21] Clearly, this behavior is very different from that seen in dehydrogenases, although we do not yet have sufficient understanding of this phenomenon to be able to predict when coupled motion will arise. As regards the primary exponent, these tend to be less inflated than secondary values even under the circumstance of significant H-tunneling (cf Tables I and III). In a paper by Grant and Klinman, calculations using the expanded Bell correction for tunneling indicate that exponents will peak for reaction coordinate frequencies somewhat greater than 1000 i cm^{-1} and then decline again toward 3.26.[7] Thus, it is conceivable that the exponential inflations predicted by Saunders will only occur when tunneling is moderate and primary isotope effects are relatively small. In an interesting theoretical analysis by Huskey, it is argued that breakdowns from the rule of the geometric mean, manifest in 'abnormal' exponents, will only occur when coupled motion is an important component of the reaction coordinate.[27] This latter point can be examined experimentally and hopefully, will be addressed in the near future.

2.5 Demonstration of Tunneling in the Flavo-Protein, Monoamine Oxidase B

In our continuing effort to examine the scope of H-tunneling in enzymic reactions, the monoamine oxidase reaction appeared to be an excellent system to pursue. This enzyme contains covalently bound flavin and hence would extend the phenomenon of H-tunneling to a new class of redox cofactor. In addition, the enzyme reacts with benzylamines, allowing the use of previously synthesized, isotopically labelled substrates. Finally, experimental evidence supports the intermediacy of radical intermediates in benzylamine oxidation, extending our studies into this class of reaction.[28]

Initially, isotope effects were measured for p-methoxybenzylamine at pH 7.5. At 25 °C, the primary k_H/k_T isotope effect is 21.4±0.8, which is comparable to a k_H/k_D of 8.4 [using eqn (2)]; this value is in good agreement with a previously determined value for k_H/k_D using stopped flow kinetics.[28] Subsequently, k_D/k_T and k_H/k_T isotope effects were measured at 2 and 43 °C, showing a trend in exponential relationships. As can be seen from Table VII, the exponents relating primary isotope effects decrease from 3.13 to 2.57 [with exponents relating secondary isotope effects decreasing from 2.79 to 1.20 (data not shown)]. This trend strongly suggests the presence of a temperature dependent commitment at pH 7.5 and thus, the need to find another set of conditions for pursuing tunneling. To search for conditions that might give isotope effects closer to intrinsic values, $^DV_{max}$ and $^D(V_{max}/K_m)$ were measured at pH 6.0, 7.5 and 9.0, showing a general increase of isotope effects as the pH was lowered. Additionally, V_{max}/K_m was found to decrease approximately 20-fold as the pH was lowered from 7.5 to 6.0, conditions expected to reduce the contribution of an external commitment to the overall rate.

Table VII. pH and temperature dependence of primary tritium isotope effects in the oxidation of *p*-methoxybenzylamine, catalyzed by monoamine oxidase B.[a]

t(°C)	k_D/k_T(obs)	k_H/k_T(obs)	Exp [b]
	pH 7.5	pH 7.5	
2.0	2.70±0.04	13.03±0.51	2.57
25.0	2.70±0.10	21.43±0.81	3.08
43.2	2.46±0.07	16.87±0.35	3.13
	pH 6.1	pH 6.1	
10.0	2.90±0.13	28.89±1.10	3.16
15.0	2.89±0.05	27.59±0.70	3.12
25.0	2.69±0.09	21.98±1.16	3.20
30.0	2.63±0.04	21.49±0.95	3.17
35.0	2.57±0.07	18.52±0.99	3.09
40.0	2.51±0.07	17.42±0.40	3.11
43.0	2.51±0.13	17.02±0.52	3.08

[a] From ref 29.
[b] Exponent relating k_D/k_T(obs) to k_H/k_T(obs).

Encouraged by these findings, the measurement of k_D/k_T and k_H/k_T was pursued in detail at pH 6.1 between 10 and 43 °C (Table VII); experiments were not carried out below 10 °C, due to an untenably slow reaction with the deuterated substrate. At 25 °C, both isotope effects and exponents are found to be very similar to pH 7.5, indicating that if commitments are present they are not altered by the large change in rate that accompanies the reduction in pH. Of particular significance is the observation that exponents do not change systematically in the 10 to 43 °C range. This shows there are no temperature dependent changes in the level of expression of intrinsic isotope effects, such that a commitment, if present, would have to be independent of temperature as well as pH. The fact that the primary exponent is unchanged, despite a combined 260-fold change in rate (20-fold due to pH and 13-fold due to temperature), supports the view that the H-transfer step is fully expressed at pH 6.1.

In this context we were surprised to find that the exponents relating secondary isotope effects all fall below 3.26. Analogous to the interpretation of isotope effects in the bovine plasma amine oxidase reaction, the small magnitude of exponents relating secondary isotope effects rules out any significant coupling of motion between the primary and secondary hydrogens of substrate. Although we do not yet have an explanation for exponents below 3.26, there are two possibilities under consideration. The first is that a small temperature *independent* commitment is present under all conditions of measurement.

(Secondary isotope effects, by nature of their smaller size, show a greater sensitivity of exponential relationships to commitments.) As the commitment neither appears to change with pH at 25 °C nor with temperature at pH 6.1, it would have to be an internal commitment related to the H-transfer step itself. A second, more speculative explanation relates to the nature of the H-transfer step itself. As described by Truhlar and co-workers, tunneling leads to corner cutting across the potential energy surface, with H and D crossing the barrier at different positions.[30] In the event of a large degree of tunneling, the position of the barrier crossing for H vs D will differ significantly. Since in the absence of coupled motion secondary isotope effects reflect changes in force constant between the ground state and point of barrier penetration, this could give rise to 'different' secondary isotope effects for D transfer (k_D/k_T) than H transfer (k_H/k_T). Under these conditions, the exponential relationship between k_D/k_T and k_H/k_T may be difficult to predict.

Arrhenius plots of the data in Table VII at pH 6.1 lead to straight lines, which upon extrapolation to infinite temperature give the Arrhenius pre-factors in Table VIII.

Table VIII: Arrhenius parameters for 1°KIE's with monoamine oxidase B.[a]

	A_L/A_T(obs)	A_L/A_T(sc)
k_H/k_T	0.13±0.03	>0.6[b]
k_D/k_T	0.52±0.05	>0.9[b]

[a] From ref 29.
[b] Semi-classical limits, from ref 26.

The values for $A_H/A_T = 0.13\pm0.03$ and $A_D/A_T = 0.52\pm0.05$ are well below the semi-classical limits and, in fact, appear quite similar to the data for bovine serum amine oxidase (Table V). In light of the possibility of a temperature independent commitment, the data were re-analyzed on the assumption of a small commitment which would restore the secondary exponent to 3.26. Using a C_f value of 0.5 changed the data very little, yielding values for A_H/A_T and A_D/A_T of 0.18 and 0.53, respectively. It should be noted that the presence of this commitment has a far greater effect on the individual isotope effect values, elevating intrinsic isotope effects above measured values in Table VII, eg to a primary k_H/k_T of 32.5 and a secondary k_H/k_T of 1.51 at 25 °C. Both of these appear outside the semi-classical range. Once again, the temperature dependence of k_D/k_T provides an excellent control, since the commitment in k_D/k_T is expected to be negligible.

In the monoamine oxidase reaction we find that the temperature dependence of secondary isotope effects does not give isotopic Arrhenius pre-factors significantly below unity ($A_H/A_T = 0.77\pm0.15$ and $A_D/A_T = 0.92\pm0.04$).

While the small deviations from unity could be due to coupled motion between the hydrogen being transferred and the secondary hydrogen, the magnitude of exponents argues against this being the case. The large secondary isotope effects (k_H/k_T(obs) = 1.34 at 25 °C, pH 6.1) are close to equilibrium values seen for $sp^2 \rightarrow sp^3$ rehybridizations. Thus, it is possible that the oxidation of substrate occurs in two steps, with an equilibrium secondary isotope effect on the first step and a large primary isotope effect on the second step. However, as a result of stopped flow studies, which fail to detect chemical intermediates in benzylamine oxidation and structure reactivity correlations, which implicate free radical intermediates,[28] the most straightforward mechanism that can be written for monoamine oxidase B is one involving a single step hydrogen atom transfer accompanied by significant tunneling.

2.6 Factors Controlling Tunneling in Enzyme Reactions

The above described evidence for H-tunneling in the reactions catalyzed by alcohol dehydrogenases from yeast and horse liver, bovine serum amine oxidase and monoamine oxidase B suggests that this may be a general property of enzyme catalyzed H-transfer process. From the outset we have been curious whether there are unique properties of enzyme catalysis which may facilitate tunneling. Features of enzymes which may qualify in this regard are:

(i) *The exclusion of bulk solvent from enzyme active sites, with the role for solvent being assumed by carefully positioned active site side chains.* For reactions in solution, solvent reorganization is generally required to satisfy changes in charge distribution at the transition state. To the extent that solvent motion accompanies H motion, the mass of the transferred particle is increased and tunneling is decreased.

(ii) *The alteration of equilibrium constants toward unity for the interconversion of bound reactants and products.* Tunneling can only occur from matched energy states in reactant and product, and hence is expected to become more probable the more closely matched energy levels are for E·S relative to E·P. It has previously been argued that matching of internal energy states is a catalytic advantage to enzymes.[31]

(iii) *The binding of reactants in close proximity with precise geometries at enzyme active sites.* Tunneling is exquisitely dependent on the ratio of reaction barrier height to width (reaction barrier curvature), such that alterations in this shape through, for example, compression of bound substrates could have a very significant effect on the tunneling probability.

Several of these points are currently under investigation, and will be addressed in part below. Additionally, we have been attempting to understand the role for thermal activation in enzyme catalyzed H-transfer reactions, given an expected temperature independence of tunneling. Measured enthalpies of activation for the H-transfer step in each enzyme system studied thus far are, indeed, sizeable: E_a = 51 kJ/mol, 56 kJ/mol and 54 kJ/mol for yeast alcohol dehydrogenase, bovine serum amine oxidase and monoamine oxidase B, respectively.

2.6.1 Nature of Thermal Activation

Insight into the role for thermal activation in enzyme catalyzed tunneling reactions is intimately linked to the boundary conditions established for the modeling of this phenomenon. Modeling can be addressed in a more straight forward manner for bovine serum amine oxidase and monoamine oxidase B than for alcohol dehydrogenase, given the absence of data implicating coupled motion in the former two systems. Bruno and Bialek have explored the properties of the bovine serum amine oxidase reaction, using a model in which *all* of the H motion is quantum mechanical. These authors have been successful in reproducing temperature dependent primary isotope effects through the use of a fluctuating potential for protein.[32] According to this view of catalysis, thermal motion in protein gives rise to continuous sampling of ground state configurations for bound reactants. At a critical configuration, the ratio of the barrier height to width is optimal for tunneling and hydrogen moves from the reactant to product well. A feature incorporated into this scheme is degenerate energy levels for E•S and E•P, although this is not essential, since a more complicated model could allow for adjustments in internal energy levels prior to the tunneling event.

While the above described model provides a satisfactory fit to the data for bovine serum amine oxidase, it is likely to be incomplete in that it does not allow for partitioning of thermal excitation into the reacting C-H bond. It is to be expected that increases in temperature will lead to bond distortions involving changes in bond angles, van der Waals' interactions and possibly excitation of the reacting C-H bond to the first excited vibrational mode. Some of the earliest efforts at modeling H-tunneling in organic reactions involved the addition of a correction factor to the reaction rate on the assumption that the quantum event would occur near the top of the classical barrier (*eg* the Bell correction[23]). Given the extreme simplicity of this type of model it is remarkable that it has been so successful in reproducing experimental phenomenon. For example, this was the approach used by Huskey and Schowen[15] in their original predictions of H-tunneling and in Saunders' calculations[6] leading to the proposed comparison of D/T and H/T isotope effects as a probe of tunneling. Of particular significance, both of these theoretical predictions were made in advance of their experimental verification in the reactions catalyzed by yeast and horse liver alcohol dehydrogenase.

More recently, Truhlar and co-workers have advanced a sophisticated approach to the computation of reaction rates and isotope effects.[30] This method involves the construction of a potential energy surface for the reaction under study, followed by the application of variational transition state theory to calculate rate parameters. Computations for a variety of H-transfer reactions reveal a phenomenon called 'corner cutting', which involves H and D crossing of the potential energy surface at some point below the top of the barrier. A unique feature of this treatment is that H and D cross the potential barrier at different positions. Recent efforts at reproducing the available data for the yeast alcohol dehydrogenase reaction suggest that observed isotope effects and their exponential relationship can be matched with variational transition state theory.[33] Although the model currently under study is restricted to 10 atoms, and cannot accommodate detailed interactions between protein side chains and bound substrates, the preliminary results are encouraging. One feature of tunneling near the top of the semi-classical barrier is the high density of states, such that tunneling is not expected to be critically dependent on degenerate ground state energy levels for the interconversion of reactant and product. This relationship between internal thermodynamics and tunneling has recently been addressed for the interconversion of aromatic benzyl alcohol substrates, catalyzed by yeast alcohol dehydrogenase.

2.6.2 Role of Internal Thermodynamics in H-tunneling

Our initial studies of tunneling in the yeast ADH reaction employed benzyl alcohol, a substrate which, by chance, was characterized by an internal equilibrium constant [K_{eq}(int)] close to unity. It therefore became important to examine alternate substrates characterized by perturbed values for K_{eq}(int). The substrate *p*-chlorobenzyl alcohol, with a value for K_{eq}(int) *ca* 10-fold less than benzyl alcohol, was an excellent candidate for an initial exploration of this issue. As summarized in Table IX, measured values for k_D/k_T and k_H/k_T at 25 °C are related by exponents which are essentially identical to the unsubstituted alcohol. Similarly, a study of temperature dependent isotope effects in the plasma amine oxidase reaction with *p*-methoxybenzyl alcohol has revealed isotopic Arrhenius pre-factors (data not shown) which are almost identical to those seen with benzylamine. These studies indicate clearly that the property of K_{eq}(int) = 1 is *not a prerequisite* for enzymatic H-tunneling.[34]

Table IX: Comparison of tritium isotope effects and their exponential relationship for the oxidation of *p*-Cl and *p*-H-benzyl alcohols by yeast alcohol dehydrogenase.[a]

	1° KIE'S			2° KIE'S		
	k_D/k_T (obs)	k_H/k_T (obs)	Exp[b]	k_D/k_T (obs)	k_H/k_T (obs)	Exp[b]
p-Cl	1.59 ±0.03	6.59 ±0.13	4.06	1.03 ±0.01	1.34 ±0.01	9.90
p-H	1.72 ±0.02	7.13 ±0.07	3.58	1.03 ±0.01	1.35 ±0.02	10.2

[a] T = 25 °C. From refs 16 and 34.
[b] Exponent relating k_D/k_T(obs) to k_H/k_T(obs).

A more quantitative evaluation of this phenomenon has been possible using multiple substrates for yeast ADH. As shown in previous studies,[12] ring substituted benzyl alcohols are characterized by almost identical rate constants in the direction of NAD$^+$ reduction (k_{ox}):

$$E \cdot NAD^+ \cdot X\text{-}C_6H_4\text{-}CH_2OH \underset{k_{red}}{\overset{k_{ox}}{\rightleftarrows}} E NADH \cdot X\text{-}C_6H_4\text{-}CHO$$

By contrast, rate constants in the direction of aldehyde reduction (k_{red}) increase in a regular fashion in proceeding from electron releasing to electron withdrawing substituents.[11] A consequence of these trends is a regular decrease in K_{eq}(int) and a concomitant increase in DG° from -6.9 kJ/mol (*p*-methoxy) to 0.2 kJ/mol (*p*-H) to 5.8 kJ/mol (*p*-chloro).[12] Both the nature of structure reactivity correlations and the small magnitude of the secondary k_D/k_T isotope effect with benzyl alcohol (k_D/k_T = 1.03 vs K_D/K_T = 1.10)[16] indicate an early transition state, closely resembling alcohol. Using the Hammond postulate to predict transition state structure, it is expected that the transition state will become more symmetrical as the rate for aldehyde reduction increases; a corollary of this prediction is that intrinsic isotope effects will increase in the order: *p*–CH$_3$O < *p*–H < *p*–Cl.

In Table X magnitudes of primary and secondary k_D/k_T isotope effects have been summarized as a function of three different ring substituents.

Table X: Correlation of intrinsic tritium isotope effects with thermodynamic parameters for the oxidation of ring substituted benzyl alcohols catalyzed by yeast alcohol dehydrogenase.[a]

	p-CH$_3$O-	*p*-H-	*p*-Cl-
1°k_D/k_T	1.94±0.06	1.73±0.02	1.59±0.03
2°k_D/k_T	1.12±0.02	1.03±0.01	1.03±0.01
ΔH° (kJ/mol)	12.9	15.6	21.2
ΔG°(kJ/mol)	-6.9	0.2	5.8

[a] From ref 34.

As noted earlier, magnitudes for k_H/k_T may contain small commitments (even for reactions where H-transfer has been concluded to be rate limiting); however, magnitudes for k_D/k_T are expected to be free of this ambiguity and are, therefore, directly ascribable to intrinsic isotope effects. It is of considerable interest that the trend in both primary and secondary isotope effects in Table X is opposite to that predicted from semi-classical considerations with isotope effects in the order: *p*-CH$_3$O > *p*-H > *p*-Cl.[30]

The relationship between tunneling and thermodynamic driving force has been considered in some detail by Bell,[23] in the context of models in which tunneling occurs near the top of the reaction barrier. While matching of ground state internal thermodynamics is not required for this type of model, the probability of tunneling is predicted to increase as the ground state energies of reactant and product become more equal; since tunneling leads to inflated magnitudes for isotope effects, these are also predicted to change in an analogous manner. The logic for this correlation derives from an analysis of the area under the reaction curve available for tunneling. Microscopic reversibility dictates that this area will be the same in both reaction directions, such that within a series of reactants those characterized by exo- and endo-energetic reaction profiles are predicted to have decreased tunneling relative to the isoenergetic state (cf Figure 4).

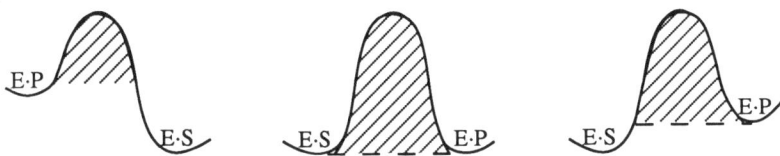

Figure 4: The area available for tunneling (hatched) is shown to depend on the energy difference between reactants and products (from ref 23).

At the time that we initiated our studies of yeast ADH we were uncertain whether a correlation between degree of tunneling and internal thermodynamics would be more likely to occur with $\Delta G°$ or $\Delta H°$. A kinetic study of k_{ox} and k_{red} was therefore conducted as a function of temperature, leading to estimates of $\Delta H°$ and $T\Delta S°$ as a function of substituent. As indicated in Table X, $\Delta H°$ (not $\Delta G°$) increases in a fashion which corresponds directly to the order of observed k_D/k_T isotope effects. From the data for ADH we conclude (i) that tunneling can occur when $\Delta H°$ is endothermic and (ii) that the degree of tunneling increases as the value of $\Delta H°$ decreases toward zero.[34] While it is premature to try to generalize beyond these studies, it will be extremely interesting to see the consequence of perturbations in internal thermodynamics (through the use of mutant enzymes as well as altered substrates) on H-tunneling in additional enzyme systems.

3   SUMMATION AND FUTURE DIRECTIONS

As shown in this report, the phenomenon of H-tunneling has now been extended to include nicotinamide-dependent reactions (ADH's), a topa-quinone-containing enzyme (the copper containing serum amine oxidase) and a flavo-protein (monoamine oxidase B). These enzymes encompass reactions involving hydride, proton and hydrogen atom activation. The techniques that have proven most fruitful involve a study of k_D/k_T and k_H/k_T isotope effects and their exponential and temperature dependencies. As described in several systems, the successful application of these probes requires that the H-transfer step be close to rate limiting. While many enzymes do not fall into this category, changes in rate determining step can be brought about through the use of alternate substrates and more recently, through site specific mutagenesis. The question of the role for thermal activation in facilitating the temperature independent quantum event is an intriguing one and model dependent. A realistic, current view involves the distribution of thermal activation into a variety of modes which leads to changes in protein geometry and distortion of the C-H bond itself. The position of barrier crossing is likely to occur near the top of the classical barrier with a very small fraction of reactants actually passing over the top of the barrier. For the future it will be extremely interesting to examine the effects of very low temperature on the ability of an enzyme to bring about H-tunneling; in principle this could be achieved in cryosolvents through the use of a pre-bound, caged reagent whose reaction would be initiated by photoactivation. Two features of enzyme reactions which may facilitate the tunneling process are the tendency of enzymes to perturb the equilibrium constant for conversion of bound substrate and product toward unity and the possible use of compression in achieving the closest possible proximity between reactants. Regarding the importance of thermodynamics in the control of tunneling, it has been shown that tunneling can occur for systems which deviate from $K_{eq}(int) = 1$. A more quantitative analysis of this phenomenon in the yeast ADH reaction indicates that tunneling may, however, peak when $\Delta H° = 0$. Ongoing studies are now focused on the potentially important role of compression in bringing about tunneling. Site specific mutagenesis offers the opportunity to modify selectively individual residues which contact bound substrates. Preliminary studies with ADH suggest that a change in the bulk of a hydrophobic residue which resides behind the reacting carbon at C-4 of the nicotinamide cofactor can alter the degree of tunneling; in particular, replacement of Val[203] by Ala has been found to reduce tunneling substantially, accompanied by a *ca.* 70-fold reduction in V_{max}/K_m.[35] This intriguing result suggests that H-tunneling may provide several orders of magnitude rate enhancement to enzyme mediated H-processes, possibly *via* an 'evolutionary fine tuning' of the catalytic potential energy surface.

REFERENCES

1   D B Northrop, *Biochemistry*, 1975, **14**, 2644.
2   M Palcic and J P Klinman, *Biochemistry*, 1983, **22**, 5957.
3   S M Miller and J P Klinman, *Biochemistry*, 1983, **22**, 3091.
4   C G Swain, E C Stiver, J R Reuwer, and L J Schaad, *J Am Chem Soc*, 1958, **80**, 5885.
5   H J Stern and R E Weston, *J Chem Phys*, 1974, **60**, 2815.
6   W H Saunders, *J Am Chem Soc*, 1985, **107**, 164.
7   K L Grant and J P Klinman, *Bioorg Chem*, 1992, **20**, 1.
8   D B Northrop and R G Duggleby, *Bioorg Chem*, 1990, **18**, 435.

9   H Eklund, B V Plapp, J-P Samama, and C-I Brandén, *J Biol Chem*, 1982, **257**, 14349.
10 J P Klinman, *CRC Crit Rev Biochem*, 1981, **10**, 39.
11 J P Klinman, *J Biol Chem*, 1972, **247**, 7977.
12 J P Klinman, *Biochemistry*, 1976, **15**, 2018.
13 K M Welsh, D J Creighton, and J P Klinman, *Biochemistry*, 1981, **19**, 2005.
14 P F Cook, N J Oppenheimer, and W W Cleland, *Biochemistry*, 1981, **20**, 1817.
15 W P Huskey and R L Schowen, *J Am Chem Soc*, 1983, **105**, 5704.
16 Y Cha, C J Murray, and J P Klinman, *Science*, 1989, **243**, 1325.
17 V C Sekhar and B V Plapp, *Biochemistry*, 1990, **29**, 4289.
18 Y Cha and J P Klinman, unpublished results.
19 D-H Park and B V Plapp, *J Biol Chem*, 1992, **267**, 5527.
20 B J Bahnson, D-H Park, K Kim, B V Plapp, and J P Klinman, *Biochemistry*, 1993, **32**, 5503.
21 K L Grant and J P Klinman, *Biochemistry*, 1989, **28**, 6597.
22 C Hartmann and J P Klinman, *Biochemistry*, 1992, **30**, 8138.
23 R P Bell, *The Tunnel Effect in Chemistry*, Chapman and Hall, New York, 1980.
24 J P Klinman, *Trends in Biochem Sci*, 1989, **14**, 368.
25 H F Koch and D B Dahlberg, *J Am Chem Soc*, 1980, **102** 6102.
26 M E Schneider and H J Stern, *J Am Chem Soc*, 1972, **94**, 1517.
27 W P Huskey, *J Phys Org Chem*, 1991, **4**, 361.
28 M C Walker, PhD Thesis, Emory University, 1987.
29 T Jonsson, D Edmondson, and J P Klinman, *Proc 11th Intl Symp Flavins and Flavoproteins*, Nagoya, Japan, in press.
30 a) B C Garrett and D G Truhlar, *J Am Chem Soc*, 1980, **102**, 2559.
    b) S C Tucker, D G Truhlar, B C Garrett, and A D Isaacson, *J Chem Phys*, 1985, **82**, 4102.
31 a) W J Albery and J R Knowles, *Biochemistry*, 1976, **15**, 5631.
    b) K P Nambiar, D H Stauffer, P A Kolodziej, and S A Benner, *J Am Chem Soc*, 1983, **105**, 5886.
32 W J Bruno and W Bialek, *Biophys J*, 1992, **63**, 689.
33 Y Kim and J P Klinman, unpublished results.
34 J Rucker, Y Cha, T Jonsson, K L Grant, and J P Klinman, *Biochemistry*, 1992, **31**, 11489.
35 B J Bahnson and J P Klinman, unpublished results.

COBALAMIN-DEPENDENT METHIONINE SYNTHASE: DISSECTION OF A LARGE PROTEIN INTO FUNCTIONAL AND STRUCTURAL DOMAINS

James T Drummond, Sha Huang, and Rowena G Matthews*

Department of Biological Chemistry and Biophysics Research Division, The University of Michigan, Ann Arbor, Michigan 48109, USA.

1   INTRODUCTION

Cobalamin-dependent methionine synthase catalyzes the transfer of a methyl group from methyltetrahydrofolate (**1**) to homocysteine (**2**), producing tetrahydrofolate (**3**) and methionine (**4**) (Figure 1). As shown there, the cobalamin prosthetic group functions as a methyl carrier in the transfer, accepting a methyl group from methyltetrahydrofolate to form methylcobalamin and then transferring the methyl group to homocysteine with the concomitant formation of enzyme-bound cob(I)alamin. In agreement with this proposed double displacement mechanism, the overall reaction proceeds with retention of stereochemistry at the transferred methyl group.[1]

The enzyme must pay a price for using the highly reactive and strongly reducing cob(I)alamin intermediate during turnover. This form of the enzyme is occasionally intercepted and oxidized to form inactive cob(II)alamin enzyme. Return of the cob(II)alamin enzyme to the catalytic cycle requires a reductive methylation.[2] This methylation requires adenosylmethionine rather than methyltetrahydrofolate as the methyl donor.[3] *In vitro*, activation can be achieved using dithiothreitol and aquocobalamin to mediate reduction or $FMNH_2$ formed by hydrogenation over platinum.[4] Alternatively, two flavoproteins isolated from crude extracts of *Escherichia coli*,[5] and referred to as the R and F proteins, catalyze the transfer of reducing equivalents from NADPH to methionine synthase. These flavoproteins have been purified to homogeneity and characterized,[5,6] and more recently their genes have been cloned and sequenced.[7,8] The R protein is a member of the family of ferredoxin oxido-reductases that accept hydride equivalents from NADPH and transfer electrons to a protein acceptor,[9] in this case the F protein.

The F protein, a flavodoxin, is an electron transferase that mediates electron transfer to methionine synthase. The catalytic cycle of methionine synthase is shown in Figure 2. The enzyme cycles in catalysis between

Cobalamin-Dependent Methionine Synthase

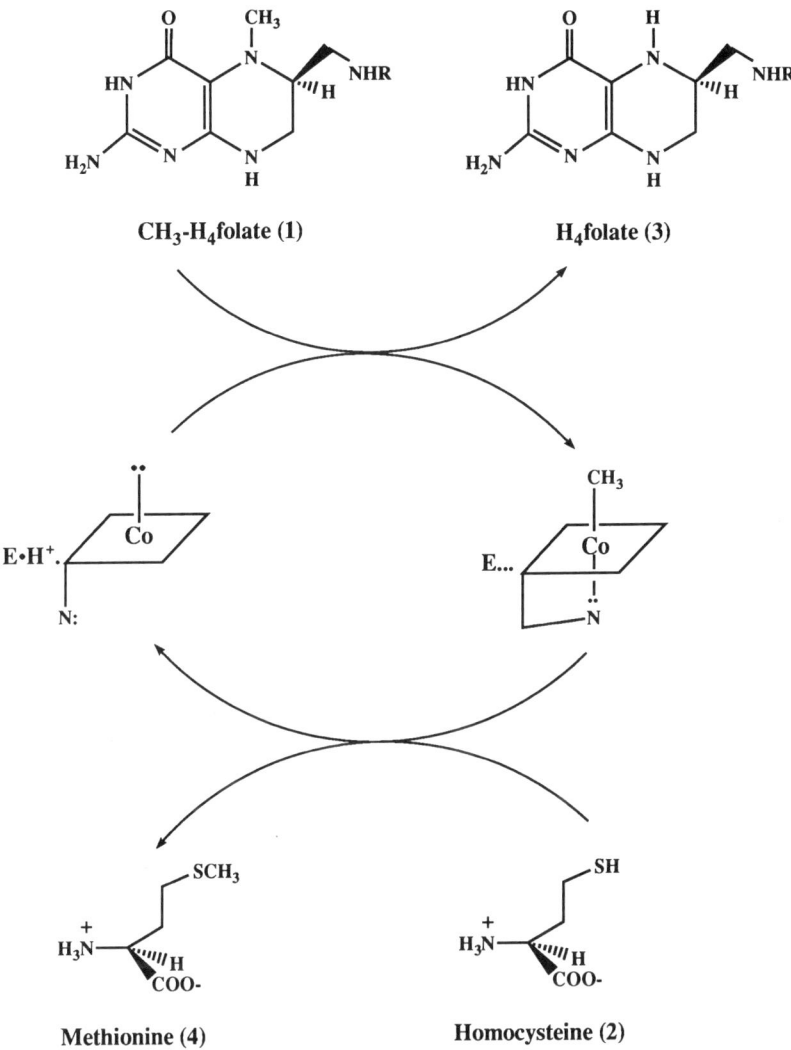

Figure 1 Reaction catalyzed by cobalamin-dependent methionine synthase

methylcobalamin and cob(I)alamin forms. However, the cob(I)alamin prosthetic group is occasionally oxidized to form inactive cob(II)alamin enzyme. It has been estimated that cob(II)alamin enzyme is formed approximately once every 100 turnovers.[6,10] These experiments were performed using NADPH and the R and F proteins to catalyze reductive activation. Our own recent experiments have measured the number of turnovers enzyme with a methylcobalamin prosthetic group undergoes before it is oxidized to cob(II)alamin under an atmosphere of argon, in the absence of AdoMet. These experiments suggest that the enzyme averages 1800 turnovers before oxidation occurs, and may reflect the potent oxygen scavenging activity of the chemical reducing system (aquocobalamin and dithiothreitol) employed.[11] Each time the inactive enzyme is generated, it must be reactivated using adenosylmethionine and the reducing system, which are present under standard assay conditions.

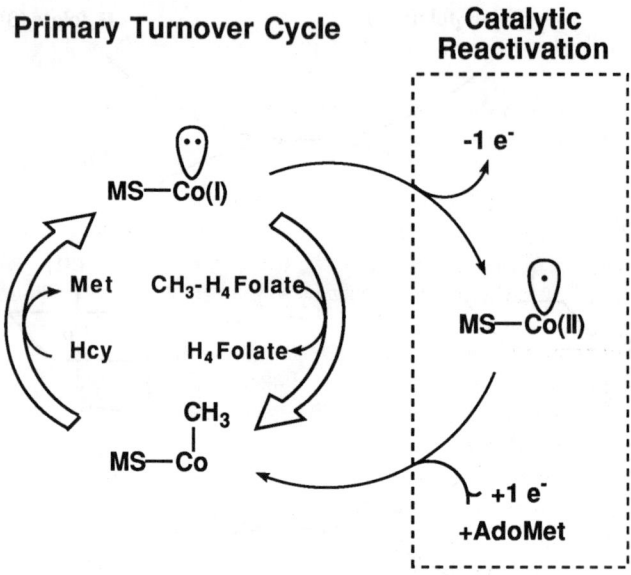

Figure 2 Schematic summary of the catalytic cycle of methionine synthase

There are several issues raised by the proposed catalytic mechanism, and our ongoing research is aimed at addressing these issues. The transfer of a methyl group from methyltetrahydrofolate to cob(I)alamin formally involves nucleophilic displacement at the α-carbon of a tertiary amine, and such a reaction is without precedent in the organic chemistry literature. Salient questions are how methyltetrahydrofolate is activated for methyl transfer, and what the mechanism of transfer might be. The second issue is why adenosylmethionine is required for reductive activation, if cob(I)alamin reacts readily with methyltetrahydrofolate during the catalytic cycle. We have suggested that the requirement for adenosylmethionine is thermodynamic, in that a highly endergonic reduction of cob(II)alamin to cob(I)alamin must be coupled to an exergonic driving reaction.[12] But if so, the mechanism by which reduction of cobalamin is coupled to methyl transfer from adenosylmethionine remains to be described. Finally, the enzyme catalyzes three different methyl transfers to and from the cobalamin prosthetic group, each with highly constraining geometric requirements for the methyl transfer. How is the successive positioning of the substrates *vis á vis* the cobalamin accomplished? What are the signals for determining which substrate is to be presented to the cobalamin? We shall begin by examining relevant model reactions for the reaction catalyzed by cobalamin-dependent methionine synthase, and then proceed to describe our own recent work on the enzymatic reaction.

2   MODEL STUDIES OF THE METHIONINE SYNTHASE REACTION

2.1   Transfer of the Methyl Group From Methylcobalamin to Homocysteine

The transfer of the methyl group from [^{13}C-methyl]methylcobalamin to thiols under anaerobic conditions has been investigated by ^{13}C-NMR.[13] The rate of methyl transfer to β-mercaptoethanol is pH dependent, and reflects the concentration of the attacking thiolate between pH 7 and 11. Such results had been predicted earlier by Schrauzer,[14] who postulated a nucleophilic attack of a thiolate anion on the methyl group of methylcobalamin. The reaction of dithiothreitol with methylcobalamin is first order in methylcobalamin and in thiol (in excess over cobalamin). No evidence for coordination of thiolate to methylcobalamin at the lower axial position is seen, even though the methyl transfer reactions are slow.[13] These results were interpreted as providing evidence for an S_N2 displacement. However, several observations were not fully consistent with this interpretation. In particular, the rate of methyl transfer was decreased only two-fold when methylcobalamin was replaced with ethylcobalamin[15], although S_N2 reactions involving alkyl transfer typically show decreases of approximately 100-fold when X-CH$_3$ is replaced by X-CH$_2$CH$_3$. A very small change in rate when methylcobalamin is substituted by ethylcobalamin would be more consistent with an S_N1 reaction, or with a very late S_N2 transition state in which the alkyl-cobalamin bond is already weakened and the alkyl group has considerable carbocation character. The rate of methyl transfer to mercaptoethanol is very slow, even at pH 9.7 and 43°C, proceeding with a rate constant of 6.5×10^{-5} M^{-1} s^{-1}.

The studies of Kräutler[16] suggest a mechanism by which the enzyme may accelerate the rate of methyl transfer. He has examined the dependence of the carbon-cobalt bond strength of methylcobinamides on the ligand in the lower axial position of the cobinamide by measuring the equilibrium shown in equation 1 in aqueous solution. His results indicate that the carbon-cobalt bond is stabilized by ~4.2 kcal/mol by coordination of the dimethylbenzimidazole moiety in the lower axial position of cobalamin.

$$CH_3\text{-cobalamin} + cob(I)inamide \leftrightarrow CH_3\text{-cobinamide} + cob(I)alamin \qquad [1]$$

These results suggest that nucleophilic attack of a thiolate on enzyme-bound methylcobalamin might be facilitated by displacement of dimethylbenzimidazole in the lower axial position of the prosthetic group, as shown below. Such a mechanism would also reduce the reorganization energy associated with the formation of cob(I)alamin, which is preferentially planar 4-coordinate.[17]

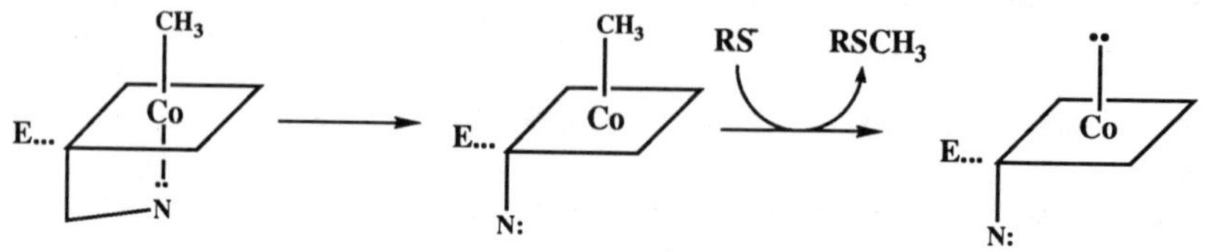

2.2 Transfer of The Methyl Group From Methyltetrahydrofolate to Cob(I)alamin

Three possible modes of activation of the N^5-methyl group of methyltetrahydrofolate have been proposed; activation by protonation to form a quaternary nitrogen (5, below), or activation by one or two electron oxidation of the folate ring to form a quinonoid methyldihydrofolate (6), or a cationic amine radical (7).

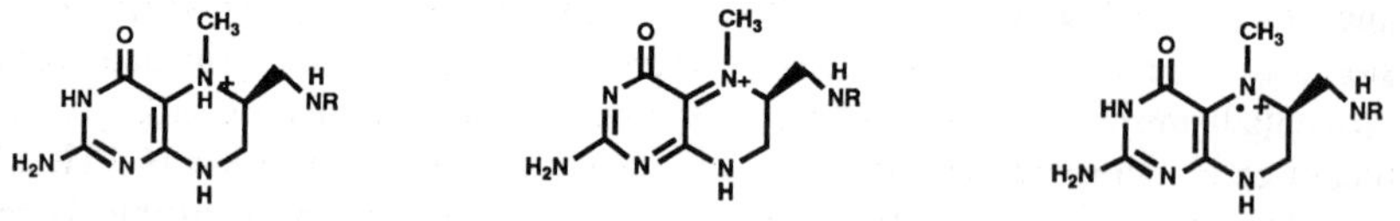

Protonated CH$_3$-H$_4$folate (5)　　Quinonoid methyldihydrofolate (6)　　Amine radical cation (7)

Dr Ellen Hilhorst[18] has recently developed model reactions to probe the occurrence of methyl transfer from one or two electron oxidized 5,6,7-trimethyltetrahydropterin to cob(I)alamin. She was unable to demonstrate methyl transfer to cob(I)alamin. She then synthesized 5,5-dimethyl-6,7-dimethyl-2-pivaloyl-5,6,7,8-tetrahydropteridinium tetrafluoroborate, and was able to demonstrate significant methyl transfer to cob(I)alamin. These studies suggest that protonation of N^5-methyltetrahydrofolate (pK$_a$ ~5.2)[19] may be the most effective mode of activation of methyltetrahydrofolate for transfer of the N^5-methyl group to cob(I)alamin. It should be noted that cob(I)alamin is a uniquely acidic strong nucleophile, with the pK$_a$ for protonation of the lone pair of electrons in the d$_{z^2}$ orbital being ≤ 1.0.[20] Thus the enzyme might readily accomplish protonation of methyltetrahydrofolate while cob(I)alamin remained unprotonated.

3 STRUCTURE OF THE ENZYME

Cobalamin-dependent methionine synthase from *Escherichia coli* is a monomeric protein with a molecular mass of 136.1 kDa.[21] It is one of the largest polypeptides synthesized in *E coli*,[22] and contains 1227 amino acid residues.[21] The *metH* gene encoding the enzyme has been sequenced,[23,24] although both published sequences contained frameshift errors that led to premature termination of the deduced amino acid sequence. The C-terminus of the protein was definitively established by peptide sequencing and electrospray mass spectrometry.[21] Limited homologies have been identified in proteins that bind cobalamins, and there is no homology between MetH and the cobalamin-independent methionine synthase from *E coli* that is encoded by the *metE* gene.[25] Our studies have suggested that the MetH enzyme is a structural mosaic of regions that each possess specific binding and catalytic properties. Our initial analyses involved characterizing fragments of the protein obtained by proteolysis of the native enzyme with trypsin. Tryptic digestion initially cleaves the enzyme C-terminal to Arg896 to yield 98 and 37 kDa peptides.[21,23] We were able to show that the cobalamin remained bound to the larger, N-terminal peptide, and to a smaller 28 kDa peptide obtained on more extensive proteolysis of the 98 kDa peptide.[23]

N-terminal sequencing of the peptides derived by proteolytic cleavage established that the cobalamin-binding region extends from residues 643 to 896 and lies at the C-terminus of the 98 kDa peptide.[21,23] If the purified enzyme is stored at room temperature in the absence of EDTA, the enzyme undergoes cleavage to yield a 28 kDa domain with similar but not identical boundaries, and this fragment has been crystallized.[26] Determination of the structure of this peptide, containing bound cobalamin, is now in progress using X-ray crystallography. Crystals of the peptide diffract to ~2.7 Å, and isomorphous platinum chloride, gold chloride and trimethyl lead derivatives have been obtained (C. Luschinsky Drennan and M L Ludwig, unpublished data).

Marsh and Holloway have succeeded in aligning the deduced amino acid sequences of the β-subunits of methylmalonyl-CoA mutase (*mutB*) from humans, mice and *Propionobacterium shermanii*, the small subunit of glutamate mutase (*mutS*) and residues 743-880 of methionine synthase (*metH*).[27] Residues 743-880 lie within the cobalamin-binding region of methionine synthase. Two short stretches of conserved sequence were identified in the alignment, a DXHXXG motif spanning residues 756-761 of MetH, and a GXX(X)IXXXXGG motif spanning residues 824-834 of MetH. Secondary structure predictions of the cobalamin-binding region of MetH using the Chou-Fasman and Garnier-Osgulthorpe-Robson algorithms predict an alternating pattern of α-helix and β-sheet,[23,27] and predictions using PredictProtein[28] indicate extensive alpha helical structure throughout the region, predicted with a very high reliability index, and interspersed loops and regions of β-sheet. Preliminary phasing of the cobalamin-

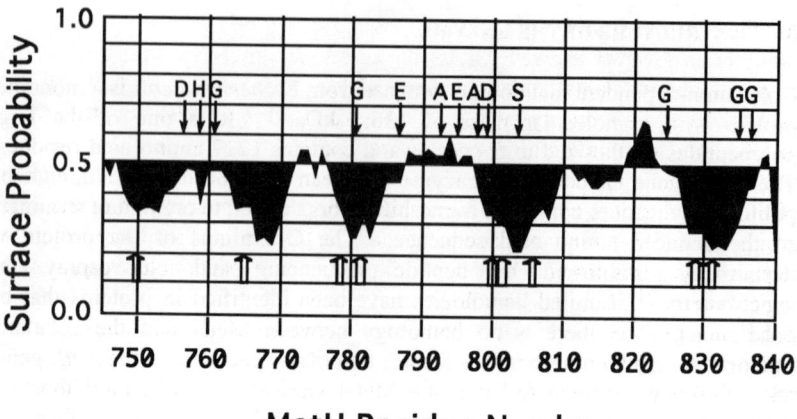

Figure 3 Conservation of residues in the cobalamin-binding region of methionine synthase with residues of the small subunit of glutamate mutase (MutS). The alignment of MetH with MutS and MutB proposed by Marsh and coworkers[27] involves an extremely hydrophobic portion of the MetH as shown in this plot of surface probability generated with the algorithm in MacVector (International Biotechnologies). Identities between MetH and MutS are highlighted by arrows here, but similarities extend to Mut B from several sources. Broad arrows below the line indicate conserved leucine, valine or isoleucine residues. The conserved residues are largely hydrophobic or acidic or are glycine residues.

binding region also indicates extensive α-helical structure, and suggests that many of the contacts to the bound cobalamin are made by the side chains of residues in helices. The 28 kDa peptide is extremely hydrophobic, as shown in Figure 3, and many of the residues that are conserved in the alignment of Marsh and Holloway are concentrated in regions of predicted high hydrophobicity and low surface probability. While the decorating side chains and the nucleotide loop of cobalamin are hydrophilic, and would be expected to be involved in hydrogen bonding interactions with polar groups in the binding pocket, the overall environment of the cobalamin prosthetic group is expected to be hydrophobic.

We have been able to locate the AdoMet binding region within the protein. When the holoprotein is subjected to ultraviolet irradiation in the presence of [^{3}H-*methyl*]-AdoMet, the label is covalently attached to the C-terminal 37 kDa domain of the protein. If this C-terminal region is generated by tryptic cleavage and then isolated, it can also be labeled by irradiation in the presence of tritiated AdoMet.[11] Since AdoMet is involved in the reductive activation of enzyme in the cob(II)alamin form, but not in catalytic turnover with substrates, this observation suggests that the function of the C-terminal region might be to catalyze reductive activation of the enzyme. Subsequent experiments showed that if enzyme containing methylated cobalamin is cleaved with trypsin, and the N-terminal 98 kDa region is isolated, this region is initially fully active in catalytic turnover.[11] However, the isolated region slowly accumulates in a form containing bound cob(II)alamin, and can no longer be reactivated in the presence of AdoMet and a reducing system. These results are certainly congruent with a role for the C-terminal region in catalyzing methyl transfer from AdoMet to the enzyme bound cobalamin in the N-terminal domain. Assuming an S_N2 transfer of the methyl group, a close and geometrically precise positioning of the methyl group of AdoMet with respect to the dz^2 orbital of cob(I)alamin should be required, and this implies a precise alignment of a portion of the C-terminal region *vis à vis* the cobalamin prosthetic group. Surprisingly, AdoMet is not a good substrate for methionine synthase. Catalysis of the reaction shown by equation 2, in the presence of saturating AdoMet, proceeds at only 0.6% of the rate of methyl transfer to homocysteine when methyltetrahydrofolate is the methyl donor.[4]

$$AdoMet + Homocysteine \rightarrow AdoHcy + Methionine \quad [2]$$

Catalysis of this reaction shows the same requirement for a reducing system seen during catalysis of methyl transfer from methyltetrahydrofolate, and presumably requires a preliminary activation of enzyme in the cob(II)alamin form by reductive methylation. Our studies indicate that the major form of enzyme during steady-state catalysis of the reaction shown by equation 2 is cob(I)alamin, suggesting that the methylation of cob(I)alamin by AdoMet is rate limiting during turnover.

One possible explanation for these observations is that the C-terminal region is not normally positioned appropriately for methyl transfer from AdoMet to the enzyme-bound cobalamin when the enzyme is cycling between methylcobalamin and cob(I)alamin forms, so that methylation by methyltetrahydrofolate is favored. Our observation that the N-terminal fragment containing bound methylcobalamin is initially fully active in catalytic turnover indicates that the binding sites for both methyltetrahydrofolate and homocysteine are located in the N-terminal region. If methyl transfers both to and from the cobalamin proceed by S_N2 mechanisms, the enzyme must alternately present bound methyltetrahydrofolate to cob(I)alamin and then remove the tetrahydrofolate product from this position prior to positioning homocysteine so as to receive the methyl group from methylcobalamin. Catalysis of this molecular juggling act may explain the remarkably large size of the methionine synthase peptide. The oxidation of cob(I)alamin to cob(II)alamin may be the signal that results in positioning elements of the 37 kDa domain adjacent to the dz^2 orbital of the cobalamin, as shown in schematic form in Figure 4. Since catalysis of reaction 2 presumably involves enzyme cycling between cob(I)alamin and methylcobalamin forms, most of the enzyme will be in an inappropriate conformation for rapid remethylation of cob(I)alamin by AdoMet.

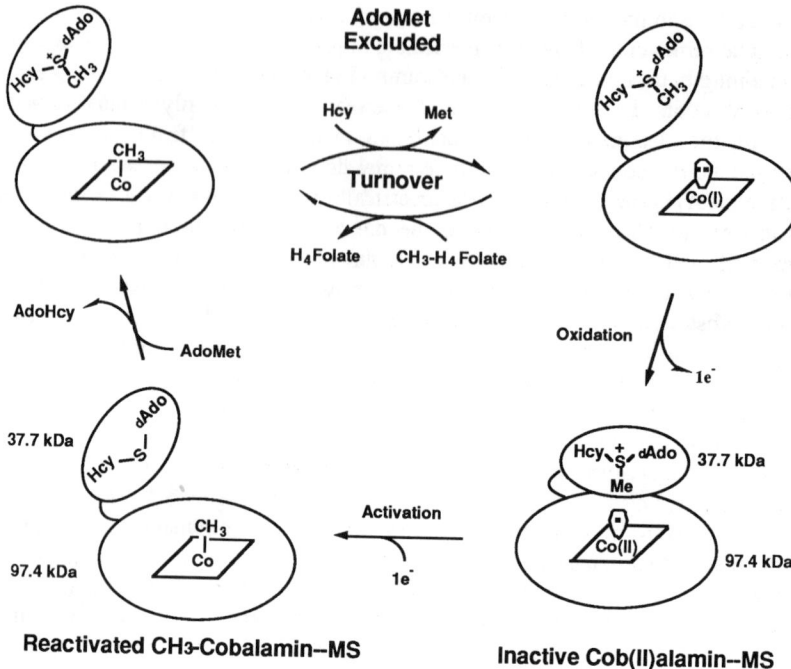

Figure 4 Proposed conformational changes associated with activation of methionine synthase suggesting a physical mechanism for exclusion of AdoMet from the catalytic site during turnover.

4 CATALYSIS OF THE PHYSIOLOGICAL REACTION

4.1 Reductive Activation of Methionine Synthase

Three enzymes are now known to be activated in reactions involving AdoMet and reduced flavodoxin. In addition to methionine synthase, these enzymes are pyruvate-formate lyase[29] and anaerobic ribonucleotide reductase (P Reichard and R G Matthews, unpublished data). In both the latter enzymes, the activation process results in the generation of a protein radical, thought to be formed following homolytic cleavage of AdoMet and abstraction of a hydrogen atom from the protein backbone to form 5'-deoxyadenosine.[29,30] We have recently examined the products formed from AdoMet during reductive activation of methionine synthase using high pressure liquid chromatography on a reversed-phase C_{18} column. Enzyme was incubated anaerobically with NADPH, AdoMet, and the R and F proteins, following the protocol developed by Huennekens and his coworkers,[31] and the formation of enzyme-bound methylcobalamin was monitored spectrophotometrically. At intervals, aliquots were removed for HPLC analysis. The formation of adenosylhomocysteine was shown to occur concomitant with the appearance of methylated enzyme. We prepared a standard curve to quantitate absorbance of adenine-containing products based on A_{254}. The formation of AdoHcy was stoichiometric with the methylation of the cobalamin prosthetic group, and no other adenine-containing products are formed during the reaction. These experimental results are fully consistent with the label transfer studies of Taylor and Weissbach, that indicated transfer of the methyl group of AdoMet to cobalamin during reductive activation,[32] and with the absorbance changes associated with reductive activation that are consistent with formation of methylcobalamin.[31] Since the product of AdoMet cleavage that is observed in pyruvate-formate lyase and the anaerobic ribonucleotide reductase is 5'-deoxyadenosine, rather than AdoHcy, our results also suggest that activation of methionine synthase differs from these other enzymes in that a radical is not generated on the protein.

We have used two different approaches to measure the reduction potential of the enzyme-bound cobalamin prosthetic group. Both approaches establish that the reduction of the prosthetic group from the cob(II)alamin to cob(I)alamin states using electrons derived from NADPH is a highly endergonic reaction, and requires coupling to an exergonic methyl transfer to proceed to any significant extent. Our first measurements of the enzyme-bound cob(II)alamin/cob(I)alamin reduction potential employed EPR measurements of the concentration of enzyme-bound cob(II)alamin, following poising of the enzyme at varied potentials in an electrochemical cell.[12] These measurements were performed in 20 mM potassium phosphate buffer, pH 7.2, containing 100 mM KCl and 500 µM TRIQUAT or methylviologen and 175 µM methionine synthase. The enzyme was poised at 25° C, and the EPR measurements were made at 77 K. Under these conditions, a midpoint potential of -526 ± 5 mV vs. the standard hydrogen electrode was measured. More recently, we have measured the reduction potential of the enzyme-bound prosthetic group by equilibration with photoreduced methylviologen at 25° C. Methionine synthase, 33 µM, was dissolved in 100 mM potassium phosphate buffer, pH 7.2, containing 100 mM KCl and 100 µM methylviologen. These measurements indicate a somewhat higher midpoint potential, -463 mV. The origin of the difference in the midpoint potentials is not understood; one possibility is that the oxidized enzyme has a tendency to dimerize at the higher concentrations of enzyme used in the EPR titrations, thus lowering the measured reduction potential. Regardless of which value we choose for the midpoint potential of the enzyme-bound cob(II)alamin/cob(I)alamin couple, equilibration of the enzyme with $NADP^+$/NADPH (which has a midpoint potential of -341 mV at pH 7.2) will result in less than 1% reduced methionine synthase.

The presence of either CH_3-H_4folate or AdoMet shifts the equilibrium distribution of cobalamin species observed during reduction by converting cob(I)alamin to methylcobalamin. The $\Delta G^{o'}$ associated with methylation of enzyme-bound cob(I)alamin by CH_3-H_4folate is estimated to be -0.09 kcal/mol, while methyl transfer to cobalamin from AdoMet is associated with a $\Delta G^{o'}$ of > - 9 kcal/mol.[12] Thus the requirement for AdoMet rather than CH_3-H_4folate for reductive methylation of enzyme in the cob(II)alamin form using NADPH as the reductant is primarily thermodynamic; the methyl donor must provide the driving force necessary to drive an endergonic reduction to completion. However, the mode by which methyl transfer from AdoMet is coupled to reduction of cob(II)alamin remains to be elucidated, and it has not been established that enzyme-bound cob(I)alamin is an intermediate in reductive activation. An alternative scenario would be a homolytic methyl transfer from AdoMet to cob(II)alamin to generation methylcobalamin and the sulfur radical cation of AdoHcy, with reduction of the radical.

It is possible that other mechanisms could be used to reduce flavodoxin more fully, particularly during anaerobic growth. The estimated midpoint potential of the flavodoxin semiquinone/hydroquinone couple at pH 7.2 is -455 mV vs. the standard hydrogen electrode,[33] and reduction of methionine synthase in the cob(II)alamin form by the flavodoxin hydroquinone should be a readily reversible reaction. Thus, it is not only necessary to exclude AdoMet from the active site during the normal catalytic turnover; it is also necessary to exclude flavodoxin. Assuming reversible electron transfer between the flavodoxin hydroquinone (FMNH-) and the cob(II)alamin prosthetic group of methionine synthase, the flavodoxin semiquinone should reoxidize the enzyme on contact. If the flavodoxin binding site involves recognition of elements from the 37 kDa domain, the mechanism proposed in Figure 4 for physical exclusion of AdoMet from the catalytic cycle would serve equally well to exclude the flavodoxin semiquinone.

4.2 Enzyme Cycles in Catalysis Between Methylcobalamin and Cob(I)alamin forms

There is clear evidence for participation of enzyme-bound cob(I)alamin in the catalytic cycle. The formation of methylcobalamin and cob(I)alamin intermediates on the enzyme was initially demonstrated spectrophotometrically.[31,32,34] More recently, enzyme-bound cob(I)alamin has been shown to be a kinetically competent intermediate.[35] Steady-state and pre-steady-state kinetic analyses indicate that the methyl transfers from CH_3-H_4folate to cobalamin and from methylcobalamin to homocysteine proceed in a ternary complex of enzyme with these two substrates, where CH_3-H_4folate is bound first and homocysteine second (Figure 5). Substrate binding occurs to enzyme in the methylcobalamin form, and the first methyl transfer must thus be from methylcobalamin to homocysteine. The second methyl transfer regenerates methylcobalamin prior to the release of methionine and H_4folate.[35] In this way the reactive cob(I)alamin state of the enzyme is protected, in that it is only generated when the methyl-donating substrate is already bound to the enzyme. Our recent studies indicate that the formation of the ternary complex and the catalysis of methyl transfer from CH_3-H_4folate to cobalamin, and from cobalamin to homocysteine are carried out by the 98 kDa N-terminal region of the protein.[11]

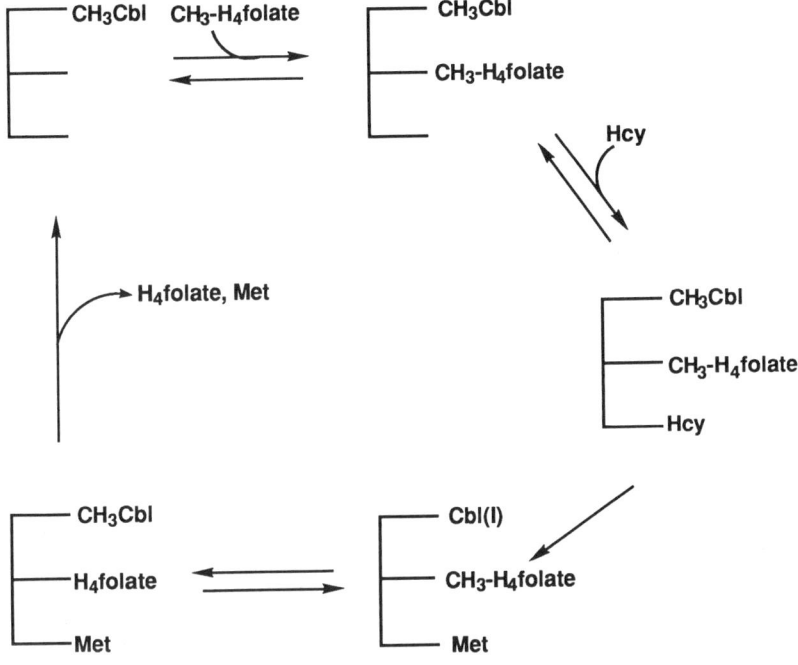

Figure 5 A kinetic scheme for catalysis by methionine synthase.

5 INACTIVATION OF METHIONINE SYNTHASE BY NITROUS OXIDE

Prolonged exposure of humans to nitrous oxide (N_2O), a commonly employed anaesthetic gas, results in development of the symptoms of cobalamin deficiency.[36,37] These symptoms include megaloblastic anaemia and a characteristic neuropathy of the spinal cord termed subacute combined spinal cord degeneration. Development of the symptoms of cobalamin deficiency following exposure to nitrous oxide was soon shown to be associated with decreases in methionine synthase activity in cells and tissues of animals and humans exposed to nitrous oxide.[38,39] Since recovery of activity required 2-3 days after termination of exposure to nitrous oxide, it was assumed that nitrous oxide leads to irreversible inactivation of methionine synthase, and recovery is associated with synthesis of new enzyme. Model studies of the interaction of cob(I)alamin with N_2O established that cob(I)alamin was oxidized and N_2 was liberated, presumably according to equations 3 and 4.[40,41] Since these studies were conducted in the presence of an excess of cob(I)alamin, the oxidation product of equation [4] was not directly characterized, but rather inferred, and only cob(II)alamin was observed.

$$\text{cob(I)alamin} + N_2O + 2H^+ \rightarrow \text{cob(III)alamin} + N_2 + H_2O \quad [3]$$
$$\text{cob(III)alamin} + \text{cob(I)alamin} \rightarrow 2\,\text{cob(II)alamin} \quad [4]$$

However, the observation that damage to the cobalamin cofactor did not occur during oxidation of cob(I)alamin by N_2O was certainly consistent with the reaction stoichiometry inferred.[41] In contrast, the studies of Allen and his coworkers[39] demonstrated that exposure of rats to nitrous oxide led to the disappearance of labeled cobalamin from the peak normally associated with methionine synthase activity during chromatography on DEAE-cellulose, and to the appearance of cobalamin analogues.

We were able to demonstrate that inactivation of methionine synthase by nitrous oxide could be accomplished *in vitro*, using purified enzyme from *E Coli*, or partially purified enzyme from pig liver.[42] Inactivation of the *E Coli* enzyme required turnover in the presence of substrates, AdoMet, and a reducing system, and appeared to result in irreversible loss of enzyme activity. When the enzyme was reisolated following inactivation, the spectrum remained characteristic of enzyme in the cob(II)alamin form, but significant losses of enzyme-bound cobalamin were observed relative to controls where the enzyme was reisolated following turnover under argon. We postulated at that time that loss of enzyme activity following exposure to nitrous oxide resulted from enzyme-catalyzed reduction of N_2O to form hydroxyl radical, as shown in equation 5. The generation of hydroxyl radical, or its equivalent, at the active site of the enzyme could readily explain the observed loss of activity and the formation of cobalamin analogues observed by Allen and his coworkers.[39]

$$\text{cob(I)alamin} + N_2O + H^+ \rightarrow \text{cob(II)alamin} + N_2 + OH\cdot \quad [5]$$

More recent studies from our laboratory strongly support this model for enzyme inactivation following exposure to nitrous oxide.[43] However, to our surprise, inactivation does not result primarily from damage to the cobalamin, but rather from damage to the protein, and particularly to the C-terminal domain of the protein. Inactive enzyme reisolated after exposure to nitrous oxide contains approximately 80% cob(II)alamin, but is unable to catalyze the reductive methylation of this cobalamin using AdoMet.[44] Since nitrous oxide must be reduced by cob(I)alamin bound to the N-terminal 98 kDa domain, and yet the hydroxyl radical is reacting with residues in the C-terminal domain of the protein, characterization of the sites of reaction will provide a first indication of the portion of the C-terminal domain that lies close in space to the cobalamin, at least during reductive activation.

REFERENCES

1 T M Zydowsky, L F Courtney, V Frasca, K Kobayashi, H Shimuzu, L-D Yuen, R G Matthews, S J Benkovic, and H G Floss, *J Am Chem Soc*, 1986, **108**, 3152.
2 J R Guest, S Friedman, and M A Foster, *Biochem J*, 1962, **84**, 93P.
3 J H Mangum and K G Scrimgeour, *Fed Proc*, 1962, **21**, 242.
4 R T Taylor and H Weissbach, *J Biol Chem*, 1967, **242**, 1502.
5 K Fujii and F M Huennekens, *J Biol Chem*, 1974, **249**, 6745.
6 K Fujii, J H Galivan, and F M Huennekens, *Arch Biochem Biophys*, 1977, **178**, 662.
7 C Osborne, L-M Chen, and R G Matthews, *J Bacteriol*, 1991, **173**, 1729.
8 V Bianchi, P Reichard, R Eliasson, E Pontis, M Krook, H. Jornvall, and E Haggard-Ljungquist, *J Bacteriol*, 1993, **175**, 1590.
9 P Hemmerich and V Massey, *Biochem Soc Trans*, 1980, **8**, 241.
10 M A Foster, M J Dilworth, and D D Woods, *Nature*, 1964, **201**, 39.
11 J T Drummond, S Huang, R M Blumenthal, and R G Matthews, *Biochemistry*, 1993, **32**, 9290.
12 R V Banerjee, S R Harder, S W Ragsdale, and R G Matthews, *Biochemistry*, 1990, **29**, 1129.
13 H P C Hogenkamp, G T Bratt, and S-z. Sun, *Biochemistry*, 1985, **24**, 6428.
14 G N Schrauzer, *Acc Chem Res*, 1968, **1**, 97.
15 H P C Hogenkamp, G T Bratt, and A T Kotchevar, *Biochemistry*, 1987, **26**, 4723.
16 B Kräutler, *Helv Chim Acta*, 1987, **70**, 1268.
17 D Lexa and J M Saveant, *J Am Chem Soc*, 1976, **98**, 2652.
18 E Hilhorst, 'Model Studies of the Methionine Synthase Reaction,' PhD Thesis, The University of Amsterdam, 1993, Centrale Drukkerij, University of Amsterdam.
19 J M Whitely, J H Drais, F M Huennekens, *Arch Biochem Biophys*, 1969, **133**, 436.
20 D Lexa and J M Saveant, *Acc Chem Res*, 1983, **16**, 235.
21 J T Drummond, R R Ogorzalek Loo, and R G Matthews, *Biochemistry*, 1993, **32**, 9282.
22 R A VanBogelen and F C Neidhardt, *Electrophoresis*, 1991, **12**, 955.
23 R V Banerjee, N L Johnston, J K Sobeski, P Datta, and R G Matthews, *J Biol Chem*, 1989, **264**, 13888.
24 I G Old, D Margarita, R E Glass, and I Saint Girons, *Gene*, 1990, **87**, 15.
25 J C Gonzâlez, R V Banerjee, S Huang, J S Sumner, and R G Matthews, *Biochemistry*, 1992, **31**, 6045.
26 C L Luschinsky, J T Drummond, R G Matthews, and M L Ludwig,

J Mol Biol, 1992, **225**, 557.
27 E N G Marsh and D E Holloway, *FEBS Lett*, 1992, **310**, 167.
28 B Rost and C Sander, *Nature*, 1993, **360**, 540.
29 J Knappe, F A Neugebauer, H P Blaschkowski, and M Gänzler, *Proc Natl Acad Sci USA*, 1984, **81**, 1332.
30 E Mulliez, M Fontecave, J Gaillard, and P Reichard, *J Biol Chem*, 1993, **268**, 2296.
31 K Fujii and F M Huennekens, 'Biochemical Aspects of Nutrition', K. Yagi, ed., Japan Scientific Societies, Tokyo, 1979, p 173.
32 R T Taylor and H Weissbach, *Arch Biochem Biophys*, 1969, **129**, 728.
33 H Vetter Jr and J Knappe, *Hoppe-Seyler's Z Physiol Chem*, 1971, **352**, 433.
34 R T Taylor and M L Hanna, *Arch Biochem Biophys*, 1970, **137**, 453.
35 R V Banerjee, V Frasca, D P Ballou and R G Matthews, *Biochemistry*, 1990, **29**, 11101.
36 H C A Lassen, E Henriksen, F Neukirch, and H S Kristensen, *Lancet i*, 1956, 527.
37 R B Layzer, *Lancet i*, 1978, 1227.
38 R Deacon, J Perry, M Lumb, I Chanarin, B Minty, M J Halsey, and J F Nunn, *Lancet ii*, 1978, 1023.
39 H Kondo, M L Osborne, J F Kolhouse, M J Binder, E R Podell, C S Utley, R S Abrams, and R H Allen, *J Clin Invest*, 1981, **67**, 1270.
40 R G Banks, R J S Henderson, and J M Pratt, *J Chem Soc A*, 1968, 196.
41 R Blackburn, M Kyaw, and A J Swallow, *J Chem Soc Faraday Trans*, 1977, **73**, 196.
42 V Frasca, B Stephenson Riazzi, and R G Matthews, *J Biol Chem*, 1986, **261**, 15823.
43 J T Drummond and R G Matthews, *Biochemistry*, 1994, **33**, 3732.
44 J T Drummond and R G Matthews, *Biochemistry*, 1994, **33**, 3742.

ACKNOWLEDGEMENTS

Research from our laboratory has been supported by NIH Grant R37 GM24908. JTD has been supported by NIH and Pharmacological Sciences Grant T32 and GM07767, by an NSF Graduate Fellowship, and by a Rackham Predoctoral Fellowship.

BIOMIMETIC REACTIONS WITH HYDROPHOBIC VITAMIN B_{12}

Yukito Murakami,* Yoshio Hisaeda, and Teruhisa Ohno

Department of Chemical Science and Technology, Faculty of Engineering, Kyushu University, Fukuoka 812, Japan

1 INTRODUCTION

Metalloenzymes are ingeniously designed natural catalysts that demonstrate specific substrate recognition, marked rate enhancement, and appropriate selection of reaction pathways. Our study is directed toward development of new catalytic systems through construction of model systems for functional simulation of metalloenzymes. Naturally occurring holoenzymes are individually composed of a specific apoprotein and an additional cofactor such as a coenzyme. An apoprotein generally provides a binding site for both specific coenzyme and substrate molecules, which is well separated from a bulk aqueous phase. As a consequence, an active site of each enzyme turns out to be sufficiently hydrophobic and hardly holds water molecules. Under such circumstances, reacting species become efficiently naked so that the reactivity is much enhanced due to thermodynamic reasons. Thus, it is quite important for construction of an artificial holoenzyme to select a relevant molecular assembly that is capable of reflecting such characteristic physical functions of apoproteins. We focussed on the catalytic functions of vitamin B_{12} enzymes in this work.

Vitamin B_{12}-dependent enzymes, involving the cobalt species as a reaction site, catalyse various isomerization reactions leading to the intramolecular exchange of a functional group (X) and a hydrogen atom between neighbouring carbon atoms (refer to eqn 1).[1,2] These reactions have attracted much attention because of their novel nature from the viewpoints of organic and organometallic chemistry. Carbon-skeleton rearrangement reactions, mediated by methylmalonyl-CoA mutase and glutamate mutase, are shown in eqns 2 and 3. Even though the real reaction mechanisms involved in the carbon-skeleton rearrangements have not been clarified up to the present time, radical mechanisms are considered to be the most plausible ones on the basis of ESR studies.[3-5]

We have been studying the functional simulation of vitamin B_{12} enzymes from two different aspects: clarification of electrochemical catalysis by a hydrophobic vitamin B_{12} in organic solvents that provide hydrophobic microenvironments; exploration of specific catalysis by an artificial holoenzyme composed of a hydrophobic vitamin B_{12} and a synthetic bilayer membrane, the latter being used as an apoprotein model.

2　Hydrophobic Vitamin B_{12}

Various cobalt complexes have been synthesized as model complexes of vitamin B_{12}.[6] However, all of those complexes cannot be qualified as favourable model complexes in the following aspects: (i) Redox behaviour of the central cobalt, which is mainly controlled by basicity of an equatorial ligand, must be similar to that for the naturally occurring vitamin B_{12}. (ii) Electronic properties must be equivalent to those of the natural B_{12}, which are provided by the corrin ring with eight double bonds and a direct bond between rings A and D. (iii) The steric effects, which are caused by a methyl moiety and a hydrogen atom at C(1) and C(19) positions in the corrin ring, respectively, and by four propionamides and three acetamides placed at the α- and β-peripheral sites, respectively, must be retained by model complexes.

The naturally occurring apoproteins, which provide relevant reaction sites for vitamin B_{12}, are considered to perform additional important roles that lead to desolvation and close association of reacting species.[7] On this ground, we have been interested in the catalytic activity of vitamin B_{12} in hydrophobic microenvironments in order to simulate the catalytic functions of holoenzymes concerned. Under such circumstances, we have prepared hydrophobic vitamin B_{12} derivatives which have ester groups in place of the peripheral amide moieties of natural vitamin B_{12}.[8-10] These modified cobalt complexes satisfy all the above requirements and are readily soluble in a wide range of organic solvents.

3　Electrochemical Reactions Mediated by Hydrophobic Vitamin B_{12}

We have previously reported on the redox chemistry of heptamethyl cobyrinate perchlorate, [Cob(II)7C_1ester]ClO_4, and pointed out that this complex is readily reduced to the univalent cobalt species of highly nucleophilic character by electrochemical means in nonaqueous media.[11] The cyclic voltammetry applied to [(CH$_3$)(H$_2$O)Cob(III)7C_1ester]ClO_4 indicated that the cobalt–carbon bond is cleaved by electrochemical reduction. On the basis of this information, catalytic cycles were established as shown in Figure 1. An alkylated complex, generated by the reaction of a univalent cobalt complex and an alkyl halide, is generally decomposed by photolysis or electrolysis to afford reduction and/or rearrangement products.[13-15]

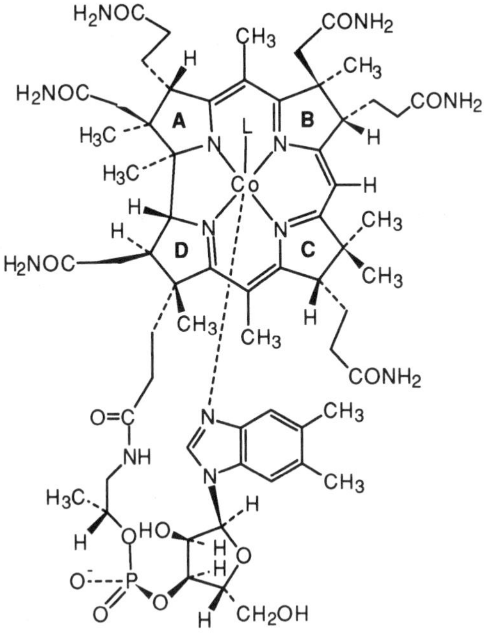

H₂NOC structure shown.

Vitamin B₁₂
(Cobalamin)

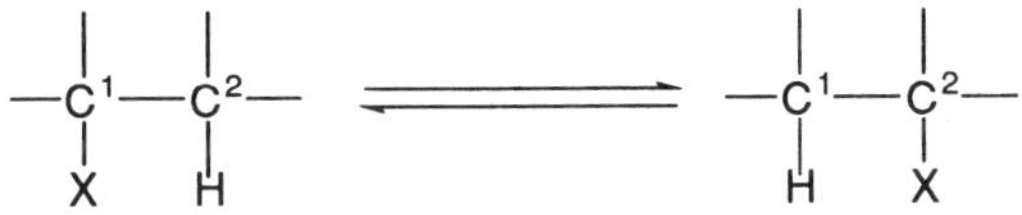

$$-\underset{X}{\overset{|}{C^1}}-\underset{H}{\overset{|}{C^2}}- \rightleftharpoons -\underset{H}{\overset{|}{C^1}}-\underset{X}{\overset{|}{C^2}}- \quad (1)$$

$$\underset{\underset{COS-CoA}{|}}{\overset{\overset{CO_2H}{|}}{H_3C-C-H}} \xrightarrow{\text{methylmalonyl-CoA mutase}} \underset{\underset{COS-CoA}{|}}{\overset{\overset{CO_2H}{|}}{H_2C-CH_2}} \quad (2)$$

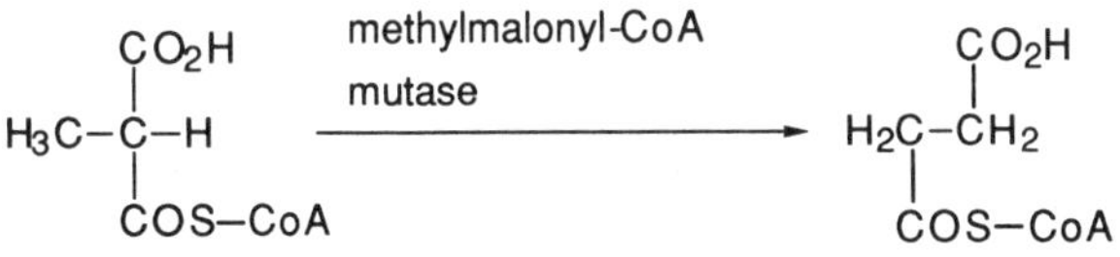

$$\underset{\underset{CH_3}{|}}{\overset{\overset{NH_2}{|}}{HO_2C-CH-CH-CO_2H}} \xrightarrow{\text{glutamate mutase}} HO_2C-CH_2-CH_2-\overset{\overset{NH_2}{|}}{CH}-CO_2H \quad (3)$$

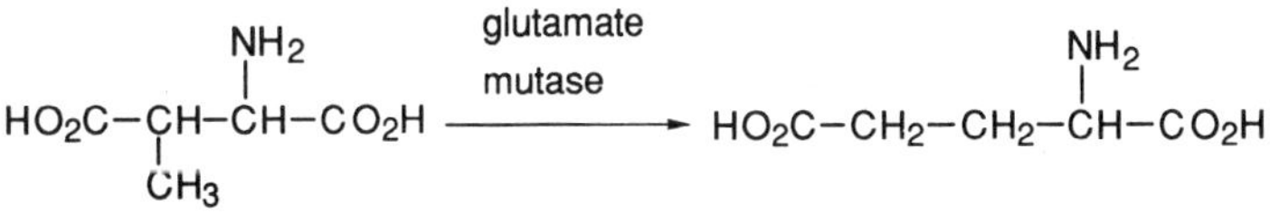

3.1 Catalytic Simulation of Methylmalonyl-CoA Mutase

We firstly adopted 2,2-bis(ethoxycarbonyl)-1-bromopropane which is considered to be a model substrate for methylmalonyl-CoA mutase. Electrolysis of the alkyl halide was carried out upon addition of [Cob(II)7C$_1$ester]ClO$_4$ in DMF under various conditions, and products were analyzed by GLC (refer to eqn 4). The catalysis was quite efficient at -2.0 V vs SCE as reflected in yields of the rearrangement product; equivalent to 100-110 times as much as a molar quantity of the hydrophobic vitamin B$_{12}$ after 2 h of the reaction.[14]

Reaction mechanisms involved in the controlled-potential electrolysis were investigated by means of electronic spectroscopy and coulometry as well as by the spin-trapping ESR technique. The results are consistent with the overall feature of electrolysis shown in Figure 2.[15] The divalent cobalt complex is first converted into the corresponding univalent cobalt species by electrochemical reduction. The alkylated complex is formed subsequently by reaction of the super-nucleophilic Co^I species with the alkyl halide. The complex is then decomposed by visible light to give the divalent cobalt species and the alkyl radical, and the latter abstracts a hydrogen atom to afford the reduction product. The alkylated complex is further reduced to the one-electron reduction intermediate in the dark. An electronic structure of the intermediate seems to be represented by two canonical forms. A proton attack on the β-carbon of the substrate induces the carbon-skeleton rearrangement, followed by the cobalt–carbon bond cleavage. On the other hand, the one-electron reduction intermediate is spontaneously decomposed to the Co^I chelate and the alkyl radical in the absence of an efficient proton source. The reduction product is mainly derived from the alkyl radical by rapid abstraction of a hydrogen atom. At -2.0 V vs SCE, the alkylated complex is converted into the two-electron reduction intermediate in the dark. This intermediate is decomposed to the Co^I chelate and an anionic species, and the rearrangement product is obtained from the latter. It needs to be noted that the simple reduction product is primarily obtained from the radical species. Since the identical radical species, which is produced by reaction of the present substrate with tributyltin hydride or by photolysis of the present substrate bound to cobaloxime, does not undergo the rearrangement reaction, both of the anionic reduction intermediates given in Figure 2 are the primary sources for the rearrangement product.

The above study is the first example of the rearrangement as catalysed by a vitamin B$_{12}$ model under electrochemical conditions. The carbon-skeleton rearrangement can be postulated to proceed via formation of the anionic intermediates. However, it is not relevant to apply these mechanisms directly to the corresponding vitamin B$_{12}$-dependent enzymic reactions, because the reduction potential as high as -2.0 V vs SCE would not be attained *in vivo*. The rearrangement reactions are generally considered to proceed via radical mechanisms *in vivo*, but the chemical nature of such radical species must be different from those in homogeneous solution. The reactivity of radical species

Biomimetic Reactions with Hydrophobic Vitamin B_{12}

[(CH$_3$)(H$_2$O)Cob(III)7C$_1$ester]$^+$: R = CH$_3$, X = CH$_3$, Y = H$_2$O

[Cob(II)7C$_1$ester]$^+$: R = CH$_3$, X = Y = none

(CN)$_2$Cob(III)7C$_3$ester : R = n-C$_3$H$_7$, X = Y = CN

[Cob(II)7C$_3$ester]$^+$: R = n-C$_3$H$_7$, X = Y = none

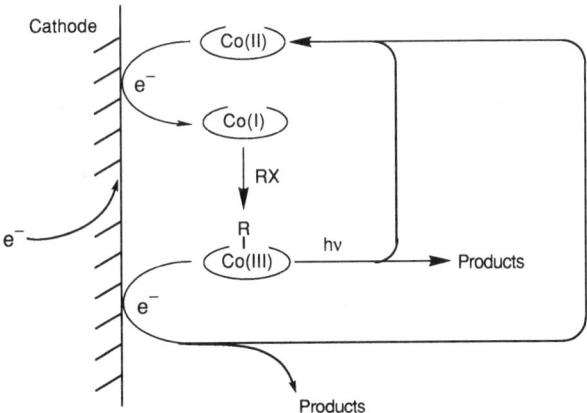

$$\underset{\text{Br}\ \ \text{CO}_2\text{C}_2\text{H}_5}{\overset{\text{CO}_2\text{C}_2\text{H}_5}{\text{H}_2\text{C}-\underset{|}{\overset{|}{\text{C}}}-\text{CH}_3}} \xrightarrow{-2.0\ \text{V vs. SCE}} \underset{\text{CO}_2\text{C}_2\text{H}_5}{\overset{\text{CO}_2\text{C}_2\text{H}_5}{\text{H}_3\text{C}-\underset{|}{\overset{|}{\text{C}}}-\text{CH}_3}} + \underset{\text{CO}_2\text{C}_2\text{H}_5}{\overset{\text{CO}_2\text{C}_2\text{H}_5}{\text{H}_2\text{C}-\text{CH}-\text{CH}_3}} \quad (4)$$

15% 80%

Fig. 1 Schematic representation of catalytic cycles

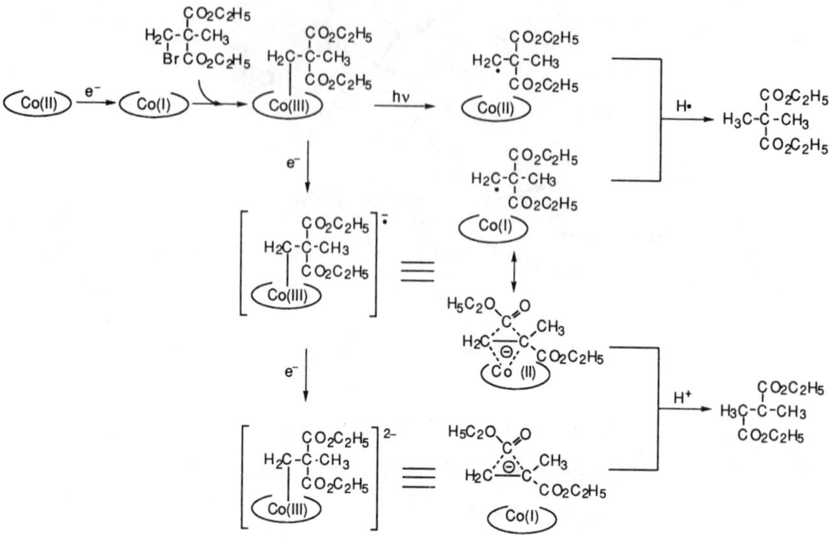

Fig. 2 Overall feature of electrochemical reaction catalysed by hydrophobic vitamin B_{12}

may be subjected to change by microenvironmental properties provided by apoproteins at the reaction sites. In this regard, the catalytic behaviour of the hydrophobic vitamin B_{12} embedded in an apoprotein model is described later.

3.2 Migratory Aptitude and Steric Effect of Functional Groups

In order to make further characterization of the catalytic proficiency of the hydrophobic vitamin B_{12} and to clarify the migratory aptitude of functional groups in the electrochemical rearrangement reaction, various substrates were also used.[14] These substrates and the corresponding products are summarized in eqns 5–7. The following aspects became apparent for the electrolysis: (i) Substrates with two electron-withdrawing groups on the β-carbon atom tend to give the corresponding rearrangement products which are derived from individual migration of the groups. (ii) Substrates with only one electron-withdrawing group on the β-carbon atom do not give the corresponding rearrangement products, except for the substrate with one thioester group. (iii) The rearrangement reaction readily proceeds under electrochemical conditions that allow the formation of anionic intermediates. In the light of the above results, the apparent migratory aptitude of electron-withdrawing groups decreases in the following sequence: $COSR \geq COR > CO_2R > CN$. Both steric bulkiness and electronic character of the migrating groups would be responsible for this tendency.

Biomimetic Reactions with Hydrophobic Vitamin B_{12}

$$\underset{\underset{Br\ \ CO_2C_2H_5}{|\ \ \ \ \ \ \ \ \ \ |}}{H_2C-C-CH_3} \xrightarrow{-2.0\ V\ vs.\ SCE} \underset{\underset{CO_2C_2H_5}{|\ \ \ \ \ \ \ \ \ \ |}}{H_3C-\underset{CN}{C}-CH_3} + \underset{\underset{CO_2C_2H_5}{|\ \ \ \ \ \ \ \ \ \ |}}{H_2C-\underset{CN}{CH}-CH_3} + \underset{\underset{CO_2C_2H_5}{|\ \ \ \ \ \ \ \ \ \ |}}{H_2C-\underset{CN}{CH}-CH_3} \quad (5)$$

30%   2%   62%

$$\underset{\underset{Br\ \ CO_2C_2H_5}{|\ \ \ \ \ \ \ \ \ \ |}}{H_2C-\underset{COCH_3}{C}-CH_3} \xrightarrow{-2.0\ V\ vs.\ SCE} \underset{\underset{CO_2C_2H_5}{|\ \ \ \ \ \ \ \ \ \ |}}{H_3C-\underset{COCH_3}{C}-CH_3} + \underset{\underset{CO_2C_2H_5}{|\ \ \ \ \ \ \ \ \ \ |}}{H_2C-\underset{COCH_3}{CH}-CH_3} + \underset{\underset{CO_2C_2H_5}{|\ \ \ \ \ \ \ \ \ \ |}}{H_2C-\underset{COCH_3}{CH}-CH_3} \quad (6)$$

3%   91%   3%

$$\underset{\underset{Br\ \ COSC_2H_5}{|\ \ \ \ \ \ \ \ \ \ |}}{H_2C-\underset{CH_3}{CH}} \xrightarrow{-2.0\ V\ vs.\ SCE} \underset{\underset{COSC_2H_5}{|}}{H_3C-\underset{CH_3}{CH}} + \underset{\underset{COSC_2H_5}{|}}{H_2C=\underset{CH_3}{CH}} + \underset{\underset{COSC_2H_5}{|}}{H_2C-\underset{CH_3}{CH_2}} \quad (7)$$

20%   25%   37%

$$\underset{\underset{Br\ \ CO_2C_2H_5}{|\ \ \ \ \ \ \ \ \ \ |}}{H_2C-\underset{CO_2R}{C}-CH_3} \xrightarrow[\text{Cobester}]{-2.0\ vs.\ SCE} \underset{\underset{CO_2C_2H_5}{|\ \ \ \ \ \ \ \ \ \ |}}{H_3C-\underset{CO_2R}{C}-CH_3} + \underset{\underset{\boxed{CO_2C_2H_5}}{|\ \ \ \ \ \ \ \ \ \ |}}{H_2C-\underset{CO_2R}{CH}-CH_3} + \underset{\underset{CO_2C_2H_5}{|\ \ \ \ \ \ \ \ \ \ |}}{H_2C-\underset{\boxed{CO_2R}}{CH}-CH_3} \quad (8)$$

$R = -C(CH_3)_3$	29 %	44 %	24 %
$R = $ cyclohexyl	24 %	47 %	22 %
$R = Ph$	29 %	43 %	19 %

(9) Cyclic α-bromo ester $\xrightarrow[\text{Cobester}]{-2.0\ vs.\ SCE}$ A + B + C

	A	B	C
n = 1 (5-membered ring)	1.5 %	43 %	0 %
n = 2 (6-membered ring)	14 %	49 %	12 %
n = 3 (7-membered ring)	3 %	44 %	Trace
n = 4 (8-membered ring)	2 %	21 %	0 %

A steric effect in the carbon-skeleton rearrangement catalysed by [Cob(II)7C$_1$ester]ClO$_4$ was investigated under the same conditions.[16] The controlled-potential electrolyses of alkyl halides having two carboxylic ester groups of different bulkiness on the same carbon atom were carried out in DMF (refer to eqn 8). As regards a correlation between bulkiness of an ester group and migratory aptitude, a smaller ester group tends to migrate to the adjacent carbon atom more readily than a larger one. This steric effect indicates that the rate-determing step in the electrolysis is not formation of the cobalt–carbon bond but rather its cleavage.

3.3 Application to Other Reactions

The electrolyses of alkyl halides with a cyclic ketone (5, 6, 7, and 8-membered ring) and an ester group were carried out in DMF in the presence of [Cob(II)7C$_1$ester]ClO$_4$ under various conditions (refer to eqn 9).[17] Products were analysed by GLC; simply reduced **A**, ring expanded **B**, and ester-migrated **C** were detected. At -2.0 V vs SCE, the major product was a ring-expansion product for all the substrates. The electrolysis plausibly proceeds as follows; the Co^{II} complex is electrochemically reduced to the Co^I species, and the corresponding alkylated complex is generated by reaction of the super-nucleophilic Co^I species with a substituted alkyl bromide. The alkylated complex is subsequently decomposed by electrolysis to afford the corresponding products, and the cobalt complex acts as a mediator repeatedly.

Strapped B$_{12}$

$$\underset{\underset{Br\;\;CO_2C_2H_5}{|\;\;\;\;\;\;\;|}}{H_2C-\underset{Ph}{\overset{|}{C}}-OCH_3} \xrightarrow[\text{Cobester}]{\text{Electrolysis}} \underset{\underset{CO_2C_2H_5}{|}}{H_3C-\underset{Ph}{\overset{|}{C}}-OCH_3} + \underset{\underset{CO_2C_2H_5}{|}}{H_2C-\underset{Ph}{\overset{|}{C}H}-OCH_3} + \cdots \quad (10)$$

D **E**

We prepared a strapped hydrophobic vitamin B_{12} in order to enhance the enantioselectivity.[18] The electrolysis of ethyl 3-bromo-2-methoxy-2-phenylpropionate with a catalytic amount of $[Cob(II)7C_3ester]ClO_4$ or strapped B_{12} was carried out at -1.8 V vs SCE (refer to eqn 10).[19] After the electrolysis for half conversion of the substrate, products, such as simply reduced **D** and phenyl-migrated **E**, were detected by GLC. The absolute configuration of **D** was analysed by HPLC. When $[Cob(II)7C_3ester]ClO_4$ was used as a catalyst, formation of the S-enantiomer prevailed over that of the corresponding R-isomer. On the other hand, the R-enantiomer was preferentially formed when strapped B_{12} was used as a catalyst. This result indicates that hydrophobic vitamin B_{12} derivatives are effective catalysts for asymmetric reactions.

4 ARTIFICIAL ENZYME COMPOSED OF HYDROPHOBIC VITAMIN B_{12} AND SYNTHETIC BILAYER MEMBRANE

It is obvious that an apoprotein plays an important role in such radical reactions as mediated by vitamin B_{12}-dependent enzymes.[7,20] However, no relevant apoprotein models which are capable of generating substrate radicals in the dark have been developed up to the present time. We have been adopting photolysis conditions to generate substrate radicals in a synthetic bilayer membrane as an apoprotein model.

4.1 Synthetic Bilayer Membranes

Characteristic features of synthetic bilayer membranes as apoprotein models are as follows.[21] (i) In view of the fact that apoproteins create relatively hydrophobic and desolvated active sites in aqueous media, analogous microenvironments can be readily provided by aggregation of peptide lipids. A hydrogen-belt domain is constructed through hydrogen-bonding interactions among the amino acid residues of lipid molecules in the bilayer domain and provides an effective reaction site. (ii) A functional group is readily introduced into the reaction site by incorporation of an appropriate amino acid residue covalently into a peptide lipid molecule. (iii) Morphological stability of the synthetic membranes is superior to that of liposomal membranes composed

$$(CH_3)_3N^+(CH_2)_5\overset{\underset{\parallel}{O}}{C}NHCH\overset{\underset{\parallel}{O}}{C}N[(CH_2)_{n-1}CH_3]_2 \ Br^-$$

with CH_2R on the CH.

$N^+C_5Ala2C_n$ (R = H)

Peptide lipid

of phospholipids, and single-walled vesicles formed upon sonication of aqueous dispersions of the synthetic lipids stay in solution over a month at least without meaningful morphological changes. (iv) Molecular motion of a guest molecule incorporated into the synthetic membrane is strongly restricted, compared to those in homogeneous solution, so as to enhance the intramolecular rearrangement reaction.

4.2 Construction of Artificial Vitamin B_{12} Enzyme

An artificial vitamin B_{12} holoenzyme is constructed with a combination of a single-walled bilayer vesicle composed of $N^+C_5Ala2C_{16}$ and the hydrophobic vitamin B_{12} [Cob(III)7C$_3$ester] as illustrated in Figure 3.[10,22] The morphological stability of the synthetic bilayer vesicle is not perturbed by noncovalent incorporation of the hydrophobic vitamin B_{12}, even though the latter has a relatively large molecular size.

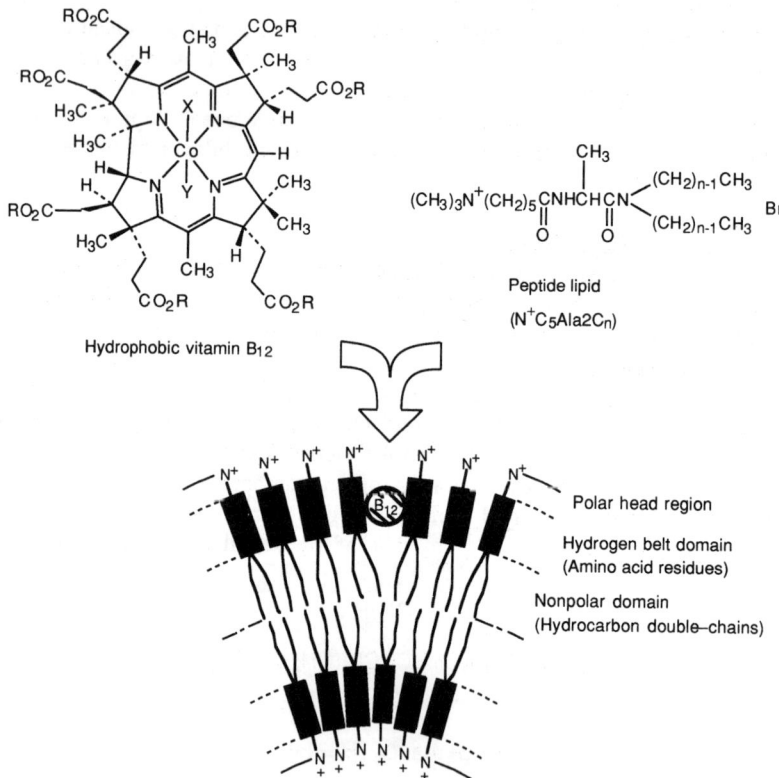

Fig. 3 Constitution of an artificial Vitamin B_{12}-dependent enzyme

4.3 Artificial Methylmalonyl-CoA Mutase

Methylmalonyl-CoA mutase catalyses interconversion of methylmalonyl-CoA and succinyl-CoA. A characteristic aspect of the substrate, methylmalonyl-CoA, is that it has two electron-withdrawing groups on the same carbon atom. We take it for granted that all compounds relevant to this structural specification are qualified as model substrates. Substrate species were bound to the hydrophobic vitamin B_{12} at one axial site of the nuclear cobalt, and the resulting alkylated complexes were noncovalently fixed in the bilayer membrane domain in aqueous media and irradiated with a 500 W tungsten lamp to result in homolytic cleavage of the Co–C bond. Some typical experimental results are shown in eqns 11–14 in comparison with the data obtained for the reactions in homogeneous solutions.[22] The apparent migratory aptitude of electron-withdrawing groups was observed to increase as follows: $CN \cong CO_2C_2H_5 < COCH_3$.

$$H_2C(COCH_3)(CO_2C_2H_5)CH_3\text{-Co}^{III} \xrightarrow{h\nu} H_3C\text{-}C(COCH_3)(CO_2C_2H_5)\text{-}CH_3 + H_2C(COCH_3)\text{-}CH(CO_2C_2H_5)\text{-}CH_3 + H_2C(COCH_3)\text{-}CH(CO_2C_2H_5)\text{-}CH_3 \quad (11)$$

	$H_3C\text{-}C(COCH_3)(CO_2C_2H_5)CH_3$	$H_2C(COCH_3)\text{-}CH(CH_3)CO_2C_2H_5$	$H_2C(CO_2C_2H_5)\text{-}CH(CH_3)COCH_3$
Methanol	70 %	Trace	8 %
Benzene	65 %	Trace	10 %
Vesicle	25 %	Trace	63 %

$$H_2C(CN)(CO_2C_2H_5)CH_3\text{-Co}^{III} \xrightarrow{h\nu} H_3C\text{-}C(CN)(CO_2C_2H_5)CH_3 + H_2C(CN)\text{-}CH(CO_2C_2H_5)CH_3 + H_2C(CO_2C_2H_5)\text{-}CH(CN)CH_3 \quad (12)$$

Methanol	87 %	0 %	0 %
Benzene	78 %	Trace	Trace
Vesicle	41 %	23 %	18 %

$$H_2C(CO_2C_2H_5)_2CH_3\text{-Co}^{III} \xrightarrow{h\nu} H_3C\text{-}C(CO_2C_2H_5)_2CH_3 + H_2C(CO_2C_2H_5)\text{-}CH(CO_2C_2H_5)CH_3 \quad (13)$$

Methanol	88 %	0 %
Benzene	82 %	1.3 %
Vesicle	75 %	9 %

$$H_2C(COSC_2H_5)(CO_2C_2H_5)\text{-CH-Co}^{III} \xrightarrow{h\nu} H_3C\text{-}CH(COSC_2H_5)(CO_2C_2H_5) + H_2C=C(COSC_2H_5)(CO_2C_2H_5) + H_2C(COSC_2H_5)\text{-}CH_2(CO_2C_2H_5) \quad (14)$$

Methanol	83 %	0 %	0 %
Benzene	80 %	Trace	Trace
Vesicle	61 %	4 %	13 %

4.4 Artificial Glutamate Mutase

Glutamate mutase mediates interconversion of methylaspartic acid and glutamic acid. Diethyl methylasparate was bound to the hydrophobic vitamin B_{12} in the manner described above, and the resulting alkylated complex underwent homolytic cleavage of the Co–C bond upon irradiation with visible light.[23,24] It is now apparent from eqn 15 that the isomerization reaction takes place only in the vesicle.

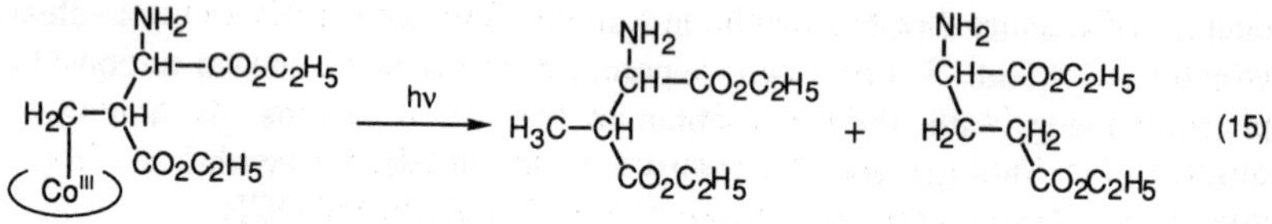

$$\underset{\substack{H_2C-CH \\ | \\ Co^{III} \quad CO_2C_2H_5}}{\overset{NH_2}{\underset{|}{CH-CO_2C_2H_5}}} \xrightarrow{h\nu} \underset{\substack{| \\ CO_2C_2H_5}}{\overset{NH_2}{\underset{|}{H_3C-CH}}} + \underset{\substack{H_2C-CH_2 \\ | \\ CO_2C_2H_5}}{\overset{NH_2}{\underset{|}{CH-CO_2C_2H_5}}} \quad (15)$$

Methanol	83 %	0 %
Benzene	78 %	Trace
Vesicle	66 %	14 %

The overall reaction sequence mediated by the artificial methylmalonyl-CoA and glutamate mutases in the single-walled bilayer vesicle is illustrated in Figure 4: a substrate ($CRXYCH_2-$) bound to a hydrophobic vitamin B_{12} undergoes the homolytic Co–C cleavage upon irradiation with visible light, the generated substrate radical is converted into the product radical, and the product radical abstracts a hydrogen atom from its vicinity to give the final product.

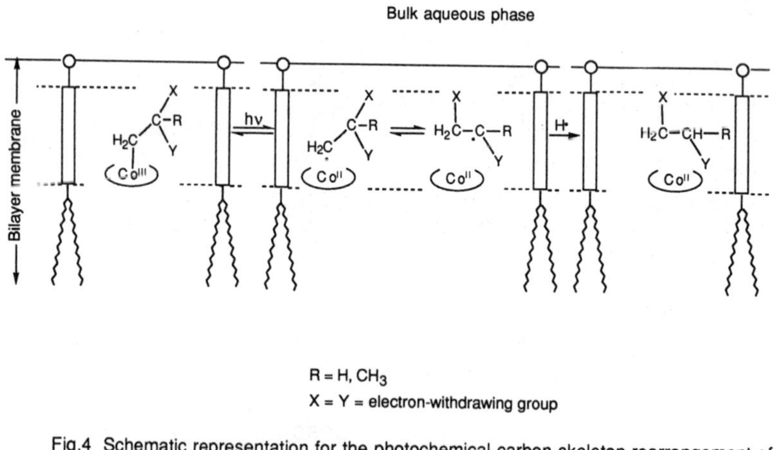

$R = H, CH_3$
$X = Y = $ electron-withdrawing group

Fig.4 Schematic representation for the photochemical carbon-skeleton rearrangement of an alkyl ligand bound to the hydrophobic vitamin B_{12}

A question arises as to the hydrogen source for formation of the product species, the final step in the overall reaction sequence mentioned above. Does a hydrogen atom come from a bulk aqueous phase or else from molecular species constituting an artificial enzyme? In order to clarify this query, product analyses were performed for the photochemical cleavage of a hydrophobic vitamin B_{12}, bearing a benzyl moiety at one axial site of the nuclear cobalt, in various deuterated media under anaerobic conditions.[25] The results obtained were as follows: (i) the solvent acts as a hydrogen source in CD_3OD to give the deuterated toluene in a 80% yield; (ii) bibenzyl is exclusively obtained in C_6D_6 which does not act as a hydrogen source; (iii) the major product in the vesicular phase with D_2O as a bulk solvent is the non-deuterated toluene. These observations indicate that a hydrogen atom is taken from the hydrophobic vitamin B_{12} or the lipid molecule and not from bulk water.

It became necessary to answer the next question as to which functional group migrates preferentially to the neighbouring carbon atom, the glycyl group or the carboxylic ester, in a reaction mediated by the artificial glutamate mutase. It became clear by utilizing a deuterated substrate that the glycyl group migrates predominantly, as observed for the corresponding enzymic reaction (refer to eqn 16).[26]

(16)

glycyl-migrated product (70 ~ 83 %)   ester-migrated product (17 ~ 30 %)

4.5 Application to Ring-Expansion Reactions

Even though naturally occurring vitamin B_{12}-dependent enzymes do not catalyse ring-expansion reactions, such reactions are mediated by the present artificial enzyme under experimental conditions comparable to those applied to the above-mentioned reactions. As shown in eqns 17 and 18, the ring-expansion reactions are much enhanced in the vesicular phase.[27]

(17)

Methanol	69 %	19 %
Benzene	57 %	28 %
Vesicle	19 %	67 %

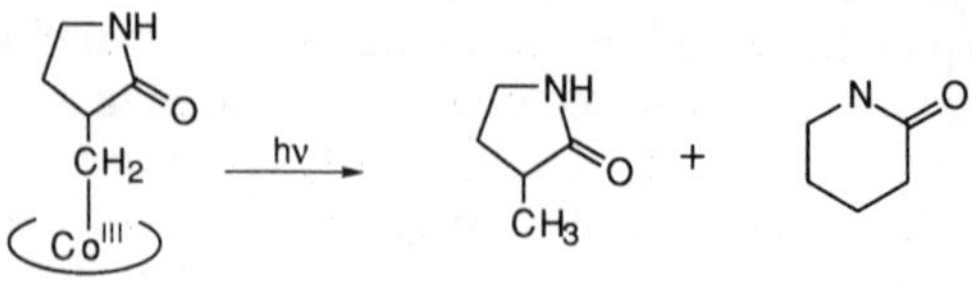

Methanol	83 %	3 %
Benzene	76 %	9 %
Vesicle	63 %	24 %

(18)

4.6 Turnover Catalysis

All the above reactions are stoichiometric rather than catalytic, even though the substrate species undergo isomerization. In order to improve upon this situation, we coupled an effective process for activation of substrates with the catalytic mediator composed of [Cob(II)7C$_3$ester]ClO$_4$ and the N$^+$C$_5$Ala2C$_{16}$ vesicle.[22,24] An appropriate amount of [Cob(II)7C$_3$ester]ClO$_4$ and a large excess of vanadium trichloride were dissolved in an aqueous medium containing the N$^+$C$_5$Ala2C$_{16}$ vesicle and a sufficient excess of a substrate. The solution was irradiated with a 500 W tungsten lamp at 20 °C under aerobic conditions. The overall reaction cycle is shown in Fig. 5. A substrate is activated by vanadium(III) ions and molecular oxygen,[28,29] and the resulting radical species undergoes coupling with [Cob(II)7C$_3$ester]$^+$ to afford the corresponding alkylated complex. The alkylated complex is subjected to homolytic cleavage to give the original substrate and the isomerized products, the former being subjected to catalysis.

In the light of our investigations, the following aspects became apparent: (i) A combination of vanadium trichloride and atmospheric oxgen abstracts a hydrogen atom from the terminal methyl group of a substrate species to form the corresponding radical species, which then undergoes reaction with [Cob(II)7C$_3$ester]$^+$ to form the Co–C bond. (ii) Since both suppression of molecular motion and desolvation effects operate on chemical species incorporated into the bilayer vesicle, [Cob(II)7C$_3$ester]$^+$ and the substrate radical, which are produced by homolytic cleavage of the Co–C bond upon photolysis, must form a tight pair. Under such conditions, the nuclear cobalt acts to promote the 1,2-migration of electron-withdrawing groups. The present artificial holoenzyme is expected to be applied to other non-enzymatic reactions that undergo similar reaction mechanisms.

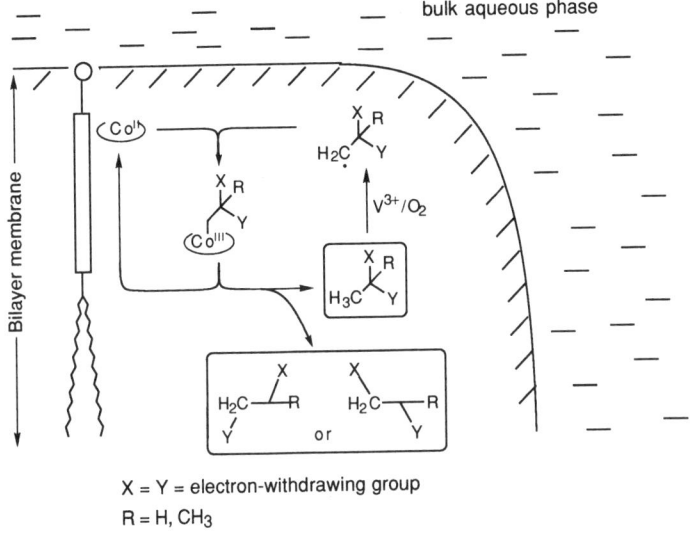

X = Y = electron-withdrawing group
R = H, CH$_3$

Fig. 5: Schematic representation of catalytic carbon-skeleton rearrangement reactions in the bilayer vesicle.

REFERENCES

1. B T Golding and D N R Rao, 'Adenosylcobalamin-dependent Enzymic Reactions', in 'Enzyme Mechanisms', eds M I Page and A Williams, The Royal Society of Chemistry, London, 1987, pp 404-428.
2. J M Pratt, 'Coordination Chemistry of the B$_{12}$ Dependent Isomerase Reactions', in 'B$_{12}$', ed D Dolphin, John Wiley, New York, 1982, Vol 1, pp 325–392.
3. Y Zhao, P Such, and J Rétey, *Angew Chem Int Ed Engl*, 1992, **31**, 215.
4. U Leutbecher, S P J Albracht, and W Buckel, *FEBS Letters*, 1992, **307**, 144.
5. C Michel, S P J Albracht, and W Buckel, *Eur J Biochem*, 1992, **205**, 767.
6. J Halpern, 'Chemistry and Significance of Vitamin B$_{12}$ Model Systems', in 'B$_{12}$', ed D Dolphin, John Wiley, New York, 1982, Vol 1, pp 501–542.
7. J. M. Pratt, *Chem Soc Rev*, 1985, **14**, 161.
8. L Werthemann, R Keese, and A Eschenmoser, unpublished results; see L Werthemann, Dissertation, ETH Zürich (Nr 4097), Juris Druck and Verlag, Zürich, 1968.
9. Y Murakami, Y Hisaeda, and A Kajihara, *Bull Chem Soc Jpn*, 1983, **56**, 3642.

10 Y Murakami, Y Hisaeda, and T Ohno, *Bull Chem Soc Jpn*, 1984, **57**, 2091.
11 Y Murakami, Y Hisaeda, A Kajihara, and T Ohno, *Bull Chem Soc Jpn*, 1984, **57**, 405.
12 Y Murakami, Y Hisaeda, T Tashiro, and Y Matsuda, *Chem Lett*, 1985, 1813.
13 Y Murakami, Y Hisaeda, T Tashiro, and Y Matsuda, *Chem Lett*, 1986, 555.
14 Y Murakami, Y Hisaeda, T Ozaki, T Tashiro, T Ohno, Y Tani, and Y Matsuda, *Bull Chem Soc Jpn*, 1987, **60**, 311.
15 Y Murakami and Y Hisaeda, *Pure Appl Chem*, 1988, **60**, 1363.
16 Y Murakami, Y Hisaeda, and T Ozaki, *J Coord Chem*, 1991, **23**, 77.
17 Y Murakami, Y Hisaeda, J Takenaka, and T Ohno, unpublished results.
18 Y Murakami, Y Hisaeda, H Kohno, T Ohno, and T Nishioka, unpublished results.
19 T Ohno, T Nishioka, Y Hisaeda, and Y Murakami, *J Mol Struct (Theochem)*, 1994, **308**, 207.
20 R G Finke, D A Schiraldi, and B J Mayer, *Coord Chem Rev*, 1984, **54**, 1.
21 Y Murakami, A Nakano, A Yoshimatsu, K Uchitomi, and Y Matsuda, *J Am Chem Soc*, 1984, **106**, 3613.
22 Y Murakami, Y Hisaeda, and T Ohno, *J Chem Soc, Perkin Trans 2*, 1991, 405.
23 Y Murakami, Y Hisaeda, and T Ohno, *Chem Lett*, 1987, 1357.
24 Y Murakami, Y Hisaeda, and T Ohno, *J Chem Soc, Chem Commun*, 1988, 856.
25 Y Murakami, Y Hisaeda, X-M Song, K Takasaki, and T Ohno, *Chem Lett*, 1991, 977.
26 Y Murakami, Y Hisaeda, X-M Song, and T Ohno, *J Chem Soc, Perkin Trans 2*, 1992, 1527.
27 Y Murakami, Y Hisaeda, T Ohno, and Y Matsuda, *Chem Lett*, 1988, 621.
28 G N Schrauzer and M Hashimoto, *J Am Chem Soc*, 1979, **101**, 4593.
29 A Maihub, J W Grate, H B Xu, and G N Schrauzer, *Z Naturforsch, Teil B*, 1983, **38**, 643.

CHASING THE ENZYMES OF ALKALOID BIOSYNTHESIS[*]

Meinhart H Zenk

*Department of Pharmaceutical Biology, University of Munich,
Karlstrasse 29, D-80333 Munich, Germany.*

1   FUNCTION AND ACTIVITY OF BENZYLISOQUINOLINE ALKALOIDS

Secondary products are the most fascinating molecules found in nature. Their extreme chemical diversity has for decades attracted the interests of chemists, pharmacists, botanists and ecologists. In the foreground was, of course, their use for mankind as medicines, poisons, spices, dyes, *etc.* But already in the 19th century there was a clear opinion influenced by Darwinism that these compounds should serve as a chemical defense for plants against their predators.[1] Under the more romantic idealistic view of nature, prevailing still at the beginning of this century, the opinion for the 'raison d'être' of these compounds changed and they were denominated as being products of the playground of evolution and the 'loot in the garbage bin of plant metabolism'.[2,3] However, in recent years we have observed a swingback in opinion and with overwhelming evidence an ecochemical function can be assigned now to secondary products in general.[4,5] Clearly the development of a secondary product improves the producer's survival fitness by interacting specifically with the physiological set up of the competing organism,[5] either as an attractant or repellent or poison.

Alkaloids comprise one of the largest groups of secondary products. Numerous alkaloids have in this context been found to exert specific plant protective influences on the producer organism.[6] Among the alkaloids the largest group are benzylisoquinolines and derivatives thereof, which comprise about 1/4, *ie* 2500 members, of all known alkaloids. These alkaloids deserve specific attention because of their medical value. Of about 29 alkaloids used as pure compounds, in Western medicine today, 6 are benzylisoquinoline derived alkaloids.[7] Specifically the most widely used plant derived natural products in therapy are the isoquinoline alkaloids codeine and morphine. More than 2% of all prescriptions in North America contain either of these two alkaloids.[8] The fascination of the complex structure of the isoquinoline alkaloids, their medical value and the attempt to come up eventually with biotechnological processes to produce these compounds *in vitro* led us to study the biosynthesis of these compounds specifically at the enzyme level.

---

[*] Dedicated to Professor Wolfgang Steglich on the occasion of his sixtieth birthday.

2   THE BIOSYNTHESIS OF ISOQUINOLINE ALKALOIDS

In the 1960s, a picture of the biosynthesis of this class of alkaloids began to emerge, somewhat speculative, based on isotopically labelled precursor feeding experiments to differentiated plants.[9] The ground work was layed down and metabolic routes to this class of alkaloids emerged.[10-13] However, the way to solve the question, how these metabolites are really formed, was to identify and characterize the enzymes involved in the synthesis of these alkaloids. This was a task that seemed at that time almost hopeless, due to the sluggish rate of expression of secondary metabolites in higher plants, which was coupled to low stationary levels of the corresponding biosynthetic enzymes and the large amounts of tannins and other phenolics which interfered with the extraction of catalytically active proteins from plants. In order to overcome this general drawback, we decided to establish plant suspension cultures and to try to use them as a source of the enzymes involved in the catalysis of isoquinoline alkaloids. The advantage would be, that the material is continuously available regardless of the season. It is undifferentiated, homogeneous and amenable to pipette transfer and the material is free of interfering micro-organisms. The most important advantage is, however, that secondary product formation can be compressed into an one to two week cultivation period, which should imply much higher levels of biosynthetic enzymes than in a differentiated field- or pot-grown plant, which may take several months to grow and produce.

The intellectual handicap at that time was the dogma that de-differentiated cell cultures are unable to form secondary products.[14] This dogma was broken only after we were able to establish a completely de-differentiated cell culture of *Morinda citrifolia*, yielding 2.5 g of anthraquinones per liter of medium which amounted to 10% secondary product, based on dry weight.[15] Subsequently, new cultures were established which produced compounds of interest. The lucky circumstance that more than 40 species of the genus *Berberis*, cultivated in our laboratory, were able to produce in the cell suspension stage copious amounts of the isoquinoline alkaloids jatrorrhizine and berberine,[16] made it possible for us to set out to study the general isoquinoline pathway leading to (*S*)-reticuline at the enzyme level.

2.1   Biosynthesis of (*S*)-Reticuline

The central intermediate for a multitude of isoquinoline alkaloids is the alkaloid (*S*)-reticuline. Reticuline is the proven precursor to such diverse groups of isoquinoline alkaloids as for instance morphinans, aporphines, rhoeadines, benzo[*c*]phenanthridines, phthalideisoquinolines, protoberberines, pavines and certain bisbenzylisoquinolines, as summarized in ref 9. The first question posed was how is (*S*)-reticuline formed in higher plants? Visually dissecting those benzylisoquinoline alkaloids known at that time, Winterstein and Trier[17] and later Robinson[18] came to the conclusion that all of these alkaloids were derived from a simple tetrahydroxylated benzylisoquinoline named norlaudanosoline.

These authors also had already suggested that this molecule might arise by condensation of dopamine with 3,4-dihydroxyphenylacetaldehyde to afford the tetraoxygenated base norlaudanosoline.[17,18] This proposal was put to test when radiolabelled tracer molecules became commercially available. Satisfactory incorporation of the labelled tracer molecule norlaudanosoline was found to occur into different classes of benzylisoquinoline alkaloids[19,20] which seemingly verified the old Winterstein and Trier[17] biogenetic hypothesis. These tracer experiments were later supported by enzymatic investigations, which showed that in cell cultures of various *Papaveraceae* an enzyme is present which stereoselectively condenses dopamine and 3,4-dihydroxyphenylacetaldehyde (as well as other substituted aldehydes) by a Pictet-Spengler-condensation to yield the (S)-configurated assumed alkaloid precursor.[21,22]

The intermediacy of the purported base norlaudanosoline in the construction of benzylisoquinoline alkaloids was seriously questioned for the first time when a common biogenetic origin was shown for the trioxygenated bases of the coclaurine type and reticuline in *Anona reticulata* plants.[23] Contradictory and incompatible results had previously been obtained after application of radiolabelled DOPA and/or dopamine to alkaloid producing plants, which were found to label exclusively the isoquinoline or 'upper half' of reticuline or reticuline-derived structures,[9,24-26] and not both halves of the benzylisoquinoline alkaloids, as it would have been expected for a tetraoxygenated precursor. The question of the pathway leading from L-tyrosine to (S)-reticuline was solved when it was demonstrated that the label from (S)-[1-^{13}C]norcoclaurine and (S)-[1-^{13}C]coclaurine was highly and nonrandomly incorporated into (S)-reticuline in benzyltetrahydroisoquinoline-producing plant cell cultures of *Berberis stolonifera, Eschscholtzia californica* as well as *Pneumus boldus*.[27] From these surprising incorporation data observed *in vivo*, it was possible to conclude that (S)-norcoclaurine is stereoselectively metabolized to (S)-reticuline *via* (S)-coclaurine, (S)-N-methylcoclaurine and (S)-3'-hydroxy-N-methylcoclaurine.[27] (S)-Norreticuline is not an intermediate in this important metabolic route.

In the meantime all enzymes involved in the transition from tyrosine to (S)-reticuline are now in hand and published.[28] The following enzymes are listed in the order of their biosynthetic sequence: (1) (S)-norcoclaurine synthase [previously called (S)-norlaudanosoline synthase],[21,22] (2) S-adenosyl-L-methionine: (R,S)-nor-coclaurine-6-O-methyltransferase (previously called (R,S)-norlaudanosoline-6-O-methyltransferase),[29] (3) S-adenosyl-L-methionine: (R,S)-coclaurine-N-methyl-transferase (previously called S-adenosyl-L-methionine: norreticuline N-methyltransferase),[30] (4) (R,S)-N-methylcoclaurine-3'-hydroxylase (most likely a phenolase),[31] (5) S-adenosyl-L-methionine: 3'-hydroxy-N-methyl-(S)-coclaurine-4'-O-methyltransferase.[28] A total of five hitherto unknown enzymes is involved in the transformation of dopamine and 4-OH-phenylacetaldehyde into (S)-reticuline (Scheme 1). Two enzymes (1 and 5) are absolutely stereoselective yielding or acting on only the (S)-configurated benzylisoquinoline alkaloid.

Three of the enzymes (1, 2, 3) had to be renamed in light of the above evidence that the trihydroxylated and not the tetrahydroxylated alkaloids as assumed previously are the true precursors. The sequence of O- and N-methylation steps is strictly governed by the sequential action of stereoselective and non-selective enzymes affording the bisphenolic dimethylether (S)-reticuline with the potential for further more complex modifications. The clarification of the (S)-reticuline pathway was the result of the combination of *in vivo* ^{13}C NMR, and *in vitro* enzymatic studies. Scheme 1 depicts the now firmly established pathway from the early precursor L-tyrosine *via* dopamine,[32] 4-hydroxyphenylacetaldehyde,[32] and norcoclaurine to (S)-reticuline.

2.2 Biosynthesis of Protoberberines

Out of the considerable number of protoberberines found in nature, berberine is no doubt the most prominent member. It is valued pharmaceutically for its antibacterial activity both as an intestinal antiseptic and to eliminate eye infections. In addition, it acts as a stomachic and anti-inflammatory agent. Berberine is produced by a considerable variety of plants.[33] *Coptis japonica* (*Ranunculaceae*) is a plant endemic to Japan which contains large amounts of berberine, palmatine, jatrorrhizine and coptisine in its rhizome. In spite of a very high content of berberine (up to 10% of dry weight) in the rhizome, this plant cannot be utilized commercially as a raw material for preparing berberine because the growth is so slow that it takes more than 5 years of cultivation to obtain only small rhizomes weighing 2 g dry weight.[34] For the first time a stable fast growing *Coptis* cell line was established which produced considerable amounts of berberine as well as other alkaloids.[34] This strain was further improved[35] and most successfully, cell suspension cultures of *C japonica* have been established which produce 7 g × l^{-1} berberine during a cultivation period of 20 days,[36] the highest yield of a secondary product thus far achieved worldwide. Using this cell strain, the production of berberine by fermentation of *C japonica* cells became one of the few successful commercial productions of a plant constituent. Previous to these activities, we had developed cell suspension culture strains of different *Berberis* species, which produced also large amounts of protoberberine alkaloids, the main constituent in this genus being jatrorrhizine and as minor alkaloids berberine, columbamine and palmatine.[16] These cultures proved to be an excellent source for the isolation of the various enzymes involved in protoberberine biosynthesis.

(S)-Reticuline is converted *via* (S)-scoulerine to various protoberberines, as has previously been determined by precurser feeding experiments using differentiated plants.[37,38] During this conversion of the benzylisoquinoline, the so-called berberine bridge is formed, *ie* the N-CH$_3$ group of reticuline is converted to carbon atom 8 of the protoberberine molecule. Rink and Böhm[39] were the first to describe a cell-free system prepared from *Macleaya microcarpa* cell suspension cultures which could convert reticuline to scoulerine. The enzyme, which was named 'berberine bridge enzyme', was found by us to occur

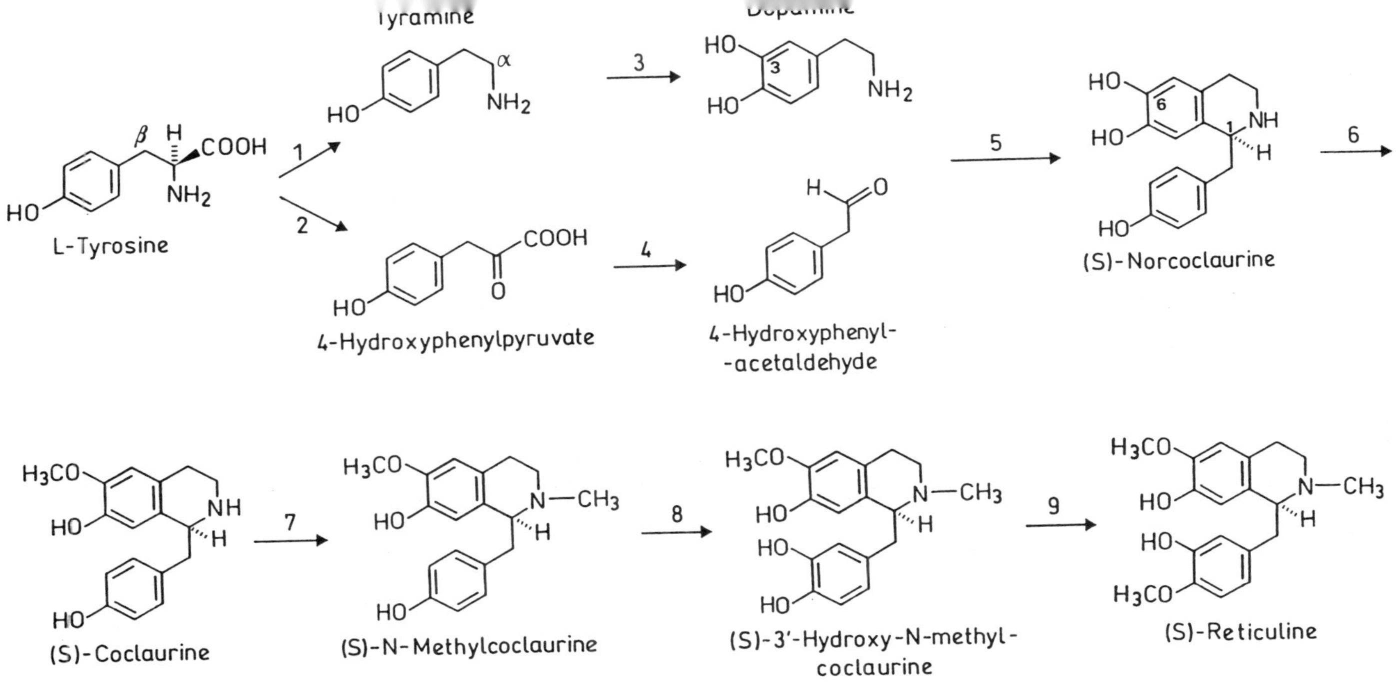

Scheme 1: The biosynthetic pathway leading from 2 molecules of L-tyrosine to (*S*)-reticuline. The individual enzymes involved are: 1 = L-tyrosine decarboxylase; 2 = L-tyrosine transaminase; 3 = phenolase; 4 = p-hydroxyphenylpyruvate decarboxylase; 5 = (*S*)-norcoclaurine synthase; 6 = S-adenosyl-L-methionine:(*R*,*S*)-norcoclaurine-6-*O*-methyltransferase; 7 = S-adenosyl-L-methionine: (*R*,*S*)-coclaurine-*N*-methyltransferase; 8 = (*R*,*S*)-*N*-methylcoclaurine-3'-hydroxylase (a phenolase); 9 = S-adenosyl-L-methionine: 3'-hydroxy-*N*-methyl-(*S*)-coclaurine-4'-*O*-methyltransferase.

in 66 samples taken from differentiated plants and from cell suspension cultures.[40] The enzyme was purified 450-fold from *Berberis beaniana* cell cultures.[40] The enzyme requires the presence of oxygen for the catalysis of the conversion of the (*S*)-enantiomer of the benzylisoquinoline (*S*)-reticuline to the corresponding tetrahydro-protoberberine (*S*)-scoulerine and releases a stoichiometric amount of H_2O_2. The DNA for this enzyme has recently been cloned from *E californica*.[41] It is a 57.352 kD enzyme (excluding carbohydrates) containing a total of 538 amino acids with a 22 amino acid containing signal peptide.[41] Most importantly, this enzyme is housed in a specific vesicle with a density of $r = 1.14$ g × ml^{-1}. Using the heterologously expressed enzyme, it is to be expected that the reaction mechanism of this enzyme will soon be solved, *ie* whether the berberine bridge is formed by an ionic or a radical mechanism.[41] In order to gain further insight into the mechanism of this transformation, the fate of a methyl group transferred from S-adenosylmethionine to scoulerine was followed.[42] It was shown that the methyl group of SAM is transferred to the nitrogen of (*S*)-norreticuline with clean inversion of configuration, which is consistent with the observation on most other methyltransferases. Chiral [*N*-CH$_3$]-labelled (*S*)-reticuline was further transformed by the action of berberine bridge enzyme and this reaction showed to operate completely stereospecifically, replacing an *N*-methyl hydrogen by the phenol group in an inversion mode. A primary kinetic isotope effect of $k_H/k_D \approx 4$ was observed. The unravelling of the steric course of berberine bridge enzyme was possible by using tritium NMR.[42]

The next step in the formation of berberine is the specific methylation of (*S*)-scoulerine at the 9-*O* position. Again from suspension cultures of several *Berberis* species an enzyme was detected and isolated which transfers the methyl group of SAM specifically to the 9-position of the (*S*)-enantiomer of scoulerine, thus producing (*S*)-tetrahydrocolumbamine.[43] The enzyme was partly enriched. It has a molecular weight of 63 kD and shows a very high degree of substrate specificity using several naturally occurring tetrahydroprotoberberines as substrates. Only scoulerine was methylated. Neither (*R*)-scoulerine nor dehydroscoulerine with its ring-C aromatized served as a substrate. The pH optimum of the enzyme is 8.9, which is identical to the pH optimum of the berberine bridge enzyme. Whether this methyltransferase does also occur in the specific vesicle is not quite clear yet. The enzyme might stick to the outside of the membrane of the vesicle or it really might be located within the vesicle.[44] This question has to await the generation of specific antibodies against this enzyme to make a clear-cut decision.

The (*S*)-tetrahydrocolumbamine thus formed is transformed to (*S*)-canadine. The enzyme responsible for the formation of the methylenedioxy-bridge is a microsomal bound cytochrome P-450 enzyme. It has been found to occur in microsomal preparations of different species of *Ranunculaceae* (*Thalictrum, Coptis*) as well as *Berberidaceae* cell cultures. The enzyme needs NADPH and O_2 as cosubstrates. The enzyme is highly specific for (*S*)-

tetrahydrocolumbamine. Its (R)-configurated isomer as well as columbamine, tetrahydrojatrorrhizine, cheilanthifoline, nandinine *etc*, do not serve as substrates for this specific enzyme. The enzyme is inhibited by specific cytochrome P-450 inhibitors including carbon monoxide. The inhibition of CO is reversed by blue light but not by red light. The pH optimum for canadine synthase is 8.5 (M Rueffer and M H Zenk, unpublished results). We had previously claimed[45] that this reaction is catalyzed by a quite different protein. Using columbamine which had been labelled in the 3' position with a [C^3H_3]-group and following the expected release of 1/3 of the tritium as an enzyme assay, enzyme activity was found in a protein which was brought to homogeneity. This protein was found to be a Fe^{2+} dependent cytosolic enzyme showing a typical protoporphyrin absorption at 408 nm (at pH 8.9) and with an unusual temperature optimum at 70 °C. Recently this enzyme turned out to be a non-specific peroxidase.[46] The previously claimed berberine synthase[45] is therefore incorrect.

The final reaction in the formation of berberine is the oxidation of the C-ring of (S)-canadine (= tetrahydroberberine). A novel oxidase was discovered in *Berberis wilsoniae* suspension cultures catalyzing in the presence of oxygen the removal of 4 hydrogen atoms from a number of tetrahydroprotoberberine alkaloids including canadine with simultaneous production of 2 moles of H_2O_2. This enzyme, (S)-tetrahydroprotoberberine oxidase, exhibits a strict specificity for the (S)-enantiomers of tetrahydroprotoberberines and benzylisoquinoline alkaloids. It shows a pH optimum at 8.9, a molecular mass of 105 kD and consists of 2 subunits each and covalently bound flavin. This catalyst is responsible for the conversion of (S)-tetrahydroprotoberberines to protoberberines including the tetrahydro-derivatives of berberine, jatrorrhizine, palmatine and coptisine. The enzyme was successfully immobilized with a half life of 200 days at room temperature.[47] It is also located in a vesicle with a specific gravity of $r = 1.14$ g × cm^{-3} as shown by direct enzymatic assay as well as immunoelectrophoresis.[48] This vesicle also contains the following protoberberine alkaloids: jatrorrhizine, palmatine, columbamine and berberine, the same alkaloids which are also found in the final storage place, the vacuole of these cells. The pH optima of 8.9 of this enzyme, as well as berberine bridge enzyme, points to the fact that within this smooth vesicle probably a rather alkaline pH persists. Since quaternary protoberberine alkaloids are known not to diffuse through membrane systems of living cells, it appears that the protoberberine alkaloids formed under the influence of (S)-tetrahydroprotoberberine oxidase are trapped within this vesicle. It has been observed by electron microscopy[48] that the small vesicles fuse with each other to form small vacuoles, which finally become confluent with the large central vacuole of the plant cell, thus releasing the vesicular protoberberines into the central vacuole of the plant cell. This is most likely the way these quaternary alkaloids are passively transported to the storage place of the plant cell.

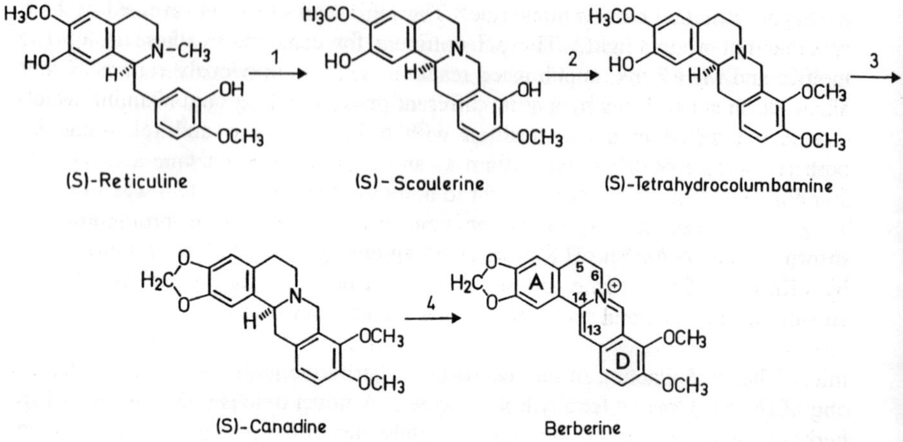

Scheme 2: The biosynthetic pathway leading from (S)-reticuline to berberine. The individual enzymes involved are: 1 = berberine bridge enzyme; 2 = S-adenosyl-L-methionine: (S)-scoulerine-9-O-methyltransferase; 3 = (S)-canadine synthase; 4 = (S)-tetrahydroprotoberberine oxidase.

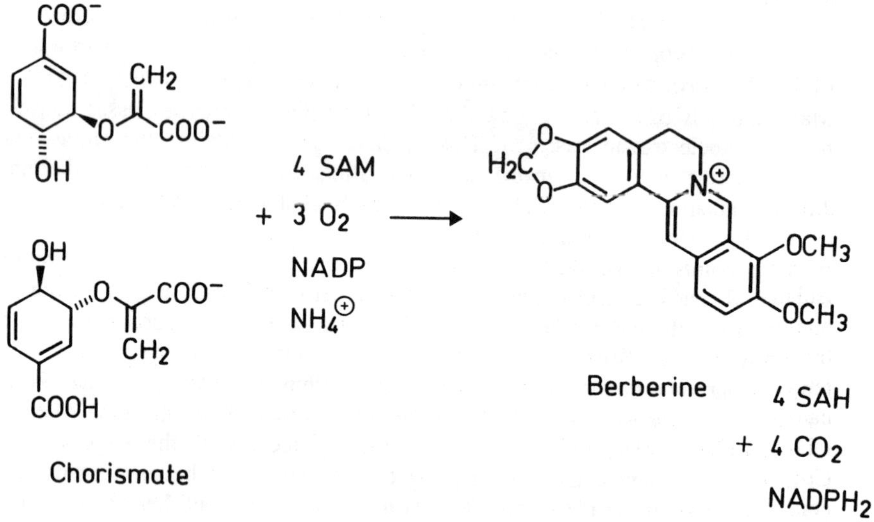

Scheme 3: Requirements for the transformation of the primary metabolite chorismate to the alkaloid berberine (SAH = S-adenosylhomocysteine).

It should be pointed out that in *Coptis japonica* a similar enzyme has been discovered: (*S*)-tetrahydroberberine oxidase (= canadine oxidase).[49,50] This enzyme is also vesicle-bound, needs the same cofactors as (*S*)-tetrahydroprotoberberine oxidase, is highly stereoselective, and produces H_2O_2 as a reaction product in addition to berberine. However, it differs considerably in molecular weight (58 kD) and in substrate specificity, in that it does not oxidize alkaloids such as (*S*)-tetrahydropalmatine or (*S*)-scoulerine. Interestingly, in two closely related plant families the penultimate enzyme, the final oxidase of the berberine biosynthetic pathway, consists of two considerably different enzymes.

The whole pathway which has been worked out now at the enzyme level is shown in Scheme 2. The branch point intermediate, (*S*)-reticuline, is transformed by catalysis of four different enzymes to berberine. At least two of these enzymes are contained in a specific vesicle, derived from the endoplasmic reticulum, while one of the enzymes (canadine synthase) is located in the microsomes of these cells. The energy requirement for the transition of chorismic acid, the branchpoint intermediate of the shikimate pathway, to berberine *via* L-tyrosine is depicted in Scheme 3. It is astonishing that very little energy is necessary to transform this primary metabolite to the alkaloid berberine. Actually the two molecules of chorismate each lose their two carboxyl groups. Two methyl groups from SAM and three molecules of oxygen from O_2 are inserted into each intermediate. One molecule of NADP and one molecule of NH_4^+ are needed for this conversion. The H_2O_2 which is produced during the formation of berberine is assumed to be split by the enzyme catalase into H_2O and O_2.

Recently it was found that berberine is obviously retained in the vacuole of the *Coptis* cells by the formation of the highly water-soluble malic acid salt.[51] Malic acid obviously acts as counterion for the quaternary and positively charged berberine in (at least) *Coptis* cell cultures.

2.3 Biosynthesis of (*R*)-Configurated Tetrahydroprotoberberines

As shown in the previous section, all tetrahydroprotoberberines which have been isolated thus far or have been established as intermediates show the (*S*)-configuration at carbon atom 14. It had been noticed, however, already in 1931 by Späth and coworkers,[52] that in *Corydalis cava* not only (*S*)-configured tetrahydroprotoberberines are present, but also (*R*)-tetrahydropalmatine, (*R*)-canadine *etc*. The absolute configuration of these compounds was established by CD-spectroscopy.[53] Bhakuni and coworkers[54,55] conducted experiments using *Cocculus laurifolius* towards the elucidation of the biosynthesis of these (*R*)-configured tetrahydroprotoberberines. These studies suggested that (*S*)-reticuline was not converted in the plant into dihydroreticuline. Further, racemization of optically active forms of tetrahydropalmatine did not take place *via* dihydrotetrahydropalmatine, as had been found to occur at the enzyme level during the formation of (R)-reticuline from (S)-reticuline in the morphinan biosynthetic pathway.[56] Clear evidence was obtained[55] that (R)-reticuline gave rise to (R)-tetrahydropalmatine. Since we had clearly shown that the berberine bridge enzyme exclusively converts (S)-reticuline to (S)-scoulerine,[40] there were two possibilities: either in these specific plants which produce the (R)-configured isoquinoline alkaloids an (R)-specific berberine bridge enzyme is present, or if this is not the case, conversion to the (R)-configured intermediates has to occur at a later stage in the biosynthesis. To resolve this discrepancy we repeated the above mentioned[54,55] feeding experiments with authentic (R)-[N-$^{14}CH_3$]reticuline to freshly cut, young *C laurifolius* branches. Absolutely no incorporation of enantiomerically pure (R)-reticuline into (R)-sinactine was observed.[57] This confirms that the berberine bridge enzyme, which was found to occur in a total of 66 plant cell cultures and whole plants, representing 5 plant families, is without exception absolutely (S)-reticuline specific.[40]

We decided, therefore, that stereochemical inversion has to occur at a later stage in the biosynthetic pathway towards (R)-configured tetrahydroprotoberberine alkaloids. *Corydalis cava* was used as the experimental plant. The berberine bridge enzyme of this particular plant was isolated and shown to be absolutely (S)-stereoselective.[57] Feeding experiments with synthetic (R)-[8-^{14}C]scoulerine showed no incorporation, while the corresponding (S)-precursors yielded 9% incorporation into (R)-stylopine, 5% into (R)-tetrahydropalmatine, and 8% into (R)-canadine. By repetition of experiments published by Holland *et al*,[26] we confirmed the incorporation of the protoberberine palmatine (lacking a stereochemical center in the C-ring) into the ^{13}C-methylated derivative, corydaline, with (14R)-chirality. Feeding experiments of [8-^{3}H]berberine and [8-^{3}H]palmatine to *C cava* plants yielded high incorporations into (R)-configured canadine and (R)-configurated tetrahydropalmatine (38% and 49%, respectively). The biosynthesis of (14R)-tetrahydroprotoberberine alkaloids requires, obviously, hydrogenation of the two double bounds at 7,8 and 13,14, the latter generating a chiral center with (R)-absolute configuration. In order to investigate the sequence of the reduction steps, a cell-free system from bulbs of *Corydalis cava* was employed. Incubation with labelled berberine showed that this compound in the presence of NADPH rapidly disappeared. An intermediate was detected, which also decreased in amount with time and, finally, (14R)-canadine appeared. The intermediate was identified as 7,8-dihydroberberine. This intermediate, when supplied in synthetic form to the enzyme mixture, was specifically reduced in the presence of NADPH to the (R)-configured canadine. Addition to the enzyme mixture of [4A-^{3}H]NADPH and [4B-^{3}H]NADPH as cofactors revealed a B-type hydrogenation. The (R)-absolute configuration of the enzyme generated products was unequivocally established by CD-spectra and enzymatic assays.[57] (R)-Tetrahydroprotoberberines are, therefore, formed *in vivo* and *in vitro* from protoberberines by a highly stereoselective reduction system as shown in Scheme 4. (R)-Tetrahydroprotoberberines are not formed from (R)-reticuline, as has been suggested previously.[54,55]

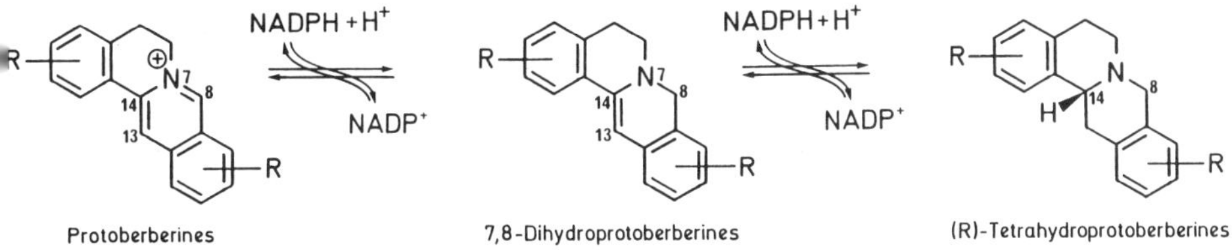

Scheme 4: Reversible enzyme system reducing protoberberine alkaloids to (R)-configured tetrahydroprotoberberine alkaloids with the intermediate formation of 7,8-dihydroprotoberberines.

2.4 Biosynthesis of Benzo[c]phenanthridines

Benzo[c]phenanthridine alkaloids are a specific group of isoquinoline alkaloids, which occur only in higher plants and are constituents mainly of the *Papaveraceae* family.[58] Three cell culture systems have been worked out, which produce oxidized benzo[c]phenanthridine alkaloids in abundance: *Eschscholtzia californica*,[59] *Papaver somniferum*,[60] and *Sanguinaria canadensis*.[61] All three systems yield copious amounts of the oxidized alkaloids when the cell suspension cultures were challenged with a fungal elicitor. Using the *E. californica* cell culture, the complete pathway leading from (S)-reticuline to the most highly oxidized member of the benzophenanthridine alkaloid family, macarpine, was worked out in our laboratory. The initial steps in the pathway were made possible by the pioneering work of Battersby[62] and Takao[63] and their coworkers. Starting with (S)-reticuline, the benzophenanthridine pathway is opened by the vesicularly contained berberine bridge enzyme which gives raise to (S)-scoulerine (step 1, Scheme 5). This alkaloid, which also is an intermediate in the protoberberine pathway, obviously has to leave the vesicle in which it was formed, diffuse within the cell towards the microsomes. Then, in the presence of NADPH and O_2 two methylenedioxy bridges are formed in two consecutive steps: (S)-cheilanthifoline from (S)-scoulerine (step 2) and (S)-stylopine from (S)-cheilanthifoline (step 3). The use of [9-OC^{3}H$_3$]- and [3-OC^{3}H$_3$]-labelled scoulerine and cheilanthifoline, respectively, offered a convenient assay for both of these enzymes.[64] Both enzymes show maximal activity at pH 8. Both enzymes were surprisingly substrate-specific. Only (S)-scoulerine and (S)-cheilanthifoline were transformed. Absolutely no reaction was observed with tetrahydroprotoberberine analogues with an (R)-configuration or any of the other derivatives tested in either enantiomerically pure (S)- or (RS)-forms. Due to the action of both cytochrome P-450 enzymes, the methylenedioxy groups are introduced first in the D-ring and later in the A-ring of the (S)-scoulerine molecule.[64] Inhibition studies clearly demonstrated that both enzymes are cytochrome P-450 proteins.[64] Both enzymes are distinct and separate as could be shown by solubilization and physical separation (L Kammerer and M H Zenk, unpublished). Stylopine and other (S)-configured tetrahydroprotoberberine alkaloids like (S)-canadine can be subjected to N-methylation at the expense of SAM which generates the cis-N-methyl-derivatives of tetrahydroprotoberberine. These have been elegantly shown to be the precursor for the benzophenanthridines in vivo.[63] The methyltransferase was partially purified, showed a pH optimum of 8.9, a temperature optimum of 40 °C, and is of about 78 kD. This enzyme (step 4) appears to be cytosolic and was named S-adenosyl-L-methionine: (S)-tetrahydroprotoberberine-cis-N-methyltransferase.[65] It opens the benzophenanthridine pathway.

The (S)-cis-N-methylstylopine acts as substrate for another cytochrome P-450 hydroxylase which oxidizes the substrate compound to protopine (step 5), the lead alkaloid of the *Papaveraceae*.[66] This microsomal cytochrome P-450 NADPH-dependent enzyme hydroxylated stereo- and regio-specifically carbon atom 14 of (S)-cis-N-methyltetrahydroprotoberberine. It is a typical monooxygenase, found in several members of the *Fumariaceae* and *Papaveraceae*, and is inhibited by cytochrome P-450 inhibitors, as well as by carbon monoxide. The enzyme was solubilized and purified 100-fold. Protopine had been recognized earlier as an important intermediate in benzophenanthridine alkaloid biosynthesis by *in vivo* studies using *Macleaya microcarpa* callus cultures.[63] The most critical step in the biosynthesis of benzophenanthridine alkaloids is the enzymatic opening of the B-ring of protopine. In an elegant piece of work, Tanahashi[67] in our laboratory succeeded in the synthesis of 6-[^{3}H]protopine which was used as a substrate for the purported hydroxylase. It was assumed that during the hydroxylation of the 6-position of protopine the ring would open and possibly rearrange either enzymatically or spontaneously. This goal was achieved by using [6-^{3}H]protopine as a substrate and measuring the release of tritium using cell-free preparations of *E californica*. Indeed a microsomal preparation could be found which catalyzed the hydroxylation of [6-^{3}H]protopine with the concomitant formation of [11-^{3}H]dehydrosanguinarine and HOT. The hydroxylation proved strictly dependent on NADPH as reduced cofactor and on molecular O_2. The monooxygenase was inhibited by the classical cytochrome P-450 inhibitors. The hydroxylase was induced by fungal elicitor about 8-fold after challenging the plant cell culture with the glucoprotein. The hydroxylase is specifically present only in those plant species which produce benzo[c]phenanthridine alkaloids in culture.[67] We assume that protopine-6-hydroxylase introduces the hydroxyl group into the protopine molecule to give (6S)-hydroxyprotopine.[68] No evidence has ever been obtained in our enzyme work for a stable intermediate between protopine and dehydrosanguinarine. Therefore we assume that a single enzymatic hydroxylation at C-6 of protopine leads to the spontaneous rearrangement of this molecule to yield dihydrosanguinarine (steps 6 and 7). Obviously the protein pocket of the active centre of the 6-hydroxylase (step 6) provokes this rearrangement, which proceeds in a quantitative manner.

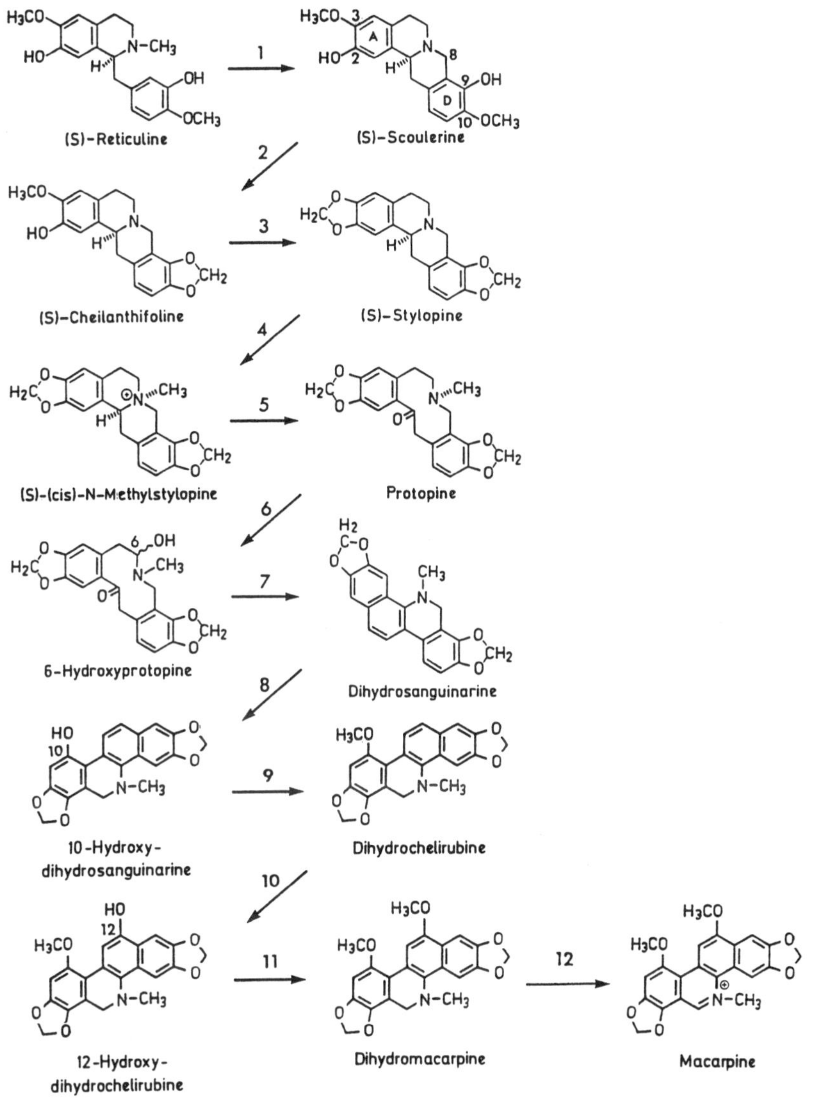

Scheme 5: The biosynthetic pathway leading from (S)-reticuline to the most highly oxidized benzo[c]phenanthridine alkaloid, macarpine. The individual enzymes involved are: 1 = berberine bridge enzyme; 2 = (S)-cheilanthifoline synthase; 3 = (S)-stylopine synthase; 4 = S-adenosyl-L-methionine:(S)-tetrahydroprotoberberine-cis-N-methyltransferase; 5 = (S)-cis-N-methyltetrahydroprotoberberine-14-hydroxylase; 6 = protopine-6-hydroxylase, step 7 is spontaneous; 8 = dihydrosanguinarine-10 hydroxylase; 9 = S-adenosyl-L-methionine: 10-hydroxydihydrosanguinarine-10-O-methyltransferase; 10 = dihydrochelirubine-12-hydroxylase; 11 = S-adenosyl-L-methionine: 12-hydroxydihydrochelirubine-12-O-methyltransferase; 12 = dihydrobenzophenanthridine oxidase.

If a plant accumulates sanguinarine than dihydrosanguinarine is subsequently oxidized by the known enzyme dihydrobenzophenanthridine oxidase to afford the fully aromatized quaternary benzo[c]phenanthridine sanguinarine.[69,61] If, however, the plant species produces macarpine, as is the case for instance in *E californica* cell cultures, two subsequent hydroxylation and two methylation steps have to follow. In subsequent work, again using *E californica* cell suspension cultures it was possible to detect two additional cytochrome P-450 enzymes: dihydrosanguinarine-10-hydroxylase (step 8) and S-adenosyl-L-methionine: 10-hydroxydihydrosanguinarine-10-*O*-methyltransferase (step 9). Both enzymes were partially characterized and found to be highly substrate specific.[70] Both enzymes are specific only for the dihydro-derivatives; the oxidized members of the sanguinarine family are not transformed. Recently, the two enzymes transforming dihydrochelirubine to macarpine were also discovered in our laboratory using suspension cultures of *Thalictrum bulgaricum* (L Kammerer, W De-Eknamkul, and M H Zenk, *Phytochemistry*, 1994, in press). The enzymes were designated dihydrochelirubine-12-hydroxylase (step 10) and S-adenosyl-L-methionine: 12-hydroxydihydrochelirubine-12-*O*-methyltransferase (step 11). Again both enzymes were found to be highly substrate specific and only the dihydro-compounds were substrates for both enzymes. The final step in macarpine synthesis is then the oxidation of dihydromacarpine to macarpine by dihydrobenzophenanthridine oxidase (step 12).[69,61] This enzyme was originally discovered in our laboratory in *E californica* suspension cultures and more recently isolated and purified to apparent homogeneity from elicited *Sanguinaria canadensis* cultures. The enzyme oxidizes a variety of dihydrobenzophenanthridine molecules and is most likely copper-containing.[61] It is only active in the presence of molecular oxygen and H_2O_2 is formed as a reaction product. This terminal reaction concludes the pathway from (S)-reticuline to macarpine which involves a total of 11 highly specific enzymes and 1 spontaneous step (step 7). Out of these 11 enzymes, 6 catalysts are cytochrome P-450 enzymes which are highly specific and seem to catalyze only one reaction each. Since the pathway from tyrosine to (S)-reticuline is also known at the enzyme level (see above), the conversion of L-tyrosine to macarpine involves a total of 20 enzymes which are now all known and at least partially characterized. Only one cDNA encoding one of these enzymes has been cloned.[41] The pathway leading to macarpine is probably the longest which has ever been elucidated at the enzyme level for any secondary product.

The knowledge of these pathways will also give a hint as to the regulation of secondary products within plants. Attention should be drawn to the recent finding that one of the signal compounds in controlling the formation of secondary products, including benzophenanthridine alkaloids, in plant cell cultures is jasmonic acid.[71] The combination of enzymology with molecular biology will be expected eventually to unravel the mode of regulation of secondary compounds.

2.5 Phenol Oxidative Coupling

The proposal of Barton and Cohen[72] correlated the structure of specific plant alkaloids in terms of the reaction mechanism. The oxidation of phenols by one electron transfer affords phenolic radicals, which, by radical pairing, form new C-C or C-O bonds either intra- or inter-molecularly. The biocatalysts involved in these transformations have until now remained obscure. The hypothesis that the ubiquitous phenol oxidase, laccase, tyrosinase, and peroxidase enzymes are of importance could be excluded due to the lack of stereo- and substrate-specificity displayed by these enzymes.[73]

Recently we presented two examples which unequivocally demonstrated that the formation of C-C and C-O bonds in the benzylisoquinoline alkaloid metabolism is catalyzed by specific cytochrome P-450 linked microsomal bound plant enzymes.

2.5.1 Intramolecular Phenolic Coupling

Barton and Cohen[72] have proposed that the crucial C12-C13 bond of morphine alkaloids can be envisaged as being formed by intramolecular phenol coupling of (R)-reticuline. Indeed, experimental proof has been obtained by employing *in vivo* experiments that (R)-reticuline is transformed to salutaridine (Scheme 6) by regioselective *para-ortho* oxidative coupling.[74] Attempts to synthesize this morphinandienone structure biomimetically have up to now resulted in rather mediocre yields.[75] An efficient chemical or enzymatic method for the phenolic oxidation of (R)-reticuline to salutaridine has been the subject of intensive research for many years. Cytochrome P-450 enzymes can function in alkaloid biosynthesis, both as oxygenases in O_2 and NADPH dependent substrate hydroxylations[66,67] and in the synthesis of C-O bonds exemplified in the formation of methylenedioxy groups from *O*-methoxyphenols.[64] This prompted us to isolate microsomal preparations from differentiated *Papaver somniferum* plants or from a thebaine producing cell suspension culture. Incubation of these microsomes together with (R)-reticuline as substrate and NADPH in the presence of oxygen yielded salutaridine.[76] The product of this transformation, salutaridine, was rigorously identified by isotope dilution analysis and derivatisation. Using (R)-[N-$^{13}CH_3$]reticuline as substrate for the microsomal conversion, the enzymatically formed product yielded the correct ^{13}C NMR spectrum for salutaridine.[77]

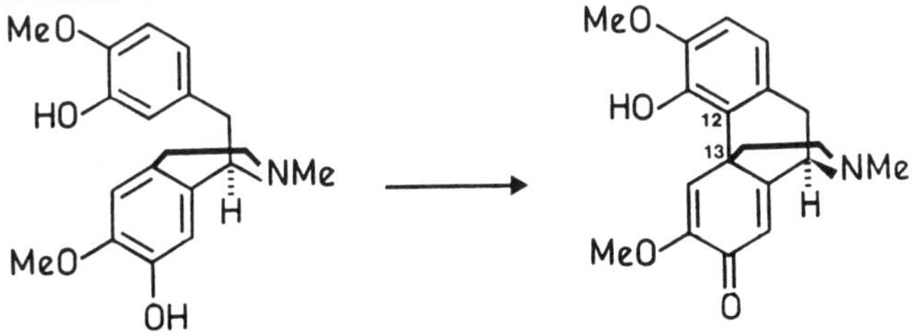

Scheme 6: Intramolecular *para-ortho* C-C coupling of (R) reticuline by the microsomal cytochrome P-450 enzyme, salutaridine synthase, to yield salutaridine.

All of these experiments demonstrated unequivocally that microsomes isolated from poppy capsules catalyzed the conversion of (*R*)-reticuline to salutaridine (Scheme 6) in the presence of NADPH. The membrane bound enzyme catalyzing this reaction has been called salutaridine synthase.[76] The synthase shows a pH optimum at 7.5 under standard assay conditions and it was surprisingly substrate specific. Only (*R*)-reticuline was transformed. Neither (*R*)-norreticuline, (*R*)-coclaurine, (*S*)-reticuline nor any of the congeners of reticuline served as substrate. Cytochrome P-450 inhibitors, such as naphthoquinones and carbon monoxide (in darkness but not in light) inhibited this phenol coupling reaction and pointed clearly towards the existence of a cytochrome P-450 dependent oxidase. This enzyme system occurs in roots, shoots, capsules of the poppy plant but not in latex. Therefore, earlier claims that the latex is the site of morphinan alkaloid biosynthesis and that peroxidative enzyme systems are responsible for the formation of salutaridine from (*R*)-reticuline have both to be refuted.[76] The C-C coupling observed here is probably of major importance also for the crucial phenol coupling steps in the biosynthesis of *Amaryllidaceae* alkaloids as well as for the formation of colchicine and its congeners, and for the *Erythrina* alkaloids, *etc.*

2.5.2 *Intermolecular Phenol Coupling*

As a case of C-O intermolecular coupling, we studied the dimerisation of 6,7,4-trioxygenated benzyltetrahydroisoquinoline bases to form the simple dimer, (*R,S*)-berbamunine.[77,78] It had previously been shown that this phenol coupling product, berbamunine, is formed in *Berberis stolonifera* cell suspension cultures.[79] The screening for more than 34 different *Berberis* suspension cultures yielded the information that *B stolonifera* is the main source of this bisbenzyltetrahydroisoquinoline alkaloid.[79] A new cytochrome P-450 enzyme isolated from this cell culture has been purified to electrophoretic homogeneity.[78] The purified hemoprotein migrated as a single band upon SDS gel electrophoresis with a minimum molecular weight of 46 kD. The purified cytochrome P-450 was successfully reconstituted with NADPH-cytochrome P-450 reductase in the presence of lipids. It displayed a maximum turnover number of 50 nmol of substrate per nmol of P-450/min. The main product of (*R,S*)-*N*-methylcoclaurine, supplied as a substrate, was the simple tail-to-tail, diaryl-ether coupled dimer, berbamunine (Scheme 7). The stoichiometry of the reaction is as follows: $2ArOH + NADPH + H^+ + O_2 \rightarrow ArO-ArOH + NADP^+ + 2H_2O$. The work presented[78] proposes that cytochrome P-450 biradical mechanisms are governed by highly regioselective processes in the biosynthesis of complex natural products. This is the first report of a cytochrome P-450 enzyme that mediates regio- and stereoselective intramolecular oxidative phenol coupling to furnish a naturally occuring dimeric compound. In this catalytic cycle, cytochrome P-450 functions as an oxidant in a bisubstrate reaction without transfer of the activated oxygen atom to either of the two chiral substrates.

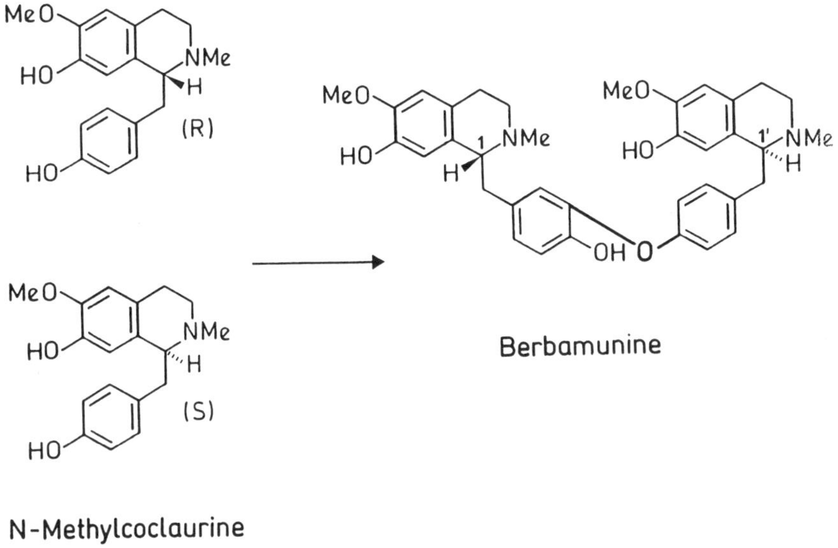

Scheme 7: Catalytic formation of berbamunine by intermolecular C-O coupling of the monomeric (R)- and (S)-N-methylcoclaurine by the microsomal cytochrome P-450 enzyme, berbamunine synthase.

We predict that a considerable number of enzymes capable of catalyzing some of the most fascinating steps in natural product biosynthesis will in future turn out to be cytochrome P-450 containing proteins. It has already been shown in the present study that by chemical synthesis it is extremely difficult to achieve C-C phenol coupling to yield the carbon skeleton of the morphinandienone alkaloids and the C-O intermolecular coupling of tetrahydrobenzylisoquinolines which are catalyzed by these interesting proteins. This type of enzyme, however, will also play a major role in the synthesis of primary metabolites such as membrane constituents, signal compounds, plant hormones, *etc.*

3 FUTURE ASPECTS OF RESEARCH IN THE ALKALOID FIELD

One of the chief goals in the future of alkaloid biosynthesis will be the elucidation of complicated individual reaction steps, catalyzed by specific enzymes, followed by working out complete pathways leading from primary metabolites to the rather elaborate alkaloidal products. Attention should be paid to the highly specific transport and accumulation mechanisms by which the alkaloids are stored in the central vacuole of the plant cell.[80] It has been shown that the uptake of alkaloids into vacuoles is carrier-mediated and the uptake mechanism is highly selective, as shown, for instance, by the exclusive preference for the (*S*)-forms of reticuline and scoulerine, while the (*R*)-enantiomers, which did not occur in the plant under investigation (*Fumaria capreolata*), were strictly discriminated. Furthermore, the uptake system could be shown to be absolutely specific for alkaloids endogenous to the plant from which the vacuoles were isolated. These results contradict the previously proposed ion-trap mechanism for alkaloid accumulation in vacuoles. It would be highly desirable to characterize biochemically and also on the molecular level these highly specific carrier molecules. It could be envisaged that, knowing these carrier systems, it would be possible to construct plants by genetic engineering capable of accumulating higher amounts of alkaloids than occur in wild-type plants.

It would be highly desirable to clone those enzymes in a given biosynthetic pathway which catalyze reactions that cannot be mimicked by synthetic organic chemistry [as for instance the conversion of (*R*)-reticuline to salutaridine]. Chemo-enzymatic biomimetic synthesis could eventually gain a major role for the production of desired alkaloids this way. Another important development requiring the knowledge of the enzymes and the corresponding genes will be the modification of alkaloid yielding plants by antisense technology. The interruption of pathways leading to undesirable gene products could either lead to completely alkaloid-free plants or specific undesired alkaloids could be eliminated by this technology. The yield of a minor alkaloid could be increased by interrupting those branches of a pathway which siphon away alkaloidal precursors from a wanted product. A great potential will be seen in cloning genes from other organisms like microbes or animals into plants in order to use the photosynthetic capability of higher plants to obtain products in a much more economic way. This could either lead to a new generation of industrial plants, producing compounds for human consumption or yield plants which are resistant against various types of pests, so that exogenous pest-control will no longer be necessary for these systems.

The knowledge of plant catalysts for the formation of secondary products and the underlying genes will eventually yield knowledge of the regulation of secondary pathways. If ever the *trans*-factors for the regulation of diverse secondary pathways become known, this knowledge could be applied by genetic manipulation to yield high alkaloid producing plants, on one side, and to develop plant cell culture fermentation to a commercial level not yet seen.

All of these considerations call for a continued effort in studying the biosynthesis of secondary plant products on the enzyme and gene level. We live in an exciting time, where the technologies exist to unravel these pathways. We are convinced that the next century will see the rise of genetically engineered industrial plants, providing plants with completely new chemical properties for agricultural use.

REFERENCES

1. E Stahl, 'Pflanzen und Schnecken', G Fischer, Jena, 1888, p 1.
2. E Haslam, *Phytochemistry,* 1977, **16**, 1625.
3. E Haslam, *Nat Prod Rep*, 1986, **3**, 217.
4. G A Rosenthal, and M R Berenbaum (eds), 'Herbivores, their interaction with secondary plant metabolites', Academic Press, Inc San Diego, 1991, Vol I and II, 2nd edition.
5. D H Williams, M J Stone, P R Hauck, and S K Rahman, *J Nat Prod,* 1989, **52**, 1189.
6. T Hartmann, in: 'Herbivores', G A Rosenthal, and M R Berenbaum (eds), Adademic Press, Inc San Diego, 1991, Vol I, p 79.
7. N R Farnsworth, in: 'Natural Products and Drug Development', P Krogsgaard-Larsen, S Brøgger Christensen, H Kofod (eds), Munksgaard, Copenhagen, 1984, pp 17.
8. N R Farnsworth, and R W Morris, *Am J Pharmacy*, 1976, **147**, 46.
9. Summarized in: I D Spenser, in: 'Comprehensive Bio-chemistry', Elsevier Press, Amsterdam, 1968, **20**, p 231.
10. D H R Barton, *Pure & Appl Chem*, 1964, **9**, 35.
11. E Leete, and J B Murrill, *Phytochemistry*, 1967, **6**, 231.
12. A R Battersby, *Proc Chem Soc,* 1963, 189.
13. A R Battersby, in: 'Oxidative Coupling of Phenols', W I Taylor and A R Battersby (eds), Marcel Dekker, Inc New York, 1967, p 119.
14. A D Krikorian, and F C Steward, 'Plant Physiology VB', Academic Press, New York, 1969, p 227.
15. M H Zenk, H El-Shagi, and U Schulte, *Planta Medica Suppl*, 1975, 79.
16. H Hinz, and M H Zenk, *Naturwissenschaften*, 1981, **67**, 620.
17. E Winterstein, and G Trier, 'Die Alkaloide', Gebrüder Bornträger, Berlin, 1910, p 157.
18. R Robinson, *J Chem Soc*, 1917, **111**, 876.
19. A R Battersby, and R Binks, *Proc Chem Soc*, 1960, 360.
20. A R Battersby, R Binks, R J Francis, D J McCaldin, and H Ramuz, *J Chem Soc*, 1964, 3600.
21. M Rueffer, H El-Shagi, N Nagakura, and M H Zenk, *FEBS Lett*, 1981, **129**, 5.
22. H-M Schumacher, M Rüffer, N Nagakura, and M H Zenk, *Planta Medica*, 1983, **48**, 212.
23. R Stadler, T M Kutchan, S Loeffler, N Nagakura, B Cassels, and M H Zenk, *Tetrahedron Lett,* 1987, **28**, 1251.
24. D S Bhakuni, A N Singh, S Tewari, and R S Kapil, *J Chem Soc, Perkin Trans I*, 1977, 1662.
25. A R Battersby, S Ruchirawat, and J Staunton, *J Chem Soc, Chem Commun*, 1974, 773.
26. H L Holland, P W Jeffs, T M Capps, and D B McLean, *Can J Chem*, 1979, **57**, 1588.

27 R Stadler, and M H Zenk, *Liebigs Ann Chem*, 1990, 555.
28 T Frenzel, and M H Zenk, *Phytochemistry*, 1990, **29**, 3505.
29 M Rueffer, N Nagakura, and M H Zenk, *Planta Medica*, 1983, **49**, 131.
30 C-K Wat, P Steffens, and M H Zenk, *Z Naturforsch*, 1986, **41c**, 126.
31 S Loeffler, and M H Zenk, *Phytochemistry*, 1990, **29**, 3499.
32 M Rueffer, and M H Zenk, *Z Naturforsch*, 1987, **42c**, 319.
33 I W Southon, and J Buckingham, 'Dictionary of Alkaloids', Chapman and Hall, London, 1989.
34 H Fukui, K Nakagawa, S Tsuda, and M Tabata, in: 'Proc 5th Int Congr Plant Tissue & Cell Culture', A Fujiwara (ed), Maruzen Co, Tokyo, 1982, pp 313.
35 E Sato, and Y Yamada, *Phytochemistry*, 1984, **23**, 281.
36 Y Fujita, and M Tabata, in: 'Plant Tissue and Cell Culture, Plant Biology', C E Green *et al* (eds), A R Liss Inc, New York, 1987, Vol 3, p 169.
37 D H R Barton, R H Hesse, and G W Kirby, *Proc Chem Soc*, 1963, 267.
38 A R Battersby, R J Francis, M Hirst, and J Staunton, *Proc Chem Soc*, 1963, 268.
39 E Rink, and H Böhm, *FEBS Lett*, 1975, **49**, 396.
40 P Steffens, N Nagakura, and M H Zenk, *Phytochemistry*, 1985, **24**, 2577.
41 H Dittrich, and T M Kutchan, *Proc Natl Acad Sci USA*, 1991, **88**, 9969.
42 T Frenzel, J M Beale, M Kobayashi, M H Zenk, and H G Floss, *J Am Chem Soc*, 1988, **110**, 7878.
43 S Muemmler, M Rueffer, N Nagakura, and M H Zenk, *Plant Cell Rep*, 1985, **4**, 36.
44 M H Zenk, M Rueffer, T M Kutchan, and E Galneder, in: 'Applications of plant cell and tissue culture', Ciba Foundation Symposium 137, John Wiley and Sons, Chichester, 1988, p 213.
45 M Rueffer, and M H Zenk, *Tetrahedron Lett*, 1985, **26**, 201.
46 W Bauer, R Stadler, and M H Zenk, *Botanica Acta*, 1992, **105**, 370.
47 M Amann, and M H Zenk, *Phytochemistry*, 1987, **26**, 3235.
48 M Amann, G Wanner, and M H Zenk, *Planta*, 1986, **167**, 310.
49 Y Yamada, and N Okada, *Phytochemistry*, 1985, **24**, 63.
50 N Okada, A Shinmyo, and Y Yamada, *Phytochemistry*, 1988, **27**, 979.
51 H Sato, G Taguchi, H Fukui, and M Tabata, *Phytochemistry*, 1992, **31**, 3451.
52 E Späth, and P L Julian, *Chem Ber*, 1931, **64**, 1131.
53 G Snatzke, J Hrebek, L Hruban, A Horeau, and F Santavy, *Tetrahedron*, 1970, **26**, 5013.
54 D S Bhakuni, S Jain, and S Gupta, *Tetrahedron Lett*, 1983, **39**, 455.
55 D S Bhakuni, S Jain, and S Gupta, *Tetrahedron Lett*, 1984, **40**, 1591.
56 W De-Eknamkul, and M H Zenk, *Tetrahedron Lett*, 1990, **31**, 4855.
57 W Bauer, and M H Zenk, *Tetrahedron Lett*, 1991, **32**, 487.
58 B D Krane, M O Fagbule, M Shamma, and B Gözler, *J Nat Prod*, 1984, **47**, 1.

59   H-M Schumacher, H Gundlach, F Fiedler, and M H Zenk, *Plant Cell Rep*, 1987, **6**, 410.
60   J M Park, S Y Yoon, K L Giles, D D Songstad, D Eppstein, D Novakovski, L Friesen, and I Roewer, *J Ferment Bioeng*, 1992, **74**, 292.
61   H Arakawa, W G Clark, M Psenak, and C J Coscia, *Arch Biochem Biophys*, 1992, **299**, 1.
62   A R Battersby, J Staunton, H R Wiltshire, R J Francis, and R Southgate, *J Chem Soc, Perkin Trans I*, 1975, 1147.
63   N Takao, M Kamigauchi, and M Okada, *Helv Chim Acta*, 1983, **66**, 473.
64   W Bauer, and M H Zenk, *Phytochemistry*, 1991, **30**, 2953.
65   M Rueffer, G Zumstein, and M H Zenk, *Phytochemistry*, 1990, **29**, 3727.
66   M Rueffer, and M H Zenk, *Tetrahedron Lett*, 1987, **28**, 5307.
67   T Tanahashi, and M H Zenk, *Phytochemistry*, 1990, **29**, 1113.
68   A R Battersby, J Staunton, M C Summers, and R Southgate, *J Chem Soc, Perkin Trans I*, 1979, 45.
69   H-M Schumacher, and M H Zenk, *Plant Cell Rep*, 1988, **7**, 43.
70   W De-Eknamkul, T Tanahashi, and M H Zenk, *Phytochemistry*, 1992, **31**, 2713.
71   H Gundlach, M J Müller, T M Kutchan, and M H Zenk, *Proc Natl Acad Sci USA*, 1992, **89**, 2389.
72   D H R Barton and T Cohen, 'Festschrift Arthur Stoll', Birkhäuser, Basel, 1957, p 117.
73   T Kametani, M Mizushima, S Takano, and F Fukumoto, *Tetrahedron*, 1973, **29**, 2031.
74   D H R Barton, G W Kirby, W Steglich, G M Thomas, A R Battersby, T A Dobson, and H Ramuz, *J Chem Soc*, 1965, 2423.
75   C Szàntay, M Bàrczai-Beke, P Péchy, G Blaskó, and G Dörnyei, *J Org Chem*, 1982, **47**, 594 and references cited therein.
76   R Gerardy, and M H Zenk, *Phytochemistry*, 1993, **32**, 79.
77   M H Zenk, R Gerardy, and R Stadler, *J Chem Soc, Chem Commun*, 1989, 1725.
78   R Stadler, and M H Zenk, *J Biol Chem*, 1993, **268**, 823.
79   B K Cassels, E Breitmaier, and M H Zenk, *Phytochemistry*, 1987, **26**, 1005.
80   B Deus-Neumann, and M H Zenk, *Planta*, 1986, **167**, 44.

ACKNOWLEDGEMENTS

Work at our laboratory was supported by the Deutsche Forschungsgemeinschaft, Bonn, (SFB 145) and the Fonds der Chemischen Industrie. Linguistic help in the preparation of this manuscript by Dr T M Kutchan is gratefully acknowledged.

MECHANISTIC AND STRUCTURAL FEATURES OF THE PICORNAVIRAL 3C PROTEASE

Racheli Kreisberg, Michael Shocken,
Dietmar Schomburg,[#] and Dorit Arad[*]

Department of Molecular Microbiology and Biotechnology, G S Wise Faculty of Science, Tel-Aviv University, 69978 Ramat Aviv, Israel.

[#] *Molecular Structure Research, G B F - Gesellschaft für Biotechnologische Forschung, Braunschweig, Germany.*

1 INTRODUCTION

Picornaviruses are single stranded positive RNA viruses that are encapsulated in a protein capsid.[1] These viruses cause a wide range of diseases in man and animals, among them common cold, poliomyelitis, hepatitis, encephalitis, meningitis and foot-and-mouth disease.[2] After inclusion into the host cell, the Picornaviral RNA is translated into a 247-kDa polyprotein that is co- and post-translationally cleaved,[2] yielding 11 mature proteins. The 2A and 3C proteolytic enzymes, which are part of the Picornaviral polyprotein, are responsible for these cleavages.[1] The 2A protease cleaves co-translationally between the structural and non-structural proteins and the 3C protease cleaves post-translationally the remaining cleavage sites except one. Having been recognized as an important protein in the Picornaviral life cycle by virtue of being responsible for its maturation, the 3C protease enzyme has been a prime target for extensive structural and mechanistic investigations during the last few years. Moreover, it has been shown that the 3C protease is also involved in the inhibition of the synthesis of cellular proteins.[3,4] The shutdown of the cellular translation provides that the Picornaviral RNA will be translated and avoids the formation of proteins that the cell needs during the viral infection.

Based on sequence homology between the 3C enzyme and trypsin-like serine proteases, Bazan and Fletterick[5] have suggested that the 3C protease is structurally related to the latter protease family, having a cysteine instead of a serine at its active site. Thus, they have classified the 3C enzyme as a 'trypsin-like cysteine protease', having His40, Asp85 and Cys147 at the catalytic site. Gorbalenya et al[6] also found that the 3C protease is homologous to the trypsin-like serine proteases, but have identified His40, Glu/Asp71 and Cys147 as the catalytic amino acids. The identification of Asp85 by Bazan and Fletterick and Glu/Asp71 by Gorbalenya et al is the result of divergent primary sequence alignments. As Bazan and Fletterick include the Picornaviral 2A proteases in their multiple sequence alignment, the identification of Asp85 is reinforced (the triad of the 2A protease consists of His20, Asp38 and Cys109). Site-directed mutagenesis experiments have confirmed that the four amino acids mentioned above are essential for the catalytic activity of the enzyme.[7]

Both Bazan and Fletterick[5] and Gorbalenya et al[6] have constructed a three-dimensional model of the 3C enzyme. The two groups have hardly dealt with the mechanistic aspects of the 3C protease. They refer to the catalytic triad of the 3C enzyme as being homologous to that of serine proteases by having a cysteine instead of a serine nucleophile. This conclusion is comprehensive regarding the fact that most textbooks that consider the mechanistic aspects of serine and papain-like cysteine proteases do not distinguish between the two mechanisms and regard them as 'very similar'.[8]

We suggest that the 3C protease constitutes a unique enzyme family which is neither a serine nor a pure cysteine protease family. The distinction is important when one wants to design specific inhibitors against the 3C protease which will affect neither the cellular serine nor the cellular papain-like cysteine proteases. Our claim is based upon the observation that, during the first step of the catalytic process, the nucleophilic attack of an oxygen (*eg* serine proteases) on a carbonyl centre in the substrate behaves mechanistically different from a nucleophilic attack by sulfur (*eg* cysteine proteases) on the same center. Investigating the enzymic apparatus through this outlook opens new horizons to the study of the 3C proteases and contributes to the basic mechanistic knowledge on serine and papain-like cysteine proteolytic enzymes.

2 RESULTS AND DISCUSSION

2.1 The Behaviour of an Oxygen Nucleophile *Versus* a Sulfur Nucleophile

Howard and Kollman[9] have reached, by combined *ab initio* and molecular mechanics calculations, a surprising conclusion which states that the potential surface for a hydroxide anion attack on a carbonyl is different from the same reaction with a sulfhydride anion. Whereas the charged tetrahedral intermediate 1 (Figure 1) is a stable species on the potential reaction surface of a hydroxyl attack on a carbonyl, the analogous species 2 formed by a sulfhydride nucleophile attack on carbonyl is not at a minimum on the potential energy

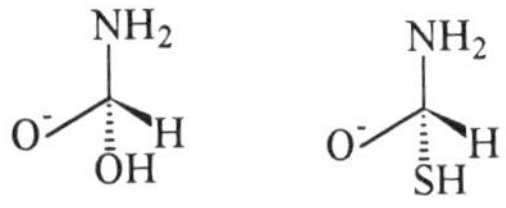

Figure 1: The products of nucleophilic attack of OH⁻ (1) and SH⁻ (2) on an amide.

surface. Thus, calculations predict that, unlike **1**, **2** is not a stable intermediate.

Regarding the fact that a sulfhydride anion is comparable in its nucleophilic properties with a hydroxide nucleophile, this finding is odd. Yet, we have repeated the calculations with the MP4/6-31+G*//MP4/6-31+* basis set and the basic outcome did not change: the charged tetrahedral intermediate (2) does not exist as a minimum on the potential energy surface.[10] The reason for this has been investigated by molecular orbital theory and will be published elsewhere.[10]

This observation, if it obtains *in vivo*, has an immediate implication for the distinction between the mechanisms of proteolytic enzymes whose catalytic site has an oxygen or a sulfur nucleophile, *eg* serine and cysteine proteases, respectively. Experimental data[11] and the theoretical investigations mentioned above[10] imply that the catalytic triads of serine and papain-like cysteine proteases, although very similar in spatial arrangement, function differently during the catalytic reaction.

The enzymatic catalysis of papain-like cysteine proteases has to proceed through an alternative path that can lead to the products, without the formation of a charged tetrahedral intermediate. Arad *et al*[11] have suggested that in papain an early protonation of the substrate by the catalytic histidine takes place prior to the nucleophilic attack by the sulfur. The nucleophilic attack of the cysteine sulfur on a protonated substrate allows the formation of a neutral tetrahedral intermediate.

This implies that the role that the catalytic histidine plays in the enzymic apparatus is different for each type of enzyme (serine or cysteine protease). In serine proteases its role is to abstract a proton from the catalytic serine residue, activating the nucleophile. Its role in cysteine proteases is to donate a proton to the substrate, enabling the nucleophilic attack. This conclusion has raised a question which relates to the role of the third member of the catalytic triad: aspartic acid in serine versus asparagine in papain-like cysteine proteases. In serine proteases the aspartate allows the formation of the serine nucleophile and in papain-like cysteine proteases asparagine stabilizes the positive charge on the protonated histidine maintaining the ability of the latter to donate its proton to the substrate.

2.2 Cellular Serine and Cysteine Enzymes *Versus* 3C Viral Proteases

Comparison of the catalytic triads of the serine, papain-like cysteine and the 3C proteases shows that the triad of the 3C protease is a hybrid between the catalytic triads of the cellular enzymes:

serine proteases:	Ser195	His57	Asp102
viral 3C proteases:	Cys147	His40	Asp85
cysteine proteases:	Cys25	His 159	Asn175

It is evident that if one agrees that the mechanisms of serine and cysteine proteases are distinct, then the mechanism of the 3C protease, based on examination of its catalytic triad, differs from both. The cysteine nucleophile in the triad of the 3C requires, as in the case of the papain-like cysteine proteases, protonation of the substrate prior to the nucleophilic attack. But the third member of its triad is an aspartic acid, as in the serine proteases. Aspartic acid in the 3C protease increases the basicity of the histidine and, thus, diminishes its ability to protonate the substrate.

In order to investigate the difference in the proton affinity of the histidine in these three enzymes, MP2/3-21G*//3-21G* *ab initio* calculations[12] of a protonated histidine model in the presence of models for aspartate and asparagine were performed. The proton affinity is estimated as the difference in the total energies of the protonated product and the unprotonated reagent. The amino acids were modelled by using 4-methyl-imidazole (4-Me-Im), acetate and acetamide to simulate histidine, aspartate and asparagine, respectively. The calculations (summarized in Table 1) show that the gas-phase proton affinities of 5-Me-Im that is hydrogen bonded to an acetamide is 14.5 kcal/mol higher compared with 5-Me-Im. As was expected, the proton affinity of 5-Me-Im which is hydrogen bonded to an aspartate is 106.9 kcal/mol higher than that of 5-Me-Im.

Table 1: The energy of protonation of 4-Me-Im, 5-Me-Im complexed with an acetamide and 5-Me-Im complexed with an acetate.

Reaction	ΔH/kcal mol^{-1}
5-Me-Im + H$^+$ → 5-Me-ImH$^+$	249.5
5-Me-Im-acetamide + H$^+$ → 5-Me-ImH$^+$-acetamide	264.0
5-Me-Im-acetate + H$^+$ → 5-Me-ImH$^+$-acetate	356.4

From these calculations it can be concluded that the proton affinity of the catalytic histidine in both the serine and the 3C proteases is higher than that of the cysteine proteases. This conclusion correlates with that of Warshal *et al*[13] who have demonstrated that the aspartate stabilizes the HisH$^+$ by electrostatic interactions in serine proteases. Mutating the aspartate to an alanine, a neutral side chain, disabled the stabilization of HisH$^+$. Analogously, an acetamide lacks the electrostatic properties of the aspartate and, thus, does not dramatically stabilize the HisH$^+$. Similar investigations of the proton affinity of Me-Im have been reported lately.[14]

The calculations, thus, emphasize the existing difference between the catalytic triad of the serine, papain-like cysteine and 3C enzymes. The ability of HisH$^+$ in cysteine proteases to protonate the substrate is possible, while being less favoured both in the serine and in the 3C proteases. In the case of the serine proteases there is no need to protonate the substrate prior to the nucleophilic attack. In the 3C protease with a cysteine nucleophile this step is crucial, yet the proton transfer process is disfavoured owing to the existence of an aspartate in its triad. The conclusion is that the 3C triad, as currently described (Cys, His, Asp[5,15]), is unable to function effectively. Several possibilities for a more effective mechanism exist:

1   The catalytic cysteine and histidine operate independently of the aspartic acid. The aspartic acid may function oppositely, *eg* switch off the activity of the histidine.

2   In the 3C protease there should be an additional amino acid responsible for the protonation of the substrate prior to the nucleophilic attack.

3   The totally conserved aspartic acid has another functional role (*eg* in the second function of the 3C enzyme: shutdown of cellular translation).

During the checking of the proof of this paper there appeared a publication by M James *et al*,[24] presenting details of the crystal structure of the 3C protease of Hepatitis A virus. Their description of the active site complements our conclusions, by revealing that the Aspartic acid 84 predicted to be part of the catalytic triad, turns to the opposite direction from the histidine by forming a strong salt bridge with Lysine 202.

2.3 Detection of an Additional Histidine by Multiple Sequence Alignments of the 3C Proteases

A multiple sequence alignment of all the 3C proteases that are available in the GENBANK[15] was performed using PileUp from the GCG program.[16] This alignment shows that the only three completely conserved amino acids in the viral 3C proteases are His40, Asp85 and Cys147. This finding opposes the outcome of the alignment of Gorbalenya *et al* which claimed that the triad of the 3C enzyme consists of His40, Glu/Asp71 and Cys147. Our multiple sequence alignment clearly shows that one Foot-and-Mouth disease virus 3C protease has an arginine at position 71 instead of an aspartic or a glutamic acid. The possibility that the residue at this position plays another functional or a structural role as suggested by recently published site-directed mutagenesis experiments,[7] is reasonable also on the basis of the analysis in the previous section.

In addition to these three residues, another histidine at position 161 was detected to be conserved in all the aligned sequences, except for two polio 3C proteases. This conservation implies that His161 is functionally important and that it may be a candidate for the protonation of the substrate, prior to the nucleophilic attack. The two exceptional cases in which no histidine exists at position 161 were analyzed on the RNA level. The RNA sequence of one of these two sequences (pollb31b) is compared to a sequence that has the His161 (poll) and shows that there is a deletion of one nucleotide upstream of the codon of His161 and another deletion of two nucleotides downstream of this codon (Figure 2).[17] This mutation causes a frame shift and a deletion of one amino acid in this polio 3C protease.

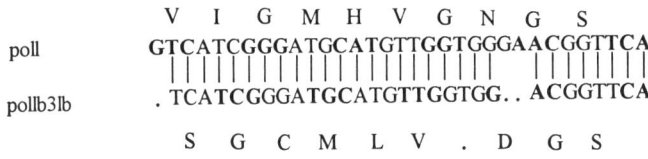

Figure 2: The nucleotides and corresponding amino acids surrounding position 161 in a 3C protease that codes for His161 and in a 3C enzyme that lacks this residue at position 161.

No information is available on the influence of this mutation on the activity of these two polio 3C enzymes that lack His161.

2.4 Molecular Modelling of the Rhoniviral Type 14 3C Protease

Three structural models of the Picornaviral 3C protease have been suggested, by Bazan and Fletterick,[5] by Gorbalenya et al[6] and by Arad et al.[18] The two former models are based upon a small homology at the primary and secondary structure level to trypsin-like serine proteases. In the present paper we present the existence of homology to the trypsin-like serine proteases that can be extended to the tertiary structure level. This was achieved by using the Eisenberg program.[19] A profile of several structurally solved enzymes was run against a database which contains all the available PDB sequences[20] and the primary sequences of all the 3C proteases. A minimum score of 8 indicates the existence of structural homology. The profile of bovine trypsin (4ptp) identifies cellular trypsin-like serine proteases with a high score. The score for the detection of bacterial trypsin-like serine proteases (3sgb and 1sga) by this profile is much lower, even though structural homology between these cellular and bacterial enzymes does exist. Therefore, the identification of the primary sequence of 3C proteases with a score of ca 4 - 5 was considered as indicative of structural homology. The results of the Eisenberg program are summarized in Table 2.

According to the detected structural homology of the 3C protease to the trypsin-like serine proteases, the latter serves as a template for the construction of a structural model of the former. The 3C protease and these serine proteases are aligned in such a way that the catalytic amino acids of the 3C protease (His40, Asp85 and Cys147) are aligned with those of the serine proteases. Moreover, the amino acids that are conserved in the Rhinoviral and Enteroviral 3C proteases are aligned with the residues that form the structural core of the trypsin-like serine proteases (data not shown). The alignment was refined with secondary structure and hydrophobicity predictions (data not shown). This alignment served as a template for the model construction of the 3C protease. Deletions were replaced by shorter loops using an expert database from the Bragi program[21] and the model was subjected to energy minimization.[22]

Table 2: The results of the Eisenberg program, using 3sgb (Streptomyces griseus protease B), 1hne (human neutophil elastase), 9pap (papaya papain), 1sbc (subtilisin Carlsberg), 4ptp (bovine trypsin), 3est (pig pancreatic elastase), 2alp (Lysobacter enzymogenes alpha-lytic proteinase), 2sga (Streptomyces griseus protease A) and 1sgt (Streptomyces griseus trypsin).

	3sgb	1hne	9pap	1sbc	4ptp	3est	2alp	2sga	1sgt	3C
3sgb	X				4.4	3.8	10.4	31.5	2.6	3.8 - 5.6
1hne		10.7	10.4							2.0 - 4.7
9pap					28.0	13.3		3.8		1.6 - 2.2
1sbc										1.1 - 1.6
4ptp	4.5	16.0		2.0		12.2		1.2		2.2 - 5.0
3est	3.3	16.0		1.6	16.4		2.2	1.9	11.1	1.9 - 2.8
2alp	9.4				1.9	2.3		8.4		
2sga	29.0				4.2		8.8		2.2	2.2 - 5.1
1sgt		7.3		2.7	14.1	8.8				1.7 - 3.1

The structural model of the Rhinoviral type 14 3C protease has a globular shape and its catalytic amino acids, His40, Asp85 and Cys147, are located in a groove that is accessible by the substrate.

2.5   His161 is Located in the Vicinity of the Catalytic Triad

The model structure of the 3C protease has revealed that His161 is located in the vicinity of the three catalytic amino acids. It should be mentioned that the location of His161 is in fact determined during the molecular modelling process: His161 is aligned with one of the amino acids that is involved in binding in the trypsin-like serine proteases. Therefore, this residue will automatically be positioned in the neighbourhood of the catalytic triad, due to the vicinity of the triad and binding site in enzymes.

The spatial arrangement of His161 in relation to the other catalytic residues is particularly interesting: His161 is located as a mirror image of His40, while Cys147 lies in the plane of symmetry (Figure 3). This observation, obtained by modelling, is confirmed by the preliminary crystallographic data on HPA.[24]

2.6   His161 Serves as a Proton Donor to the Substrate

The additional histidine, His161, is not only located near the active site, but is also close to the carbonyl of the substrate (see next section). Together with the fact that His40 is hydrogen bonded to Asp85, a combination that makes His40 a stronger base and a weaker proton donor, it is credible to speculate that the His161 may serve as a proton donor to the substrate. It should be mentioned that

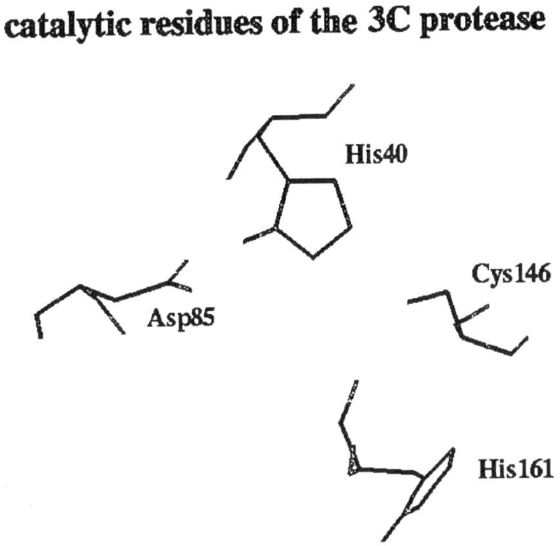

catalytic residues of the 3C protease

Figure 3: The catalytic triad of the 3C protease that consists of His40, Asp85, Cys147 and His161.

the source of the proton that protonates His161 is unknown. It can be derived either from a charged side chain, which has not yet been detected, or from the solvent.

Figure 4 displays one of the mechanisms that may operate in the 3C protease. Four catalytic amino acids participate in this mechanism.

The suggested mechanism is based on the fact that aspartic acid 85 reduces the protonation ability of histidine 40: in the resting state Cys147 is ionized and both His40 iand His 161 are protonated. The substrate is protonated by His161, which is more feasible, and the nucleophilic attack forms a neutral tetrahedral intermediate. This species breaks down into a thioacylenzyme and an amine.

2.7 Reconstruction of a 3C Protease Substrate

A model of the interaction between the 3C protease and its substrate may indicate whether the residues that are involved in this interaction are located close enough to allow protonation of the carbonyl of the substrate by His161. This raises the need for a substrate of 3C protease whose three-dimensional structure has been experimentally determined. However, no such structure is yet available. In order to overcome this problem, we have reconstructed a substrate of the Rhinoviral type 14 3C protease. The reconstruction is based on the fact that the Xray structures of the Rhinoviral type 14 structural proteins have been determined.[23]

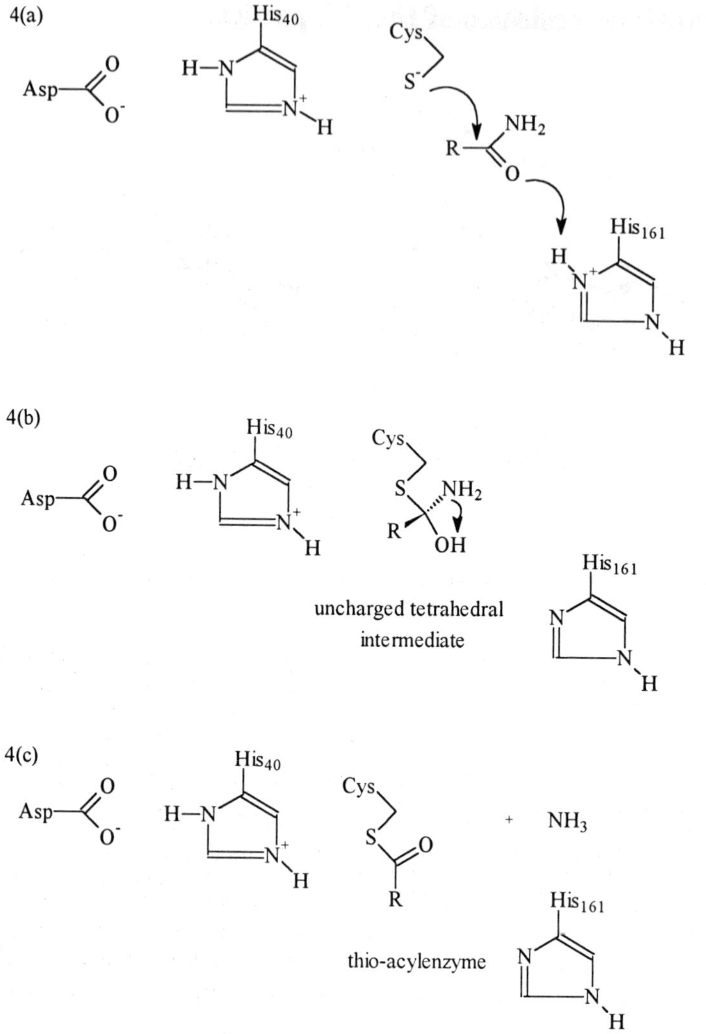

4(a)

4(b)

uncharged tetrahedral
intermediate

4(c)

thio-acylenzyme

Figure 4: A possible mechanism of the 3C protease in which His161 participates.

The cleavage site between two of the four structural proteins (1B and 1C) is hydrolyzed by the 3C protease. Therefore, reconstructing the peptide bond between the C-terminus ending with Gln of the 1B protein and the N-terminus ending with Gly the 1C protein forms the 1B/1C cleavage site (Figure 5).

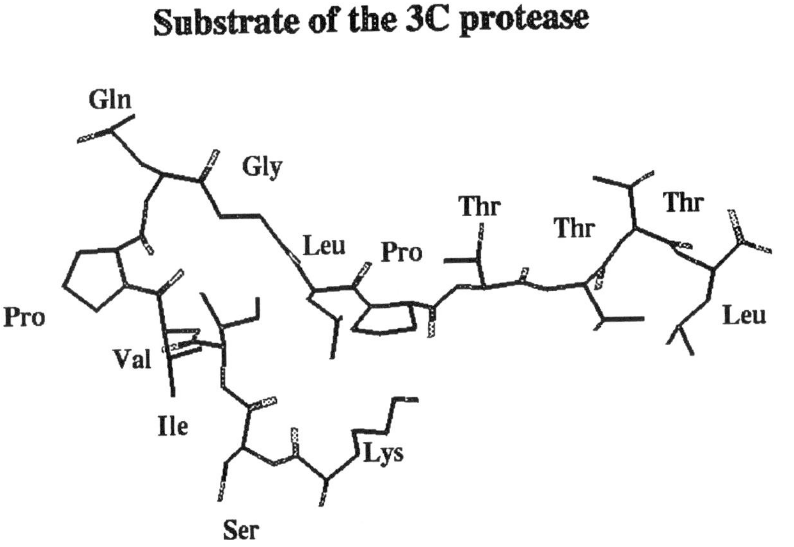

Substrate of the 3C protease

Figure 5: The reconstructed substrate of the 3C protease, which mimics the 1B/1C cleavage site.

The assumption is that the conformation of the mature 1B and 1C proteins is similar to that of the 1BC precursor in general and that the conformation of the free termini resembles that of the reconstructed 1B/1C site in particular. We are aware of the fact that the conformations of the termini of these two proteins may be different from their conformation in the precursor. Two facts permit us to regard this construct as a credible substrate of the 3C protease that mimics the 1B/1C cleavage site: (1) the N-terminus of 1C is composed of an exceptionally large number of prolines that rigidify the structure; this may hint that this domain separates the proteins whose cleavage site is being recognized and hydrolyzed by the 3C protease; (2) a survey of structural patterns with similar sequences to the 3C cleavage sites reveal a typical loop which resembles the structure of the reconstructed substrate.

A close look reveals that there is a possibility that the relatively conserved Gln at the P1 position of the substrate can form an internal hydrogen bond (*ie* the side chain NH_2 group is hydrogen bonded to the oxygen of the carbonyl of the scissile peptide bond) (Figure 6).

In such a way, Gln at P1 may facilitate the polarization of the carbonyl in a similar way that the oxy-anion hole operates in the serine protease family.[12]

The reconstructed substrate was docked into the 3C protease in such a way that the Cys147 nucleophile relates to the scissile carbonyl of the reconstructed substrate, in the same way that the nucleophile of the *Streptomyces griseus*

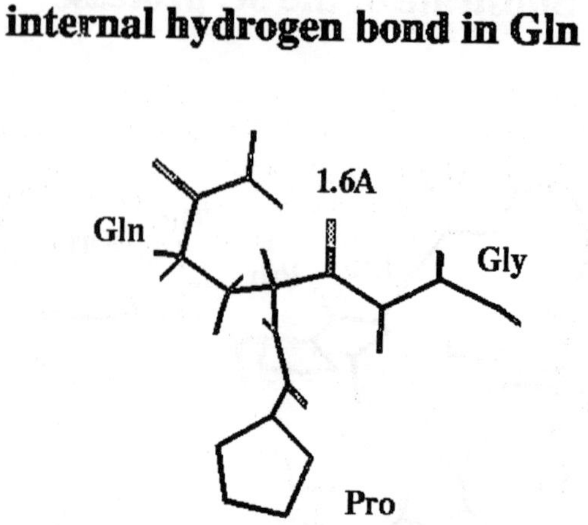

internal hydrogen bond in Gln

Figure 6: The side chain of Gln that is hydrogen-bonded through its amine to the carbonyl oxygen of the scissile bond.

protease B, a bacterial trypsin-like serine protease, does towards the turkey ovomucoid third domain inhibitor (3sgb). The complex 3C protease-reconstructed substrate was submitted to energy minimization. The minimized complex reveals the orientation of His161 towards the scissile bond and the amino acids of the 3C protease which seem to be involved in binding the substrate.

2.8 Orientation of His161 Towards the Reconstructed Substrate
The docking of the reconstructed substrate into the 3C protease shows that His161 can in fact protonate the carbonyl of the scissile bond. The protonation will allow the nucleophilic attack of Cys147 to form an uncharged tetrahedral intermediate (for a full description of the mechanism, see Figure 4).

2.9 Binding Site of the 3C Protease Based upon the Docking
The docking of the reconstructed substrate into the 3C enzyme can detect the amino acids of the 3C protease that are involved in substrate binding. These amino acids include:

Thr19, Thr20, Lys22, Glu24, Phe25
Ala41, Gln42, Pro43, Asp45
Glu80, Lys81, Phe82
Thr142, Lys143, Thr144, Gly145, Gln146, Gly148
His161, Val162, Gly163, Gly164

The amino acids that are conserved in the Rhinoviral and Enteroviral 3C proteases (data not shown) are underscored.

The specificity of the 3C protease to Gln at P1 may be determined by Thr142 that forms a hydrogen bond with this residue in the complex. The participation of Thr142 in binding of the substrate is reinforced by the models that were suggested by Bazan and Fletterick[5] and by Gorbalenya et al.[6] These two groups suggest that Thr142 and His161 are part of the binding site.

3.0 Inhibitor Design

Serine and papain-like cysteine proteases work through similar mechanisms, yet a delicate distinction between them[9] provides a way to inhibit only one type of enzyme. The problem of finding an inhibitor that will specifically inhibit one enzyme is even more acute for the viral 3C proteases. The human body contains vital serine and papain-like cysteine enzymes which we do not want to inhibit when designing a viral protease inhibitor as an anti-viral drug. The fact that there is a distinction between serine, papain-like cysteine and 3C proteases is encouraging in this regard. One of our goals is thus using the mechanistic concepts and designing a specific inhibitor for the viral enzyme which will not inhibit the cellular enzymes.

References

1. A C Palmenberg, *Ann Rev Microb*, 1990, **44**, 603.
2. M J H Nicklin, H Toyoda, M G Murray, and E Wimmer, *Biotechnology*, 1986, **4**, 36.
3. M Tesar and O Marquardt, *Virology*, 1990, **174**, 364.
4. O Marquardt, *FEMS Microbiology Letters*, 1993, **107**, 279.
5. J F Bazan and R J Fletterick, *Proc Natl Acad Sci*, 1988, **85**, 7872.
6. A E Gorbalenya, A P Donchenko, V M Blinov, and E V Koonin, *FEBS Letters*, 1989, **243**, 103.
7. K M Kean, M T Howell, S, Grunert, M Girard, and R J Jackson, *Virology*, 1993, **194**, 360.
8. A Fersht, *Enzyme Structure and Mechanism*, London, 2nd edition, W H Freeman and Company, New York, 1977, 413.
9. A Howard and P A Kollman, *J Am Chem Soc*, 1988, **110**, 7195.
10. M Shocken and D Arad, *J Am Chem Soc*, submitted for publication.
11. D Arad, R Langridge, and P A Kollman, *J Am Chem Soc*, 1990, **112**, 491.
12. M J Frisch, M Head-Gordon, G W Trucks, J B Foresman, H B Schlegel, K Raghavachari, M A Robb, J S Binkley, C Gonzalez, D J Defress, D J Fox, R A Whiteside, R Seeger, C F Melius, J Baker, R L Martin, L R Kahn, J J P Stewart, S Topiol, and H A Pople, 1990, Gaussian, Inc, Pittsburgh, PA, USA.

13 A Warshal, G Naray-Szabo, F Sussman, and J-K Hwang, *Biochemistry*, 1989, **28**, 3629.
14 A A Bliznyuk, H F Schaefer III, and I J Amster, *J Am Chem Soc*, 1993, **115**, 5149.
15 J W Fickett, Los Alamos National Laboratory, Los Alamos, New Mexico, 87545, USA, *Trends Biochem Sci*, 1986, **11**, 190.
16 J Devereux, P Haeberli, and O Smithies, *Nucleic Acids Research*, 1984, **12**, 387.
17 B L Semler, C W Anderson, N Kitamura, P F Rothberg, W L Wishart, and E Wimmer, *Proc Natl Acad Sci USA*, 1981, **78**, 3464.
18 D Arad, R Kreisberg, and M Shochen, *J Chem Inf Comput Sci*, 1992, **33**, 345.
19 J U Bowie, R Luthy, and D Eisenberg, *Science*, 1991, **253**, 164.
20 F C Bernstein, T F Koetzle, G J B Williams, E F Meyer Jnr, M D Brice, J R Rogers, O Kennard, T Shimanouchi, and M Tasumi, *J Mol Biol*, 1977, **122**, 535.
21 D Schomburg and J Reichelt, *J Mol Graphics*, 1988, **6**, 161.
22 B R Brooks, R E Bruccoler, B D Olafson, D J States, S Swaminathan, and M Karplus, *J Comp Chem*, 1983, **4**, 187.
23 E Arnold and M G Rossmann, *Acta Crystallogr*, Sect A, 1988, **44**, 270.
24 M Allaire, M M Chernaia, B A Malcolm, and M N G James, *Nature*, 1994, **369**, 72.

Acknowledgements

This research was supported by the Lady Davis Fellowship Trust, Jerusalem, Israel (to D A), the Bertha (Hartz) Axel Fellowship (to R K) and the Ministry of Science and the Ministry of Absorption (to M S).

NEIGHBOURING GROUP PARTICIPATION: A MODEL FOR ENZYMIC CATALYSIS?

Keith Bowden

*Department of Chemistry and Biological Chemistry,
University of Essex, Wivenhoe Park, Colchester, Essex CO4 3SQ, UK.*

1 INTRODUCTION

A substituent can affect rates and equilibria by a number of means. For substituents non-proximate to the reaction centre, such effects are mainly considered to be electronic in nature. However, for proximate substituents, these electronic effects can be accompanied by steric effects and/or neighbouring group participation. The latter occurs when a substituent interacts directly with the reaction centre through partial or complete bonding.[1] If such participation gives rise to an enhanced rate, the substituent is considered to provide anchimeric assistance. Such behaviour is often referred to as intramolecular, as opposed to intermolecular, catalysis when detected in reactions such as the hydrolysis of esters.[2]

Intramolecular reactions often proceed very much faster than the corresponding intermolecular reactions. The source of this acceleration remains a subject of debate.[3,4] A number of explanations have been advanced for this phenomenon, some of which appear inter-related.[3] Furthermore, these rate increases have been considered to be of special significance to enzymic catalysis in which substrate and catalytic group are brought together in a stereochemically exact manner.

However, such intramolecular effects can be demonstrated in studies of equilibria, as well as in rate processes. If the effects apparent in the equilibria reactions can be related to those in the rate process reactions, it cannot be argued that any 'special' effects[4] associated with the formation of the transition state from the initial state could be the source of such rate enhancements.

2 RESULTS AND DISCUSSION

It has been possible to conduct six different studies[5-9] of rate and equilibrium processes which involve neighbouring group participation and intramolecular

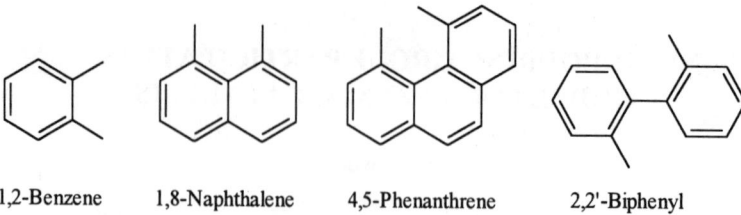

| 1,2-Benzene | 1,8-Naphthalene | 4,5-Phenanthrene | 2,2'-Biphenyl |

Figure 1: Template used.

catalysis using a series of simple templates. These templates are shown in Figure 1. They are the 1,2-benzene, 1,8-naphthalene, 4,5-phenanthrene, and 2,2'-biphenyl. All these can, in principle, be related to a simple benzene intermolecular system. The model reactions to be considered in this study are the following:

(i) The alkaline hydrolysis of esters by neighbouring carbonyl (formyl) groups. This proceeds by addition of hydroxide to the neighbouring carbonyl group, followed by intramolecular nucleophilic attack.[5,9] This is shown in Figure 2 below, as I.

(ii) The base-catalyzed cyclization of esters by neighbouring carbon acid (acetyl) groups. This proceeds by base-catalyzed ionization of the neighbouring carbon acid, followed by intramolecular nucleophilic attack by the enolate anion.[6] This is shown in Figure 2 below, as II.

(iii) The base-catalyzed cyclization of diacetyl compounds. This proceeds by base-catalyzed ionization of the neighbouring acetyl group, followed by intramolecular attack by the enolate anion and subsequent dehydration.[7] This is shown in Figure 2 below, as III.

(iv) The intramolecular Cannizzaro reaction of diformyl compounds. This proceeds by hydration and ionization, followed by intramolecular hydride transfer.[8] This is shown in Figure 2 below, as IV.

(v) The ring-chain tautomerism of formyl-carboxylic acids. This is the equilibrium between the chain formyl-carboxylic acid and the ring hydroxy-lactone. This is shown in Figure 2 below, as V.

(vi) The hydration of diformyl compounds. This is the equilibrium between the diformyl compound and the cyclic monohydrate.[9,10] This is shown in Figure 2 below, as VI.

In Table 1 are shown the relevant rate and equilibrium constants for each template and for those for the corresponding intermolecular reactions. For the latter, estimates are usually based on extrapolations or model reactions. The use of the latter, in particular, may result in some uncertainty in the constants. However, as the magnitudes of the rate and equilibrium constant ratios for intra-

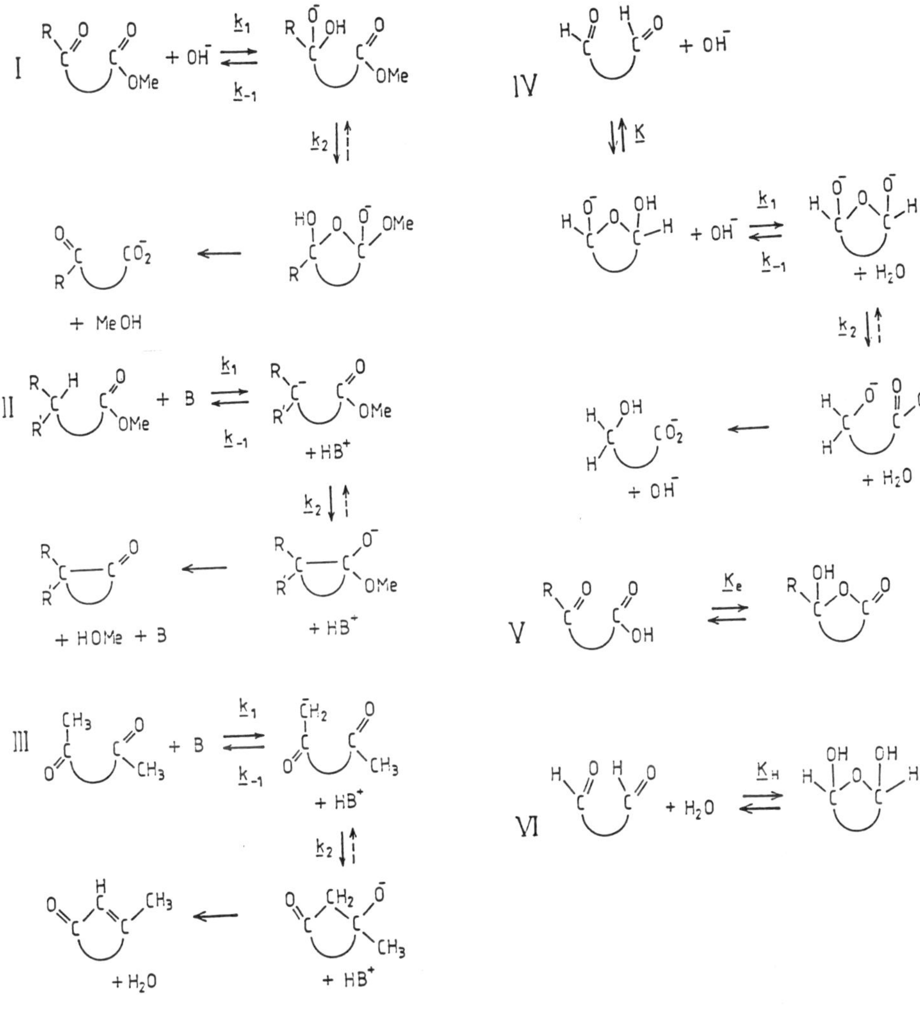

Figure 2: Intramolecular reactions studied.

to inter-molecular reactions are very large, this is less important. For ester hydrolysis, the alkaline hydrolysis of methyl benzoate is used as a model reaction.[5] For the Cannizzaro reaction, the reactions of the first three entries in Table 1 refer to the second-order reaction with the mono-anion of the cyclic hydrate as the initial state.[8] The following two entries refer to third- and fourth-order reactions, respectively, with aldehydes as the initial states. As a model reaction for the ring-chain tautomerism for intermolecular addition, the addition of water to benzaldehyde[11] will give an upper limit, the addition of benzoic acid to benzaldehyde being very much less favourable.[12]

Table 1: Rate of equilibrium constants for the intra- and inter-molecular reactions.

Link	I Ester hydrolysis,[a] $10^3 k_2/dm^3 mol^{-1} s^{-1}$	II Cyclisation of esters,[c] $10^3 k_1/s^{-1}$	III Cyclisation of diacetyl compounds, $10^3 k_1/s^{-1}$	IV Cannizzaro reaction,[e] $k_2/dm^3 mol^{-1} s^{-1}$	V Ring-chain tautomerism,[i] K_e	VI Cyclic monohydration,[k] K_h
1,2-Benzene	271,000	0.0082	0.683	0.0086	4.6	0.124
1,8-Naphthalene	35,500	15.5	417	36.6	1,200	>25
4,5-Phenanthrene	7,470	-	98.5	6.40	3,000	16
2,2'-Biphenyl	19.0	-	27.0	$[1.2]^{f,g}$	<0.04	<0.04
None	8.50^b	-	-	$[3 \times 10^{-4}]^{f,h}$	$<2 \times 10^{-4}$ [j]	-

[a] in 70% (v/v) dioxane-water at 20°C. [5,9]
[b] Methyl benzoate. [5]
[c] In 36.3 mole % methanolic DMSO containing 0.035 M NaOMe at 30°C. [6]
[d] In 40.0 mole % aqueous DMSO containing 0.011 M Me$_4$NOH at 30°C. [7]
[e] In 70% (v/v) dioxane-water at 60°C. [8]
[f] In water at 40°C. [8]
[g] dm^{6}mol^{-2}s^{-1}.
[h] dm^{12}mol^{-3}s^{-1}.
[i] In 80% (w/w) 2-methoxyethanol-water at 25°C. [9,10]
[j] In water at 25°C (see text). [9]
[k] In 33 mole % aqueous dioxane at 60°C. [8,9]

Table 2: Rate and equilibrium constant ratios for the intra- and inter-molecular reactions.[a]

Link	Reaction	k/k_0 or K/K_0					
	I	II	III	IV	V	VI
1,2-Benzene	1.0	1.0	1.0	1.0	1.0	1.0
1,8-Naphthalene	1.31×10^{-1}	1900	611	4300	261	>200
4,5-Phenanthrene	2.76×10^{-2}	-	144	740	650	130
2,2'-Biphenyl	7.01×10^{-5}	-	39.5	$[1.0]^b$	<0.01	<0.3
None	3.13×10^{-5}	-	-	$[2.5 \times 10^{-4}]^b$	$<4 \times 10^{-5}$	-

[a] For details see Table 1.
[b] Relative to the 2,2'-biphenyl system.

In Table 2 are shown the rate and equilibrium constant ratios for the six reactions. There are, of course, important differences in the actual structures of the reaction types and sometimes variations in the nature of the rate-determining steps for the rate processes. Considering this, there is a remarkable similarity in the trends observed in the ratios. This is, normally, 1,8-naphthalene > 4,5-phenanthrene > 1,2-benzene > 2,2'-biphenyl >> none. Thus, the same general effects are noted in the equilibrium process reactions as in the rate process reactions.

The large enhancements observed for the intramolecular reactions appear to result, in the main, from two sources.[13] The first is an entropy effect and arises from the proximity of the interacting groups. This results from the great advantage of having linked together the reacting groups in close proximity as opposed to having to bring two reactant species together for reaction to occur. The second is steric strain. The latter can be either increased or induced in the 'chain' system; but can be relieved by reaction to form the 'ring' system. Ring strain and resonance will play an important part in determining the relative reactivity of the systems employing the four templates shown in Figure 1. Thus, the 1,2-benzene, 1,8-naphthalene, 4,5-phenanthrene and 2,2'-biphenyl reaction systems studied here involve 4-, 6-, 7-, and 8-membered rings, respectively. The relative reactivity pattern observed here is that generally expected for reactions involving the formation of such rings.[14] As such, these results are in accord with the considerations[3] that no 'special' effects are required to account for the rapid reactions observed in intramolecular rate processes.

A second approach to the understanding of the importance of such factors using this type of model can now be considered. Methyl 2-acetylbenzoate rapidly undergoes alkaline hydrolysis, employing neighbouring group participation by the carbonyl group, by the pathway shown in Figure 3, as I. Methyl 8-acetyl-1-naphthoate also undergoes alkaline hydrolysis, but reacts by attack on the ester group by the enolate carbanion, generated by rate-determining proton abstraction from the acetyl group by hydroxide anion, by the pathway shown in Figure 3, as II. The dichotomy displayed by these closely related esters clearly indicates the specific control of reaction afforded by orientation and environment of the assisting group. For the 8-acetyl-1-naphthoate, the carbanion is generated with favourable orientation for substitution and is closely proximate to the ester carbonyl-carbon. However, catalysis by the carbonyl group has distinct spatial requirements, unlike the carbon-acid process. Carbonyl participation involves severe crowding in the 8-acetyl-1-naphthoate, compared with the 2-acetylbenzoate. Furthermore, the 2-acetylbenzoate does not have the favourable orientation, in terms of both angle and distance, for the carbon-acid path to be favoured. The carbonyl participation route can be considered to simulate an esterase, while the carbon-acid participation route simulates aldolases and related enzymes, especially those enzymes catalyzing Claisen condensations or retrocondensations.[5] Thus, control of reaction type can be achieved by the employment of these factors.

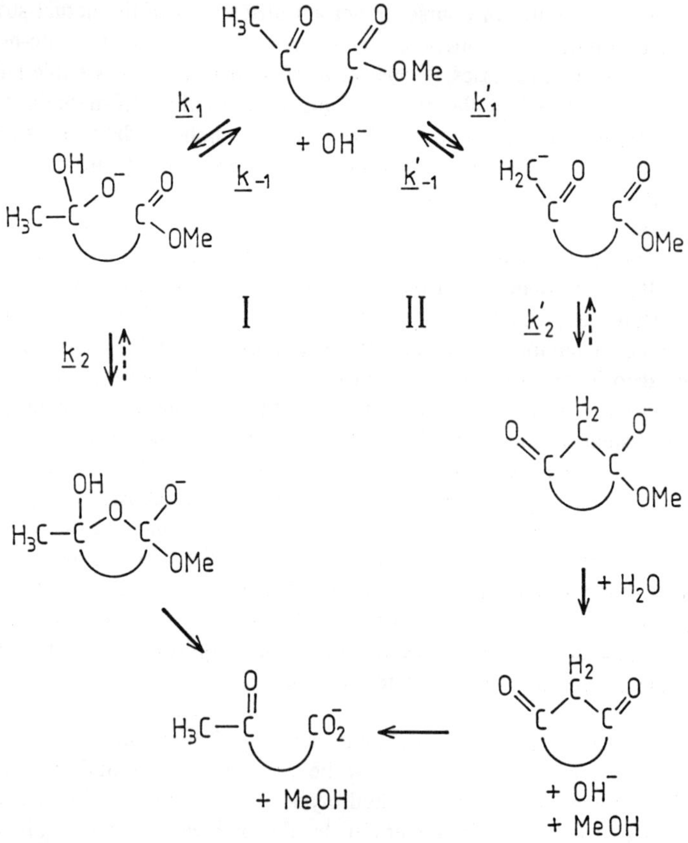

Figure 3: Mechanistic dichotomy for participation.

3 CONCLUSIONS

The importance of proximity (entropy), steric strain, and orientation (both angle and distance) can be demonstrated in both rate and equilibrium processes. No 'special' effects appear to be required for intramolecular catalysis in such systems as those considered here.

REFERENCES

1. B Capon, *Quart Rev*, 1964, **18**, 45; B Capon, and S P McManus, *Neighbouring Group Participation*, Plenum Press, New York, 1976.
2. A J Kirby and A R Fersht, *Progr Bioorg Chem*, 1971, **1**, 1.
3. M I Page and W P Jencks, *Gazz Chim Ital*, 1987, **117**, 455 and references therein.
4. F M Menger, *Acc Chem Res*, 1985, **18**, 128 and references therein.
5. K Bowden, *Adv Phys Org Chem*, 1993, **28**, 171 and references therein.
6. K Bowden and M Chehel-Amiran, *J Chem Soc, Perkin Trans* 2, 1986, 2035.
7. K Bowden and A Brownhill, unpublished results.
8. K Bowden, F A El Kaissi, and N S Nadvi, *J Chem Soc, Perkin Trans* 2, 1979, 642; M R Abbaszadeh and K Bowden, *J Chem Soc, Perkin Trans* 2, 1990, 2081; K Bowden, F A El Kaissi, and R J Ranson, *J Chem Soc, Perkin Trans* 2, 1990, 2089; F Anvia and K Bowden, *J Chem Soc, Perkin Trans* 2, 1990, 2093; K Bowden, A M Butt, and M Streater, *J Chem Soc, Perkin Trans* 2, 1992, 567.
9. K Bowden and C S Manning, *J Chem Res (S)*, 1994, 28.
10. K Bowden and G R Taylor, *J Chem Soc (B)*, 1971, 1390; K Bowden and A M Last, *J Chem Soc, Perkin Trans* 2, 1973, 1144.
11. P Greenzaid, *J Org Chem*, 1973, **38**, 3164.
12. E G Sander and W P Jencks, *J Am Chem Soc*, 1968, **90**, 6154.
13. K Bowden and A M Last, *Can J Chem*, 1971, **49**, 3887.
14. L Mandolini, *Adv Phys Org Chem*, 1968, **22**, 1 and references therein.

ACKNOWLEDGEMENTS

The author wishes to thank his students and ex-students who contributed to the studies underlying this work.

ENZYME-CATALYSED HYDROXYLATIONS OF AROMATIC SUBSTRATES: STEREOCHEMICAL AND MECHANISTIC ASPECTS

Derek R Boyd,[a] Narain D Sharma,[a] and Howard Dalton[b]

[a]*School of Chemistry, Queen's University of Belfast, Belfast BT9 5AG, UK.*
[b]*Department of Biological Sciences, University of Warwick, Coventry CV4 7AL, UK.*

1   INTRODUCTION

The enzyme-catalysed oxidation of arenes can occur by two major pathways. In eucaryotes (animals, fungi and plants) monooxygenase enzymes catalyse the epoxidation of arenes. Thus, arene oxides (and the derived *trans*-dihydrodiols and phenols) have been isolated from polycyclic aromatic hydrocarbons (*eg* naphthalene[1]) and their aza-analogues (*eg* quinoline[2]) using mammalian liver enzymes. The mechanism of arene epoxidation in the presence of cytochrome P-450 enzyme systems (monooxygenases) has been rigorously investigated over the past decade and chemical model systems have been developed.

In procaryotes (bacteria) the dioxygenase-catalysed oxidation of arenes has been found to proceed *via* the initial formation of *cis*-dihydrodiol metabolites, followed by enzyme-catalysed dehydrogenation to yield catechols and ring cleavage products. The development of mutant strains of bacteria (lacking the dehydrogenase enzyme) by Gibson and coworkers[3] has allowed the *cis*-dihydrodiol metabolites of a wide range of arenes to be isolated. Although the dioxygenase enzymes responsible for *cis*-dihydrodiol formation in bacteria have been purified, the mechanism of this reaction remains unknown at present (Scheme 1).

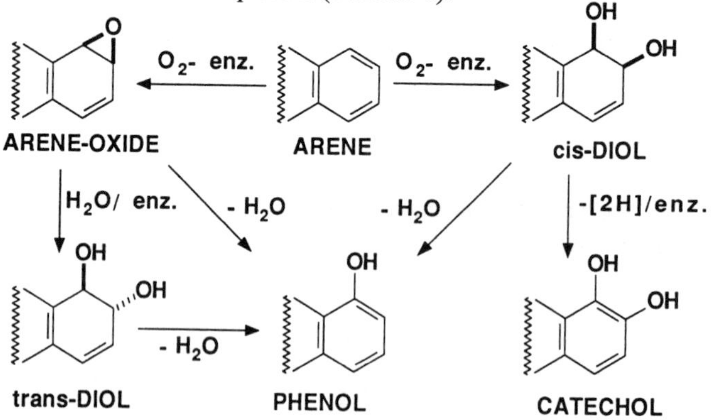

Scheme 1: Oxidative metabolism of arenes.

2 DISCUSSION

As a prelude to investigating the mechanism of *cis*-dihydrodiol formation a comprehensive programme has been undertaken to determine both structure and stereochemistry of arene metabolites produced by mutant strains of the soil bacterium *Pseudomonas putida*.

Biotransformation of benzene using *P putida* UV4 gives a single *cis*-dihydrodiol, *cis*-1,2-dihydroxy-1,2-dihydrobenzene. By contrast there are six possible isomeric *cis*-dihydrodiol metabolites from monosubstituted or 1,4-disubstituted (R ≠ R') benzenes and twelve possible isomeric *cis*-dihydrodiols from 1,2- or 1,3-disubstituted (R ≠ R') benzenes. In practice, the dioxygenase enzyme system in *P putida* UV4 appears to be both regioselective and stereoselective. Thus, oxidation of a wide range of monosubstituted benzene substrates gave only 2,3-*cis*-dihydrodiols in *P putida* UV4 (Scheme 2).

Generally applicable stereochemical methods for the determination of the enantiomeric excess (% ee) and absolute configuration of *cis*-dihydrodiol metabolites have now been developed. The direct formation of diMTPA ester derivatives of *cis*-dihydrodiols was precluded as a measure of enantiopurity due to instability. The latter problem was circumvented by the initial formation of the more stable cycloadducts formed with the dienophile 4-phenyl-1,2,4-triazoline-3,5-dione. The resulting formation of a single isomer due to *cis*-cycloaddition relative to the diol group, followed by diMTPA formation, provided an NMR (^{1}H and ^{19}F) method for ee determination (using both *R* and *S* forms of MTPA chloride).[4] While this method has proved to be applicable to more than twenty examples of *cis*-dihydrodiols derived from mono- and disubstituted arenes, instability and steric hindrance proved to be a problem with a small proportion of these metabolites. Chiral stationary phase (CSP) HPLC has recently provided a more convenient and sensitive alternative method for enantiopurity determination.[5] The general applicability of the CSP-HPLC method has been established by its success in more than fifteen examples.

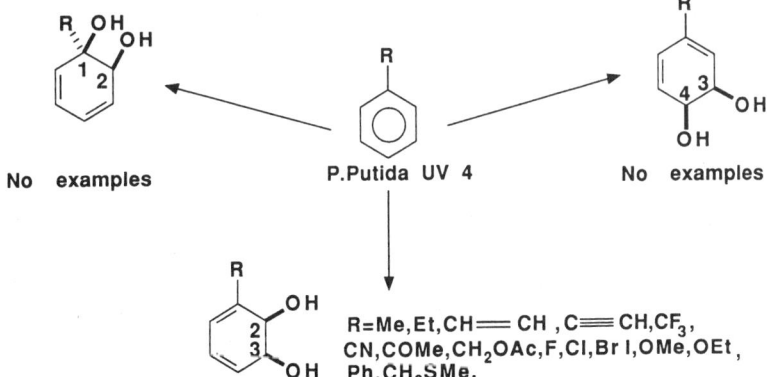

Scheme 2: Oxidation of monosubstituted benzenes by *Pseudomonas putida* UV4.

The crystalline diMTPA derivatives obtained from 4-phenyl-1,2,4-triazoline-3,5-dione adducts of *cis*-dihydrodiols have been examined by X-ray crystallography and thus the absolute configurations have been unequivocally established.[4] NMR spectral characteristics of the latter diMTPA esters have also been used as a less rigorous method for absolute configuration determination. Circular dichroism (CD) spectroscopy has now been developed as a more convenient, sensitive and general method for assigning absolute configurations to the parent *cis*-dihydrodiols.[6] The preferred conformation of the skew diene moiety in *cis*-dihydrodiol metabolites of mono-substituted benzenes was determined by the nature of the substituent. Thus, intra-molecular hydrogen bonding between a fluorine substituent and a neighbouring *pseudo*equatorial OH group resulted in a P configuration for the diene. Conversely, larger substituents (eg an iodine atom) resulted in the proximate OH group being *pseudo*axial and the diene group having an M configuration (Scheme 3).

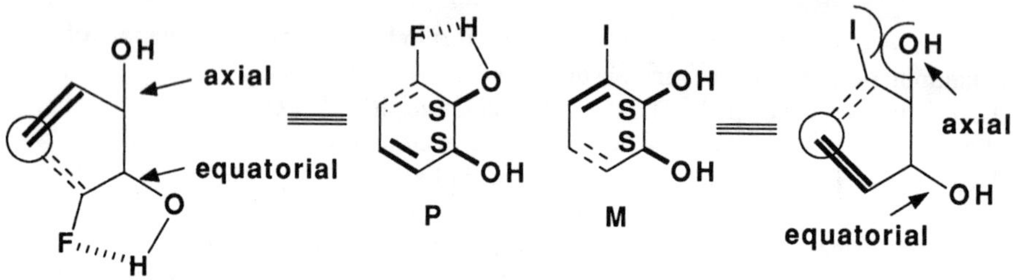

Scheme 3: Configurations of the diene moieties of *cis*-dihydrodiol metabolites.

The combination of ^{1}H and ^{19}F NMR spectroscopy and X-ray crystallography for a range of diMTPA esters, allied to the much more sensitive and generally applicable CSP-HPLC and CD methods, should now be applicable to the stereochemical assignment of many *cis*-dihydrodiol metabolites derived from intact cell or pure enzyme oxidation of mono- or disubstituted arenes.

Monosubstituted benzene substrates (with the exception of fluorobenzene) were found to yield *cis*-dihydrodiols of identical absolute configuration and enantiopurity (> 99% ee). The fluorine atom is closer in size to a hydrogen atom than the other substituents present on the *cis*-dihydrodiols. It has been observed[5] that the differential in substituent size at the 1- and 4-positions of benzene substrates has a marked effect upon facial selectivity during *cis*-dihydrodiol formation. Thus, a simple model based upon the stereodirecting effect of the larger substituent (L) at position 1 (relative to the smaller substituent S at position 4) allows a reliable prediction of absolute configuration for the *cis*-dihydrodiol product from *P putida* (Scheme 4).

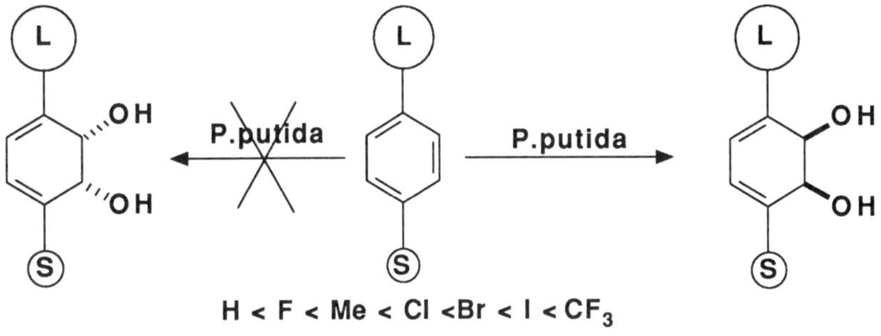

Scheme 4: Predictive model for the oxidation of 1,4-disubstituted benzenes by *P putida*.

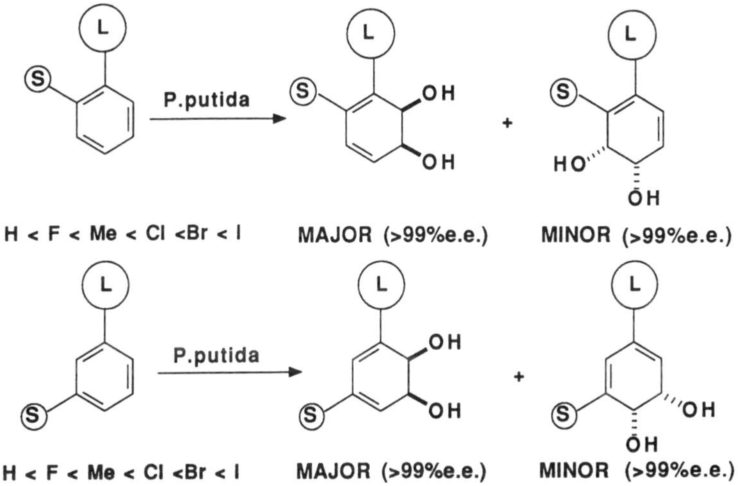

Scheme 5: Predictive models for the oxidation of 1,2- and 1,3-disubstituted benzenes by *P putida*.

A similar directing effect has been found during the formation of *cis*-dihydrodiols from 1,2- and 1,3-disubstituted benzene substrates. Thus, the larger group L was found to direct oxidation preferentially to a proximate bond (regiodirecting effect) and to a particular stereoheterotopic face (stereodirecting effect) in a predictable manner (Scheme 5).

Cis-Dihydroxylation of a carbocyclic ring in the polycyclic arene substrates naphthalene, quinoline, quinoxaline, benzofuran, benzo[*b*]thiophene, dibenzofuran and dibenzothiophene using *P putida* UV4 gave in each case a single enantiomer of identical absolute configuration.[7,8] The position of the heteroatom (N, O, S) in the heterocyclic ring appeared to influence regioselectivity during *cis*-diol formation although no clear trend was observed (Scheme 6).

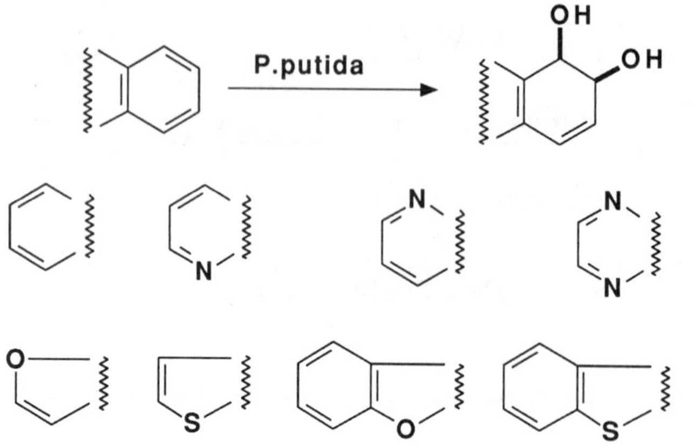

Scheme 6: Oxidation of heterocycles by *Pseudomonas putida* UV4.

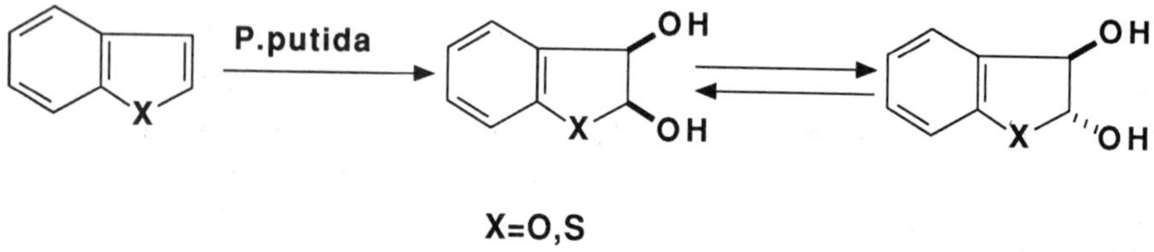

Scheme 7: Oxidation of benzofuran and benzo[*b*]thiophene by *Pseudomonas putida* UV4.

Oxidation of the heterocyclic ring of quinoline and quinoxaline gave phenolic metabolites (possibly *via* unstable *cis*-dihydrodiol intermediates). The latter proposal is supported by the isolation of *cis*-dihydrodiol metabolites resulting from oxidation of the heterocyclic rings of both benzofuran and benzo[*b*]thiophene. These heterocyclic *cis*-dihydrodiols were found to spontaneously isomerize to the corresponding *trans*-dihydrodiols, presumably *via* the open chain aldehyde (Scheme 7).

On the basis of the preferred absolute configurations from more than thirty *cis*-dihydrodiol metabolites obtained from mono- and disubstituted benzenes, and bicyclic and tricyclic arene substrates a composite picture may be built up. This is consistent with the steric differential models proposed and gives a crude picture of the active site of the dioxygenase enzyme system.

The metabolism of polycyclic arene substrates by *P putida* was found to yield either *cis*-dihydrodiol or phenol metabolites. When bicyclic arenes bearing a benzylic methylene group in the adjacent ring were used as substrates, monohydroxylation at the benzylic position was found to occur in all cases using *P putida* UV4. The benzylic hydroxylation was generally found to occur specifically at one prochiral hydrogen atom yielding a single enantiomer of R configuration (Scheme 8).

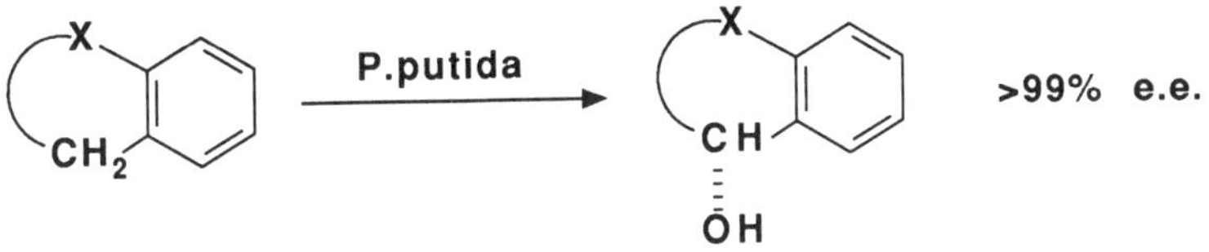

Scheme 8: Oxidation of benzocycloalkenes and derivatives by *P putida*.

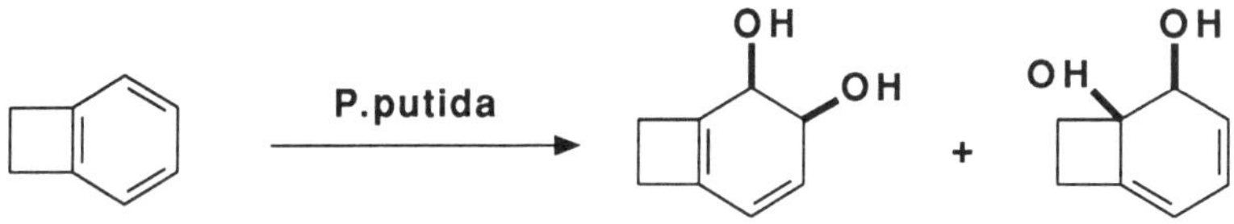

$X = (CH_2)_2, (CH_2)_3, (CH_2)_4, (CH_2)_5$

$X = (CH_2)O, (CH_2)_2O, (CH_2)S, (CH_2)_2S$

$X = (CH_2)CH=CH, (CH_2)_2CH=CH, (CH_2)_3CH=CH$

Scheme 9: Oxidation of benzocyclobutene by *P putida*.

The exception to the trend in Scheme 8 was found to be benzocyclobutene which gave the *S* configuration preferentially (20% ee).[9] Benzocyclobutene also proved to be exceptional in yielding *cis*-dihydrodiol metabolites. The small ring size thus appears to make benzocyclobutene behave rather like *ortho*-xylene in yielding *cis*-dihydrodiol metabolites from *P putida* (Scheme 9).

Evidence of further oxidation of bicyclic benzylic alcohols to yield the corresponding ketones was obtained (presumably *via* a dehydrogenase enzyme). A range of bicyclic benzylic alcohol enantiomers of opposite configuration was obtained by semi-preparative CSP-HPLC separation from racemic samples. When individual enantiomers were in turn used as substrates for *P putida* UV4, an unusual example of enantioselectivity was observed. Thus, using the *S* enantiomer of benzocyclobutan-1-ol as substrate a triol was produced from dihydroxylation of the neighbouring bond on the opposite face as the directing hydroxyl group. By contrast the *R* enantiomer was selectively oxidized to benzocyclobutanone and other products (Scheme 10).

The benzylic hydroxyl group present on 1-indanol, 3-hydroxy-2,3-dihydrobenzofuran and 3-hydroxy-2,3-dihydrobenzo[*b*]thiophene was similarly found to have a profound influence on the nature of metabolites formed. Thus, one alcohol enantiomer appeared to be oxidized exclusively to yield an all *cis* triol while the other enantiomer was dehydrogenated to yield a ketone. The triol metabolites from 3-hydroxy-2,3-dihydrobenzofuran and 3-hydroxy-2,3-dihydrobenzo[*b*]-thiophene were too unstable to be isolated and spontaneously dehydrated to yield *cis*-dihydrodiols (Scheme 11).

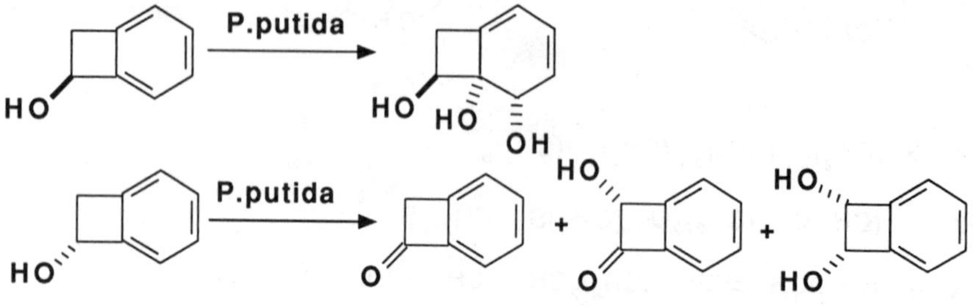

Scheme 10: Oxidation of benzocyclobutan-1-ol by *P putida*.

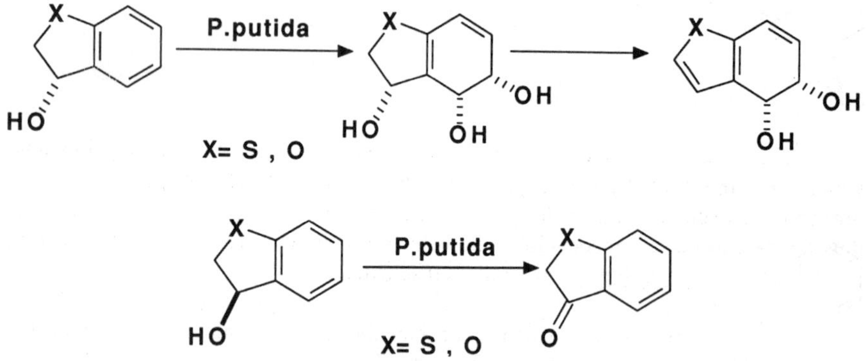

Scheme 11: Oxidation of 3-hydroxy-2,3-dihydrobenzofuran and 3-hydroxy-2,3-dihydrobenzo[*b*]thiophene by *P putida*.

A similar alkane → benzylic alcohol → *cis*-triol → *cis*-diol sequence was postulated when 1,2- and 1,4-dihydronaphthalene were used as substrates for *P putida*.[10] Thus, using ^{2}H-labelled dihydronaphthalene substrates the isolated *cis*-1,2-dihydroxy-1,2-dihydronaphthalene metabolites showed a labelling pattern consistent with the formation of unstable all *cis*-triol intermediates. Use of individual enantiomers of the arene hydrates (1-hydroxy-1,2-dihydronaphthalene and 1-hydroxy-1,4-dihydronaphthalene) as substrates again appeared to show that the *R*-enantiomers were selectively oxidized to the all *cis*-triol intermediates, whereas the *S*-enantiomers gave 1-naphthol (Schemes 12 and 13).

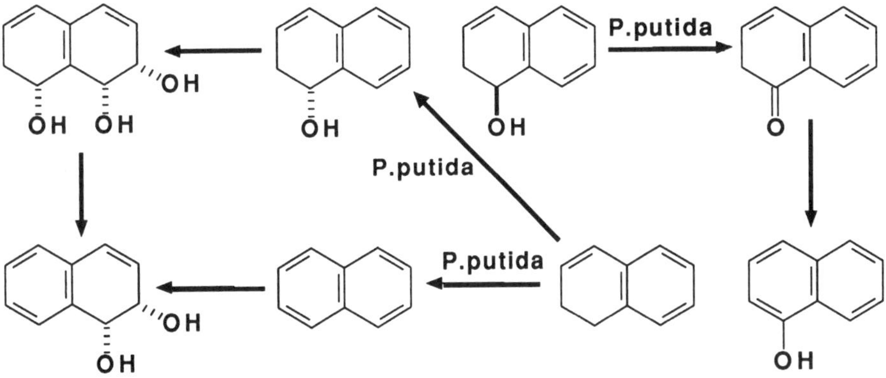

Scheme 12: Oxidation of 1,2-dihydronaphthalene by *P putida*.

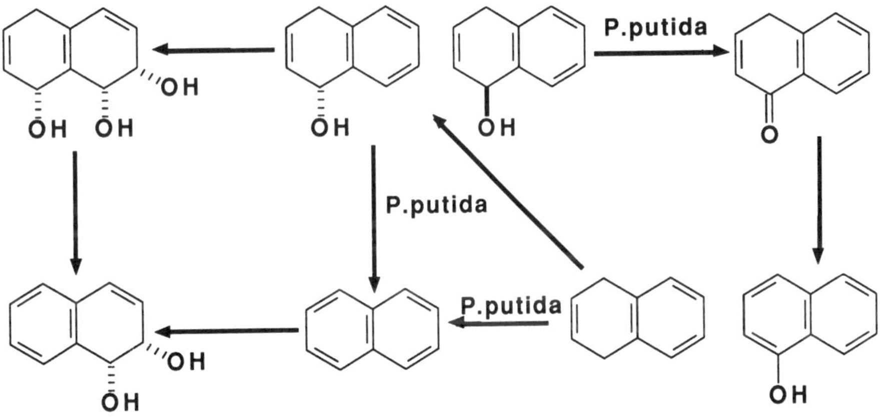

Scheme 13: Oxidation of 1,4-dihydronaphthalene by *P putida*.

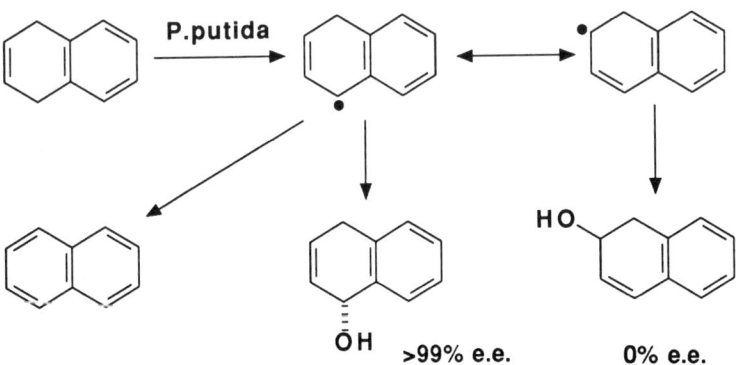

Scheme 14: Postulated intermediate radicals in the oxidation of 1,4-dihydronaphthalene by *P putida*.

Previous studies[11] using a purified dioxygenase enzyme from *P. putida* suggested that benzylic hydroxylation proceeded *via* a carbon-centred radical intermediate. A similar benzylic radical could account for the formation of the arene hydrates 1-hydroxy-1,4-dihydronaphthalene and 2-hydroxy-1,2-dihydronaphthalene from 1,2-dihydronaphthalene (Scheme 14). Similarly, the formation of 2-hydroxy- and 4-hydroxythiachromene and the isolation of the corresponding thiacoumarin and thiachromone from thiachromene substrate can be rationalized in terms of benzylic radical formation. The stereochemistry of benzylic hydroxylation was studied using individual enantiomers of 4-^{2}H labelled chroman. 4-*R*-Hydroxychroman was produced in enantiopure form with total retention of configuration (Scheme 15).

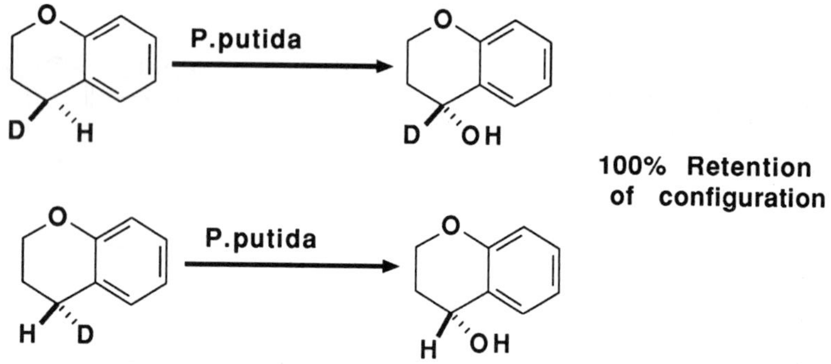

Scheme 15: Oxidation of 4-^{2}H labelled chroman by *P putida*.

These preliminary results obtained with intact cells of *P putida* UV4 are consistent with Gibson's proposal[11] for carbon-centred radical formation as the initial step in benzylic hydroxylation.

The synthetic potential of enantiopure hydroxylation products produced by *P putida* UV4 has been greatly facilitated by the recent commerical availability of several *cis*-dihydrodiols. As a result, a marked increase in the number of target molecules produced from *cis*-dihydrodiols has been reported[12] over the past three years. The stability of such *cis*-dihydrodiols is clearly of importance in synthetic applications. Kinetic studies have been carried out on the aromatization of seventeen *cis*-dihydrodiols obtained by enzymatic and chemoenzymatic methods from mono-substituted arenes. The remarkable difference in stability of the 2,3-*cis*-dihydrodiols is exemplified by those of ethoxybenzene and diphenylsulfone where the rate of aromatization of the latter is 10^6-fold less than the former.[13]

3 CONCLUSIONS

(i) The development of new methods for the determination of absolute configuration and enantiomeric excess of cis-dihydrodiol enantiomers should greatly facilitate future studies with pure dioxygenases.
(ii) The directing effects of substitutents during cis- dihydrodiol formation appear to be dominated by steric considerations. Thus, regio- and facial selectivity will be determined mainly by the nature of the larger substitutent.
(iii) The enantioselective effect on a single benzylic alcohol configuration to yield a cis,cis-triol enantiomer has been established.
(iv) Superimposition of a range of stereochemically defined cis-dihydrodiol metabolites has allowed preliminary predictions to be made about the shape of the active site of the dioxygenase enzyme present in *P putida*.

REFERENCES

1. D M Jerina, J W Daly, B Witkop, P Zaltzman-Nirenberg, and S Udenfriend, *Biochemistry*, 1970, **9**, 147.
2. S K Agarwal, D R Boyd, H P Porter, W B Jennings, S J Grossman, and D M Jerina, *Tetrahedron Lett*, 1986, **27**, 4253.
3. D T Gibson, V Subramanian, B D Ensley, W Reineke, and S H Safe in *Microbial Degradation of Organic Compounds*, ed D T Gibson, Marcel Dekker, New York, 1984, Ch 7, pp 181-252.
4. D R Boyd, M R J Dorrity, M V Hand, J F Malone, N D Sharma, H Dalton, D J Gray, and G N Sheldrake, *J Am Chem Soc*, 1991, **113**, 666.
5. D R Boyd, N D Sharma, M V Hand, M R Groocock, N A Kerley, H Dalton, J Chima and G N Sheldrake, *J Chem Soc, Chem Commun*, 1993, 974.
6. D R Boyd, H Dalton, and A F Drake, manuscript in preparation.
7. D R Boyd, N D Sharma, R Boyle, B T McMurray, T A Evans, J F Malone, H Dalton, J Chima, and G N Sheldrake, *J Chem Soc, Chem Commun*, 1993, 49.
8. D R Boyd, N D Sharma, R Boyle, R A S McMordie, J Chima, and H Dalton, *Tetrahedron Lett*, 1992, **33**, 1241.
9. D R Boyd, N D Sharma, P J Stevenson, J Chima, D J Gray, and H Dalton, *Tetrahedron Lett*, 1991, **32**, 3887.
10. D R Boyd, R A S McMordie, N D Sharma, H Dalton, P Williams, and R O Jenkins, *J Chem Soc, Chem Commun*, 1989, 339.
11. L P Wackett, L D Kwart, and D T Gibson, *Biochemistry*, 1988, **27**, 1360.
12. S M Brown and T Hudlicky in *Organic Synthesis: Theory and Applications*, Vol II, JAI Press Inc, Greenwich, Connecticut, 1992, 113 and refs therein.
13. D R Boyd, J Blacker, B Byrne, H Dalton, M V Hand, S C Kelly, R A More O'Ferrall, S N Rao, N D Sharma, and G N Sheldrake, *J Chem Soc, Chem Commun*, 1994, 313.
14. D R Boyd, M V Hand, N D Sharma, J Chima, H Dalton, and G N Sheldrake, *J Chem Soc, Chem Commun*, 1991, 1630.

IRON-SULFUR AND FLAVIN-DEPENDENT DEHYDRATIONS IN ANAEROBIC BACTERIA

Klaus Bendrat, Ulrich Eikmanns, Antje E M Hofmeister, Anne-Grit Klees, Uta Müller, Uwe Scherf, and Wolfgang Buckel*

Philipps-Universität, Laboratorium für Mikrobiologie, Fachbereich Biologie, 35032 Marburg, Germany.

1 INTRODUCTION

Anaerobic bacteria have the remarkable capacity to eliminate water from a whole series of hydroxyacyl-CoA derivatives ranging from 2-hydroxyacyl-CoA (Figure 1, eqn 1) and 3-hydroxyacyl-CoA (eqn 2) to 4-hydroxybutyryl-CoA (eqn 3) and 5-hydroxyvaleryl-CoA (eqn 4; for a review of such dehydrations see ref 1). Only the mechanism of the enzymic dehydration of 3-hydroxyacyl-CoA to 2,3-enoyl-CoA (eqn 2) is reasonably well understood. The C-H bond at the α-position, which has to be cleaved during this dehydration, is activated by the electron withdrawing thiolester. In the other substrates the corresponding C-H bond is located either at the β- or γ-position and hence is not activated. In this paper we describe four enzymes catalysing such unusual dehydrations, probably involving novel types of chemistry.

2 (R)-2-HYDROXYACYL-COA DEHYDRATASES

All known 2-hydroxyacyl-CoA dehydratases catalyse the reversible elimination of water from the (R)-enantiomer to the corresponding (E)-2,3-enoyl-CoA in a *syn*-manner (Figure 1, eqn 1). Examples are the dehydration of (R)-lactyl-CoA to acrylyl-CoA[2,3] (R = H) and (R)-2-hydroxybutyryl-CoA to crotonyl-CoA[4] (R = CH_3) which are catalysed by the same enzyme from *Clostridium propionicum*. In our laboratory another example, the reversible dehydration of (R)-2-hydroxyglutaryl-CoA to (E)-glutaconyl-CoA (R = CH_2-COO^-), has been studied extensively. The reaction is a key step in the fermentation of glutamate to ammonia, carbon dioxide, acetate, butyrate and hydrogen by certain strict anaerobic bacteria, as are *Acidaminococcus fermentans* and *Fusobacterium nucleatum*. The substrate (R)-2-hydroxyglutaryl-CoA is generated *in vivo* from acetyl-CoA and (R)-2-hydroxyglutarate in a reaction catalysed by glutaconate CoA-transferase. Owing to a lack of specificity of the enzyme the other

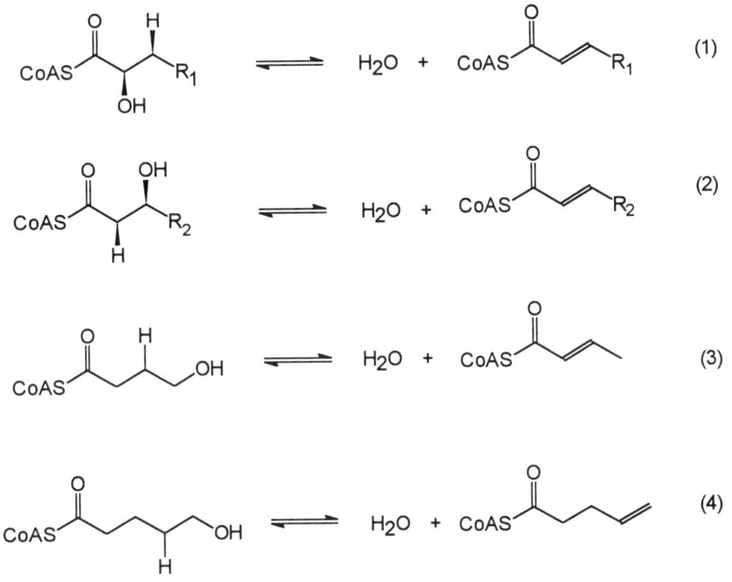

Figure 1: Dehydrations of hydroxyacyl-CoA derivatives found in anaerobic bacteria.
(R_1 = H, CH_3, CH_2-COO^-, phenyl; R_2 = H, alkyl)

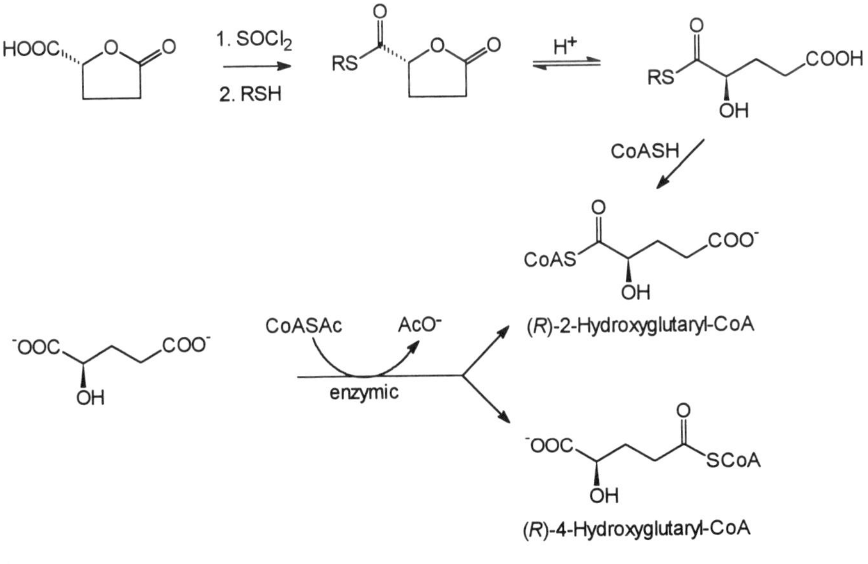

Figure 2: Formation of (R)-2-hydroxyglutaryl-CoA, the substrate of the dehydration to glutaconyl-CoA.
[R = CH_3-$(CH_2)_6$-NH-CH_2-CH_2-]

possible isomer, (R)-4-hydroxyglutaryl-CoA, is also formed, though at a lower rate. The isomers have been separated by HPLC and isolated in pure form. Furthermore, (R)-2-hydroxyglutaryl-CoA has been synthesized in a chemically defined way. Only the 2-isomer is attacked by the dehydratase and transformed to glutaconyl-CoA[5] (Figure 2).

The extremely oxygen sensitive (R)-2-hydroxyglutaryl-CoA dehydratase system from *A fermentans* consists of two components, the actual dehydratase, a heterotetramer, $\alpha_2\beta_2$ (α 54 kDa, β 42 kDa),[6] and an activator, a homodimer, γ_2 (γ 27 kDa). The genes of the three subunits are clustered together in the order hgd*CAB*. No sequence similarity to any other protein was detected. The dehydratase component was readily purified from cell extracts of *A fermentans*. It contains 8 mol Fe and 8 mol inorganic sulfur,[6] as well as reduced riboflavin and flavin mononucleotide (FMN). Homogenous activator was only obtained after overexpression of *hgdC* in *Escherichia coli*. Although under air the activator had a half-life of seconds (1% residual activity after exposure to air for 60 s), there was no difference in activity whether the expression was performed under anaerobic or aerobic growth conditions. However, harvesting of the cells and purification of the enzyme had to be performed under strict anaerobic conditions.[7] The reason for this extreme oxygen sensitivity is probably an additional iron-sulfur cluster that was recently detected in the activator (4Fe and 3S / homodimer).

(R)-2-Hydroxyglutaryl-CoA dehydratase from *F. nucleatum* consists of only one component, which is composed of three subunits $\alpha\beta\gamma$ (α 49 kDa, β 39 kDa, γ 24 kDa).[8] It appears likely that the γ-subunit has the function of the activator in the enzyme system from *A fermentans*. The other properties of the fusobacterial enzyme are similar to those of the dehydratase component from *A fermentans*. The heterotrimer (112 kDa) contains about 0.5 mol riboflavin, as well as about 4 mol Fe and 4 mol inorganic sulfur, but no FMN. The enzyme is also extremely sensitive towards oxygen (6% residual activity after exposure to air for 60 s). However, the oxidation of the reduced riboflavin in the enzyme requires about an hour. Hence, the part of the enzyme responsible for activation, most probably the γ-subunit, represents the extremely oxygen sensitive part, whereas the actual dehydratase appears to be more resistant towards oxygen.

For catalysis the dehydratase component from *A fermentans* has to be activated by incubation with the activator (dehydratase/activator approx 100/1), 5 mM $MgCl_2$, the reducing agent Ti(III)citrate (280 μM) and catalytic amounts of ATP (150 μM). The fusobacterial enzyme is activated under identical conditions but without additional activator. Reversible, transient inactivation of both enzyme systems occurred in the presence of the oxidants 4-nitrobenzoate, 2-nitrophenol, 3-nitrophenol, 4-nitrophenol or chloramphenicol (all at concentrations $\geq$ 1 μM). Under these conditions 4-nitrobenzoate was reduced to 4-aminobenzoate. Irreversible inactivation was observed with 6 μM nitrite and

30 μM hydroxylamine. The radical scavenger hydroxyurea (5 mM), however, did not inactivate the enzymes. It has been shown that ATP is hydrolysed to ADP and P_i during 'activation'. No phosphorylation or adenylation of either the enzymes or the enzyme-bound riboflavin was observed.[8]

Any speculation on the mechanism of (R)-2-hydroxyglutaryl-CoA dehydratase has to account for the stereochemical results obtained with the closely related (R)-lactyl-CoA dehydratase from C. propionicum.[3,4] During the dehydration of defined mixtures of (2R,3R) and (2R,3S)-2-hydroxy[3-^{3}H]butyryl-CoA to crotonyl-CoA the expected amounts of ^{3}HOH were formed, being consistent with a syn-elimination of water. Upon further incubation, however, no more ^{3}HOH was released indicating the absence of any racemisation at C-3.[4] Furthermore, hydration of [3-^{2}H$_1$]acrylyl-CoA in ^{3}HOH yielded (3R)-[3-^{2}H$_1$,^{3}H]lactyl-CoA of 53% enantiomeric excess.[3] According to these data a radical abstraction of the unactivated hydrogen with the formation of a methine or methylene radical at C-3 of either substrate appears unlikely, since there is evidence from coenzyme B_{12}-dependent reactions that these radicals are subject to rapid racemisation.[9] In accordance with these considerations Brunelle and Abeles proposed an alternative mechanism in which an as yet unknown metal is inserted into the unactivated C-H bond without the involvement of a radical.[3]

A two-step ionic mechanism, which considers the presence of iron-sulfur clusters and reduced flavin in the enzyme, proposes in the first step the replacement of the hydroxyl group of the substrate by the reduced flavin in an S_N2 reaction to yield enzyme-bound glutaryl-CoA. Thereby the nucleophile could be either the hydride or carbon 4a of the reduced flavin. In the second step, dehydrogenation by the now oxidized flavin leads to the product glutaconyl-CoA and regenerates the reduced flavin (Figure 3).[4,8] The S_N2 displacement of the hydroxyl group by a hydride is facilitated by the electron withdrawing thiolester and probably by binding the hydroxyl group to a specific iron of the iron-sulfur cluster which could act as a Lewis acid as observed in aconitase.[10]

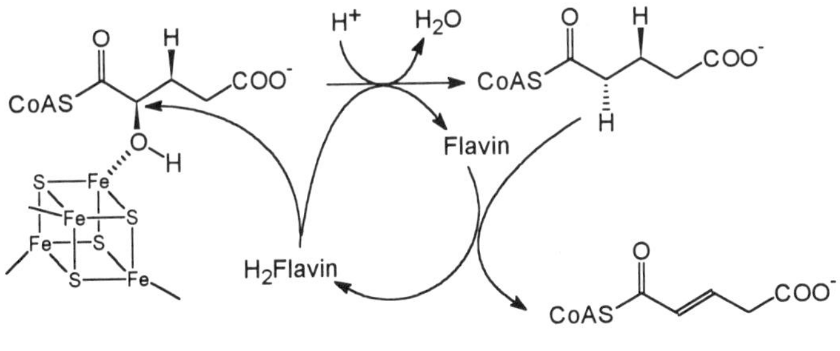

Figure 3: Hypothetical mechanism for the dehydration of (R)-2-hydroxyglutaryl-CoA to glutaconyl-CoA All reactions are reversible.

Figure 4: A model for the reductive displacement of a hydroxyl group.[11]

The reduction of dimethyl 2-hydroxyglutarate *O-p*-toluenesulfonate to dimethyl glutarate by sodium cyanoborohydride with inversion of configuration may serve as a chemical model[11] (Figure 4). Hence, binding of a hydroxyl group to an iron-sulfur cluster might convert the hydroxyl group into a similar good leaving group as does tosylation. The second step of the overall dehydration is a reaction catalysed by the well known acyl-CoA dehydrogenases, in which the unactivated hydrogen at C-3 is removed by the flavin as a hydride rather than as a proton. However, the dehydratase did not catalyse the oxidation of glutaryl-CoA to glutaconyl-CoA with an artificial electron acceptor (ferricenium ion).

A reaction related to the dehydration of 2-hydroxyacyl-CoA is the overall *anti*-elimination of phosphate from 5-enoylpyruvyl-shikimate 3-phosphate to the diene chorismate catalysed by chorismate synthase (Figure 5). Interestingly, for activity the enzyme also requires a reduced flavin (FMNH$_2$) which participates in the catalytic cycle. Furthermore, the absorbance spectrum of the intermediate resembles that of a C4a-adduct.[12] Chorismate synthase contains no iron-sulfur cluster, which is not required for catalysis, because phosphate is a much better leaving group than a hydroxyl ion or water.

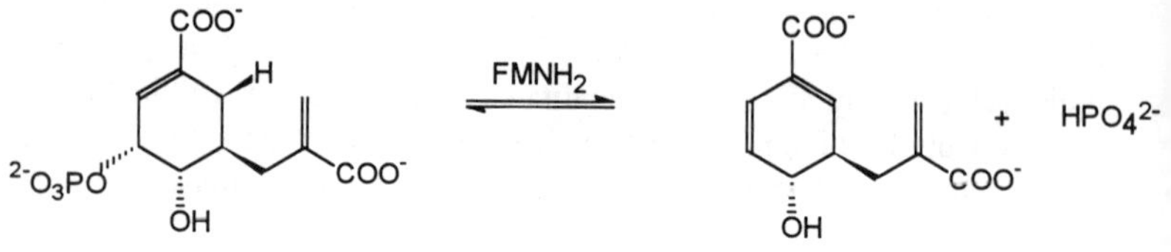

Figure 5: Reaction catalysed by chorismate synthase.

3 4-HYDROXYBUTYRYL-COA DEHYDRATASE

The reversible dehydration of 4-hydroxybutyryl-CoA to crotonyl-CoA (Figure 1, eqn 3), a step in the pathway of the fermentation of 4-aminobutyrate to ammonia, carbon dioxide, acetate and butyrate by *Clostridium aminobutyricum*,[13] poses a similar mechanistic problem. However, 4-hydroxybutyryl-CoA dehydratase from this organism is less complex than the (*R*)-2-hydroxyacyl-CoA dehydratases.[14] The dark brown enzyme is a homotetramer, α_4 (α 56 kDa) containing about 2 mol FAD, 16 mol Fe and 16 mol inorganic sulfur. The dehydratase also catalyses the stereospecific removal of one of the two hydrogens at carbon 3 of 4-hydroxybutyryl-CoA,[15] as well as the complete isomerisation and hydration of vinylacetyl-CoA to crotonyl-CoA and 4-hydroxybutyryl-CoA, respectively. When the equilibrium is reached, K_{eq} = [crotonyl-CoA]/[4-hydroxybutyryl-CoA] = 4.2, vinylacetyl-CoA cannot be detected any more. Activation of the enzyme by ATP is not required. Upon exposure to air both activities, the dehydratase as well as the isomerase, are transiently increased by about 20%, followed by irreversible inactivation yielding a yellow protein. Reduction of the active brown dehydratase by sodium dithionite leads to an inactive enzyme which is reactivated by oxidation with ferricyanide. Cyclopropylcarboxyl-CoA, which might be derived by cyclisation of 4-hydroxybutyryl-CoA, is not transformed to crotonyl-CoA by the enzyme.

A hypothetical mechanism analogous to that proposed for the dehydration of 2-hydroxyacyl-CoA (Figure 3) accounts for the fact that FAD, present in the active enzyme, is in the oxidized state. Hence, the first step should be an oxidation followed by a reduction. Thus, 4-hydroxybutyryl-CoA should be oxidized to yield $FADH_2$ and enzyme-bound 4-hydroxycrotonyl-CoA, a vinylogous 2-hydroxyacyl-CoA. The latter might be reduced by a hydride derived from $FADH_2$ in an S_N2' reaction to yield vinylacetyl-CoA which finally isomerises to crotonyl-CoA. Again the iron-sulfur cluster may serve as a Lewis acid facilitating the departure of the hydroxyl group, whereas the enoyl-CoA moiety favours the nucleophilic displacement of the hydroxyl group by a hydride[14] (Figure 6).

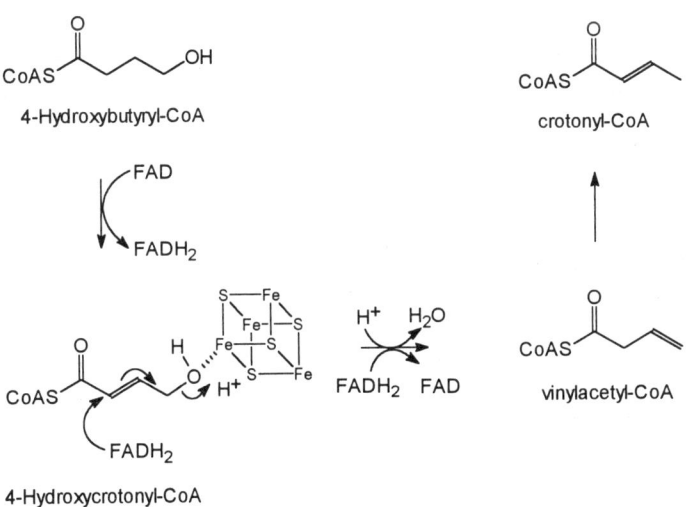

Figure 6: Hypothetical mechanism of 4-hydroxybutyryl-CoA dehydratase. All reactions are reversible.

4 5-Hydroxyvaleryl-CoA Dehydratase

The reversible dehydration of 5-hydroxyvaleryl-CoA to 4-pentenoyl-CoA (Figure 1, eqn 4) requires the cleavage of a C-H bond at the γ-position of the chain. However, in this case the transient introduction of a double bond solves the problem by forming 5-hydroxy-2-pentenoyl-CoA, a vinylogous 3-hydroxyacyl-CoA, as intermediate. Subsequent dehydration yields 2,4-pentadienoyl-CoA which is finally reduced to 4-pentenoyl-CoA (Figure 7).

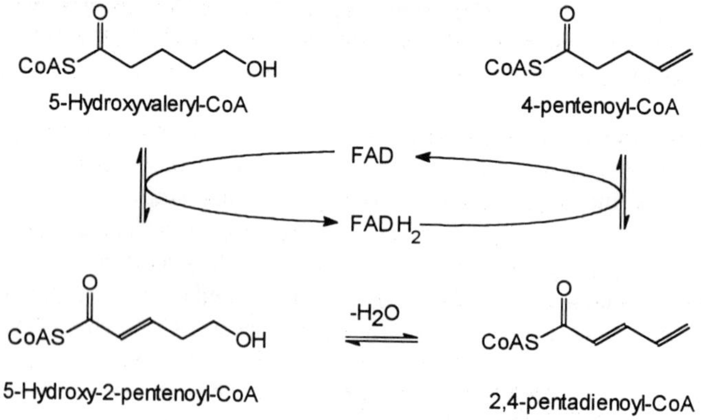

Figure 7: Reactions catalysed by 5-hydroxyvaleryl-CoA dehydratase.

The properties of a crystalline 5-hydroxyvaleryl-CoA dehydratase purified from *C aminovalericum* completely matched this proposal. The green homotetrameric enzyme α_4 (α 42 kDa) contains FAD (1 mol/subunit) and has dehydratase as well as acyl-CoA dehydrogenase activity. Furthermore, in the presence of the electron acceptor ferricenium ion, 5-hydroxy-2-pentenoyl-CoA and 2,4-pentadienoyl-CoA, rather than 4-pentenoyl-CoA, were produced from 5-hydroxyvaleryl-CoA. By conducting the hydration of 4-pentenoate in ^2H_2O in the presence of acetyl-CoA and a CoA-transferase, optically active 5-hydroxy(2,4-$^2H_2$)valerate was obtained, consistent with a transient, stereospecific formation of a double-bond between C-2 and C-3.[16]

These results led to a revision of the pathway of δ-aminovalerate fermentation by *C aminovalericum* proposed by Barker *et al*.[17] In agreement with these authors the amino acid is converted to 5-hydroxyvaleryl-CoA *via* glutaric semialdehyde and 5-hydroxyvalerate. However, dehydration of 5-hydroxyvaleryl-CoA in the presence of an as yet unknown electron acceptor yields 2,4-pentadienoyl-CoA rather than 4-pentenoyl-CoA which may not be generated *in vivo*. In a consecutive step, 2,4-pentadienoyl-CoA is reduced to 3-pentenoyl-CoA by a green FAD-containing enzyme which has also been purified from *C aminovalericum*.[18] Final isomerisation yields 2-pentenoyl-CoA which disproportionates to acetate, propionate and valerate.

5 CONCLUSIONS

The dehydrations of hydroxyacyl-CoA derivatives can be divided into two mechanistically different groups. The dehydrations of 3- and 5-hydroxyacyl-CoA are apparently simple acid base catalyses with activation *via* a transiently introduced double bond as necessary. Hence, the enzymes either are devoid of any prosthetic group or contain only FAD. In contrast, the dehydrations of 2- and 4-hydroxyacyl-CoA, in which unactivated C-H bonds have to be cleaved, require enzymes with flavins and iron-sulfur clusters. Interestingly, in the 2-hydroxyacyl-CoA dehydratases the riboflavin occurs in the reduced form whereas the FAD of 4-hydroxybutyryl-CoA dehydratase has to be in the oxidized state in order to generate an active enzyme. The importance of the oxidation states for activity suggests a redox role for the flavins in the dehydrations. Furthermore, the dehydration of 2-hydroxyacyl-CoA may be initiated by a reduction, whereas that of 4-hydroxybutyryl-CoA by an oxidation.

REFERENCES

1. W Buckel, *FEMS Microbiol Reviews*, 1992, **88**, 211.
2. R D Kuchta and R H Abeles, *J Biol Chem*, 1985, **260**, 13181.
3. S L Brunelle and R H Abeles, *Bioorg Chem*, 1993, **21**, 118.
4. A Hofmeister and W Buckel, *Eur J Biochem*, 1992, **206**, 547.
5. A-G Klees and W Buckel, *Biol Chem Hoppe-Seyler*, 1991, **372**, 319.
6. G Schweiger, R Dutscho, and W Buckel, *Eur J Biochem*, 1987, **169**, 441; R Dutscho, G Wohlfahrt, P Buckel, and W Buckel, *Eur J Biochem*, 1989, **181**, 741.
7. K Bendrat, U Müller, A-G Klees, and W Buckel, *FEBS Lett*, 1993, **329**, 329.
8. A-G Klees, D Linder, and W Buckel, *Arch Microbiol*, 1992, **158**, 294.
9. J Rétey and J A Robinson, 'Stereospecificity in organic chemistry and enzymology', Verlag Chemie, Weinheim, 1982, p 201.
10. H Beinert and M C Kennedy, *Eur J Biochem*, 1989, **186**, 5.
11. C P Whitman, G Hajipour, R J Watson, W H Johnson, M E Benbenek, and N J Stolowich, *J Am Chem Soc*, 1992, **114**, 10104.
12. M N Ramjee, S Balasubramanian, C Abell, J R Coggins, G M Davies, T R Hawkes, D J Lowe, and R N F Thorneley, *J Am Chem Soc*, 1992, **114**, 3151.
13. J K Hardman and T C Stadtman, *J Biol Chem*, 1963, **238**, 2088.
14. U Scherf and W Buckel, *Eur J Biochem*, 1993, **215**, 421.

15 P Willadsen and W Buckel, *FEMS Microbiol Lett*, 1990, **70**, 187.
16 U Eikmanns and W Buckel, *Eur J Biochem*, 1991, **197**, 661.
17 H A Barker, L D'Ari, and J Kahn, *J Biol Chem*, 1987, **262**, 8994.
18 U Eikmanns and W Buckel, *Eur J Biochem*, 1991, **198**, 263.

ACKNOWLEDGEMENTS

This work was supported by grants from the Deutsche Forschungsgemeinschaft and the Fonds der Chemischen Industrie.

ENZYME ENGINEERING BY *IN VITRO* SELECTION USING GENETIC AND ORGANIC TOOLS

Patrice Soumillion, Pascale Sartiaux, Michèle Bouchet,
Jacqueline Marchand-Brynaert, and Jacques Fastrez*

*Laboratoire de Biochimie Physique et des Biopolymères,
Université Catholique de Louvain, Place L Pasteur, 1, Bte, 1B,
B1348 Louvain-la-Neuve, Belgium.*

1   INTRODUCTION

Enzymes are highly complex organic molecules capable of catalysing a large variety of reactions. A considerable effort has been devoted during the last decades to the understanding of the structure-function relationship of these catalysts. Sometimes, in favourable circumstances, it is possible to predict the changes that should be introduced in the amino acid sequence of a protein to affect its properties in a desirable way. Most of the time, however, despite the impressive progress in the last few years (for reviews see references 1 and 2), no reliable prediction can be made. It remains for instance difficult to design mutations that will lead to significant improvements in thermal stability or to alter drastically the specificity. In the absence of the required knowledge, enzymologists have used the tricks of Nature that have been unravelled by the geneticists: mutate randomly the gene coding for the enzyme to be engineered and then screen or select for the organisms producing the improved enzyme (for a review see reference 3).

An extremely elegant technique has also been introduced a few years ago to create enzymes *de novo* by eliciting the formation of antibodies against transition state (TS) analogues.[4-6] Again, in this approach, it is necessary to screen or select for the antibodies with the best catalytic activities. This is *a fortiori* true if the biocatalyst needs further engineering.

There is a need for efficient screening or selection techniques. In a screening technique, a large collection of micro-organisms producing variants of the mutated enzymes are analysed. The process is quite laborious. In the best cases where a very sensitive detection test is available, the enzymatic activity can be observed directly on individual colonies of the micro-organism plated on a Petri dish at a density of 10^3 colonies per plate. To analyse 10^6 colonies, 10^3 plates have to be looked at.

A selection technique is in principle far more powerful. Here, the enzyme to be engineered is essential for the survival of the organism. An improvement in the properties of the protein can confer a biological advantage under the specific conditions imposed on the culture. Then the microorganisms producing the improved enzyme will grow faster and after a few generations, they will be dominant in the population. By definition this *in vivo* selection technique is applicable only for engineering enzymes whose new properties are beneficial from the biological point of view.

We will describe here a new method for *in vitro* selection of engineered enzymes. This method uses basically two tools: a relatively simple genetic tool that has been introduced recently and an organic tool whose characteristics can be designed by the organic chemist to open new possibilities for a selection process. As the selection is done *in vitro*, the requirement for biological advantage disappears.

2   RESULTS

The genetic tool is described in Figure 1. It consists of a virus that can infect the common bacteria *Escherichia coli* and replicates in them. During the extracellular phase of its life cycle, this virus, classically called a bacteriophage, or more simply a phage, is a long filamentous particle. The genome of the phage is encapsulated as a single stranded DNA inside a capsid made up of proteins. About 2300 copies of a small protein (the product of gene 8 or g8p) make up the body of the long cylinder; other proteins close the two ends of the cylinder: one of them seen as a knob at the top of the phage is of interest for this work. It is the product of gene 3 (in short g3p) (for a review on the biology of the filamentous phages see reference 7).

With the techniques of DNA recombination, it is relatively easy to introduce a foreign DNA sequence coding for a peptide or a protein into the normal phage's genome in such a way that when the genes of the coat are expressed for viral morphogenesis the g3p will be replaced by a fusion protein. This technology was initiated by G Smith.[8,9] In this construction a foreign enzyme can be directly connected by a peptide bond to the normal g3p. The stop codon, that normally specifies when the synthesis of the protein sequence is complete, is replaced by a few codons that will direct the continuation of the biosynthesis; after these codons comes the DNA sequence coding for g3p. In this way, we will have what can be called a phage-enzyme. This is an entity in which there is a physical association between an enzyme displayed on the surface of the phage and its encapsulated gene in a supramolecular association. The advantage of this genetic tool is that if a single phage infects an *E coli* cell in a culture, it will replicate; its progeny, after having been exported in the medium, will also infect other cells so that

Enzyme Engineering by In Vitro Selection Using Genetic and Organic Tools   151

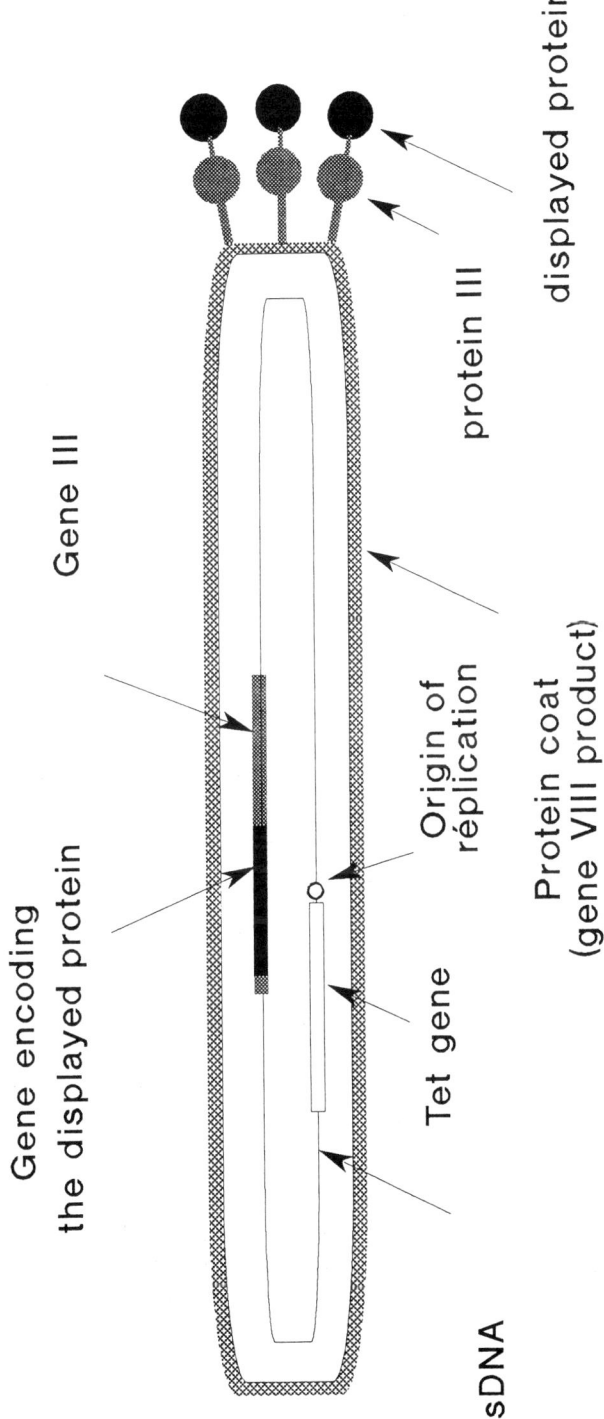

Figure 1: Phage-enzyme combination for use as a genetic tool.

an enormous amplification of the genome will ensue. This confers an extreme sensitivity of detection.

Many proteins have been displayed on the surface of a phage, and examples are given in Table 1.

Table 1: Examples of proteins displayed on phages

Protein Displayed	Reference
Antibody or Fab fragment	26, 27
α-Subunit ecto domain of IgE receptor	28
Human growth hormone	29
Enzyme inhibitor	30
Alkaline phosphatase	31
Trypsin	32

The function of an antibody, a hormone or an inhibitor, is simply to recognise its specific ligand and to bind it so that a physiological response is triggered. The phages displaying proteins of that kind (phage-receptors) can be relatively easily selected from a mixture by an affinity chromatography technique. The ligand is immobilised on a solid support. A solution of a mixture of phages is loaded on the column, but only the phages displaying a protein with an affinity for the ligand are immobilised. After extensive washing, the phages with the expected properties are eluted. The eluted solutions are used to infect an *E coli* culture for amplification. To obtain the best binders of a mixture, the selection is repeated a few times. Even if the enrichment in good binders is relatively low in a single selection experiment, it increases exponentially with the number of runs, so that one can efficiently select the best receptors (for examples see the references in Table 1).

To apply the same efficient *in vitro* selection techniques in enzymology, a basic change has to be introduced in the selection principle. Selection must be done **not for binding but for catalysis.** We do that by first labelling the phages bearing active enzymes so that they can later be selected by affinity chromatography. The labelling is done with an organic tool that we will call a label. This label is a bifunctional organic compound. It features two heads: one is a suicide inhibitor (also called suicide substrate or mechanism based inhibitor) of the enzyme to be selected (Figure 2). A suicide inhibitor is a relatively unreactive molecule whose inhibitory capacity becomes activated by the target enzyme itself through a pathway that specially uses the enzyme's mechanism (for a review see reference 10). The second head of the label is a ligand that can be recognised by an immobilised receptor. In this work we use biotin as the ligand because it forms very tight

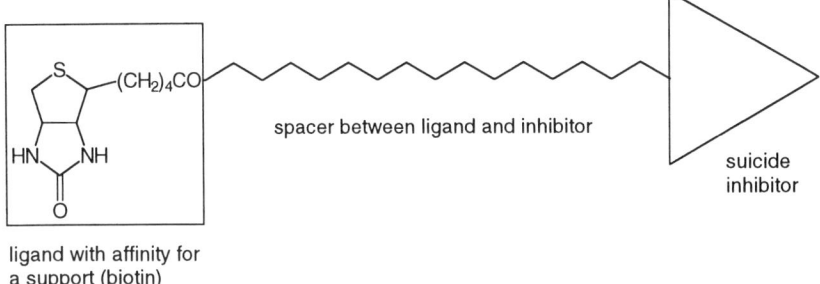

Figure 2: Bifunctional compound for labelling phages bearing active enzymes.

complexes with proteins like avidin or streptavidin. The biotin-avidin immobilisation technology is well established (for a review see reference 11). The space between the two heads can contain an easily cleavable bond like a disulfide bridge (see below).

On incubation of a mixture of phages with the label, if the mechanism based inhibitor is perfectly designed, only the phages displaying active enzymes will become labelled and consequently selectable. As the label forms a covalent bond with the inhibitor and the biotin-streptavidin complex is extremely tight ($K_D = 10^{-15}$ M^{-1}), the elution of bound phages can only be done by cleavage of a chemical bond. We use two methods, reduction of the disulphide of the spacer or very specific proteolytic cleavage of a peptide sequence introduced between the displayed enzyme and g3p. The recovered phages are replicated to amplify the signal before analysis.

We have cloned the genes of two proteins in the phage fd (using the fd-DOG1 modification engineered at the MRC in Cambridge[12]). The first protein is the RTEM-β-lactamase. This enzyme hydrolyses β-lactam antibiotics very efficiently and as such confers on the bacteria producing it a resistance to them. The X-ray structure of the enzyme has been determined:[13-14] it is an ellipsoid. The carboxyl terminal part of the sequence is organised as a helix whose carboxylate function is solvent accessible so that it was expected that a fusion protein with g3p (whose structure is not known) could be constructed. To avoid unfavourable contacts between the two globular proteins, a heptapeptide is introduced as the connection. It contains the sequence Ile-Glu-Gly-Arg that is recognised by a very specific protease of the blood clotting cascade: factor Xa. This factor Xa site will be used to disconnect the phage from the column after immobilisation.

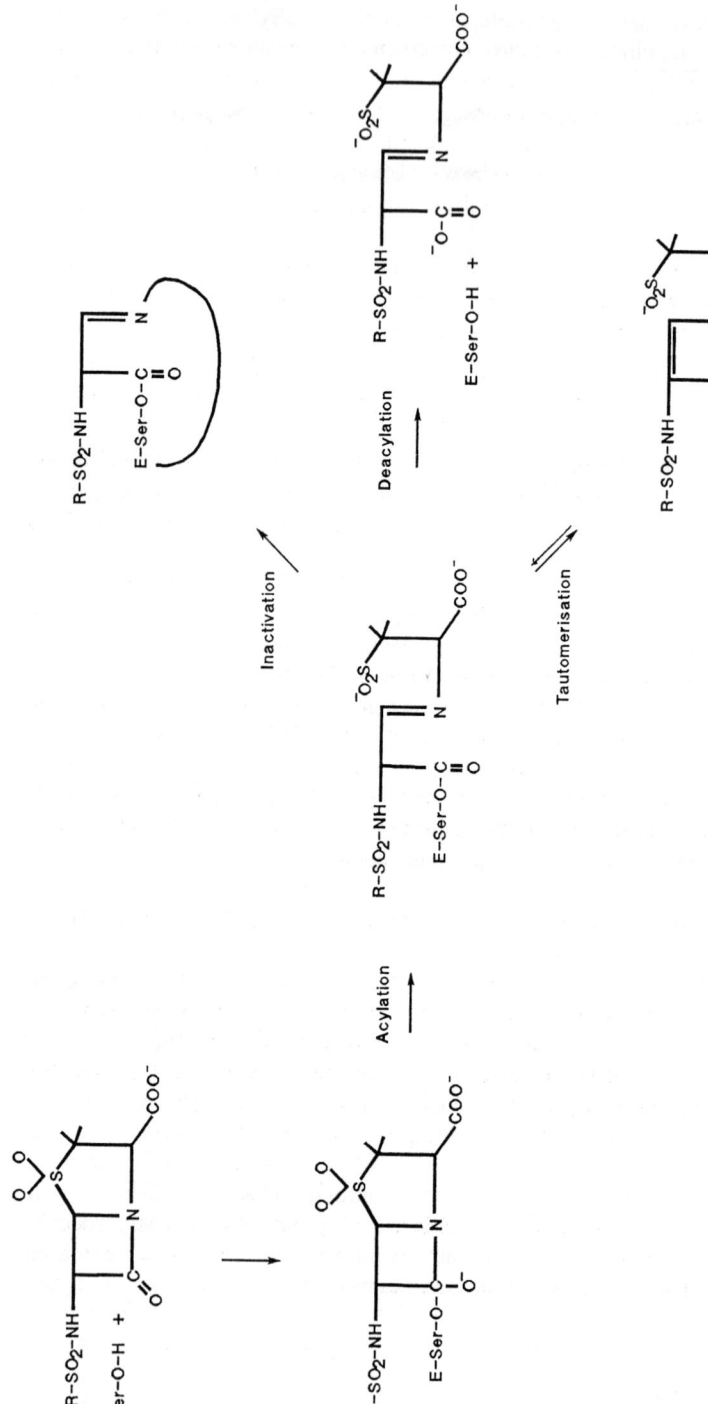

Figure 3: Mechanism of inhibition of a β-lactamase (E-Ser-OH) by a penicillin sulfone suicide inhibitor (refs 19 and 20).

To prove that it is possible to select specifically for catalysis and not simply substrate binding, we have constructed two phage-enzymes. The first one [fd-Bla(+)] displays the wild type β-lactamase that presumably will be active on this phage; the second one, fd-Bla(-), is a mutant β-lactamase in which serine-70, the essential nucleophile in the active site, has been mutated to alanine. This mutation destroys the catalytic activity of the enzyme without impairing the binding affinity for its substrate.[15,16]

The β-lactamase phage-enzyme fd-Bla(+) shows all the expected properties. The number of phages produced per *E coli* cell is large enough to allow a direct detection of the β-lactamase activity on colonies growing on plates with a sensitive chromogenic substrate: nitrocefin. This shows that the β-lactamase is active and affords an *in vivo* **screening** for colonies hosting fd-Bla(+) versus fd-Bla(-). The production of fd-Bla(+) also confers ampicillin resistance to the phages harbouring colonies: this allows an *in vivo* **selection**. After isolation of these phages, it is possible to measure the β-lactamase activity of the phage directly in solution. The specific activity corresponds to what would be observed if each phage would display about 4 copies of the enzyme, in rather satisfactory agreement with expectation. It is clear that the enzymes are covalently attached to the phages, as in an ultrafiltration of the solution, on a membrane that allows soluble β-lactamase to go through but not fd-phages, no enzymatic activity is found in the ultrafiltrate. The sensitivity of the enzyme to irreversible inhibition by a suicide inhibitor, free or attached to the biotin ligand, is very similar to that of the free enzyme.[17]

In its normal mechanism, β-lactamase hydrolyses penicillin derivatives in a multistep process with transient formation of a covalent intermediate, an acyl-enzyme resulting from serine 70 attack on the β-lactam carbonyl.[18] The interaction of the enzyme with a mechanism based inhibitor is more complex. A beautiful and detailed analysis of the inactivation pathway by inhibitors similar to the compound that we have used has been completed by J Knowles and his collaborators[19,20] (Figure 3). In the inhibition mechanism, the triggering event that transforms the suicide substrate into an irreversible inhibitor is the normal opening of the β-lactam ring. The acylation leads in a stepwise or concerted manner to a β-elimination, affording an acyl-enzyme that can either deacylate quickly or tautomerise to a more stable acyl-enzyme or react with a nucleophile (a lysine) in the active site to block completely the enzyme.

A label featuring a sulfone derivative of penicillin and a biotin moiety connected by a spacer including a disulfide bridge (Figure 4) has been prepared. The suicide substrate head in that label differs from a normal substrate in two respects. The sulfur of the penicilloyl moiety is oxidised to the sulfone level. This creates a good leaving group. The exocyclic nitrogen which is normally protected as an amide bond is included in a sulfonamide function. This increases the acidity of the C_6-H bond, again favouring a β-elimination of the sulfinic acid.

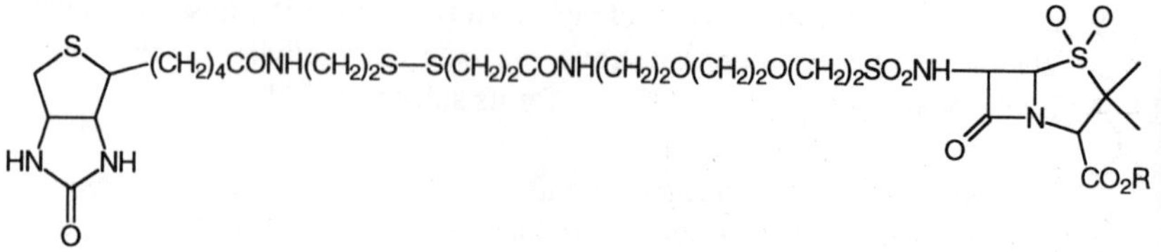

Figure 4: Label derived from biotin and penicillin sulfone.

Mixtures of fd-Bla(+) and fd-Bla(-) were incubated with the label. After elimination of excess inhibitor, streptavidin coated magnetic beads were added to the phages; then the immobilised phages were recovered either by proteolytic elution with factor Xa or by reduction with dithiothreiotol. From the ratios between the active [fd-Bla(+)] and inactive phages [fd-Bla(-)] in the initial mixture and in the eluted solution, an enrichment factor (E) is obtained. When the elution is done by proteolysis, $E = 35 \pm 15$, when it is done by reduction with DTT, $E = 9 \pm 3$. The difference between these two numbers is related to a non-specific labelling by the inhibitor. Indeed, penicillins are known to react non-specifically with proteins (these reactions are thought to be responsible for the observed immune responses[21]). The penicillin sulfones are even better acylating agents; they can react with the major coat protein (g8p) or with g3p. The products of these reactions are only eluted with DTT, not with factor Xa.[22,33]

The second phage enzyme that we have constructed displays subtilisin. The first purpose of the experiment was to show that this protein behaves like the free enzyme when attached to the phage. Subtilisin is produced in the bacterium as a pre-pro-subtilisin. The pre-peptide ensures secretion of the enzyme by the bacterium. The pro-sequence is required to guide the proper folding of the protein. In its absence, the enzyme is unable to reach the correct tertiary structure. After having catalysed the folding, the pro-sequence is removed by intramolecular proteolysis leaving a fully active subtilisin.[22]

We have indeed observed that, under suitable conditions, maturation of pro-subtilisin into subtilisin also occurs on the phage. This increases our confidence that many phage displayed enzymes will behave like the corresponding free enzymes.

For the selection of active phages, another principle of mechanism based inactivation can be envisaged. Serine proteases are known to be

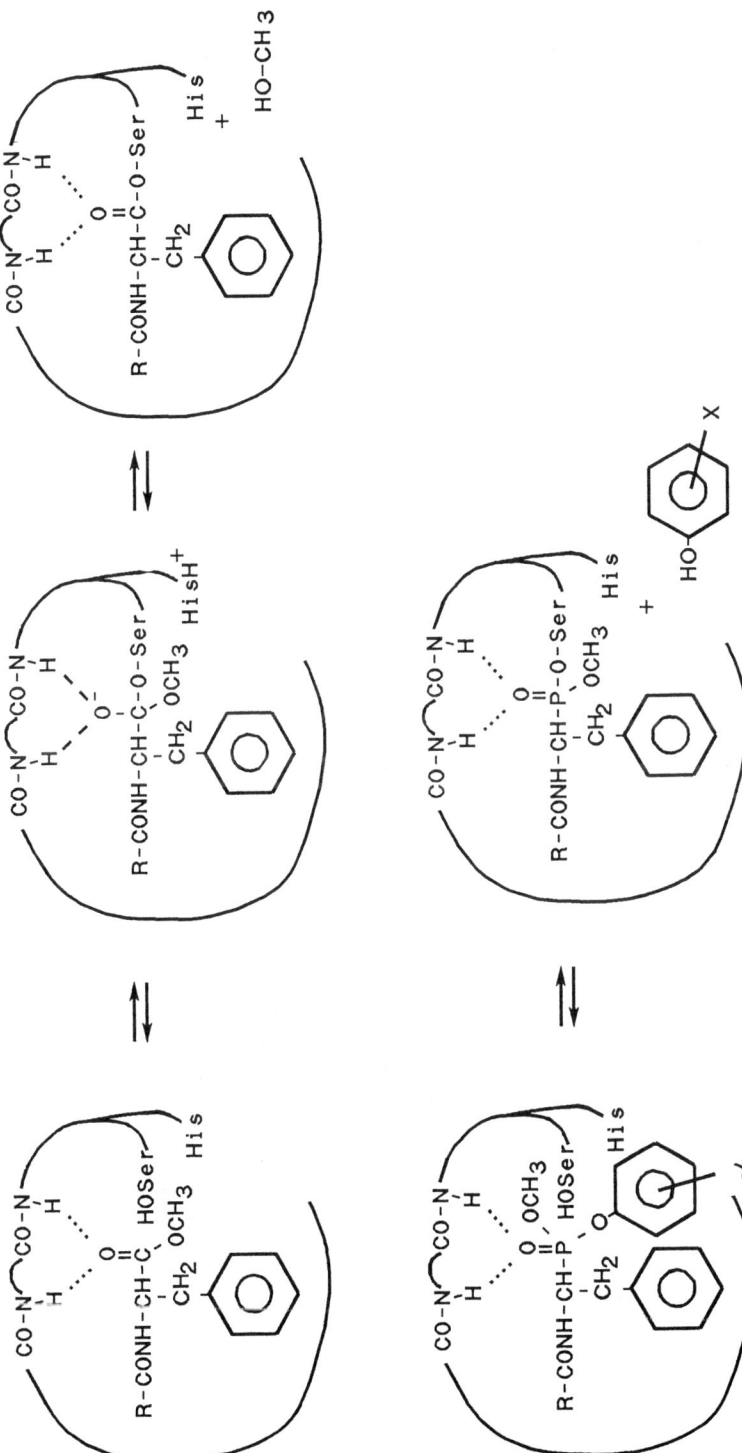

Figure 5: Inhibition of a serine protease by a phospho-amino acid ester, showing the similarity between the phosphoryl enzyme and the tetrahedral intermediate from protease attack on an amino acid ester.

inhibited by esters of phospho-amino acids.[23,24] These phosphorylate the essential serine to afford a very stable phosphoryl enzyme which mimics the tetrahedral transition state of the acylation reaction by an ester or an amide (Figure 5). Enzymes stabilise transition states by being more complementary to them than to the ground states of the substrates. Accordingly, a covalent transition state analogue is a good irreversible inhibitor. It has further been shown that phosphorylation recruits most of the substrate's features essential for acylation.[25]

3   DISCUSSION

The data presented show that it is possible to display active enzymes on the surface of phages and to select the phages displaying them from a mixture where inactive mutants are predominant. As the inactive phages display proteins that remain able to bind their substrate, the selection process is for catalytic activity versus binding.

The enrichment factors are lower than hoped for. This may be a consequence of the high reactivity of the penicillin sulfone used as a suicide substrate. This compound may react non specifically with amino acid side chains of the viral proteins; it may also label the active site residues without having been activated by the enzyme. Better designed inhibitors could avoid these reactions. This is, however, not a major limitation of the approach. To select from a highly diverse library containing millions of mutants, the selection procedure will simply have to be repeated four to five times.

The approach is likely to be of rather general interest for protein engineering. Suicide inhibitors have been designed for many classes of enzymes. In using them, one directly selects for catalytic efficiency.

This technique will open new avenues in protein engineering. For instance, by random mutagenesis on the gene coding for a displayed enzyme, it will become possible to select for mutations that stabilise the enzyme. This will be feasible because the phages are quite resistant to denaturation, so that selection can be run under conditions where the phages are still viable, while the wild type enzyme is denatured and some stabilised mutants are active. The technology will also be used to facilitate the changes in enzymatic specificity that appear to be difficult to design by simple site directed mutagenesis. Here a combination of directed and random mutageneses combined with *in vitro* selection is expected to be rather efficient. More ambitious projects like the creation of new catalytic activities might at some stage also take advantage of the technology.

It is clear from the description of the method that the contribution of physical organic chemists, synthetic organic chemists or bioorganic chemists in the design and implementation of the suicide inhibitors, in collaboration with enzymologists, is vital for the success of the approach.

REFERENCES

1. W M Atkins and S G Sligar, *Curr Opin Struct Biol*, 1991, **1**, 611.
2. Ch Wilson and D A Agard, *Curr Opin Struct Biol*, 1991, **1**, 617.
3. M J Zoller, *Curr Opin Struct Biol*, 1991, **1**, 605.
4. F Kohen, J B Kim, H R Lindner, Z Eshhar, and B S Green, *FEBS Letters*, 1980, **111**, 427.
5. R A Lerner and S J Benkovic, *Bio Essays*, 1988, **9**, 107.
6. P G Schultz, *Science*, 1988, **240**, 426-433.
7. P Model and M Russel, Filamentous Bacteriophage in *The Bacteriophages II*, R. Calendar Ed, Plenum, New York, 1988.
8. G P Smith, *Science*, 1985, **228**, 1315.
9. S F Parmley and G P Smith, *Gene*, 1988 **73**, 305.
10. M A Ator and P Ortiz de Montellano, 'Mechanism-based (suicide) enzyme inactivation', in *The Enzymes* 3rd Edn, Vol 19, D S Sigman, & P D Boyer eds, Academic Press, 1990, pp 213-282.
11. M Wilchek and E A Bayer, *Anal Biochem*, 1988, **171**, 1.
12. T Clackson, H R Hoogenboom, A R Griffiths, and G Winter, *Nature*, 1991, **352**, 624.
13. C Jelsch, F Lenfant, J M Masson, and J P Samama, *FEBS Letters*, 1992, **299**, 135.
14. N C J Strynadka, H Adachi, S E Jensen, K Johns, A Sielecki, C Betzel, K Sutoh, and M N G James, *Nature*, 1992, **359**, 700.
15. L J Mazzella, S Pazhanisamy, and R F Pratt, *Biochem J*, 1991, **274**, 855.
16. G Dalbadie-McFarland, J J Neitzel, and J H Richards, *Biochemistry*, 1986, **25**, 332.
17. G I Dmitrienko, C R Copeland, L Arnold, M E Savard, A J Clarke and T Viswanatha, *Bioorg Chem*, 1985, **13**, 34.
18. J-M Ghuysen, *Ann Rev Microbiol*, 1991, **45**, 37.
19. J Fisher, R L Charnas, S M Bradley, and J R Knowles, *Biochemistry*, 1981, **20**, 2726.
20. J R Knowles, *Acc Chem Res*, 1985, **18**, 97.
21. A L DeWeck and G Blum, *Internat Arch Allergy Appl Immunol*, 1965, **27**, 221.
22. M Inouye, *Enzyme*, 1991, **45**, 314.
23. P A Bartlett and L A Lamden, *Bioorg Chem*, 1986, **14**, 356.

24. J Fastrez, L Jespers, D Lison, M Renard, and E Sonveaux, *Tetrahedron Lett*, 1989, **30**, 6861.
25. I M Kovach, M Larson, and R L Schowen, *J Am Chem Soc*, 1986, **108**, 5490.
26. J McCafferty, A D Griffiths, G Winter, and D J Chiswell, *Nature*, 1990, **348**, 552.
27. C F Barbas III, A S Kang, R A Lerner, and S J Benkovic, *Proc Natl Acad Sci, USA*, 1991, **88**, 7978.
28. M W Robertson, *Prot Eng*, 1993, **6**, 73.
29. S Bass, R Greene, and J A Wells, *Proteins: Struct Funct Genet*, 1990, **8**, 309.
30. B L Roberts, W Markland, A C Ley, R B Kent, D W White, S K Guterman, and R C Ladner, *Proc Nat Acad Sci, USA*, 1992, **89**, 2429.
31. J McCafferty, R H Jackson, and D J Chiswell, *Prot Eng*, 1991, **4**, 955.
32. D R Corey, A K Shiau, Q Yang, B A Janowski, and C S Craik, *Gene*, 1993, **128**, 129.
33. P Soumillion, L Jespers, M Bouchet, J Marchand-Brynaert, G Winter, and J Fastrez, *J Mol Biol*, 1994, **237**, 415.

ISOTOPE-AIDED NMR STUDIES OF PROTEIN-LIGAND INTERACTIONS

James Feeney

Laboratory of Molecular Structure, National Institute for Medical Research, The Ridgeway, Mill Hill, London NW7 1AA, UK.

1 INTRODUCTION

Over the last two decades there has been a great deal of interest in the application of high resolution NMR spectroscopy to studies of interactions of small molecules with proteins. Such interactions are very important in many areas of biology, for example in complexes of enzymes with substrates, receptors with drugs and transcription factors with DNA duplexes. NMR studies can provide information about the *specificity* of the interaction and in some cases the interacting groups on the ligand and the protein can be identified. The technique is particularly useful for detecting the ionisation state of a group on the protein or the ligand and monitoring how this is perturbed on formation of the complex. NMR can also provide information about the *solution conformations* of the bound ligand and the protein. It is an excellent method for detecting and monitoring multiple conformational states in solution. Several cases have been documented where different conformations give rise to separate superimposed NMR spectra. In such cases, one can monitor the equilibrium between the different forms by measuring the intensities of the signals in the different spectra. NMR can also be used to characterise *dynamic processes* in protein ligand complexes such as sidechain and backbone motions of the protein, dynamic processes within the ligand itself (for example, aromatic ring flipping) and rates of breaking and formation of hydrogen bonds.

In order to extract this wealth of information about protein-ligand complexes, it is first necessary to make detailed assignments of the signals to specific nuclei in the protein and the ligand. This is often difficult because of the complexity of the protein spectra which contain many overlapping broad signals, the line widths increasing as the molecular weight increases. However, recent advances in NMR technology and methodology, combined with the use of isotopically labelled materials, have provided dramatic improvements in spectral resolution which make it increasingly easier to tackle such problems.

1.1 Isotopic Labelling

Table I indicates some of the types of isotopic labelling which have been used for simplifying NMR studies of protein-ligand complexes. One method of isotopic labelling involves partial deuteriation of the molecules with the aim of simplifying the complex ^{1}H spectrum. For example, this can be achieved by introducing specifically deuteriated amino acids into the protein. By selective deuteriation, the ^{1}H spectrum is considerably simplified and by comparing the spectra of the non-deuteriated and selectively deuteriated samples it is possible to detect and assign those ^{1}H signals that are present in the former but absent in the latter. Another approach is to deuteriate all of the protein residues either completely or to a high level (~ 70%). Samples where the protein is completely deuteriated are useful for detecting the ^{1}H signals of the bound ligand and the conformation of the bound ligand can sometimes be obtained from measurements of intramolecular NOE effects in such cases.[1] Samples with random, partial deuteriation are also useful in that their spectra have signals with simpler multiplet structures and narrower line widths (lower dipolar interactions).[2] In all cases the deuterium is introduced using biosynthetic methods.[2,3] Studies of selectively deuteriated proteins have been undertaken since the earliest days of biological NMR[1-6] and several examples will be considered later (see Table II).

Another form of labelling involves specific enrichment of either the ligand or the protein with an isotope amenable to direct NMR detection. The chosen isotope is one with a low natural abundance (for example, ^{13}C or ^{15}N) such that the nuclei from the enriched sites give large signals which are easy to detect and assign. The advantage of this direct approach is that the spectra are very simple and can often be examined using 1D NMR techniques. Such signals are ideal for monitoring dynamic processes which require complex line shape analysis or for relaxation time studies which involve careful measurements of signal intensities. This direct labelling procedure has been used extensively and several examples are indicated in Table III. In the third approach, specific or general ^{13}C or ^{15}N labels are introduced and then the ^{13}C or ^{15}N signals are detected indirectly, and the ^{1}H signals from protons attached to the isotopic labels are detected selectively. The full potential of this powerful method of spectral simplification has been realised only relatively recently with the advent of the appropriate multi-dimensional NMR techniques. Several 3D and 4D NMR experiments have been devised which provide NMR signals characterised by 3 or 4 different nuclear frequencies.[7-11] These signals can readily be resolved for proteins as large as 20 kDa.

Table I

Methods of isotopic labelling

Specific and general deuteriation
- direct detection of remaining ^{1}H signals
- simplifies ^{1}H spectrum
- assigns the missing ^{1}H signals
(see examples in Table II)

Specific enrichment
·direct detection of isotopic label, eg. ^{3}H, ^{13}C, ^{15}N, ^{19}F of ligands
(see examples in Table III)

Specific and general ^{13}C and ^{15}N labelling
·selective detection of ^{1}H signals from protons bonded to X nucleus and indirect detection of X nucleus signals using various X-^{1}H correlation experiments such as:

(i) indirect detection of X nucleus signals *via* attached protons (2D HMQC[11]).

(ii) 2D X-^{1}H correlation experiments with NOESY or TOCSY relay (2D HMQC.NOESY or HMQC.TOCSY[12]).

(iii) 3D ^{1}H-X-^{1}H correlation experiments (3D NOESY.HMQC, COSY.HMQC or TOCSY.HMQC[7]).

(iv) 3D ^{1}H-^{15}N correlation experiments (3D HMQC.NOESY.HMQC[13,14]).

(v) 3D experiments for side-chain resonance assignments (3D HCCH.COSY and HCCH.TOCSY[8]).

(vi) 3D and 4D triple resonance experiments for backbone resonance assignments (3D HNCA[7,15]).

(vii) 3D and 4D triple resonance experiments for correlating backbone and side-chain resonances[10]).

2 DIHYDROFOLATE REDUCTASE

One of our major interests has been the study of antifolate drugs binding to the enzyme dihydrofolate reductase (DHFR). The aim of these studies is to understand, at the molecular level, some of the factors which control the specific interactions between the ligands and the protein. DHFR catalyses the reduction of dihydrofolate (I) to tetrahydrofolate (II) using NADPH as coenzyme.[16]

DHFR is an important enzyme in the cell because the product of its reaction, tetrahydrofolate, is required to produce thymidine precursors for DNA synthesis. Antifolate drugs act by inhibiting the enzyme in malignant or parasitic cells and include several clinically useful agents such as trimethoprim (antibacterial), methotrexate (antineoplastic) and pyrimethamine (antimalarial). Trimethoprim, for example, binds to bacterial DHFR three thousand times more tightly than it does to the mammalian enzyme. Clearly, understanding the specificity of such binding could have implications for improved drug design. We have carried out extensive NMR studies on complexes of *L.casei* DHFR with substrates and antifolate drugs and by using this information in conjunction with available crystallographic data[17] we have been able to examine the specificity of binding, the presence of multiple conformational states and various dynamic processes in the complexes.[18-29] The relatively small size of the protein, (Mr 18,300), and the availability of isotopically labelled samples and modern 2D and 3D NMR techniques have allowed us to obtain extensive ^1H, ^{15}N and ^{13}C resonance assignments for several protein-ligand complexes.

The *L.casei* strain grows on a defined medium of amino acids and this is convenient for biosynthetic incorporation of selectively labelled amino acids. We also have available an over-expressing strain of *E.coli* into which the *L.casei* DHFR gene has been cloned. This grows on minimal media containing ammonium salts and glucose as the major sources of nitrogen and carbon respectively: it is ideal for biosynthetic incorporation of $^{15}N/^{13}C$ (using ^{15}N-ammonium chloride and uniformly ^{13}C-labelled glucose).

2.1 Assignment of Protein Resonances

Our most detailed assignment studies have been on the complex of *L.casei* DHFR with methotrexate in aqueous solution. There are more than 800 different types of hydrogen nuclei in the complex and these give rise to more

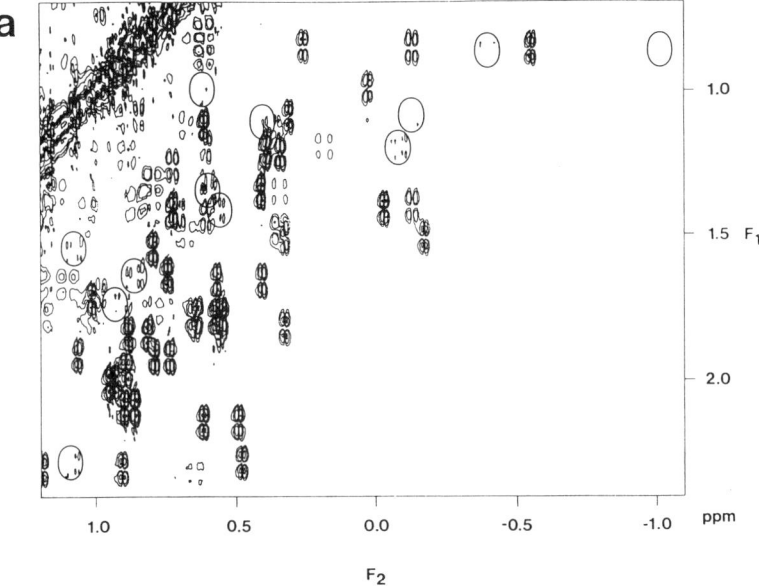

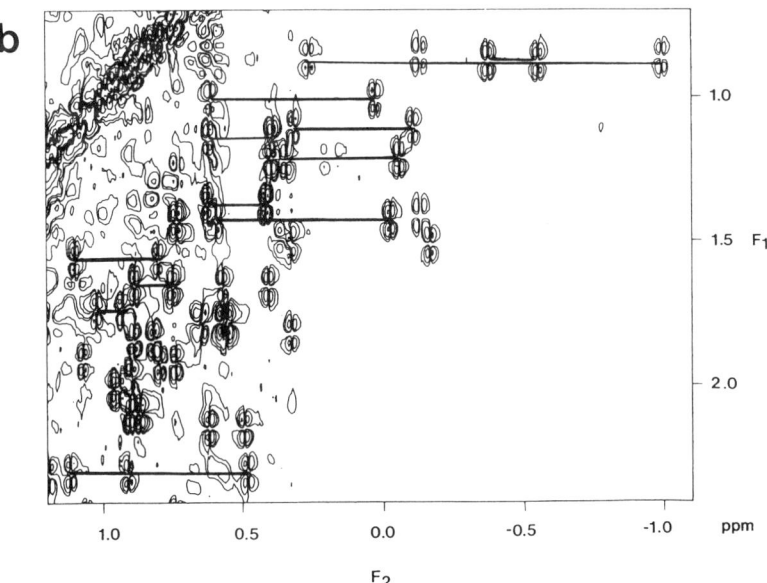

Fig.1. The high field 'aliphatic' region of the 2D DQF.COSY ^{1}H contour plot for the dihydrofolate reductase.methotrexate complex recorded at 308K (a) selectively deuterated enzyme incorporating $(2.S,4R)[5,5,5-^2H_3]$leucine (**1**) (b) non-deuteriated enzyme. The positions of the leucine cross-peaks involving the 4-pro-*R* methyl group are circled in (a) and the methyl pairs from each leucine are joined with a line in (b).
(Used with permission from Ostler *et al.*, 1993[33]).

TABLE II

NMR Studies on *L.casei* DHFR samples containing isotopically labelled amino acids

Sample	Experiment
Deuteriated DHFRs	
[^{2}H-αCH]-valine	2D COSY experiments used to identify α-CH ^{1}H chemical shifts in DHFR.MTX complex (46)
[^{2}H$_6$-γCH$_3$]-valine	1D and 2D COSY ^{1}H experiments used to identify Val γ-CH$_3$ ^{1}H chemical shifts in DHFR.TMP complex (32)
[^{2}H$_3$-4-Pro-R CH$_3$]-leucine	2D COSY ^{1}H experiments used to make stereospecific assignments for 4-Pro-R and 4-Pro-S-leucine CH$_3$ resonances (33)
[^{2}H$_2$-3′,5′]-Tyr plus His, Trp and Phe with all aromatic protons replaced by ^{2}H	^{1}H NMR experiments used to identify Tyr 2′,6′-proton signals in complexes of DHFR with folate, folinic acid, MTX, TMP, 2,4-diaminopyrimidine, p-nitrobenzoyl-L-glutamate (5)

TABLE II / cont.

Selectively deuterated aromatic amino acids (I to IV in Fig.2)	1D ^{1}H experiments for assignments of Tyr 2',6'-^{1}H, Phe-2',3',5',6'-^{1}H and His 4'-^{1}H signals 2D COSY spectrum used to identify Phe 2',3',5',6'-^{1}H signals 2D NOESY experiments used to identify unambiguous NOEs in spectra simplified by deuteriation (35,36)
Uniformly ^{15}N-labelled DHFR	3D TOCSY.HMQC, NOESY.HMQC and HMQC.NOESY.HMQC experiments used to simplify the spectra in the amide NH region (13,30). Possible to make backbone NH and α CH signal assignments using sequential assignment methods for DHFR.MTX complex (30).
Uniformly ^{15}N/^{13}C-labelled DHFR	3D HNCA experiments to determine backbone NH and α CH assignments using scalar coupling connections only: assignments determined for 158 of 162 residues (41). 3D HCCH.COSY and HCCH.TOCSY experiments used to determine the sidechain resonance assignments for most of residues in DHFR.MTX complex (31).

than 3000 detectable cross-peaks in the 2D NOESY spectrum. By using isotopic labelling techniques in combination with 2D and 3D NMR experiments it has proved possible to obtain almost all of the protein signal assignments for the DHFR-MTX complex.[30,31] In the earlier stages of the assignment work several selectively deuteriated DHFRs were prepared and examined: details of these are summarised in Table II.

2.1.1 *Assigning Protein Signals to Amino Acid Residue Types*

The first stage in the assignment procedure is to assign the signals to amino acid residue types (eg. Phe, Leu, Val). Selective deuteriation is a useful method for making such assignments unambiguously in large proteins. For example, by examining the NMR spectra of DHFR prepared with γ-Me deuteriated valine we could assign the valine methyl ^{1}H resonances.[32] Stereospecific assignments have also been made using this approach by examining stereospecifically deuteriated amino acids incorporated into the protein. This is illustrated in Fig.1 which shows the 2D COSY spectra for the methotrexate complex formed with the normal enzyme (Fig.1a) and with the enzyme containing [4-Pro-R-CH$_3$]-deuteriated leucine (Fig.1b).[33] Comparison of the spectra reveals the positions of the cross-peaks which are present in the spectrum of the complex formed with the normal enzyme and absent in the spectrum of the complex with the deuteriated protein. The stereospecific methyl assignments for 12 of the 13 leucines were made in this way (for one of the leucines the two methyl groups have protons with the same chemical shift): subsequently, specific assignments to particular leucine residues were made using 3D HCCH.COSY and HCCH.TOCSY experiments.[31]

In other experiments we have obtained spectral simplification by incorporating several deuteriated amino acids simultaneously into the protein. For example, by examining a DHFR sample where all aromatic ring protons except those for the Tyr 2',6'-protons had been replaced by deuterium one could easily assign the Tyr 2',6'-proton signals.[5] Preparing deuteriated aromatic amino acids is relatively straightforward.[34]

In one of the deuteriated DHFR samples we have introduced fully deuteriated tryptophan and partially deuteriated tyrosine, histidine and phenylalanine (see Fig.2).[35,36] In this sample, phenylalanine is the only aromatic amino acid which will give rise to signals showing scalar couplings and these can be easily detected in the 1D (Fig.2) and 2D COSY spectrum (Fig.3). This mixture of amino acids was chosen to have sufficient deuteriation to simplify the spectra but to retain enough protons to provide NOE information for assignment purposes. Fig.4 shows a comparison of part of the 2D NOESY spectra of the selectively deuteriated and normal enzyme in their complexes with methotrexate. The dramatic simplification in the spectrum is obvious. This can be illustrated by considering the NOE cross-peaks involving the Leu 113 methyl signal (indicated on spectrum). In the NOESY spectrum of the

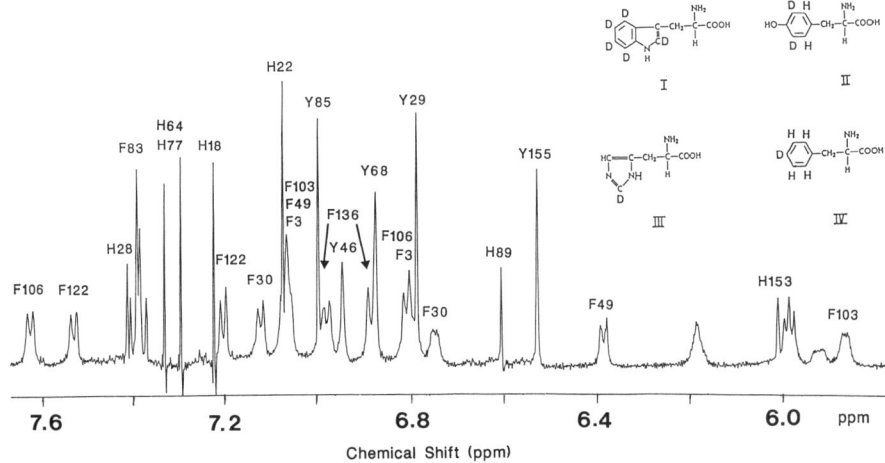

Fig.2. The aromatic region of the resolution-enhanced ^{1}H spectrum of the selectively deuteriated dihydrofolate reductase.methotrexate complex recorded at 600 MHz. 1.3 mM solution in D$_2$O buffer.
(Used with permission from Birdsall et al., 1990 [35]).

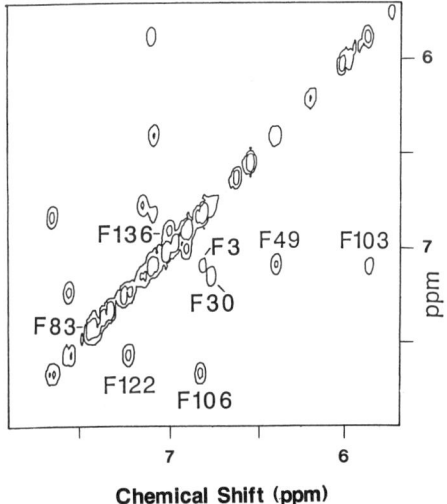

Fig.3. The aromatic region of the absolute-value ^{1}H-COSY spectrum for the dihydrofolate reductase.methotrexate complex prepared from the selectively deuteriated enzyme incorporating the selectively deuteriated amino acids (I-IV). 500 MHz at 308 K and pH 6.5.
(Used with permission from Birdsall et al., 1990[35]).

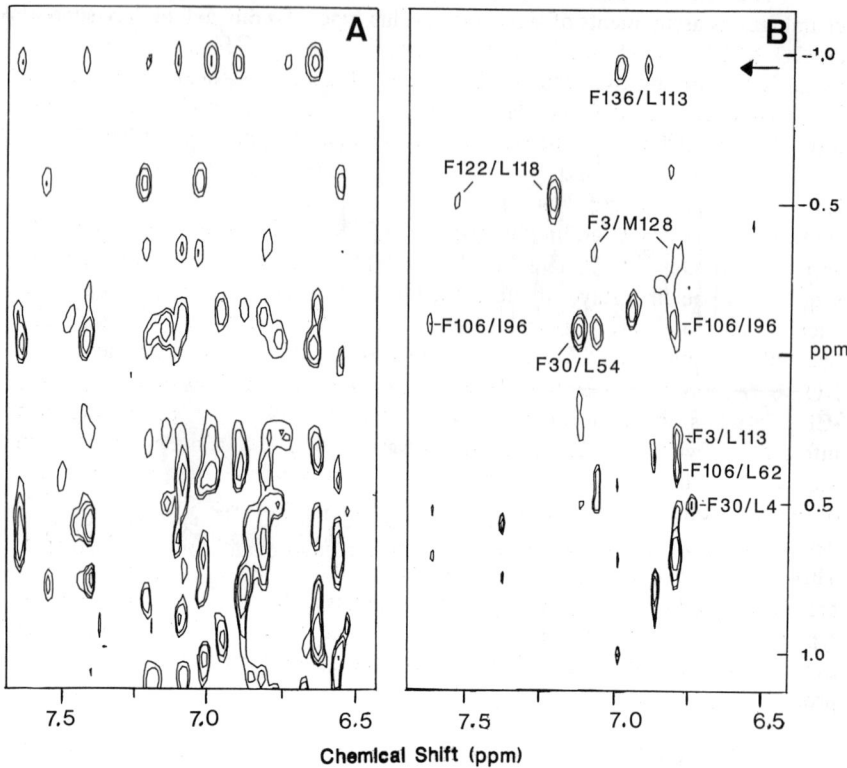

Fig.4 Part of the ^{1}H-NOESY spectrum of the dihydrofolate reductase.methotrexate complex. Spectrum (A) was obtained with non-deuteriated enzyme while spectrum (B) is the same region obtained with the deuteriated enzyme.
(Used with permission from Birdsall et al., 1990[35]).

complex formed with the normal enzyme these methyl protons give rise to many cross-peaks with aromatic protons but in the simplified spectrum from the selectively deuteriated sample, only one set of NOE connections to aromatic protons is detected: these are from the δ_1 and δ_2 protons of Phe 136.

The selective deuteriation experiments discussed here allow us to make unambiguous assignments of signals to residue type. To this extent they support the 2D NMR methods traditionally used for assigning signals to residue type based on identifying characteristic COSY and TOCSY cross-peak patterns based on scalar interactions.[37] For large proteins, the latter methods become more difficult to apply and the deuteriation approach becomes more important.

2.1.2 Backbone Assignments

Wüthrich[37] has described how one can use COSY and NOESY experiments to assign the backbone α-CH and NH resonances to specific residues in the protein sequence. The first stage of the procedure is to assign all the sidechain and backbone protons according to residue type (for example, Gly, Val, Ala). The backbone proton signals are then assigned to specific residues by identifying NOE connections between the NH proton of one residue and the NH, α-CH or β-CH$_2$ protons of adjacent residues in the sequence and combining this information with knowledge of the independently determined amino acid sequence. At this stage, the observed NOE patterns can be interpreted in terms of the common secondary structure elements found in proteins and thus used to determine where the helical and β-sheet regions are located in the sequence. The NOESY spectra also provide information about all pairs of protons which are close together (≤ 0.5 nm) in the molecule and distance constraints of this type can be used in conjunction with distance geometry or restrained molecular dynamics calculations to calculate the full three dimensional structure of a small protein.[38-40]

Using Wüthrich's approach, it proved possible to make sequential assignments for about 40% of the residues in *L.casei* DHFR. In order to obtain all the backbone assignments for a protein of this size it is necessary to resort to ^{15}N and ^{13}C labelling of the protein using samples uniformly labelled with 95% ^{15}N and ^{13}C. Table II summarises the experiments carried out on these labelled proteins. Using uniformly ^{15}N labelled DHFR in combination with 3D TOCSY.HMQC and NOESY.HMQC experiments, it was possible to make backbone NH and α-CH ^{1}H assignments for 146 of the 162 residues in dihydrofolate reductase.[30] These 3D experiments are analogues of ^{1}H-^{1}H COSY, TOCSY and NOESY experiments designed to give spectra where the signals are dispersed into a third dimension according to the ^{13}C or ^{15}N chemical shift frequencies. This approach provides a dramatic simplification in the spectra by reducing the overlap of signals. For example, the 3D NOESY.HMQC spectrum obtained from a uniformly ^{15}N labelled protein

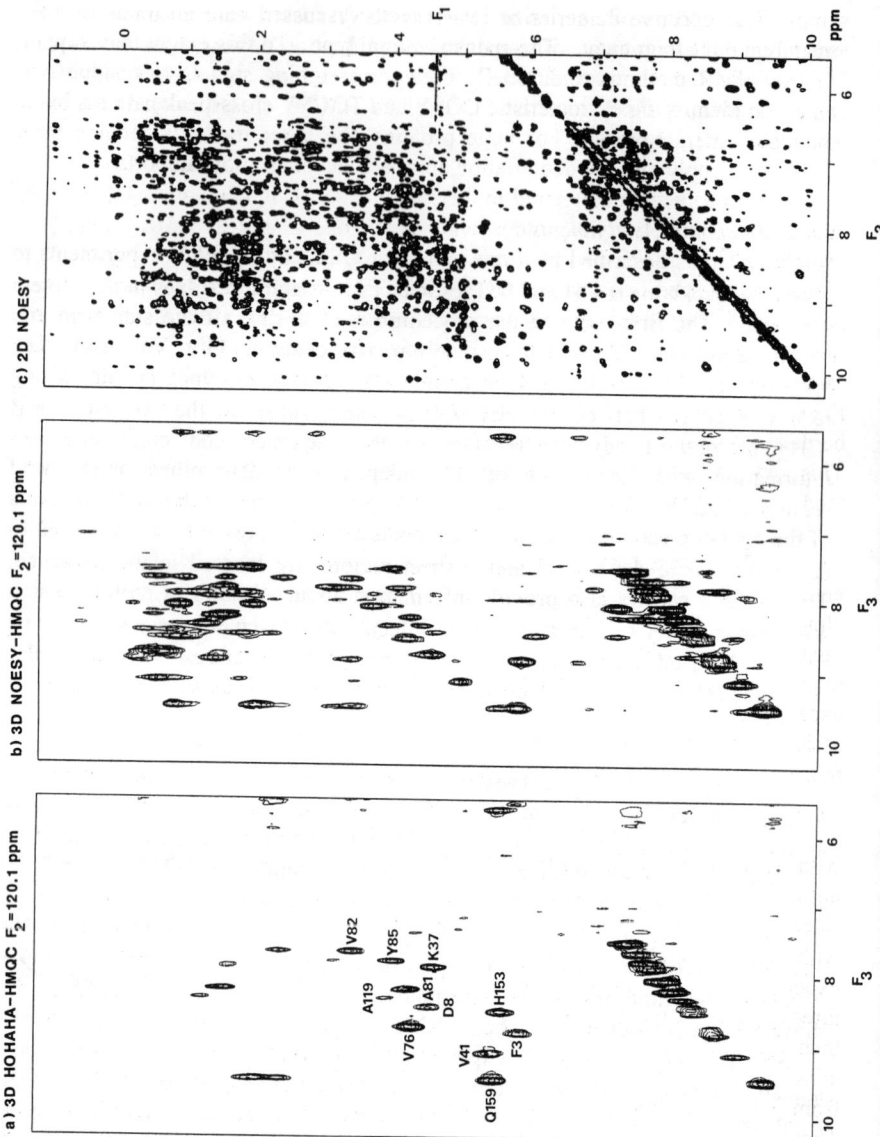

Fig.5 Comparison between a 2D NOESY spectrum and identical region from F_1-F_3 cross sections of 3D ^{15}N/^{1}H NOESY-HMQC and HOHAHA-HMQC experiments for the uniformly ^{15}N-labelled DHFR-MTX complex. In the F_2 (^{15}N) slice from HOHAHA-HMQC spectrum the assignments are given for the NH-α-CH cross peaks. The F_1-F_3 cross sections shown are among the most complicated but still contain cross peaks from only 11 backbone amide protons, with no NH shift degeneracy. The 3D NOESY-HMQC slice is clearly much simpler and more amenable to analysis than the corresponding region from the 2D NOESY spectrum.
(Used with permission from Carr et al., 1991[30]).

sample is composed of a series of ^{1}H-^{1}H NOESY spectra each characterised by a different ^{15}N frequency. The spectral simplification achieved can be seen in Fig.5 where the complex 2D NOESY spectrum from unlabelled DHFR is compared with a slice from the 3D NOESY.HMQC spectrum from the ^{15}N labelled protein.

One problem encountered in the 3D NOESY.HMQC method is that NOE connections between NH protons with degenerate chemical shifts cannot be detected. This problem has been overcome[13,14] by applying a modified sequence (referred to as an HMQC.NOESY.HMQC sequence) such that cross-peaks are characterised by only one ^{1}H frequency and two ^{15}N frequencies. In this case, NOE connections can be detected between NH protons which have the same ^{1}H chemical shift but different ^{15}N chemical shifts. This is illustrated in Fig.6 where the NH - NH NOE connections between His 28 and Tyr 29 could be detected despite their having almost degenerate NH ^{1}H chemical shifts.[13,30] This approach is particularly useful for helical proteins where the NH protons tend to resonate over a smaller chemical shift range.

More recent experiments, using the uniformly ^{13}C/^{15}N labelled DHFR in combination with 3D HNCA experiments,[7,15] have enabled us to obtain an almost complete list of backbone assignments for the complex of DHFR with methotrexate.[31] The latter method gives cross-peaks which depend on scalar interactions involving backbone ^{13}C and ^{15}N nuclei and proves to be an excellent method for making reliable backbone assignments. Backbone assignments for 158 of the 162 residues in DHFR have been obtained by using this approach.[41]

The ^{15}N/^{13}C enriched DHFR sample has also been used in combination with 3D HCCH.COSY and HCCH.TOCSY experiments[8,9] (the ^{15}N labelling is not required for these experiments) to provide an essentially complete list of the sidechain resonance assignments for ^{1}H and ^{13}C nuclei in the complex of DHFR with methotrexate (31). In the HCCH.COSY and HCCH.TOCSY experiments[8,9] the magnetisation is first transferred from a proton onto its directly bound carbon by an INEPT type pulse sequence. The magnetisation is then transferred to the adjacent carbon via the carbon-carbon scalar coupling constant in a COSY or TOCSY type transfer and ultimately it is transferred from the carbon onto its directly bonded proton by a reverse INEPT experiment. These experiments give very efficient transfer of the magnetisation at all stages of the process because of the large ^{13}C-^{1}H and ^{13}C-^{13}C coupling constants and they can give useful correlation information for proteins as large as 30kDa.[10,42] The cross-peaks in these spectra are characterised by one ^{13}C frequency and two ^{1}H frequencies, the latter defining the normal COSY or TOCSY experiment, depending on the sequence used (see Fig.7).

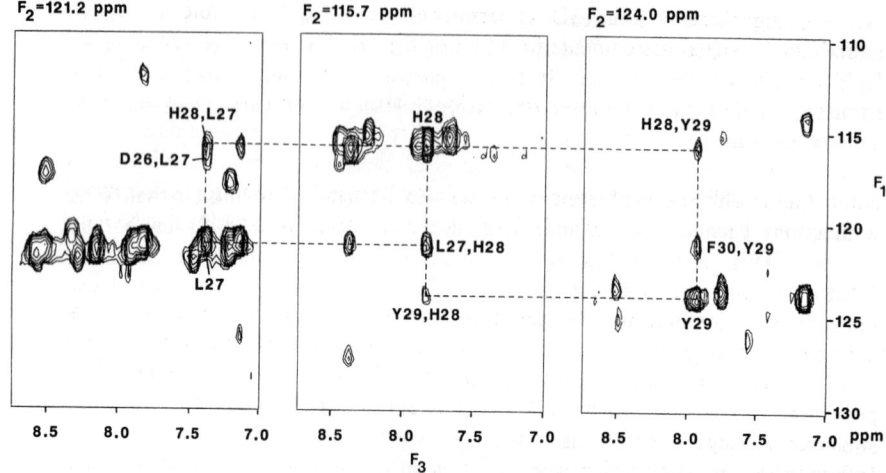

Fig. 6 F_2 (^{15}N) slices from a (^{15}N/^{1}H) HMQC-NOESY-HMQC spectrum of DHFR-MTX taken at the amide ^{15}N shifts of Leu 27, His 28 and Tyr 29. The labelled cross peaks arise from the sequential NH-NH NOEs linking neighbouring residues in the sequence. In the HMQC-NOESY-HMQC experiment, cross peaks arising from NH-NH NOEs are characterized by the ^{15}N shifts of both amide nitrogens and the ^{1}H shift of the amide proton to which NOE transfer has occurred. Hence, the experiment can detect NOEs between NHs with identical or quite similar ^{1}H shifts.
(Used with permission from Carr et al., 1991[30]).

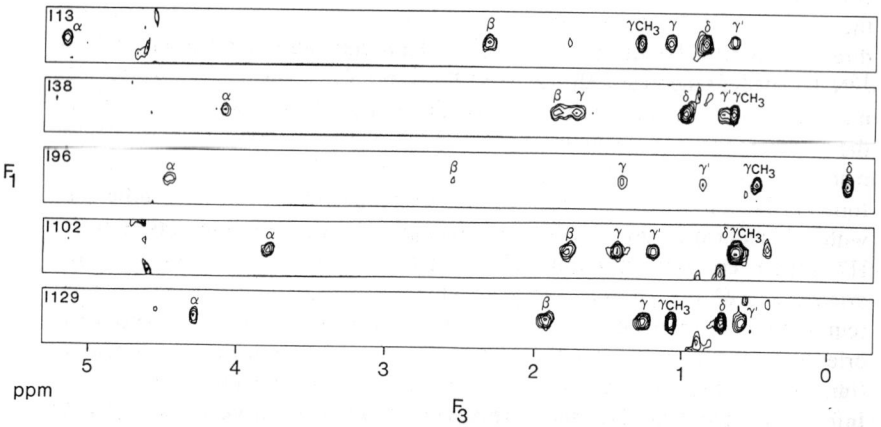

Fig. 7 A series of F_1/F_3 strips from the HCCH-TOCSY spectrum of the DHFR-MTX complex taken at the δ carbon shifts (F_2) of I13 (15.11 ppm), I38 (16.23 ppm), I96 (11.20 ppm), I102 (10.64 ppm) and I129 (16.79 ppm). The labelled cross peaks correspond to through bond correlations spanning the full length of the isoleucine sidechains, from the δ methyl protons (F_1) to γ methyl, γ, γ', β and α protons (F_3).
(Used with permission from Soteriou et al., 1993[31]).

Using these various 3D experiments we have been able to obtain unambiguous signal assignments for most of the ^{1}H, ^{13}C and ^{15}N nuclei in the DHFR.MTX complex:[31] the secondary structure has been defined[30] and detailed calculations of the three dimensional structure are currently underway. The availability of an essentially complete list of resonance assignments is a basic prerequisite for any structural studies on the enzyme in its various complexes. The assigned protein signals can be used to monitor interactions with assigned nuclei in the ligands from measurements of intermolecular NOEs.

2.1.3 Assignment of Ligand Resonances

For ligands which bind very tightly to the enzyme ($K_a > 10^8$ M^{-1}), the classical method of making the resonance assignments is to use isotopically labelled analogues (^{2}H, ^{3}H, ^{13}C, ^{15}N) in combination with various experimental NMR procedures. Table III indicates the various isotopically labelled substrate and substrate analogues which have been examined in complexes with DHFR. Information about ionisation states, interactions, multiple conformations and dynamic processes in the complexes have been extracted from the spectra: details of the results of such studies are included in Table III. Selectively deuteriated ligands can help in making ^{1}H resonance assignments if we compare ^{1}H spectra of the complexes formed with deuteriated and non-deuteriated ligands. We have used this approach to make the assignments of the NADP$^+$ nicotinamide ring protons in complexes of DHFR.NADP$^+$ formed with various selectively deuteriated NADP analogues.[43] A more direct method is to use ^{3}H NMR to examine tritiated ligands bound to DHFR. For example, we have examined ^{3}H-7,3',5'-folic acid in its complexes with DHFR and NADP$^+$ and confirmed the presence of interconverting conformational states by monitoring the assigned ^{3}H-7 tritium resonances nuclei (the tritium chemical shifts are directly related to ^{1}H chemical shifts).[44] DHFR complexes formed with ^{13}C or ^{15}N labelled ligands can be examined by direct detection of the ^{13}C or ^{15}N nucleus. Because of the low natural abundance of these nuclei, the signals from the nuclei at enriched positions are easily detected and the assignments can be made directly. Using these approaches, three pH-dependent conformational forms of bound folate have been detected in the complex of labelled folic acid with NADP$^+$ and *L.casei* DHFR.[29,45] Intermolecular NOEs between the folate H7 proton and protein protons have provided evidence for two different orientations of the pteridine ring. One of these is the correct orientation required for the catalytic reduction (the active conformation), whereas the other orientation is similar to that found in complexes of DHFR with methotrexate (turned-over by 180° compared with that of the active conformation). Information about the tautomeric and ionisation states of the folate pteridine rings in the different forms has also been obtained from studies of ^{13}C chemical shifts measured for bound ^{13}C-labelled folates.[45]

For ligands which have protons directly attached to ^{13}C or ^{15}N it is possible to detect selectively the protons attached to the labelled nucleus by

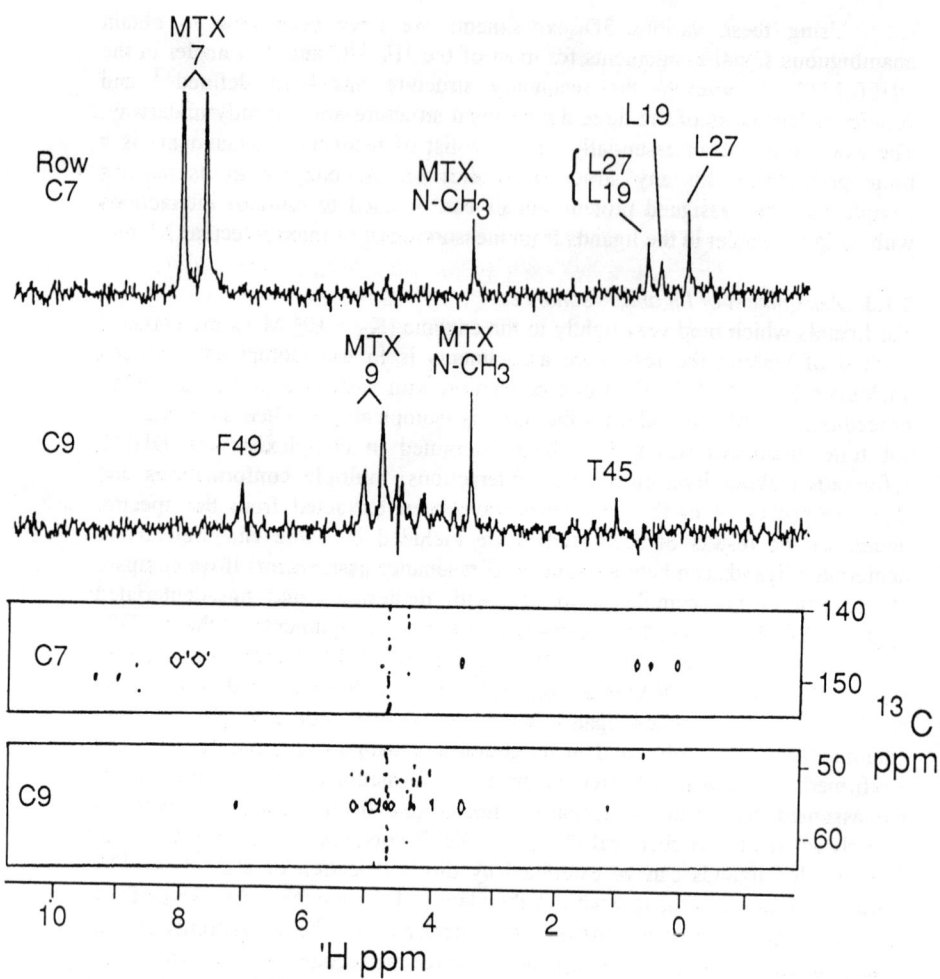

Fig.8 Part of the two-dimensional ^{1}H-^{13}C HMQC-NOESY experiment of the binary complex of *L. casei* DHFR with [4,7,-8a,9-^{13}C]methotrexate at 308 K and pH 6.5 showing the regions containing C7 and C9. The upper traces are the rows at the frequency of C7 and C9 showing the connections to their attached protons and NOEs to signals to nearby residues. The H7 signal appears as a doublet because no ^{13}C decoupling in F_2 was used.
(Used with permission from Cheung *et al.*, 1993[45]).

using appropriate pulse sequences (for example, HMQC experiments). 2D and 3D extensions of these experiments (for example, NOESY.HMQC) allow selective detection of the NOEs between ligand protons on ^{13}C or ^{15}N and neighbouring protons in the protein.[12] This is particularly valuable in cases where the protein resonance assignments are known since the intermolecular NOEs provide crucial distance constraints between the nuclei on the ligand and those on the protein. This is illustrated in Fig.8 which shows the HMQC.NOESY spectrum of ^{13}C labelled methotrexate at C7 in its complex with DHFR:[45] the NOEs measured from the pteridine H7 proton to CH$_3$ protons of Leu 19 and Leu 27 of DHFR confirm that the methotrexate pteridine ring has the same orientation in solution as that observed in the crystal structure studies of Bolin and coworkers[17] on the complex *L.casei* DHFR.NADPH.MTX. Clearly, the combined use of labelled ligands and multi-dimensional NMR techniques provides a powerful means of measuring intermolecular NOEs in protein-ligand complexes.

2.1.4 *Multiple Conformations*
NMR is a very powerful method for detecting multiple conformations and several examples have been encountered in our studies on DHFR complexes (see Table III). It is important to characterise such multiple conformations because each conformation potentially offers a new starting point for inhibitor design. It is worth noting that the presence of multiple conformations will also complicate any study of structure / activity relationships. In several cases the observed multiple conformations result from a flexible ligand binding in different ways but essentially occupying the same binding site. The most notable examples of this are provided by the ^{19}F studies on fluoronitropyrimethamine where two rotational isomers have been found to bind to the enzyme[52] and the multinuclear NMR studies on the complexes of folate with DHFR[43-45] discussed above.

2.1.5 *Dynamic Processes*
In Table III it is seen that NMR can provide detailed information about various dynamic processes within the complexes. For example, in the case of the complex of DHFR with trimethoprim it has been possible to show that the benzyl ring flips 125 times during the lifetime of the complex (0.5s) at 298 K. Furthermore, the hydrogen bond between the trimethoprim N1 proton and the Asp 26 carboxylate is broken and reformed 17 times during the lifetime of the complex.[26] These insights into the detailed dynamic behaviour of the complex were obtained from line shape analysis of ^{13}C and ^{15}N signals from isotopically enriched trimethoprim complexed to the enzyme.

TABLE III

NMR studies of *L.casei* DHFR with isotopically labelled ligands

Sample	Experiment
Trimethoprim analogues	
[2,5-^{13}C]-trimethoprim	^{13}C-2 chemical shift shows that N1 of TMP is protonated when bound to DHFR (47).
[4,6-^{13}C]-trimethoprim	Multiple ^{13}C signals show presence of two conformations in DHFR.TMP.NADP$^+$ complex (48).
[7,4'-methoxy-^{13}C]-trimethoprim	^{13}C-7 T$_1$ relaxation studies on DHFR.TMP give correlation time for protein (14 x 10^{-9} sec) and ^{13}C-methoxy T$_1$ indicates dihedral angle changes of ± 25° to ± 35° on sub-nanosecond time scale (26). 2D NOESY.HMQC experiments used to obtain ^{1}H-^{1}H NOEs between ligand and protein (Birdsall *et al.*, unpublished results).
[3',5'-methoxy-^{13}C]-trimethoprim	Analysis of ^{13}C line shapes as function of temperature gives rates of ring flipping (250s^{-1} at 298K) in DHFR.TMP complex (26).

TABLE III / cont.

Sample	Experiment
[3',5'-methoxy-^{13}C]-brodimoprim	^{13}C line shape analysis as a function of temperature gives rate of ring flipping (48).
[1,3,2-amino-^{15}N]-trimethoprim	N-1 ^{15}N chemical shift for DHFR.TMP complex indicates that N1 is protonated (80 ppm difference between protonated and non-protonated states) (49). N-1 proton signal detected and line width measurements in DHFR.TMP give rates of breaking and reforming hydrogen bond (26,49). 2D HMQC.NOESY gives ^{1}H-^{1}H NOEs between protein and ligand (Martorell, Birdsall and Feeney, unpublished results).
Methotrexate analogues [2,4a,6-^{13}C]-methotrexate	C-2 ^{13}C chemical shifts show that MTX N1 is protonated in DHFR.MTX complex (50,51).
[4,7,8a,9-^{13}C]-methotrexate	2D HMQC.NOESY gives ^{1}H-^{1}H NOEs between ligand and protein in DHFR.MTX complex (45).

TABLE III / cont.

Sample	Experiment
[3',5'-difluro]-methotrexate	Analysis of ^{19}F line shape as a function of temperature gives rates of ring flipping in DHFR.MTX (7 x 10^{3}s^{-1} at 298 K) and DHFR.MTX.NADPH (20 x 10^{3}s^{-1} at 298 K) complexes (28).
Folic acid analogues	
[4,6,8a-^{13}C]-folic acid	Multiple ^{13}C signals shows three pH dependent conformations in DHFR.Folate and DHFR.Folate.NADP$^+$ complexes (45). Tautomeric and ionisation states of bound folate determined from ^{13}C chemical shifts (45).
[2,4a,7,9-^{13}C]-folic acid	As above. 2D HMQC.NOESY gives ^{1}H-^{1}H NOEs for H7: these characterise two different orientations of pteridine ring (45).
[7,9,3',5'-^{3}H]-folic acid	^{3}H NMR shows multiple signals in DHFR.Folate.NADP$^+$ complex confirming multiple conformations (44).

TABLE III / cont.

Sample	Experiment
Pyrimethamine analogues [3'-nitro-4'-fluoro]-pyrimethamine	Multiple ^{19}F signals in complex with DHFR indicates two bound conformations corresponding to rotational isomers. Ternary complex with NADP$^+$ changes specificity of pyrimethamine rotamer binding (52).
Coenzyme analogues [^{13}CO]-NADP$^+$	Differences in ^{13}C line widths between free and bound NADP$^+$ in DHFR.NADP$^+$ complex gives dissociation rate constant (55 ± 10 s^{-1} at 283 K) (53). Three pH dependent conformations of the DHFR.Folate.NADP$^+$ complex were first detected using this ligand (29).
Selectively deuteriated coenzymes [4-^{2}H]-NADP$^+$ [6-^{2}H]-NADP$^+$	Selectively deuterated analogues allow assignments of corresponding protons in complexes with DHFR (43).
(4R)-[4-^{2}H$_1$]-NADPH	^{1}H NMR on tetrahydrofolate produced by reduction of folate with DHFR and selectively deuteriated NADPH used to determine stereochemistry of reduction (54,55).

3 CONCLUSION

The wide range of NMR studies involving complexes formed using isotopically labelled dihydrofolate reductase and its inhibitors described here illustrates the enormous potential of this approach. The spectral simplification provided by these methods often allows us to extract detailed information about interactions, conformations and dynamic processes within the complexes.

REFERENCES

1. V L Hsu and I M Armitage, *Biochemistry*, 1992, **31**, 12778.
2. D Le Master, *'Methods in Enzymology'*, N J Oppenheimer and T L James, 1989.
3. H L Crespi and J J Katz, *Nature*, 1969, **224**, 560.
4. J L Markley, I Putter, and O Jardetzky, *Science*, 1968, **161**, 1249.
5. J Feeney, G C K Roberts, B Birdsall, D V Griffiths, R W King, P Scudder, and A S V Burgen, *Proc Roy Soc London B*, 1977, **196**, 267.
6. C H Arrowsmith, J Carey, L Treat-Clemons, and O Jardetzky, *Biochemistry*, 1989, **28**, 9610.
7. M Ikura, L E Kay, and A Bax, *Biochemistry*, 1990, **29**, 4659.
8. A Bax, G M Clore, P C Driscoll, A M. Gronenborn, M Ikura, and L E Kay, *J Magn Reson*, 1990, **87**, 620.
9. S W Fesik and E R P Zuiderweg, *Quart Rev Biophys*, 1990, **23** (2), 97.
10. G M Clore and A M Gronenborn, *Progress in NMR Spectroscopy*, 1991, **23**, 43.
11. A M Gronenborn, A Bax, P T Wingfield, and G M Clore, *FEBS Letters*, 1989, **243**, 93.
12. S W Fesik, E R P Zuiderweg, E T Olejniczak, and R T Gampe Jr, *Biochem Pharmacol*, 1990, **40**, 161.
13. T A Frenkiel, C J Bauer, M D Carr, B Birdsall, and J Feeney, *J Magn Reson*, 1990, **90**, 420.
14. M Ikura, A Bax, G M Clore, and A M Gronenborn, *J Am Chem Soc*, 1990, **112**, 9020.
15. L E Kay, M Ikura, R Tschudin, and A Bax, *J Magn Reson*, 1990, **89**, 496.
16. R L Blakley, *'Folates and Pterins'*, Wiley, New York, 1985.
17. J T Bolin, D J Filman, D A Matthews, R C Hamlin, and J Kraut, *J Biol Chem*, 1982, **257**, 13650.
18. J Feeney, 'NMR Spectroscopy in Drug Research', Alfred Benzon Symposium No 26, Copenhagen, 1988.
19. J Feeney, *Biochem Pharmacol*, 1990, **40**, 141.
20. J Feeney, *'Bioorganic Chemistry in health care and technology'*, Plenum Press, 1991.

21 J Feeney, 'Proceedings of Palio Workshop on Molecular Structure, Dynamics and Recognition', 1991.
22 J Feeney and B Birdsall, *'NMR of Macromolecules: A practical approach'*, Oxford University Press, 1993.
23 D J Antonjuk, B Birdsall, A S V. Burgen, H T A Cheung, G M Clore, J Feeney, A Gronenborn, G C K Roberts, and W Tran, *Brit J Pharmacol*, 1984, **81**, 309.
24 B Birdsall, J Feeney, C Pascual, G C K Roberts, I Kompis, R L Then, K Muller, and A Kroehn, *J Med Chem,* 1984, **23**, 1672.
25 A Gronenborn, B Birdsall, E I Hyde, G C K Roberts, J Feeney, and A S V Burgen, Molecular Pharm, 1981, 20, 145.
26 M S Searle, M J Forster, B Birdsall, G C K Roberts, J Feeney, H T A Cheung, I Kompis, and A J Geddes, *Proc Natl Acad Sci USA*, 1988, **85**, 3787.
27 S J Hammond, B Birdsall, M S Searle, G C K Roberts, and J Feeney, *J Mol Biol,* 1986, **188**, 81.
28 G M Clore, A M Gronenborn, B Birdsall, J Feeney, and G C K Roberts, *Biochem J*, 1984, **217**, 659.
29 B Birdsall, J Feeney, S J B Tendler, S J Hammond, and G C K Roberts, *Biochemistry,* 1989, **28**, 2297.
30 M D Carr, B Birdsall, J Jimenez-Barbero, V I Polshakov, C J Bauer, T A Frenkiel, G C K Roberts, and J Feeney, *Biochemistry*, 1991, **30**, 6330.
31 A Soteriou, M D Carr, T A Frenkiel, J E McCormick, C J Bauer, D Sali, B Birdsall, and J Feeney, *J Biomol,* 1993, **3**, 535.
32 M S Searle, S J Hammond, B Birdsall, G C K Roberts, J Feeney, R W King, and D V Griffiths, *FEBS Letters*, 1986, **194**, 165.
33 G Ostler, A Soteriou, C M Moody, J A Khan, B Birdsall, M D Carr, D W. Young, and J Feeney, *FEBS Letters*, 1993, **318**, 177.
34 D V Griffiths, J Feeney, G C K. Roberts, and A S V Burgen, *Biochim Biophys Acta,* 1976, **446**, 479.
35 B Birdsall, J R P Arnold, J Jimenez-Barbero, T A Frenkiel, C J Bauer, S J B Tendler, M D Carr, J A Thomas, G C K Roberts, and J Feeney, *Eur J Biochem,* 1990, **191**, 659.
36 J Feeney, B Birdsall, J Akiboye, S J B Tendler, J Jimenez-Barbero, G Ostler, J R P Arnold, G C K Roberts, A Kühn, and K Roth, *FEBS Letters*, 1989, **248**, 57.
37 K Wüthrich, *'NMR of Proteins and Nucleic Acids'*, John Wiley and Sons, 1986.
38 T F Havel and K Wüthrich, *J Mol Biol,* 1985, **182**, 281.
39 R Kaptein, E R P Zuiderweg, R M Scheek, R Boelens, and W F van Gunsteren, *J Mol Biol*, 1985, **182**, 179.
40 G M Clore, A M Gronenborn, A T Brunger, and M Karplus, *J Mol Biol*, 1985, **186**, 435.

41   A Soteriou, M D Carr, C J Bauer, T A Frenkiel, B Birdsall, J E McCormick, and J Feeney, unpublished results.
42   S Grzesiek, H Döbeli, R Gentz, G Garotta, A M Labhardt, and A Bax, *Biochemistry*, 1992, **31**, 8180.
43   B Birdsall, A M Gronenborn, E I Hyde, G M Clore, G C K Roberts, J Feeney, and A S V Burgen, *Biochemistry*, 1982, **21**, 5831.
44   N Curtiss, S Moore, C Gibson, J R Jones, J Bloxidge, B Birdsall, and J Feeney, *Biochem J*, In press.
45   H T A Cheung, B Birdsall, T A Frenkiel, D D Chau, and J Feeney, *Biochemistry*, 1993, **32**, 6846.
46   J Feeney, B Birdsall, G Ostler, M D Carr, and M Kairi, *FEBS Letters*, 1990, **272**, 197.
47   G C K Roberts, J Feeney, A S V Burgen, and S Daluge, *FEBS Letters*, 1981, **131**, 85.
48   H T A Cheung, M S Searle, J Feeney, B Birdsall, G C K Roberts, L Kompis, and S J Hammond, *Biochemistry*, 1986, **25**, 1925.
49   A W Bevan, G C K Roberts, J Feeney, and I Kuyper, *Eur Biophys J*, 1985, **11**, 211.
50   L Cocco, J P Groff, C Temple Jnr, J A Montgomery, R E London, N S Matwiyoff, and R L Blakley, *Biochemistry*, 1981, **20**, 3972.
51   B Birdsall, J Andrews, G Ostler, S J B Tendler, W Davies, J Feeney, G C K Roberts, and H T A Cheung, *Biochemistry*, 1989, **28**, 1353.
52   B Birdsall, S J B Tendler, J R P Arnold, J Feeney, R J Griffin, M D Carr, J A Thomas, G C K Roberts, and M F G Stevens, *Biochemistry*, 1990, **29**, 9660.
53   J L Way, B Birdsall, J Feeney, G C K Roberts, and A S V Burgen, *Biochemistry*, 1975, **14**, 3470.
54   P A Charlton, D W Young, B Birdsall, J Feeney, and G C K Roberts, *J Chem Soc Chem Commun*, 1979, 922.
55   P A Charlton, D W Young, B Birdsall, J Feeney, and G C K Roberts, *J Chem Soc Perkin Trans 2*, 1985, 1349.

ACKNOWLEDGEMENTS

The work described here has involved the collaboration of many colleagues. In particular, I would like to acknowledge the collaborations with Berry Birdsall, Gordon Roberts, Mark Carr, Alice Soteriou, Andrew Cheung, Tom Frenkiel and Chris Bauer. I thank Andrew Lane and Tom Frenkiel for their comments on this article.

SEQUENTIAL ELECTRON TRANSFER AND BASE CATALYSIS IN THE N-DEALKYLATION OF AMINES BY CYTOCHROME P450 ENZYMES

F Peter Guengerich

*Department of Biochemistry and Center in Molecular Toxicology
Vanderbilt University Nashville, TN 37232-0146, USA.*

1   INTRODUCTION

The cytochrome P450 (P450) enzymes are the catalysts involved in the oxidation of the majority of drugs, pollutants, pesticides, and carcinogens as well as endogenous compounds such as steroids, alkaloids, and eicosanoids. The gene family is found throughout nature, and within a single species more than 40 different P450 proteins can be found.[1,2]

Most of the P450 enzymes are mixed-function oxidases; *ie*, they catalyze reactions of the stoichiometry

$$NAD(P)H + O_2 + R \rightarrow NAD(P)^+ + H_2O + RO$$

where R is a hydrocarbon, amine, sulfide, olefin, or other atom that can be oxidized. Elucidation of the catalytic mechanisms of these haemoproteins has been a subject of great interest. Biomimetic model studies have been of some use; the conclusions drawn from P450 research appear to apply to other monooxygenases. This report will consider a particular set of oxidation reactions catalyzed by P450s, the oxidation of amines.

2   SEQUENTIAL ELECTRON TRANSFER IN P450 REACTIONS

A number of lines of investigation support the view that the active form of P450 includes the formal entity $(FeO)^{3+}$, which transfers oxygen to a substrate.[3-6] In the case of the N-dealkylation reactions, carbinolamine intermediates are implicated from the results of ^{18}O labelling studies.[7]

Several results support the view that the oxidation of amines by P450s involves sequential 1-electron transfer. When 4-alkyl-1,4-dihydropyridines are oxidized by P450s, alkyl radicals can be trapped and characterized.[8]

Cycloalkylamines rearrange to give products rationalized by 1-electron chemistry;[9] further, even strained aliphatic hydrocarbons with low oxidation potentials show such rearrangements.[10] The low kinetic hydrogen isotopes seen in N-dealkylation reactions are not consistent with hydrogen atom abstraction but are compatible with 1-electron oxidation of the amine.[11,12] Rates of N-demethylation (V_{max}) of a series of p-substituted N,N-dimethylanilines can be compared with Hammett σ factors to obtain a ρ value of -0.6 with rat P450 2B1,[13] a value consistent with ring stabilization of a positively charged intermediate.

We extended the original study with the p-substituted N,N-dimethylanilines to a larger set and also measured the 1-electron $E_{1/2}$ for each compound. The parameter log k_{cat} was plotted against ΔG (calculated from $E_{1/2}$), and the data could be fitted to two derivatives of the basic Marcus equation, the Rehm-Weller equation:

$$\Delta G^{\ddagger} = \Delta G°/2 + [(\Delta G°/2)^2 + (\lambda/4)^2]^{1/2}$$

and Agmon-Levin equation:

$$\Delta G^{\ddagger} = \Delta G° + \lambda/4\ln2 \, \ln[1 + e^{-4\ln2(\Delta G°/\lambda)}]$$

Analysis of the fitted curves with either equations yielded values of ~23 kcal mol^{-1} for λ, the exchange factor, and ~1.8 V, the apparent $E_{1/2}$ for 1-electron oxidation of the N,N-dialkylamines by the putative $(FeO)^{3+}$ entity.[14]

Further considerations indicated that the apparent $E_{1/2}$ could be considered to be composed of two parts:

$$E_{1/2 \, app} = E_{1/2} (\text{intrinsic}) + E_{cf}$$

where the coulombic factor, E_{cf}, can be expressed as

$$E_{cf} = +14.4/r_{1,2} \cdot D \text{ (expressed in V)}$$

where the value +14.4 is derived from $(Z_1 - Z_2 - 1)e^2 f$, $r_{1,2}$ is the distance between the two sites in this outer sphere electron transfer, and D is the dielectric constant. Considerations of reasonable values for these parameters indicate that the P450 protein moiety could readily enhance the $E_{1/2, \, intrinsic}$ by +1.0 V.[14]

3 COMPARISON OF P450S AND PEROXIDASES

The P450s and peroxidases both catalyze amine dealkylations and are thought to use the same formal $(FeO)^{3+}$ moiety. Several mechanistic possibilities exist (Figures 1 and 2).

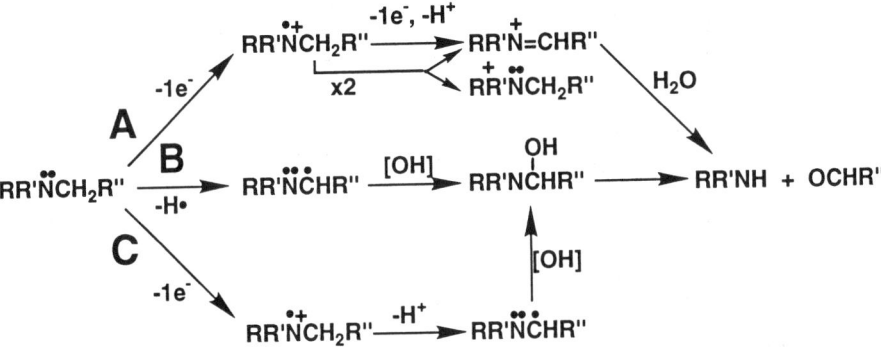

Figure 1: Possible mechanisms of N-dealkylation by haemoproteins. Path A is excluded for P450s on the basis of ^{18}O labelling studies.[7]

P450s show low kinetic hydrogen isotope effects in such reactions but many peroxidases do not.[11] We repeated the earlier findings and extended the types of approaches and studies with different enzymes (Table 1).[15]

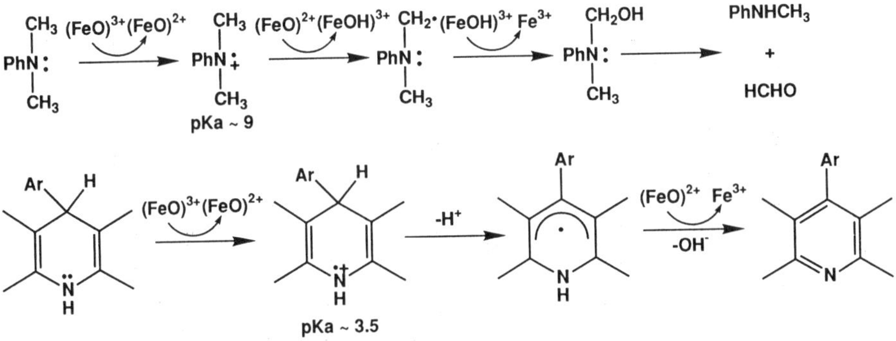

Figure 2: Postulated mechanisms of oxidation of PhNMe$_2$ and 1,4-dihydropyridines.[8,15,16]

Table 1: Kinetic hydrogen isotope effects for N-demethylation of $PhNMe_2$

System	V_{max} nmol product formed min^{-1} (nmol heme)$^{-1}$	D_V	K_m (μM)	$D_{(V/K)inter}$	$D_{(V/K)intra}$	$T_{(V/K)}$
Rat liver microsomes, NADPH	19.2 ± 0.5	1.1 (0.9-1.2)	15 ± 10	1.2 (0.9-1.6)	1.4 ± 0.1	2.2 ± 0.1
P450 2B1, reductase, NADPH	18.9 ± 0.2	1.0 (0.9-1.1)	45 ± 2	1.8 (1.2-2.5)	1.4 ± 0.1	3.4 ± 0.1
Chloroperoxidase, H_2O_2	$9,480 \pm 420$	1.0 (0.9-1.0)	144 ± 27	2.9 (2.2-4.1)	2.0 ± 0.1	3.2 ± 0.2
Prostaglandin H synthase, H_2O_2	57.1 ± 2.1	2.5 (2.3-2.6)	130 ± 13	5.9 (4.7-7.4)	3.5 ± 0.3	8.1 ± 0.1
Haemoglobin C_2H_5OOH	148 ± 12	2.4 (2.2-2.7)	287 ± 80	2.9 (1.8-4.9)	5.2 ± 0.2	13.6 ± 0.5
Horseradish peroxidase, C_2H_5OOH	$11,800 \pm 520$	6.7 (6.0-7.6)	824 ± 81	3.2 (2.2-4.6)	5.3 ± 0.1	12.7 ± 0.5
Horseradish peroxidase, H_2O_2	$15,800 \pm 1,040$	8.1 (7.1-9.1)	$1,310 \pm 140$	4.4 (2.7-6.0)	4.7 ± 0.1	8.4 ± 0.2
FeTPP-Cl, PhIO	4.7	1.2				1.6 ± 0.1
FeF$_5$TPP-Cl PhIO	10.7	2.8				2.9 ± 0.1
MnTPP-Cl, PhIO	8.0	2.1				1.8 ± 0.1
CrTPP-Cl, PhIO	0.7	2.2				2.9 ± 0.1

Table 2: Kinetic hydrogen isotope effects for oxidation of dihydropyridines

System	Substrate	V_{max} nmol product formed min^{-1} (nmol heme)$^{-1}$	D_V	K_m (μM)	$D_{(V/K)}$	$T_{(V/K)}$
Horseradish peroxidase, H_2O_2	Nifedipine	9.9 ± 0.1	1.2 (1.1-1.2)	39 ± 1	1.5 (1.2-1.8)	1.9 ± 0.1
Horseradish peroxidase, H_2O_2	1,4-Dihydro-2,6-dimethyl-4-phenyl-3,5-pyridine-dicarboxylic acid dimethyl ester	4.0 ± 0.3	1.9 (1.6-2.2)	54 ± 8	2.3 (1.4-3.9)	2.2 ± 0.2
Prostaglandin H synthase, H_2O_2	Nifedipine	124 ± 9	2.1 (1.8-2.6)	86 ± 11	1.6 (0.9-2.6)	2.0 ± 0.2
Prostaglandin H synthase, H_2O_2	1,4-Dihydro-2,6-dimethyl-4-phenyl-3,5-pyridine-dicarboxylic acid dimethyl ester	546 ± 68	1.4 (1.1-1.8)	145 ± 27	1.4 (0.8-2.5)	28 ± 0.2
P450 2B1, reductase, NADPH	Nifedipine	1.4	1.6		2.0	1.4
Rat liver microsomes, NADPH	1,4-Dihydro-2,6-dimethyl-4-phenyl-3,5-pyridine-dicarboxylic acid dimethyl ester	2.2	1.0		0.9	1.4
Haemoglobin H_2O_2	Nifedipine	3.4 ± 0.6	1.3 (0.9-1.7)	46 ± 14	1.2 (0.9-1.7)	1.5 ± 0.1
Haemoglobin H_2O_2	1,4-Dihydro-2,6-dimethyl-4-phenyl-3,5-pyridine-dicarboxylic acid dimethyl ester	5.5 ± 0.6	1.3 (0.9-1.8)	72 ± 10	1.4 (0.7-2.9)	1.6 ± 0.2

Table 3: Comparison of rates of formation of aminium radicals and HCHO from aminopyrine and N,N-dimethylthioanisole

Enzyme and Oxidant	Aminopyrine			N,N-Dimethylthioanisole	
	Rate of aminium radical formation $\mu M\ min^{-1}$	Rate of HCHO formation $\mu M\ min^{-1}$	Ratio	Rate of aminium radical formation $\Delta A_{650}\ min^{-1}$	Rate of HCHO formation $\mu M\ min^{-1}$
Horseradish peroxidase (0.41 μM)					
H_2O_2	41.3	18.9	2.2	1.15	16.0
C_2H_5OOH	28.7	10.5	2.7	0.05	3.0
Haemoglobin (0.31 μM)					
H_2O_2	16.4	6.3	2.6	0.09	2.3
C_2H_5OOH	27.8	11.2	2.5	0.31	5.6
Prostaglandin H synthase (2.1 μM)					
H_2O_2	25.1	17.7	1.4	0.25	6.9
Chloroperoxidase (3.2 nM)					
H_2O_2	<0.2	2.0	<0.1	0.04	3.0
C_2H_5OOH	<0.2	9.7	<0.02	<0.05	4.1
Microsomes (0.62 μM P450)					
NADPH, O_2	<0.2	6.7	<0.03	<0.02	7.5
P450 2B1 (0.50 μM)					
NADPH, O_2	<0.2	4.3	<0.05	<0.02	6.6

Although hydrogen atom transfer has been proposed as a mechanism for the high isotope effects seen for peroxidases,[11] there is also considerable evidence to support the view that electron transfer is operative in peroxidase-catalyzed N-dealkylation and that this transfer occurs via the heme edge.[4] We considered the possibility that the high isotope effects could be the result of a lack of base catalysis by the shielded $(FeO)^{3+}$ entity in peroxidase (Fig. 3).

As a corollary, then, P450s and biomimetic models show low isotope effects because base catalysis by the $(FeO)^{2+}$ entity is involved. To test this hypothesis, we considered the kinetic isotope effects involved in dehydrogenation of 1,4-dihydropyridines, a process analogous to N-dealkylation.[8,16] It is known that N,N-dialkylaniline aminium radicals have high pK_as (~9) but dihydropyridine aminium radicals have low pK_as (~3.5).[17] We found that all of the enzymes examined, including peroxidases and haemoglobin, showed low isotope effects (Table 2).[15]

If this view of the involvement of base catalysis is correct, then one might expect cation radicals to accumulate in those cases where high isotope effects are seen (ie no base catalysis is available). This indeed was the case when the formation of two different coloured aminium radicals was examined (Table 3).[15]

On the basis of these results, a set of proposed pictures of the active sites of hemoproteins is presented in Figure 3.

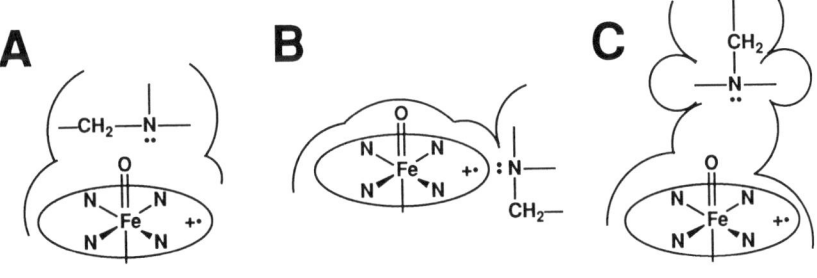

Figure 3: Interactions of amines with (A) P450s and (B,C) peroxidases.[15]

4 FORMATION AND DECOMPOSITION OF N,N-DIALKYLANILINE N-OXIDES BY P450S

If aminium radicals are intermediates in N-dealkylation reactions, then one might expect a small but finite rate of recombination (rebound) with the FeO entity to yield N-oxides or hydroxylamines, even in the presence of extensive base catalysis to produce a rearrangement by abstraction of the α-proton. However, clear evidence for this possibility in P450 reactions has not been presented. We developed procedures for the careful analysis of N-oxides of N,N-dialkylanilines by P450s, utilizing HPLC, $TiCl_3$ reduction, and more HPLC.[18] We were then able to determine that P450 2B1 formed one N-oxide per 940 N-demethylations of N,N-dimethylaniline (Table 4).

The ratio of N-dealkylation:N-oxide formation could be varied considerably as the substituents in the basic N,N-dimethylaniline structure were varied. These reactions are all resistant to catalase and superoxide dismutase, arguing against reactions due to free H_2O_2. With N,N-dimethyl-2-aminofluorene, the ratio of the two reactions was as low as 6:1. However, in contrast to the patterns seen with N-dealkylation,[13,14] there was no clear trend for p-substitution and substitution of an N-ethyl for N-methyl did not have a dramatic effect (Table 4).

Table 4: Rates of N-dealkylation and N-oxygenation of N,N-dialkylaniline derivatives by PB450 2B1

Substrate	Rate		Ratio
	N-Dealkylation	N-Oxygenation	
	nmol product formed min^{-1} (nmol P450)$^{-1}$		
N,N-Dimethylaniline	34.7 ± 0.2	0.040 ± 0.008	940
4-Methyl-N,N-dimethylaniline	39.8 ± 3.1	0.097 ± 0.017	400
4-Cyano-N,N-dimethylaniline	18.0 ± 1.9	0.183 ± 0.087	98
N-Ethyl-N-methylaniline	42.7 ± 2.2 (demethylation) 2.7 ± 0.2 (deethylation)	0.051 ± 0.006	890
N,N-Diethylaniline	25.5 ± 0.1	0.025 ± 0.004	1020
[$^2H_{10}$]N,N-Diethylaniline	7.9 ± 0.2	0.007 ± 0.005	1100
N-Phenylpyrrolidine	8.2 ± 1.0	0.057 ± 0.018	140
N,N-Dimethyl-2-aminofluorene	3.1 ± 0.7	0.49 ± 0.06	6

These and other considerations suggest that direct recombination of the $(FeO)^{2+}$ entity and aminium radical does not occur to give the N-oxides.[18] An alternative possibility involves further 1-electron transfer to give an $(FeO)^{3+}$:R_2N^+ pair that collapses, a possibility that has some credence in the comparison of products formed with those predicted by Hückel calculations[19] (Figure 4).

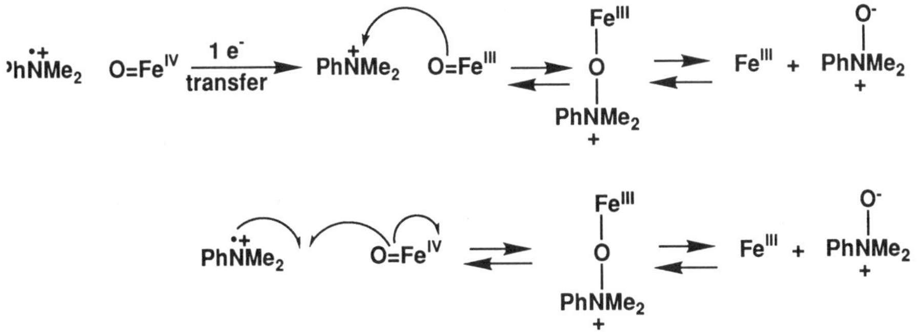

Figure 4: Possible mechanisms of amine N-oxygenation.[18]

The N,N-dialkylaniline N-oxides themselves are relatively stable in the presence of P450s. Although they have been used with biomimetic models by others and shown to undergo heterolytic scission to serve as oxygen surrogates,[20,21] the rates are rather low. P450 2B1 slowly decomposed the N,N-dimethylaniline N-oxides to give monoalkylamines but not dialkylamines. This result and the lack of support of other reactions are interpreted as evidence that the decomposition of the N-O bond is predominantly homolytic[13,18] (Figure 5).

The reductions of the N-oxides to the dialkylamines are faster than the internal decomposition reactions. This reaction is probably best explained by Polonovski chemistry.[18]

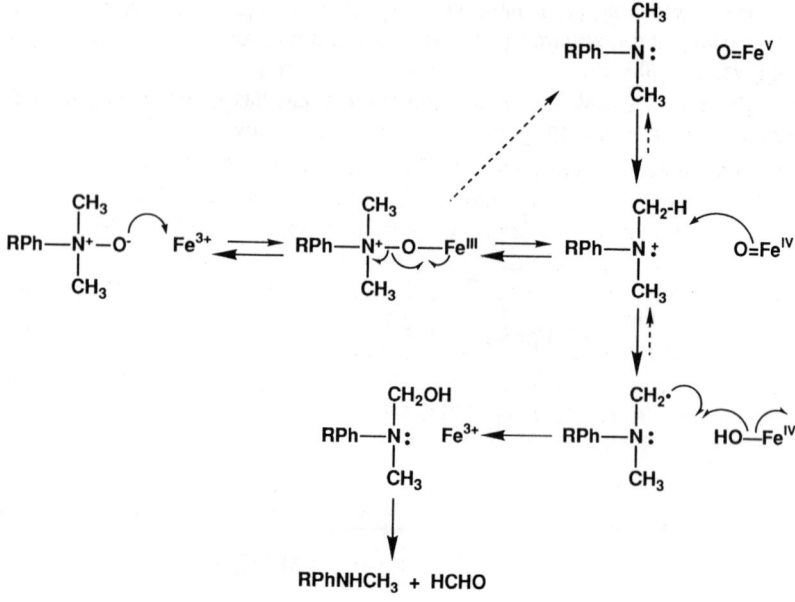

Figure 5: Possible pathways of deoxygenation of N-oxides by P450 2B1.[13,18]

REFERENCES

1. F P Guengerich, *J Biol Chem*, 1991, **266**, 10019.
2. D R Nelson, T Kamataki, D J Waxman, F P Guengerich, R W Estabrook, R Feyereisen, F J Gonzalez, M J Coon, I C Gunsalus, O Gotoh, K Okuda, and D W Nebert, *DNA Cell Biol*, 1993, **12**, 1.
3. F P Guengerich and T L Macdonald, *Acc Chem Res*, 1984, **17**, 9.
4. P R Ortiz de Montellano, *Acc Chem Res*, 1987, **20**, 289.
5. F P Guengerich and T L Macdonald, *FASEB J*, 1990, **4**, 2453.
6. F P Guengerich and T L Macdonald, Sequential electron transfer in oxidation reactions catalyzed by cytochrome P-450 enzymes. In: *Advances in Electron Transfer Chemistry, Vol 3*, edited by P S Mariano, JAI Press, Greenwich, CT, 1993, pp 191-241.
7. R E McMahon, H W Culp, and J C Occolowitz, *J Am Chem Soc*, 1969, **91**, 3389.
8. O Augusto, H S Beilan, and P R Ortiz de Montellano, *J Biol Chem*, 1982, **257**, 11288.
9. A Bondon, T L Macdonald, T M Harris, and F P Guengerich, *J Biol Chem*, 1989, **264**, 1988.
10. R A Stearns and P R Ortiz de Montellano, *J Am Chem Soc*, 1985, **107**, 4081.
11. G T Miwa, J S Walsh, G L Kedderis, and P F Hollenberg, *J Biol Chem*, 1983, **258**, 14445.

12. J P Shea, S D Nelson, and G P Ford, *J Am Chem Soc*, 1983, **105**, 5451.
13. L T Burka, F P Guengerich, R J Willard, and T L Macdonald, *J Am Chem Soc*, 1985, **107**, 2549.
14. T L Macdonald, W G Gutheim, R B Martin, and F P Guengerich, *Biochemistry*, 1989, **28**, 2071.
15. O Okazaki and F P Guengerich, *J Biol Chem*, 1993, **268**, 1546.
16. F P Guengerich and R H Böcker, *J Biol Chem*, 1988, **263**, 8168.
17. J P Dinnocenzo and T E Banach, *J Am Chem Soc*, 1989, **111**, 8646.
18. Y Seto and F P Guengerich, *J Biol Chem*, 1993, **268**, 9986.
19. G J Hammons, F P Guengerich, C C Weis, F A Beland, and F F Kadlubar, *Cancer Res*, 1985, **45**, 3578.
20. T C Bruice, *Acc Chem Res*, 1991, **24**, 243.
21. D C Heimbrook, R I Murray, K D Egeberg, S G Sligar, M W Nee, and T C Bruice, *J Am Chem Soc*, 1984, **106**, 1514.

ACKNOWLEDGEMENT

This work was supported in part by USPHS grants CA44353 and ES00267.

STRUCTURE AND MECHANISM OF PORPHOBILINOGEN DEAMINASE

Peter M Jordan *

School of Biological Sciences, Queen Mary & Westfield College, Mile End Road, London E1 4NS, UK.

1 INTRODUCTION

The biosynthesis of uroporphyrinogen III from the reactive 2-aminoketone, 5-aminolaevulinic acid, is accomplished in three enzymic reactions common to all living systems (for a review see Jordan 1991).[1] Two molecules of 5-aminolaevulinic acid are first dimerized in a Knorr reaction, catalysed by the 5-aminolaevulinic acid dehydratase, to give the trisubstituted pyrrole, porphobilinogen.[2] Porphobilinogen is subsequently transformed into uroporphyrinogen III in two stages, catalysed by porphobilinogen deaminase and uroporphyrinogen III synthase (cosynthase), in which a 1-hydroxymethylbilane, preuroporphyrinogen, is the key intermediate. These stages are shown in Scheme 1.

2 THE DISCOVERY OF PREUROPORPHYRINOGEN

Porphobilinogen deaminase (also called 1-hydroxy-methylbilane synthase) catalyses the deamination and polymerisation of four molecules of porphobilinogen to give preuroporphyrinogen. We first discovered that preuroporphyrinogen is the product of the deaminase using ^{13}C NMR by incubating regiospecifically labelled [11-^{13}C]-porphobilinogen (• in Scheme 1) with homogeneous porphobilinogen deaminase in the NMR spectrometer.[3] This resulted in the transient appearance of complex ^{13}C signals at $\delta = 23$ ppm and $\delta = 55$ ppm integrating in the ratio 3:1 and pointing to a 1-hydroxymethylbilane structure. The preuroporphyrinogen rapidly cyclised, without rearrangement, to give uroporphyrinogen I with a single signal at $\delta = 22$ ppm. Most significantly, when we incubated preuroporphyrinogen with highly purified uroporphyrinogen cosynthase, uroporphyrinogen III was the exclusive product indicating that preuroporphyrinogen was the long sought substrate for the cosynthase.[4] Preuroporphyrinogen was subsequently synthesised chemically[5] and shown unambiguously to be a 1-hydroxymethylbilane with chemical and enzymic properties identical to the intermediate isolated from our earlier enzymic experiments. This eliminated previous suggestions that preuroporphyrinogen was a 1-aminomethylbilane or an *N*-alkylporphyrinogen.

---

* Current Address: Department of Biochemistry, University of Southampton, Bassett Crescent East, Southampton SO16 7PX, UK.

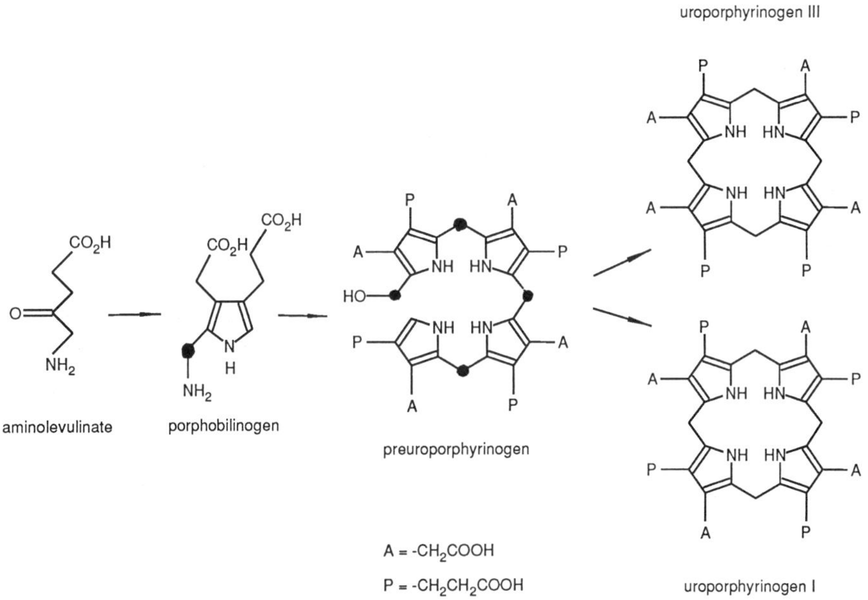

Scheme 1: The biosynthesis of uroporphyrinogen III from 5-aminolaevulinic acid.

No successful biomimetic synthesis of porphobilinogen has been achieved from 5-aminolaevulinic acid, due to symmetrical dimerisation and oxidation to give a pyrazine. However, porphobilinogen is able readily to form tetrapyrroles non-enzymically, polymerizing in dilute acid to give the four isomeric uroporphyrinogens in almost quantitative yield.[6] The observed ratio of uroporphyrinogens I, II, III and IV of 1:1:4:2 has been interpreted as indicating that the condensation between either the substituted or the unsubstituted pyrrole α-positions with the aminomethyl substituent occur with equal probability.

3   CLONING AND OVER-EXPRESSION OF *ESCHERICHIA COLI* PORPHOBILINOGEN DEAMINASE AND THE DISCOVERY OF THE DIPYRROMETHANE COFACTOR

A large amount of information has accumulated on the enzymes responsible for the biosynthesis of uroporphyrinogen III.[1] Whereas the transformation of preuroporphyrinogen into uroporphyrinogen III catalysed by uroporphyrinogen synthase still remains an enigma, recent progress in molecular biology and enzymology has assisted in revealing the intimate secrets and intricacies of the porphobilinogen deaminase catalysed reaction. Most significant, was the first characterisation, molecular cloning and overexpression of the *hemC* gene encoding *Escherichia coli* porphobilinogen deaminase that allowed the isolation of milligramme amounts of pure enzyme.[7] This led us to the discovery that native porphobilinogen deaminase possesses a novel dipyrromethane system covalently attached to the enzyme. We had suspected the presence of a dipyrrole system from previous protein chemistry studies with the deaminase from *Rhodobacter sphaeroides* but the large amounts of the *E coli* enzyme available allowed this work to be confirmed and greatly extended.[8] We named the dipyrrole system the dipyrromethane cofactor.[8] The development of a procedure, in our laboratory, to label the dipyrromethane cofactor specifically with ^{14}C indicated that the cofactor remained completely intact during catalytic turnover and that it was acting as a primer for the polymerisation,[8] in a similar fashion to that found in other polymerases.

4 ENZYME INTERMEDIATE COMPLEXES WITH ONE, TWO, THREE AND FOUR MOLECULES OF SUBSTRATE COVALENTLY LINKED TO THE ENZYME

The mechanism by which porphobilinogen deaminase, a relatively small enzyme of Mr 35,000, catalyses such a complex tetrapolymerisation reaction is of enormous interest and importance in the field of macromolecular assembly in bioorganic chemistry. Not only does the enzyme active site contain the necessary catalytic machinery to link the pyrrole units together but it must also possess the molecular architecture necessary to manoeuvre the polypyrrole chain as it is extended at the catalytic site. In addition a mechanism must exist for the release of the 1-hydroxymethylbilane product, preuroporphyrinogen, at the appropriate time. Some of the questions about the enzyme mechanism were answered by the isolation and characterisation of enzyme intermediate complexes with one, two and three pyrrole units covalently bound to the dipyrromethane cofactor, designated ES, ES_2 and ES_3.[9,10] The final complex, ES_4, is too unstable to isolate but can be detected by electro-spray mass spectrometry.[11] The four substrate molecules are attached sequentially to the dipyrromethane cofactor starting with the porphobilinogen unit that becomes the ring A in uroporphyrinogen, followed by rings B, C and D.[12,13] Only when all four pyrrole units have been attached to the dipyrromethane cofactor is preuroporphyrinogen released by hydrolytic cleavage.[10] The stages in the assembly of preuroporphyrinogen are shown in Scheme 2.

The sequential tetrapolymerisation reaction may be inhibited by the novel suicide inhibitor, 2-bromoporphobilinogen.[10] This substrate analogue is recognised by the enzyme and is deaminated normally. The resulting azafulvene (or its equivalent) then reacts with the α-position of the pyrrole ring of the

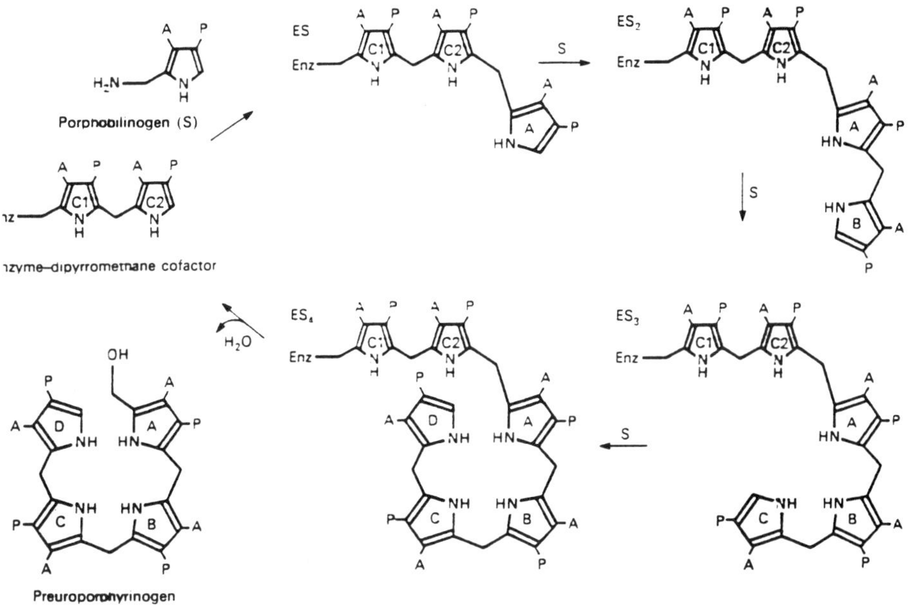

Structure and Mechanism of Porphobilinogen Deaminase

Scheme 2: The tetrapolymerisation of porphobilinogen attached to the dipyrromethane cofactor of porphobilinogen deaminase involving enzyme intermediate complexes ES, ES_2, ES_3, and ES_4.

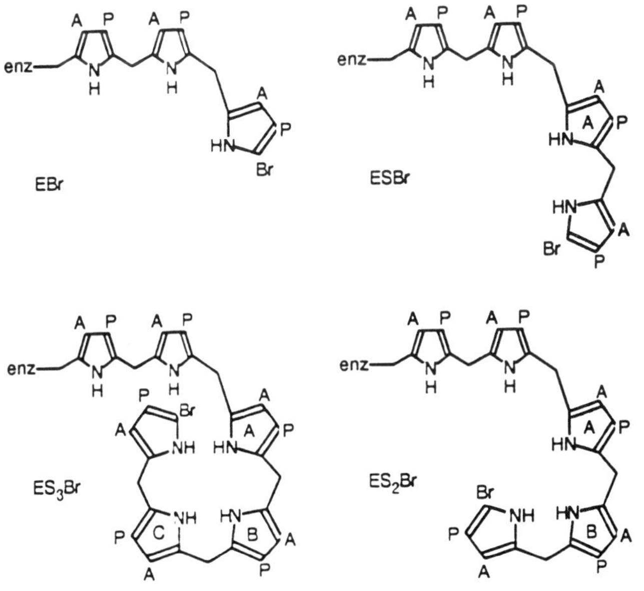

Scheme 3: The reaction of the suicide inhibitor 2-bromoporphobilinogen with porphobilinogen deaminase to give chain termination complexes.

intermediate complex. However the presence of a bromine atom at the normally free terminal α-position (see Scheme 3) prevents any further reaction with substrate (or inhibitor) and the enzyme is consequently completely inactivated. The studies with 2-bromoporphobilinogen greatly assisted in our understanding of the tetrapolymerization reaction.[10]

5   X-RAY STRUCTURE OF *E COLI* PORPHOBILINOGEN DEAMINASE

We realized several years ago that a full understanding of how this remarkable reaction is catalysed by the deaminase requires a detailed knowledge of the structure of the protein. We thus set out to overexpress and crystallize the porphobilinogen deaminase from *E coli*[14] and to determine the X-ray structure. For this we collaborated with Dr Wood and Professor Blundell at Birkbeck College, London.[15] The efforts have recently been rewarded with the first X-ray structure of any enzyme catalysing a reaction in the tetrapyrrole biosynthesis pathway.[16] The structure has been determined at a resolution of 1.7 Å.

Porphobilinogen deaminase from *E coli* is composed of three α/ß domains each consisting of about 100 amino acids[16] (Figure 1). Domains 1 and 2 are topologically related to one another, each having four parallel and one antiparallel ß-strands that form a sheet. α-Helices flank each face of these sheets. Domain 3 consists of a ß-sheet composed of three ß-strands with three α-helices interacting with one face of the sheet. The catalytic cleft is located between domains 1 and 2 into which a loop from domain 3 extends carrying the dipyrromethane cofactor, linked covalently to cysteine-242. Residues 48-58 are poorly defined in the X-ray structure and are thought to form a mobile lid over the catalytic cleft. The topology of domains 1 and 2 are closely related to a group of binding proteins that include transferrins and group II periplasmic receptors.[17]

6   KEY AMINO ACIDS IN THE CATALYTIC CLEFT

Central to the functioning of the enzyme is the presence in the catalytic cleft of a semicircle of invariant positively charged amino acid residues including arginines 11, 131, 132, 149, 155, 176 and lysine 83 (Figure 2). These residues form a positively charged surface and provide the binding sites not only for the negatively charged acetate and propionate side chains of the dipyrromethane cofactor, but also for the substrate and the polypyrrole chain. Site directed mutations of any of these residues have dramatic effects on enzyme function including inability to assemble the cofactor, depressed substrate binding and inhibition of the chain elongation process.[18,19] Table 1 shows the effects of substituting these invariant arginines for histidine.

Table 1: Effect of mutations of invariant arginine residues to histidine on the ability of the *E coli* porphobilinogen deaminase to assemble the dipyrromethane cofactor and to catalyse the chain elongation reaction.[18]

Mutation Affected	Specific Activity	K_m	Presence of Cofactor	Enzymic Stage Affected
Wild type	43	17	+	None
R11H	0.1	nd	+	E→ES
R131H	nd	nd	-	cofactor assembly
R132H	nd	nd	-	cofactor assembly
R149H	11.1	200	+	ES→ES$_2$
R155H	0.5	nd	+	E→ES ES$_3$→product
R176H	6.0	30	+	ES→ES$_2$ ES$_2$→ES$_3$

Specific enzyme activity is expressed as µmole uroporphyrinogen formed/hour/mg protein. K_m is in µM. The stage of the enzyme reaction affected by the mutation was determined by FPLC.[10] The presence of cofactor was determined by reaction with Ehrlich's reagent.[18] nd = not determinable.

7 THE CATALYTIC REACTION

The *X*-ray structure has also given clues as to how catalysis may be accomplished. Aspartate-84, an invariant residue present in all deaminases, is positioned in such a way that it can interact, by hydrogen bonding, with the hydrogen atoms attached to the nitrogens of the pyrrole rings of the cofactor and the substrate. This provides a means to stabilize the positive charges that develop on the pyrrole nitrogen atoms during C-C bond formation (Scheme 4). Mutation of aspartate-84 to glutamate reduces the k_{cat} by over two orders of magnitude whereas alanine and asparagine mutants are devoid of activity.[20] Interestingly, these two latter mutants are still able to assemble the dipyrromethane cofactor (see below) suggesting that the mechanism of cofactor assembly may be different from that of the substrate tetrapolymerisation reaction.[20]

During the tetrapolymerisation reaction it is envisaged that substantial movement occurs between domains 1 and 2 to permit the access of four molecules of substrate to the active site. The substrate binding site involves predominantly domain 1, whereas the cofactor site (C) is confined largely to

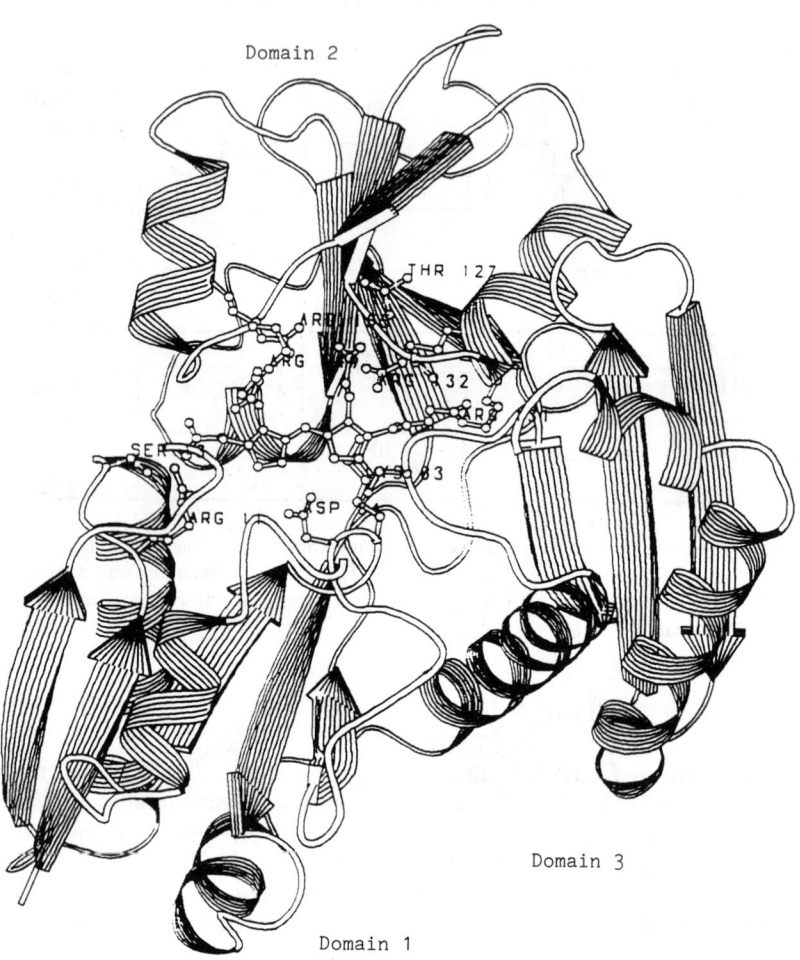

Figure 1: Ribbon representation of the 3-dimensional structure of *Escherichia coli* porphobilinogen deaminase. The catalytic site is located between domains 1 and 2. The dipyrromethane cofactor is attached to domain 3 through cysteine 242 (Louie *et al* 1992).[16] The structure shows the dipyrromethane cofactor in its oxidised form in which the C2 ring of the dipyrromethane cofactor is thought to occupy the site normally reserved for the substrate.

Structure and Mechanism of Porphobilinogen Deaminase

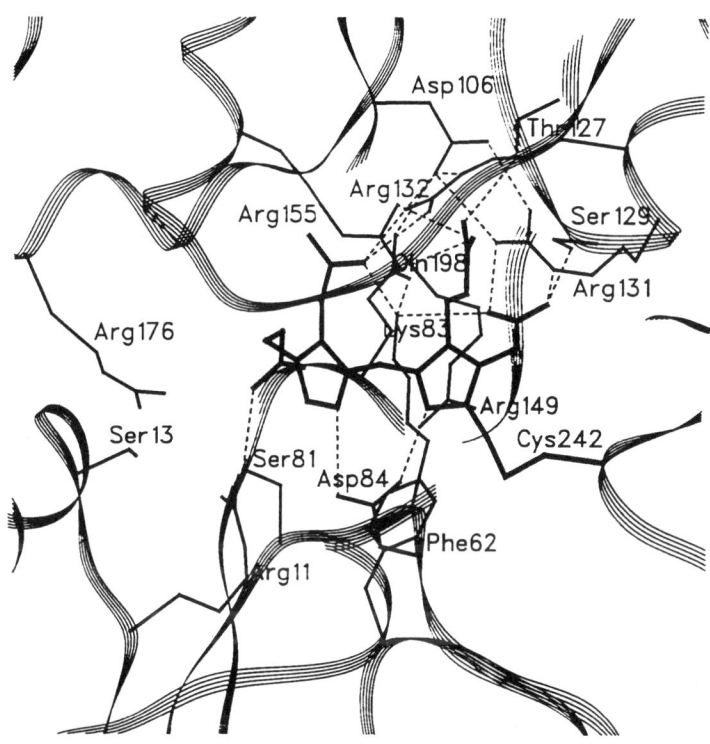

Figure 2: The catalytic site of *Escherichia coli* porphobilinogen deaminase (Louie *et al* 1992).[16] The acetate and propionate side chains of the dipyrromethane cofactor interact with several invariant arginines and lysines. Aspartate-84 interacts with the NH group of the pyrrole nitrogens of the cofactor. The vacant substrate binding site is made up from arginine 11, 149 and 155. It is also possible for aspartate-84 to interact with the pyrrole NH of the substrate (not shown).

domain 2. The location of the active site between domains 1 and 2, which are connected to one another by flexible hinge regions is beautifully designed to allow appreciable movement to occur during the polymerisation. Such flexibility is operative in the related periplasmic binding proteins.[17] Evidence for the progressive change in the conformation of the deaminase during the polymer chain elongation process has come from the observation that cysteine-134 becomes progressively more susceptible to modification with N-ethylmaleimide as the enzyme binds substrate to form ES, ES_2 and ES_3.[10,16] Cysteine-134 is located between the domains 2 and 3 suggesting that domain 3 also changes position with respect to domain 2 during the chain elongation process.

The absence of good X-ray data for the sequence from residues 48-58, that forms a 'lid' thought to loop over the catalytic cleft, suggests that this part of the enzyme is highly mobile and may play a major part in catalysis. Work is in progress to determine the role of this loop.

8   THE ROLE OF THE ENZYME IN CATALYSIS

The fact that porphobilinogen polymerizes non-enzymically with great facility in dilute acid to yield uroporphyrinogens indicates that this pyrrole unit possesses the necessary intrinsic chemistry for the condensation reaction. In the deaminase catalysed reaction the absence of obvious catalytic groups, except for aspartate-84, suggests that a primary role of the enzyme is to position precisely the nucleophilic α-position of one pyrrole ring close to the electrophilic *exo*-methylene of the azafulvene (or its equivalent) of the other reacting ring to allow the natural chemistry to express itself optimally. The X-ray structure strongly suggests that the positively charged arginines at positions 11, 149 and 155 that are thought to make up the substrate binding site (S) and arginines 131, 132, 155 and lysine 83 that make up the cofactor binding site (C) are key groups in this respect. In its carboxylate form, the bifurcated aspartate-84 could stabilise the positively charged pyrrole nitrogens on the reacting rings (Scheme 4). In its carboxylic acid form, aspartate-84 could also act to protonate the leaving group ammonia and function as a base in the final deprotonation of the α-position (Scheme 4). Although this would require the pKa of the aspartate to be some four pH units above that in free solution, its location in a hydrophobic environment under phenylalanine 60 is significant.[16] It is envisaged that after C-C bond formation, the loss of the α-hydrogen and the change from sp^3 to sp^2 hybridization provides possible impetus for the translocation of the pyrrole chain, leaving the (S) site vacant for the next molecule of substrate. After the fourth molecule of substrate (the D ring) has reacted the A ring must be able to gain access to the (S) site to allow bond cleavage and reaction with water to form the hydroxymethyl group of preuroporphyrinogen. It is interesting that the enzyme can catalyse deamination of porphobilinogen as well as

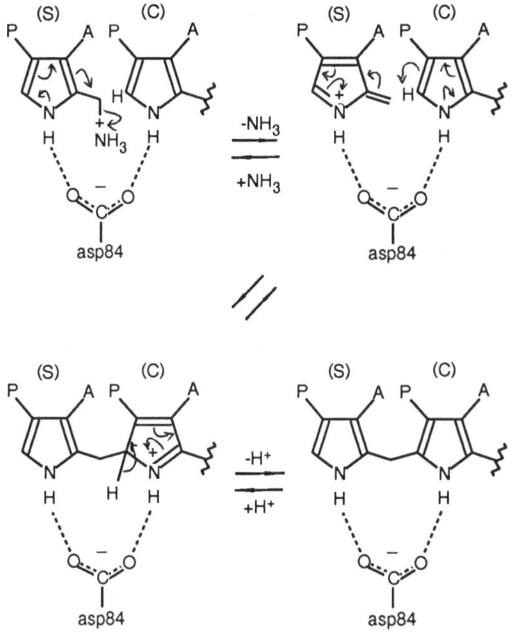

Scheme 4: The condensation reaction between porphobilinogen and the α-position of an enzyme-bound pyrrole. The catalytic aspartate-84 stabilizes the positively charged nitrogen of the azafulvene (or its equivalent) and facilitates deamination. The reaction of the nucleophilic α-position is also facilitated by aspartate-84 which stabilizes the tautomeric form.

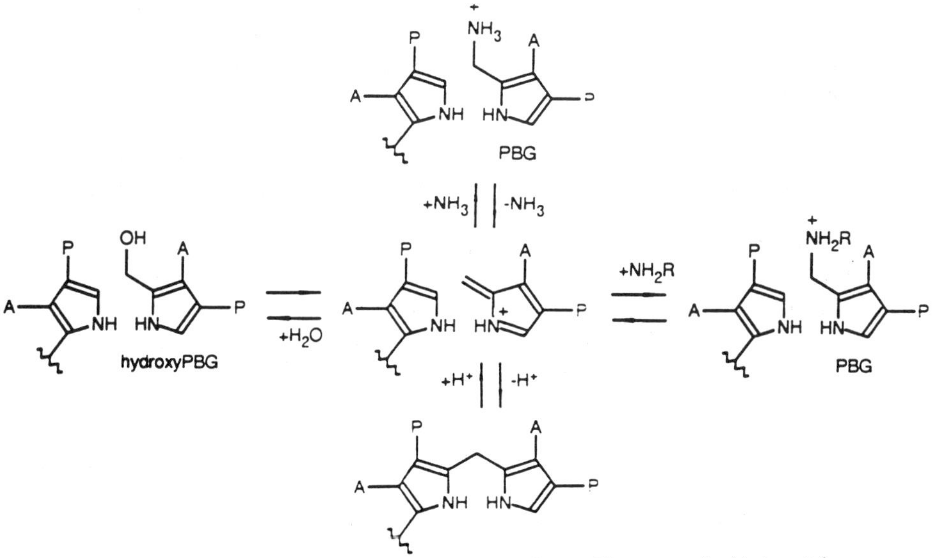

Scheme 5: Reactions catalysed by porphobilinogen deaminase. The enzyme is able to catalyse loss of ammonia or water and the addition of ammonia and water, with the azafulvene as the most likely common intermediate.

addition of ammonia and analogous bases to the enzyme intermediate complexes[21,22] to release pyrrole intermediates. Furthermore the hydroxymethyl-form of porphobilinogen acts as a good substrate[23] indicating that the enzyme can catalyse dehydration as well as hydration reactions (Scheme 5).

9 THE CHEMISTRY OF PORPHOBILINOGEN AND THE INFLUENCE OF THE ENZYME

One of the most remarkable properties of porphobilinogen deaminase is its ability to self-assemble its own dipyrromethane cofactor.[10,24] Studies using site directed mutagenesis have revealed that substitution of arginines 131 and 132 for histidine[18] or leucine[19] prevent totally the cofactor assembly process and inactive apo-enzymes result. Conversely, site-directed mutants in which the catalytic aspartate-84 is mutated to alanine or asparagine appear to be able to assemble the cofactor but are totally inactive in the tetrapolymerisation reaction.[20] These findings suggest that reaction of the porphobilinogen units destined for the dipyrromethane cofactor with the enzyme may occur by a different mechanism from the tetrapolymerisation reaction and that the former process may not require aspartate-84. This could also explain why the cofactor, once assembled, is not subject to catalytic turnover.[8]

The conformation of porphobilinogen in solution and its stereoelectronic implications may hold the key to many of its reaction characteristics both in non-enzymic and enzymic terms. NMR studies (Evans *et al* 1985),[25] suggest that at physiological pH porphobilinogen exists as an intramolecular ion pair in which the protonated aminomethyl group forms a ring with the negatively charged acetate side chain. This conformation is similar to the conformationally restricted porphobilinogen lactam and would place the leaving group in a poor position for deamination. Breaking the ion pair, either by protonation of the acetate as occurs in the non-enzymic acid catalysed polymerisation (Scheme 6a), or binding the acetate group to a positively charged arginine as is the case for the enzyme-catalysed reaction (Scheme 6b), is likely to facilitate greatly the deamination reaction by allowing the CH_2-NH_3^+ group to adopt a conformation perpendicular to the ring. It is thus possible that the process of binding of porphobilinogen to the apoenzyme provides major assistance in deamination to the azafulvene and that C-alkylation, first of cysteine-242, and then of the C1 ring of the cofactor by a second azafulvene molecule furnishes the dipyrromethane system of the native holoenzyme. Such an enzyme facilitated reaction may explain how porphobilinogen deaminase site-directed mutants lacking the catalytic aspartate-84 are able to assemble the cofactor but are unable to catalyse the formation of preuroporphyrinogen.

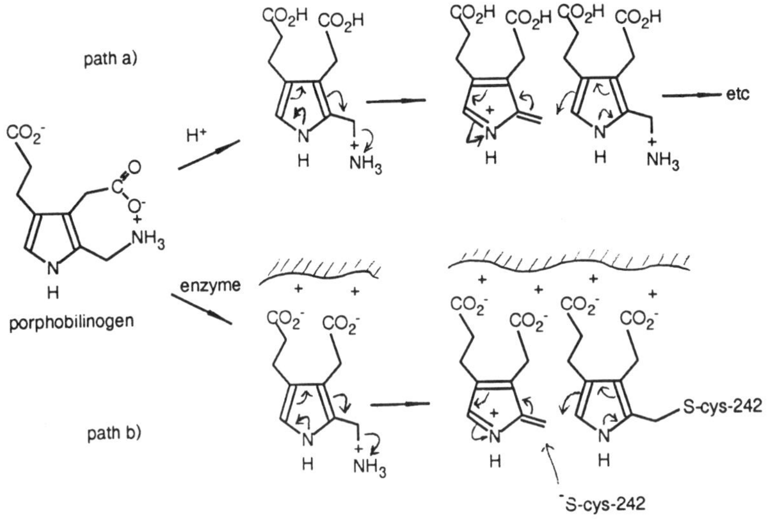

Scheme 6: Conformation of porphobilinogen. Postulated effects of a) acid and b) enzyme binding in the activation of the leaving group.

10   SUMMARY

Although the *X*-ray structure of the *E coli* porphobilinogen deaminase has answered many of the questions about the functioning of the deaminase, many more have been raised that will occupy us with exciting experiments for some time into the future. The determination of the *X*-ray structure of the enzyme intermediate complexes, currently under way, will provide us with crucial information about the way the enzyme accommodates the polypyrrole chain during the reaction and how the chain termination reaction occurs after four molecules of substrate have been linked to the dipyrromethane cofactor. In this respect the glutamate-84 mutant that forms exceptionally stable enzyme-intermediate complexes will be a crucial protein for future study.

REFERENCES

1. P M Jordan, (Ed) 'New Comprehensive Biochemistry', *Elsevier*, Amsterdam, 1991, **Vol 19**.
2. P M Jordan and J S Seehra, *FEBS Letts*, 1980, **114**, 283.
3. G Burton, P E Fagerness, S Hosozawa, P M Jordan, and A I Scott, *J Chem Soc, Chem Commun*, 1979, 202.

4   P M Jordan, G Burton, H Nordlov, M M Schneider, L Pryde, and A I Scott, *J Chem Soc, Chem Commun*, 1979, 204.
5   A R Battersby, C J R Fookes, K E Gustafson-Potter, G W J Matcham, and E McDonald, *J Chem Soc, Chem Commun*, 1979, 1155.
6   D Mauzerall, *J Am Chem Soc*, 1960, **82**, 2601.
7   S D Thomas and P M Jordan, *Nucl Acids Res*, 1986, **14**, 6215.
8   P M Jordan and M J Warren, *FEBS Letts*, 1987, **225**, 87.
9   A Berry, P M Jordan, and J S Seehra, *FEBS Letts*, 1981, **129**, 220.
10 M J Warren and P M Jordan, *Biochemistry*, 1988, **27**, 9020.
11 R T Aplin, J E Baldwin, C Pichon, C A Roessner, A I Scott, C J Schofield, N J Stolowich, and M J Warren, *Bioorg Med Chem Lett*, 1991, **1**, 503.
12 P M Jordan and J S Seehra, *FEBS Letts*, 1979, **104**, 364.
13 A R Battersby, C J R Fookes, G W J Matcham, and E McDonald, *J Chem Soc, Chem Commun*, 1979, 539.
14 P M Jordan, S D Thomas, and M J Warren, *Biochem J*, 1988, **254**, 427.
15 P M Jordan, M J Warren, B I A Mgbeje, S P Wood, J B Cooper, G Louie, P Brownlie, R Lambert and T L Blundell, *J Mol Biol*, 1992, **224**, 269.
16 G V Louie, P D Brownlie, R Lambert, J B Cooper, T L Blundell, S P Wood, M J Warren, S C Woodcok and P M Jordan, *Nature (London)*, 1992, **359**, 33.
17 G V Louie, *Current Opinion in Structural Biology*, 1993, **3**, 401.
18 P M Jordan and S C Woodcock, *Biochem J*, 1991, **280**, 445.
19 M Lander, A R Pitt, P R Alefounder, D Bardy, C Abell, and A R Battersby, *Biochem J*, 1991, **275**, 447.
20 S C Woodcock and P M Jordan, 1993 (in press).
21 J Pluscec and L Bogorad, *Biochemistry*, 1970, **9**, 4736.
22 R C Davies and A Neuberger, *Biochem J*, 1973, **133**, 471.
23 A R Battersby, C J R Fookes, G W J Matcham, E McDonald, and K E Gustafson-Potter, *J Chem Soc, Chem Commun*, 1979, 316.
24 A I Scott, K R Clemens, N J Stolowich, P J Santander, M D Gonzalez, and C A Roessner, *FEBS Letts*, 1989, **242**, 319.
25 J N S Evans, P E Fagerness, N E Mackenzie, and A I Scott, *Magnetic Resonance in Chemistry*, 1985, **23**, 939.

ACKNOWLEDGEMENTS

This work was funded by the SERC MRI and ARFC. I am grateful to Drs Warren and Woodcock at QMC London, Drs Wood, Louie, Cooper, Brownie, Lambert and Professor Blundell at Birkbeck, London and Professor M Akhtar at Southampton University for stimulating discussions. Professor A I Scott kindly provided the 2-bromoporphobilinogen.

ORGANOMETALLIC B_{12}-CHEMISTRY

Bernhard Kräutler

*Institute of Organic Chemistry, University of Innsbruck,
Innrain 52a, A-6020 Innsbruck, Austria.*

1   INTRODUCTION

Among the known metal containing natural cofactors, the B_{12}-coenzymes stand out due to their structural features as cobalt-corrins, as well as their biological organometallic chemistry, both uncovered by *X*-ray structural analysis in the laboratory of Dorothy Hodgkin.[1,2]

The dependence of the biofunctionally important reactivities of the B_{12}-coenzymes on their unique structural characteristics is still insufficiently understood,[3-5] as is the detailed rôle of coenzyme B_{12} (**1**, adenosylcobalamin, see Figure 1) itself as cofactor in a series of enzymatic reactions.[6]

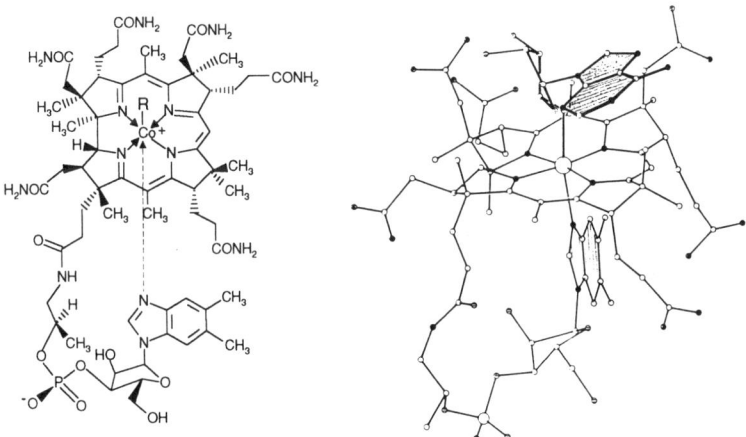

Figure 1: Left: structural formulae of coenzyme B_{12} (**1**, adenosylcobalamin, R = 5'-deoxyadenosyl), methylcobalamin (**2**, R = methyl), Co(II)-cobalamin (**3**, R= e⁻), vitamin B_{12} (**4**, cyanocobalamin, R = CN); Right: three dimensional structure of the coenzyme **1** from *X*-ray analysis.[2]

These complex processes are believed to be triggered by an enzyme activated homolysis of the organometallic bond of the bound coenzyme **1** to produce a 5'-deoxyadenosyl radical.[6,7] As pointed out by Halpern, coenzyme B_{12} indeed appears well adapted to function as a 'reversible source of alkyl radicals': thermally activated homolysis of the Co-C-bond of **1** also occurs readily in solution.[7,8].

A second, no less important biological rôle of the B_{12}-derivatives involves the catalysis of enzymatic methyl group transfer reactions *via* methyl-Co(III)-corrins, such as methylcobalamin (**2**).[9,10] Here the relevant reactivity may be the ease of methylation and demethylation at the corrin-bound cobalt-centre in (formally) nucleophilic displacement reactions.[4b,9]

This report will be concerned mainly with a detailed investigation on structure and reactivity in organometallic reactions of coenzyme B_{12}-derivatives, with structurally varied axial ligands [the nucleotide base on the lower (α)-face, an organic ligand on the upper (β)-face of the corrin-bound cobalt-centre].

2 EFFECTS OF THE NUCLEOTIDE FUNCTION

A structural characteristic of coenzyme B_{12} **1**, of methylcobalamin **2**[10] and other cobalamins[2] is their metal-coordinating 5,6-dimethylbenzimidazole nucleotide function. Other 'complete' corrins also occur naturally, that differ from the cobalamins by the constitution of their nucleotide 'bases' (see Figure 2).[4b,11] This structural variability of the nucleotide functions may be rationalized on the basis of their biosynthetic availability primarily, rather than the (potentially different) reactivity of the 'complete' corrins.[12]

The unique intramolecular coordination of the nucleotide base may influence by 'electronic' and by 'steric' effects the reactivities of the cobalamins and other 'complete' corrins in organometallic and other reactions.[4,13-16] Schrauzer,[13a,15] Pratt[16] and others[14b,c] found several nucleotide containing 'complete' organocobalamins to undergo spontaneous dealkylation in aqueous solutions considerably more readily than the 'incomplete' organo-cobinamide analogues. This pointed to a 'labilizing effect' of the axial nucleotide coordination on the *trans*-axial organometallic Co-C-bond.

Unexpectedly a stabilizing 'electronic' effect of the base coordination on the strength of the organometallic bond in 'simple' organo-Co(III)-corrins was indicated: studies of methyl transfer equilibria between methylcobalamin **2**, Co$_\beta$-methyl-cobinamide, Co(II)-cobalamin **3** and Co(II)-cobinamide revealed the homolytic (Co-C)-bond dissociation energy [(Co-C)-BDE] of the Co-CH$_3$-bond of the 'complete' **2** to be larger by 2.5 ± 0.5 kcal mol^{-1} (10.5 ± 2.1 kJ mol^{-1}) than

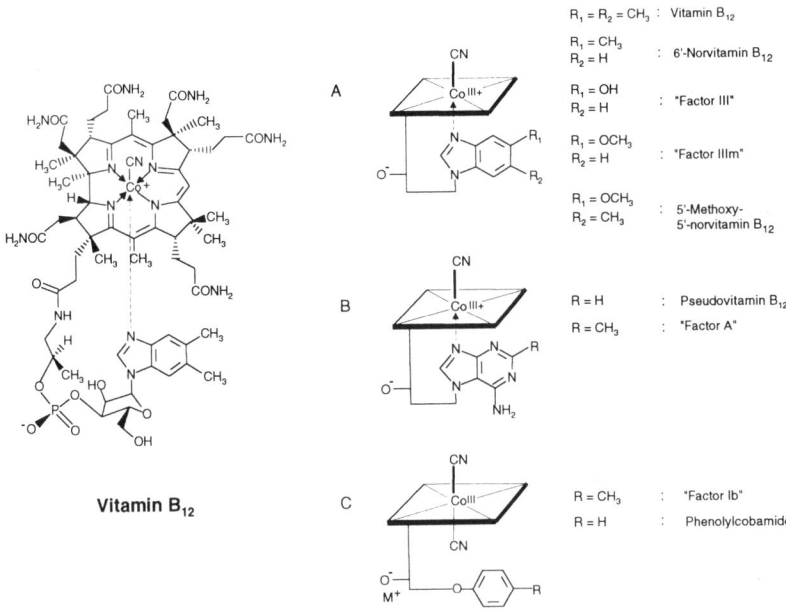

Figure 2: Structural formulae of cyano-Co(III)-forms of natural 'complete' corrins from bacterial sources (A: benzimidazolyl-cobamides; B: N(7)-purinyl-cobamides, C: dicyano-phenolyl-cobamidates).[4b,12]

in the 'incomplete' Co$_\beta$-methyl-cobinamide.[14a] Likewise, in a series of organometallic 'B$_{12}$ - models' an increase of the (Co-C)-BDE has also been noted with increasing basicity of the *trans*-axial ligand.[5,17]

However, kinetic investigations on the thermolysis of coenzyme B$_{12}$ (**1**, adenosylcobalamin) and of the analogous nucleotide free Co$_\beta$-adenosyl-cobinamide, led Finke *et al* to estimate the (Co-C)-BDE for the coenzyme to be smaller by 4.5 ± 3.8 kcal mol^{-1} (18.8 ± 15.9 kJ mol^{-1}) than that of the 'incomplete' Co$_\beta$-adenosyl-cobinamide.[14b] Thermodynamic considerations also pointed to a modest weakening effect of the nucleotide coordination on the strength of the organometallic bond in the coenzyme **1**.[14a]

A major part of the observed labilizing effect of the intramolecular nucleotide coordination in cobalamins and other 'complete' corrins with bulky organometallic groups may therefore be caused 'sterically', eg by an 'upward-folding' of the corrin ligand: the spacial requirements of the bulky nucleotide base upon intramolecular coordination to the 'lower' face of the corrin-bound cobalt-centre may be met better by a nonplanar, 'upward-bent' corrin ligand, as pointed out by Lenhert originally.[1,2] A recent systematic 'factor analysis' confirmed the upward-flexing of the corrin ligand as the major discernible factor of conformational variability of natural corrins.[18] Along these lines, Schrauzer et al[13a,15] have interpreted the observed labilizing effects of the nucleotide coordination on various organocobalamins as 'mechanochemical', since they are induced by conformational motions of the corrin ligand.[15b] Indeed, extensive structural work with various 'B$_{12}$-models'[19] by Marzilli and by Randaccio and their coworkers[13b,20] has shown significant deformation from planarity, and steric response of the ligand in these non-corrinoid compounds to the bulk of the axial ligands. The inherent non-planarity and the presumed ease of deformation of the corrin ligand has been proposed by Geno and Halpern[5] as a rationâle for Nature's preference of the corrin-ligand over the 'simpler' porphyrin ligand in the coenzyme B$_{12}$.

An *X*-ray analysis of the three-dimensional structure of Co(II)-cobalamin 3 revealed a large structural similarity of this Co(II)-corrinoid with the Co(III)-corrin part of the coenzyme 1 ('folding' of corrin ligand, position of the nucleotide base).[21] Accordingly, 'mere' better binding by the protein of the separated homolysis fragments, rather than the intact coenzyme, was suggested as a means for the hypothetical protein- and substrate-induced activation of the homolysis of the (Co-C)-bond of the coenzyme 1,[21] without the involvement of a 'conformational distortion' of the corrin ligand.[7,13b]

In order to gain insight into structural effects of the axial nucleotide coordination on the 'upward-folding' of the corrin ligand we set out to test experimentally the suggested effect of the bulk of the nucleotide base in a 'complete' vitamin B$_{12}$-analogue. The desired information appeared accessible by comparison of the structures of vitamin B$_{12}$ (4, Co$_\beta$-cyano-5',6'-dimethylbenzimidazolyl-cobamide)[2] and the non-natural 'complete' corrin Co$_\beta$-cyano-imidazolyl-cobamide 5.[22] The corrinoid 5 was synthesized by 'guided biosynthesis'[23] with the help of a culture of *Propionibacterium shermanii*, and it was isolated in crystalline form.[24]

The spectroscopic examination of the 'complete' corrinoid 5 in aqueous solution (by UV/vis-, CD-, IR-, (FAB)MS-, one- and two-dimensional ^{1}H- and ^{13}C-NMR-spectroscopy) confirmed the structural identity of this 'complete'

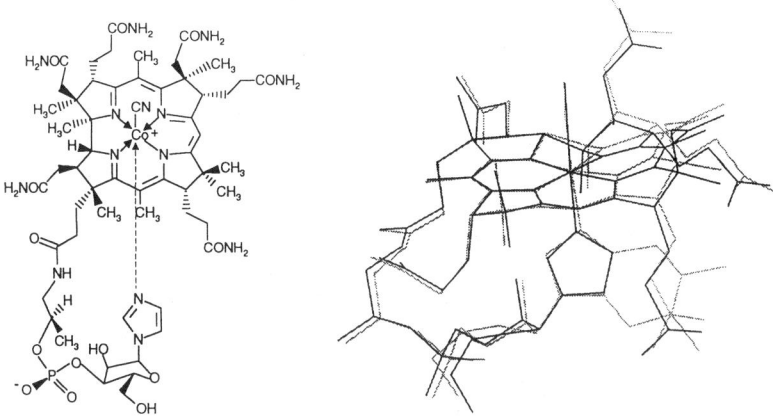

Figure 3: Left: Structural formula of Co$_\beta$-cyano-imidazolyl-cobamide **5**; Right: Superposition of three dimensional structures from X-ray analysis of vitamin B$_{12}$ (**4**, ·····) and of the imidazolyl-cobamide (**5**, ———).[24]

corrin[22] as a Co$_\beta$-cyano-imidazolyl-cobamide (see Figure 3). The spectra indicated a comparable mode of axial base coordination and did not reveal any significant differences in the build-up of the corrin ligand in **5** and in vitamin B$_{12}$ **4**.[25]

More subtle structural differences of the corrin ligand in the imidazolyl-cobamide **5**, when compared to that in vitamin B$_{12}$ **4**, were revealed by the X-ray analysis (see Figure 3). In **5** the 'folding' is reduced to 11.7°, while it amounts to 17.7° in ('wet crystals', 18.7° in 'dry crystals' of) **4**.[2] In **5**, the coordination of the imidazole-base occurs with similar bond angles at nitrogen (128° vs. 127°), but with considerably differing ones in the benzimidazolyl-cobamide **4** (123° vs. 131°).

The X-ray analytical data are compatible with a considerable degree of steric strain in the benzimidazolyl-cobamide **4** (and in the other cobalamins)[2] that is relieved in the imidazolyl-cobamide **5** and that originates in the mutual steric interaction between the 'Northern' part of the corrin ligand and the cobalt-coordinated nucleotide base. Interestingly, the 'upward-folding' in the cobalamins appears to come about mainly from an increased upward bending of the 'Northern' part of the corrin-ligand, as judged from the position of the axial ligands with respect to the 'Northern' and the 'Southern' sections of the corrin-ring (see Figure 3).[24] The indicated bias of the conformational deformation of the corrin ligand presumably can be traced back to the covalent attachment of the nucleotide loop at ring D in the 'Southern' section of the corrin-ligand. Such an uneven increase of the 'upward-folding' of the corrin-ligand due to the nucleotide coordination would produce or increase steric strain (if any) in the 'Northern' section of 'base-on' organocobalamins mainly, compared to nucleotide-free ('incomplete') or 'base-off' forms of 'complete' organocorrins. In coenzyme B_{12},[1,2] as well as in the other organocorrins, whose molecular structures have been determined so far by X-ray[26] and/or NMR-analysis,[26,27] the organic β-ligands occupy space in the 'Southern' sphere of the β-face (*eg* in the coenzyme 1, see Figure 1, but see also below). The anticipated 'upward-folding' of the corrin ligand due to the intramolecular nucleotide coordination would therefore be expected to produce a large 'mechanochemical' effect for those organocorrins, in which the organic β-ligands tend to occupy space in the 'Northern' sphere of the β-face (for an example, see below).

3   CONFORMATIONAL EFFECTS OF THE ORGANIC LIGAND

For reasons delineated above, as well as for the purpose of an increased predictability of the outcome of B_{12}-catalyzed enantioselective reactions[28] and for a better understanding of the coordination properties of (and steric and stereoelectronic effects in) organometallic B_{12}-derivatives,[4,27d] information on structural properties of organocorrins is of considerable interest.

In collaboration with the laboratory of C Kratky (University of Graz, Austria), we have set out to explore in a systematic fashion the characteristic structural properties of organocorrins related to coenzyme B_{12}.[29]

In 5'-(2'-deoxy)-adenosylcobalamin (deoxy-coenzyme B_{12}, **6**, see Figure 4), the hydroxyl group at the 2'-position of the β-bound adenosyl group of the coenzyme 1 is replaced by hydrogen. This coenzyme B_{12}-analogue has been found by Hogenkamp *et al*[30] to hydrolyze considerably faster than the coenzyme 1 itself, and has recently been subjected to an ^{1}H-NMR-analysis.[31] We have prepared a crystalline sample of this light-sensitive, organometallic B_{12}-derivative by a newly developed electrochemical route[32] from Co$_β$-methylhydroxyl-cob(III)alamin and 5'-tosyl-2'-deoxy-adenosine and characterized it thoroughly by high-field NMR-analysis in aqueous solution, as well as by a single crystal X-ray analysis.[29] These analyses, in short, showed a conformation of the organic β-ligand in the coenzyme analogue **6** that is superimposable to a remarkable extent on that of the coenzyme 1 itself.[2]

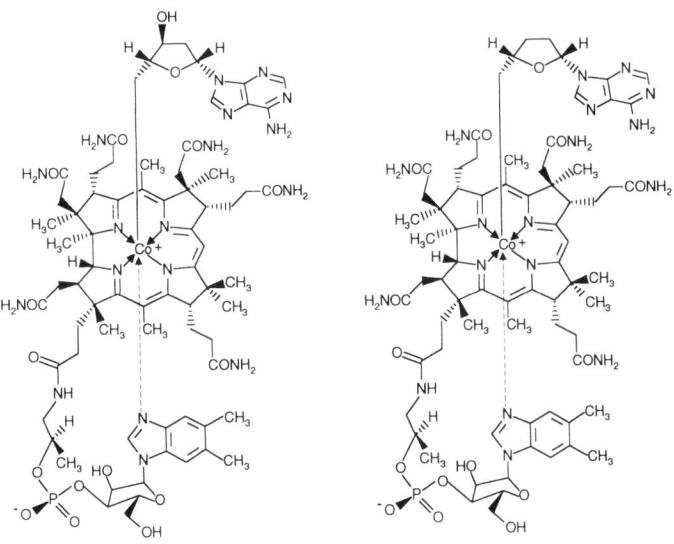

Figure 4: Structures of organocobalamins: Left: deoxy-coenzyme B_{12}, **6**; Right: dideoxy-coenzyme B_{12}, **7**.

In view of the results obtained with **6**, the extensive change in the conformation of the organic ligand in **7**, brought about by the removal of the second hydroxyl group from the adenosyl ligand of the coenzyme **1**, was unexpected. The dideoxy-coenzyme B_{12} (**7**, 5'-(2',3'-dideoxy)-adenosylcobalamin), also prepared electrochemically and obtained in crystalline form, was indicated already by the absence of an upfield shifted $H_3(C12\beta)$-singlet in its ^{1}H NMR-spectrum to differ considerably from coenzyme B_{12} in the orientation of the organometallically bound group (see Figure 5). Complete assignment of the ^{1}H- and ^{13}C-NMR spectra, as well as high-field ROESY-experiments, allowed an extensive analysis of the structure of the dideoxy-coenzyme **7** in aqueous solution: in agreement with its *X*-ray analytically determined structure in a single crystal (see Figure 6), the (predominant) conformation of **7** in solution has the adenine group positioned on top of ring D (instead of ring C, as in **1**).[29] The dideoxy-ribose ring is placed close to ring A in **7**,[29] while in **1** and in **6** the (deoxy)-ribose units are situated 'South' above the bridge between rings C and D.[2,29]

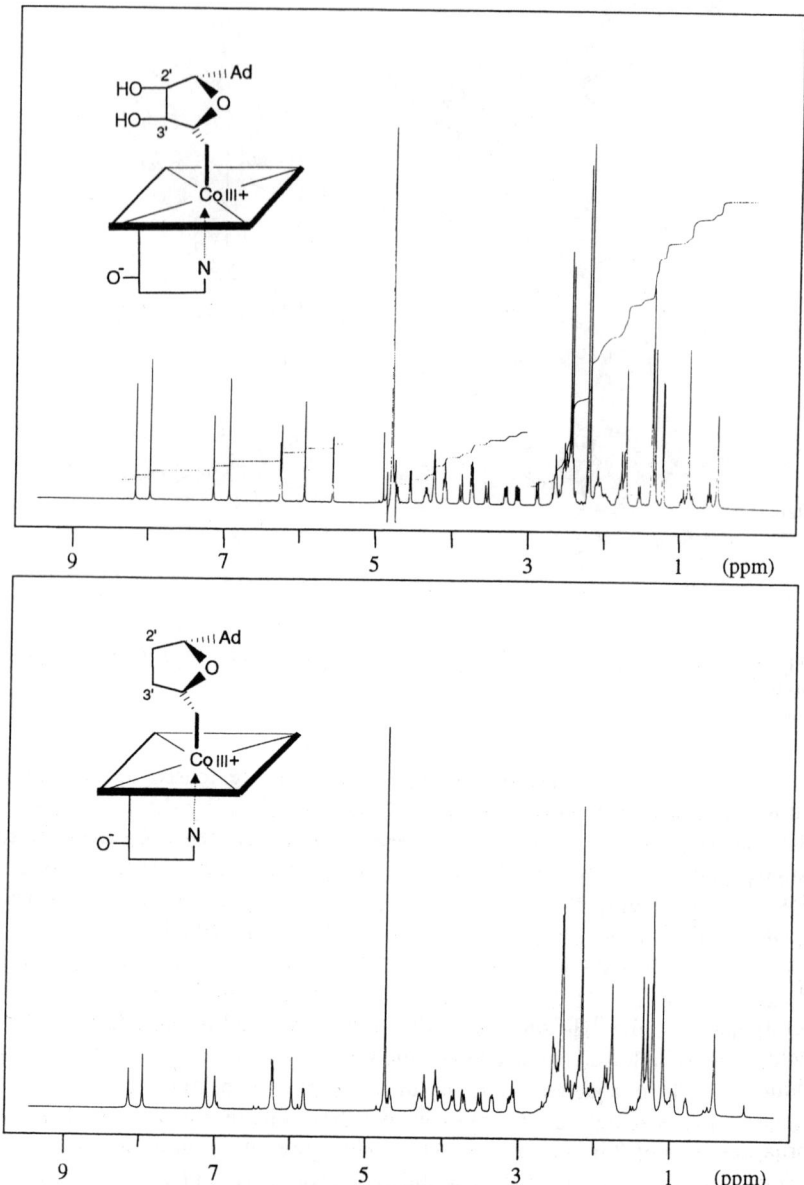

Figure 5: 400 MHz ^{1}H NMR spectra of coenzyme B_{12} (**1**, top) and of the dideoxycoenzyme B_{12} **7** (bottom) in D_2O.[29]

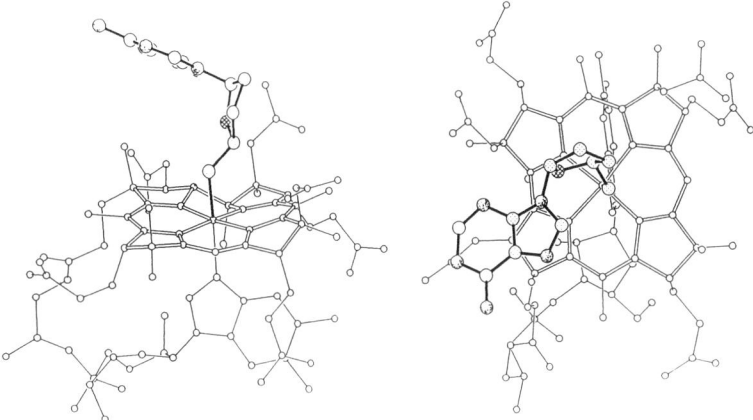

Figure 6: Three-dimensional structure of the dideoxy-coenzyme B_{12} **7**, from X-ray analysis. Left: projection from 'Western' side; Right: axial projection.[29]

The dideoxy-coenzyme **7** (see Figure 6),[29] as well as the slightly less complex organocobalamin **9** (*S*-tetrahydrofurfuryl-cobalamin, see below), represent first examples of organocobalamins, in which the organic ligand is bound to the cobalt-centre in an unprecedented conformation and resides near ring A of the corrin ligand.[29]

Before arriving at the unexpected structural result for **7**, we examined the structure of the two diastereomeric tetrahydrofurfuryl-cobalamins **8** and **9**. In these coenzyme B_{12}-analogues, only an oxacyclopentyl-methyl ligand is bound to cobalt instead of the 5'-deoxyadenosyl group of the coenzyme **1** itself. A *ca* 1:1 diastereomeric mixture of the coenzyme analogues **8** and **9** had been prepared earlier.[33] Using diastereomerically purified tetrahydrofurfuryl-(*S*)-camphorsulfonates[34] the electrochemical preparation of the practically diastereomerically uniform, crystalline *R*-tetrahydrofurfurylcobalamin **8** and *S*-tetrahydrofurfurylcobalamin **9,** and their thorough spectrosopic analysis was achieved (for their ^{1}H NMR spectra, see Figure 7):[30] exploratory two-dimensional NOESY-spectroscopic analysis of **8** and **9** in neutral D_2O-solution suggested the (2'*R*)-tetrahydrofurfuryl ligand in **8** to reside predominantly in proximity of rings C and D, similar to the situation in the coenzyme **1**; for the (*S*)-isomer **9** the information from NOESY spectra was less clear. By single crystal X-ray analyses of the diastereomeric tetrahydrofurfurylcobalamins the absolute configuration (*R* or *S*) at the 2'-carbon of the tetrahydrofurfuryl ligands was clarified.

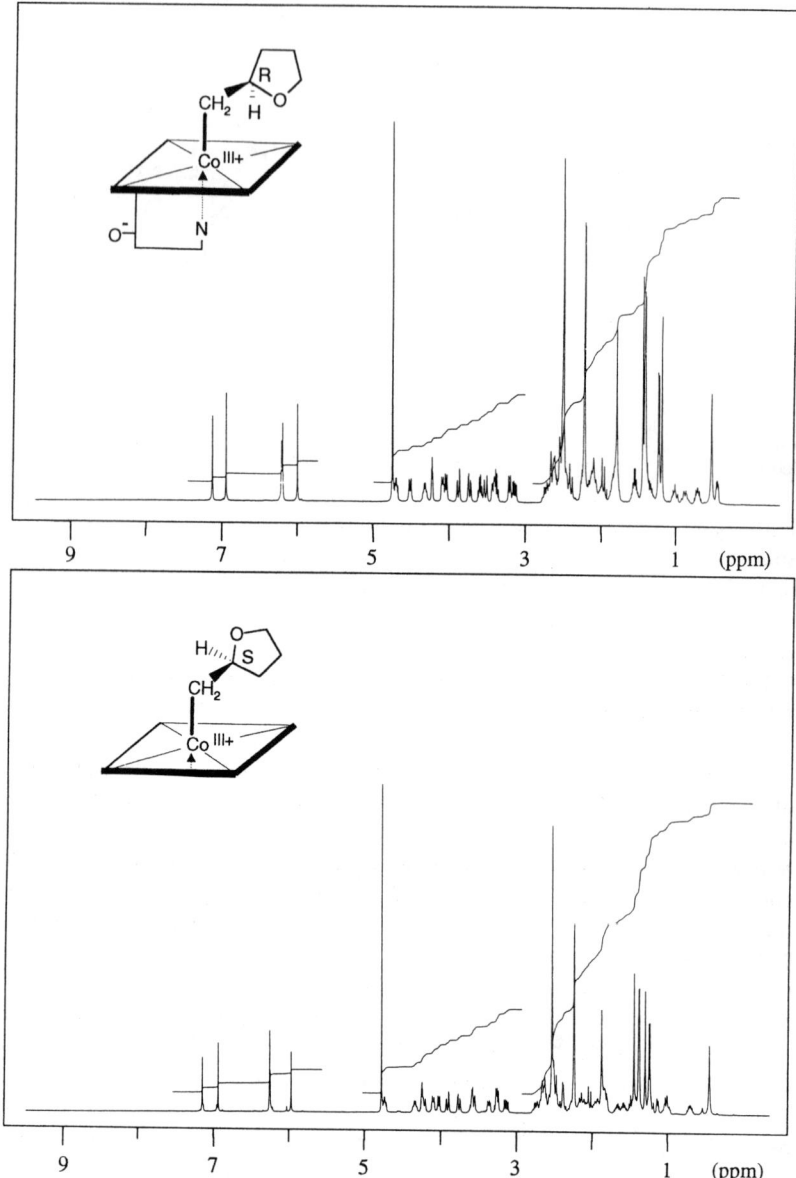

Figure 7: 400 MHz ^{1}H NMR spectra of *R*-tetrahydrofurfurylcobalamin (**8**, top) and of *S*-tetrahydrofurfurylcobalamin (**9**, bottom) in D_2O.[29]

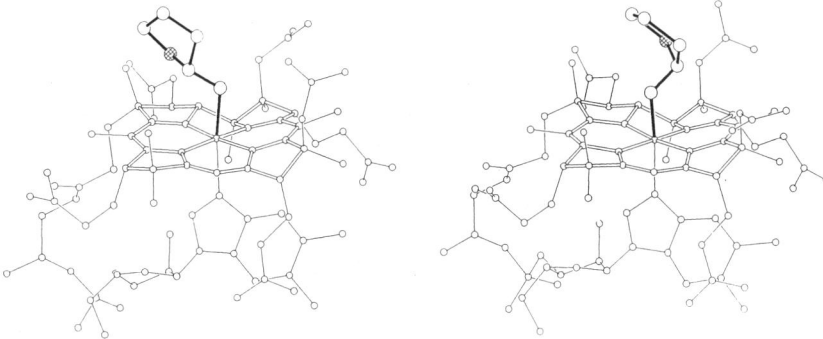

Figure 8: Three dimensional structures of tetrahydrofurfuryl-cobalamins. Left: *R*-tetrahydrofurfuryl-cobalamin **8**; Right: *S*-tetrahydrofurfurylcobalamin **9**.[29]

Further, the organocobalamins **8** and **9** could be structurally characterized in the solid state with respect to conformational properties of the metal-bound organic ligand.[29] In the diastereomeric pair **8/9** the conformation relative to the corrin ring of the tetrahydrofurfuryl ligand proved to be remarkably dependent upon the configuration at the Cβ2-position: For the *R*-isomer **8** the organic ligand turned out to be oriented towards the 'Southern' *meso*-position of the corrin ligand, *ie* roughly in an earlier encountered conformation. On the other hand, the data revealed the tetrahydrofurfuryl group in **9** to be situated in the 'Western' sphere of the β-face, exhibiting a nearly eclipsed arrangement of the (Cβ1-Cβ2)-bond of the organic ligand and of the (Co-N20)-bond (N20-Co-Cβ1-Cβ2 torsion angle: -16.7°). The corrin **9** proved to be the first organocobalamin with such an orientation of the cobalt-bound organic ligand. In its orientation relative to the corrin-ring, the (2'*S*)-tetrahydrofurfuryl group of **9** also provided a far-reaching precedence for the conformation of the organic ligand of the dideoxy-coenzyme **7**.[29] (C4' of the dideoxy-ribosyl unit of **7** has the same absolute configuration as C2' of the tetrahydrofurfuryl ligand of **9**).

In the solid state, the organo-cobalamins **6**, **7**, **8** and **9** all exhibit a large (Co-Cβ1-Cβ2)-bond angle at the metal-bound saturated carbon of their organometallic functionality (*eg* 118.9° in **8**, 130.2° in **9**). Similar bonding angles have been determined in three other *X*-ray analytically studied organo-cobalamins,[2,26] as well as in the coenzyme B_{12} itself, for which this large angle (122.2°) was interpreted as being due to steric strain.[2]

In the organo-cobalamins **6**, **7**, **8** and **9**, the (Co-Cβ1-Cβ2-O) torsion angles were determined from the *X*-ray data to be close to 90° (*eg* 86.5° in **8**,

90.4° in **9**), again similar to the situation in the coenzyme **1** (74.3°). It is the ring methylene group that is placed in an *anti*-conformation with respect to the metal in **8** and in **9**, rather than the similarly bulky, but more electronegative ring-oxygen. Accordingly, a stereoelectronic preference of a *trans*-antiplanar arrangement cobalt-carbon-carbon-oxygen is not indicated by the X-ray data. Such an arrangement would be kinetically relevant for the acid-induced decomposition of the coenzyme **1** and of other organo-corrins bearing an oxygen functionalitiy in the β-position of the organic ligand,[35] as well as in the reductive, B_{12}-catalyzed formation of alkenes from β-hydroxybromoalkanes.[36] A *trans*-antiplanar arrangement Co-Cβ1-Cβ2-oxygen, stereoelectronically favouring acid-induced (Co-C)-bond heterolysis, would be strained in the coenzyme **1** due to steric repulsion between the corrin ring and the adenosyl hydroxyl groups. By hindering such a conformation, the latter presumably may contribute sterically to the remarkable and biologically important resistance of coenzyme B_{12} with respect to acid-induced cleavage of their organometallic bond.

In contrast to the coenzyme **1**, the tetrahydrofurfurylcobalamins **8** and **9** and the coenzyme analogue **7** are indeed remarkably acid sensitive and decompose readily with the formation of aquocobalamin at room temperature: in 0.1 M aqueous phosphate buffer, *eg* at pH 3.9, **8** and **9** decompose with a half life ($t_{1/2}$) of *ca* 8 resp 5 min, *ie ca* 100 times faster than the dideoxy-coenzyme **7** and *ca* 10^4 times faster than the coenzyme **1** itself ($t_{1/2}$= *ca* 46 000 min).[37]

In summary, specific 'remote' constitutional changes in the nucleotide base and in the organometallic functionality of five vitamin B_{12}- and coenzyme B_{12}-analogues have been analysed for 'steric' (and possibly 'stereoelectronic') effects on structure and reactivity, to explore some basic conformational characteristics of organometallic B_{12}- derivatives.

REFERENCES

1   (a) D C Hodgkin, 'Vitamin B_{12} and Intrinsic Factor', H C Heinrich (ed), Proceedings of the 1st Europ Conf on Vitamin B_{12} and Intrinsic Factor, F Enke Verlag, Stuttgart, 1957, p 31. (b) P G Lenhert and D C Hodgkin, *Nature* (London), 1961, **192**, 937, (d) P G Lenhert, *Proc Roy Soc* (London) 1968, **A303**, 45.

2   Reviewed in *eg*: (a) J P Glusker, 'B_{12}', D Dolphin, ed, Wiley & Sons, New York, 1982, Vol 1, p 23. (b) M Rossi and J P Glusker, 'Molecular Structure and Energetics', J F. Liebman, and A Greenberg, eds, Vol X, VCH-Publishers, Weinheim (FRG), 1988, p 1.

3   A Eschenmoser, *Angew Chem* 1988, **100**, 5; *Angew Chem Intl Ed Engl*, 1988, **27**, 5.

4 (a) J M Pratt, 'B$_{12}$', D Dolphin, ed, J Wiley & Sons, New York, 1982, Vol 1, p 325. (b) B Kräutler, in 'The Biological Alkylation of Heavy Elements', P J Craig and F Glockling, eds, The Royal Society of Chemistry, London, Special Publication No. 66, 1988, p. 31.
5 M K Geno and J Halpern, *J Am Chem Soc*, 1987, **109**, 1238.
6 See eg (a) R H Abeles, 'Biological Aspects of Inorganic Chemistry', Wiley, New York, 1977, p 245. (b) BT Golding and DNR Rao, in 'Enzyme Mechanisms', M I Page and A Williams, eds, The Royal Society of Chemistry, London, 1987, p.404.
7 (a) J Halpern, *Science* 1985, **227**, 869. (b) J Halpern, S-H Kim, and T W Leung, *J Am Chem Soc*, 1984, **106**, 8317.
8 (a) R G Finke and B P Hay, *Inorg Chem*, 1984, **23**, 3043. (b) B P Hay and R G Finke, *J Am Chem Soc*, 1986, **108**, 4820.
9 R G Matthews, R V Banerjee, and S W Ragsdale, *BioFactors*, 1990, **2**, 147.
10 M Rossi, J P Glusker, L Randaccio, M F Summers, P J Toscano, and L G Marzilli, *J Am Chem Soc*, 1985, **107**, 1729.
11 (a) W Friedrich, 'Vitamin B$_{12}$ und verwandte Corrinoide', Vol III/2 of Fermente, Hormone und Vitamine', R Ammon and W Dirscherl, eds, Georg Thieme Verlag, Stuttgart, 1975, p 25.
12 E Stupperich, H J Eisinger, and B Kräutler, *Eur J Biochem*, 1989, **186**, 657.
13 (a) G N Schrauzer, J H Grate, M Hashimoto, and A Maihub, 'Vitamin B$_{12}$', B Zagalak and W Friedrich, eds, Proceedings of the Third European Conference, W de Gruyter, Berlin, 1979, p 511. (b) N Bresciani-Pahor, M Forcolin, L G Marzilli, L Randaccio, M F Summers, and P J Toscano, *Coord Chem Rev*, 1985, **63**, 1.
14 (a) B Kräutler, *Helv Chim Acta*, 1987, **70**, 1268; (b) B P Hay and R G Finke, *J Am Chem Soc*, 1987, **109**, 8012; (c) K Brown and H B Brooks, *Inorg Chem* 1991, **30**, 3420.
15 (a) J H Grate and G N Schrauzer, *J Am Chem Soc*, 1979, **101**, 4601; (b) G N Schrauzer and J H Grate, *ibid*, 1981, **103**, 541.
16 S M Chemaly and J M Pratt, *J Chem Soc Dalton Trans*, 1980, 2274.
17 F T T Ng, G L Rempel, C Mancuso, and J Halpern, *Organometallics*, 1990, **9**, 2762.
18 V B Pett, M N Liebman, P Murray-Rust, K Prasad, and J P Glusker, *J Am Chem Soc*, 1987, **109**, 3207.
19 (a) G N Schrauzer and J Kohnle, *Chem Ber*, 1964, **97**, 3056; (b) G Costa, *Pure Appl Chem*, 1972, **30**, 335.
20 (a) L Randaccio, N Bresciani-Pahor, E Zangrando, and L G Marzilli, *Chem Soc Rev*, 1989, **18**, 225; (b) W O Parker, Jr, E Zangrando, N Bresciani-Pahor, P A Marzilli, L Randaccio, and L G Marzilli, *Inorg Chem*, 1988, **27**, 2170.

21 B Kräutler, W Keller, and C Kratky, *J Am Chem Soc*, 1989, **111**, 8936.
22 G Eberhard, H Schlayer, H Joseph, E Fridrich, B Urz, and O Müller, *Biol Chem Hoppe-Seyler*, 1988, **369**, 1091.
23 Early results concerning 'guided biosynthesis' of 'complete' corrins are mentioned in (a) J M Pratt, 'Inorganic Chemistry of Vitamin B_{12}', Academic Press, London, 1972; (b) E Stupperich, I Steiner, and M Rühlemann, *Anal Biochem*, 1986, **155**, 365.
24 B Kräutler, R Konrat, E Stupperich, G Färber, and C Kratky, submitted for publication.
25 (a) T G Pagano and L G Marzilli, *Biochemistry*, 1989, **28**, 7213; (b) W Eisenreich and A Bacher, *J Biol Chem*, 1991, **266**, 23840.
26 (a) N W Alcock, R M Dixon, and B T Golding, *J Chem Soc Chem Commun*, 1985, 603; (b) T G Pagano, L G Marzilli, M M Flocco, C Tsai, H L Carrell, and J P Glusker, *J Am Chem Soc*, 1991, **113**, 531.
27 (a) M F Summers, L G Marzilli, and A Bax, *J Am Chem Soc*, 1986, **108**, 4285 (b) A Bax, L G Marzilli, and M F Summers, *J Am Chem Soc*, 1987, **109**, 566; (c) T G Pagano, P G Yohannes, B P Hay, J R Scott, R G Finke, and L G Marzilli, *J Am Chem Soc*, 1989, **111**, 1484; (d) R J Anderson, R M Dixon, and B T Golding, *J Organomet Chem*, 1992, **437**, 227.
28 P Bonhôte and R Scheffold, *Helv Chim Acta*, 1991, **74**, 1425.
29 G Färber, K Gruber, R Konrat, B Kräutler, C Kratky, J Maynollo and M Puchberger, publication in preparation.
30 H P C Hogenkamp and T G Oikawa, *J Biol Chem*, 1964, **239**, 1911.
31 H Yan, H Sun, H Chen, and W Tang, *Spectroscopy Letters*, 1993, **26**, 319.
32 M Puchberger, T Derer, J Maynollo, and B Kräutler, unpublished results.
33 S-H Kim, H L Chen, N Feilchenfeld, and J Halpern, *J Am Chem Soc*, 1988, **110**, 3120.
34 S-H Kim and K Stockhausen, *J Med Chem*, 1979, **22**, 1475.
35 H P C Hogenkamp, 'B_{12}', D Dolphin, ed, Wiley & Sons, New York, 1982, Vol 1, p 295.
36 R Scheffold, S Albrecht, R Orlinksi, H-R Ruf, P Stamouli, O Tinembart, L Walder, and C Weymuth, *Pure & Appl Chem*, 1987, **59**, 363.
37 M Puchberger and B Kräutler, unpublished results.

ACKNOWLEDGEMENTS

I would like to thank Doz Dr Christoph Kratky, Dr Gerald Färber and Karl Gruber (University of Graz, Austria), Dr Erhard Stupperich (University of Ulm), for their decisive collaborative work, Dr Robert Konrat, Dr Michael Puchberger and Josef Maynollo (University of Innsbruck) for their important experimental contributions, which were financially supported in part by the Austrian Science Foundation.

HYDROLYTIC CATALYSTS AS METALLOENZYME MODELS

Paolo Scrimin, Paolo Tecilla, and Umberto Tonellato*

*Centro CNR Meccanismi di Reazioni Organiche,
Dipartimento di Chimica Organica, Universita' di Padova, 35131 Padova, Italy.*

1   INTRODUCTION

Many hydrolytic metalloenzymes[1] have evolved to hydrolyze some of the most important molecules of life, from proteins to DNA. The ions of the 'essential' metallic elements function as enzyme cofactors or prosthetic groups and their activity may be the result of: (a) their inherent activity in catalyzing a chemical reaction magnified by the enzyme protein; (b) the bringing together, through complexation and activation, of the substrate and the functional groups in the active site of the enzyme; (c) their activity as Lewis acids at some point in the catalytic cycle. Over the years, a number of hydrolytic metalloenzyme models have been synthesized through elegant designs and have been carefully investigated.[2] Much of the interest for the study of reactions in metalloaggregates is also due to a continuously refined definition[3] of the structure and mode of action of carboxypeptidase A (CPA), the best known of the many zinc proteases.

2   FUNCTIONALIZED LIGAND SURFACTANTS AS ENZYME MODELS

Simple models cannot compete with the natural biological systems where many variable parameters are simultaneously operative and it is not surprising that the major difference between enzymes and their models is their reactivity and selectivity. Only large macromolecules such as the proteins, sometimes with the need of nonprotein adjuncts, can carry enough molecular information both from the point of view of thermodynamic efficiency and substrate recognition. Proteins play a crucial role in the enzymic activity by providing a precise three-dimensional pattern, the needed flexibility, and the proper environment at the active site.[1] A major drawback in the design of a model is the lack of an effective protein part which would be quite difficult or impossible to consider. This is one of the reasons that stimulated the study of amphiphilic functionalized models in their aggregate forms (micelles, vesicles).[4] Such assemblies may mimic, although to a quite limited extent, some of the functions of the protein portion of the enzymes: they bind hydrophobic substrates, attract ionic species, provide at the reaction site a distinct, more hydrophobic environment than in the bulk solution, and may show interdependency of group behaviour.[5]

2.1 Metallomicelles as Hydrolytic Catalysts

Recently, several research groups[6,7] including ours,[8] have synthesized and investigated the catalytic properties of transition metal ion ligand surfactants which are catalysts of the hydrolytic cleavage of activated esters. In our early studies[8a,b] of metallomicellar systems we focussed our attention on amphiphilic molecules of general structure **1** (Chart 1) featuring a chelating subsite (a 2,6-disubstituted pyridino residue), a hydroxy function bound into the proximity of the chelation site, a paraffinic chain, and, in some case, a quaternary ammonium

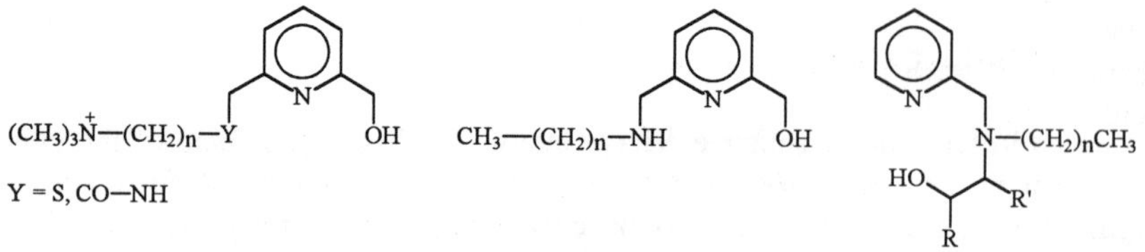

Chart 1

group to ensure solubility in water. Using *p*-nitrophenyl picolinate (PNPP) as a substrate and Cu(II) as the metal ion at relatively low concentration (0.4-0.6 mM), and at neutral pHs, more than a million-fold rate enhancements over hydrolysis in the absence of the metal ion were observed. The reaction path was shown to involve the formation of a ternary complex (ligand, metal ion, substrate) followed by a transacylation process and subsequent hydrolysis of the intermediate as illustrated in Scheme 1. The metal ion acts as a template in the cleavage of the esters and the process is a really catalytic one with fast turnover rates. Micellar amphiphilic ligands like **1** are virtually inert in the case of esters of acids devoid of chelating functions, such as the amino acids, and unable therefore to form the ternary complexes of the type shown in Scheme 1. More lipophilic ligand surfactants such as **2** and **3** (Chart 1, n =11 or 15) have been

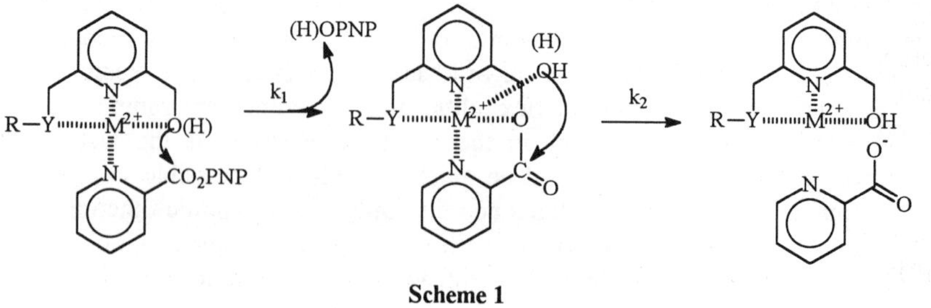

Scheme 1

shown to be more versatile ligand catalysts. They are not dispersible in neutral water; however, they chelate transition metal ions [particularly Cu(II)] with a very large complexation constant and the complexes form highly hydrophobic aggregates:[8c,d] they are effective in the cleavage not only of esters of α-amino acids but also of carboxylic and phosphoric acids. Remarkable rate accelerations were observed in the presence of the micellized complex 2·Cu(II) in the hydrolysis at pH = 6.3 of activated (*p*-nitrophenyl) esters of picolinic (over 10^6), acetic (up to 6×10^3), hexanoic (up to 4.5×10^3), dodecanoic (up to 1.7×10^4), and diphenylphosphoric (up to 2.6×10^4) acids. The mode of action in the case of non-binding substrates is different from that of amino acid derivatives: the key role is played by the micelles which strongly bind the hydrophobic substrates and bring them into the proximity of the reactive chelation site of the surfactant ligand. Moreover, ligands 3 were synthesized as pure enantiomers and were found to be enantioselective catalysts for the cleavage of chiral activated α-amino acid esters.[8e,9]

We recently addressed the following questions: (a) are metalloaggregates made of 2 effective catalysts in the cleavage of unactivated esters? (b) which factors are at the source of the enantioselection in the cleavage of chiral activated esters of α-amino acids in metallomicelles made of chiral surfactants of general structure 3? This paper presents and discusses the results obtained, addressing the two issues above by employing Cu(II) complexes with ligand surfactants 2 and 3 as micellar catalysts and the esters of α-amino acids.

3 THE LEAVING GROUP EFFECT IN THE CLEAVAGE OF PICOLINATE ESTERS EMPLOYING LIGANDS 2

In most esterase model studies, the substrates are *p*-nitrophenyl esters. These are reactive substrates and their hydrolysis is easy to follow spectrophotometrically. However, the enzyme's 'authentic' substrates are not *p*-nitrophenyl derivatives; moreover, the results observed employing esters with such good leaving groups must be taken with caution since the structural requirements of a catalyst for activated esters are not always the same as those for unactivated esters or amides.[10,11] Our investigation on the leaving group effect was also stimulated by the report by Fife and Przystas[12] that the Ni(II) or Cu(II) promoted alkaline hydrolysis of picolinate esters is virtually insensitive to the nature of the leaving group as its pK_a changes from 4 to 12. This set our hopes on finding out whether our metalloaggregates are effective also in the hydrolysis of simple alkyl esters or amides as real biomimetic models and as catalysts in their own right.

We carried out a kinetic study using ligands 2a-c and measured the rate of cleavage of picolinates 4 a-l (see Chart 2, the pK_a values of ROH are within parentheses) under a variety of conditions. Here we present and discuss the kinetic data obtained for the conditions: pH = 6.3 using MES (morpholinoethanesulphonate) buffer, 25 °C, from measurement:

(a) in pure buffer (in the case of the slowest substrates, the rate constants were obtained by extrapolation from data measured at higher temperatures);
(b) in the presence of Cu(II) only [addition of Cu(II) up to 3×10^{-4} M linearly increases the rate without evidence of any trend towards saturation conditions];
(c) in the presence of Cu(II)·2a;
(d) in the presence of non micellar Cu(II)·2b;
(e) in the presence of Cu(II)·2c.

In the presence of ligands 2a and 2b, [measurements under (c) and (d), see above], the cleavage of the most reactive esters (4a-4j) as followed spectrophotometrically in the range 260-270 nm, showed a biphasic behaviour. It was shown that the fast process can be ascribed to the cleavage of the ester with formation of the transacylation intermediate (see Scheme 1) which is then hydrolyzed to water in a slower process. The rate constant for the slower process is virtually identical for all the substrates (0.2 ± 0.1 s^{-1}) under the conditions used.

The results are shown in Figure 1 in terms of a Brønsted plot,[13] log k_ψ vs pK_a of the conjugate acid of the leaving group. The k_ψ values are those of the fast process where the biphasic behaviour was observed. Besides the kinetic benefits resulting upon addition of the metal ion and of the complexes of metal ion with the ligands, the plots of Figure 1 indicate that: (a) in the absence of metal ions, the plot is linear and the slope (β) is ca -0.25; (b) in the presence of Cu(II) the plot shows two straight lines with a break at around the pK_a value of 11-12: the slope changes dramatically from ca. 0 for the 'good' leaving groups to -0.5 for the poorer ones; (c) in the presence of Cu(II)·2a the β values are ca -0.3 and -0.65 and the change in slope is observed at pK_a = ca 9; (d) in the presence of Cu(II)·2b the β values are -0.3 and -0.5 with a break at pK_a = ca 11; (e) in the presence of Cu(II)·2c the observed rate constants (not shown in the Figure) are only slightly lower than those in the presence of Cu(II) only.

The observation of nonlinear Brønsted plots of the type shown in Figure 1 is classic evidence[14] of a change in the rate-determining step (in simple nucleophilic esterolyses from formation to breakdown of a tetrahedral intermediate) and the break in the plot is normally taken as an indication that the pK_{nuc} = pK_{lg}. Thus, in the presence of Cu(II) alone, the nucleophilic species may be assumed to be a metal ion coordinated water molecule with an apparent pK_a = 11-12, and the mode of action is that suggested (see Scheme 2, A) for the metal ion promoted esterolysis, probably through a Lewis acid and a metal hydroxide mechanism. In the presence of the micellar complex with 2a and 2b,

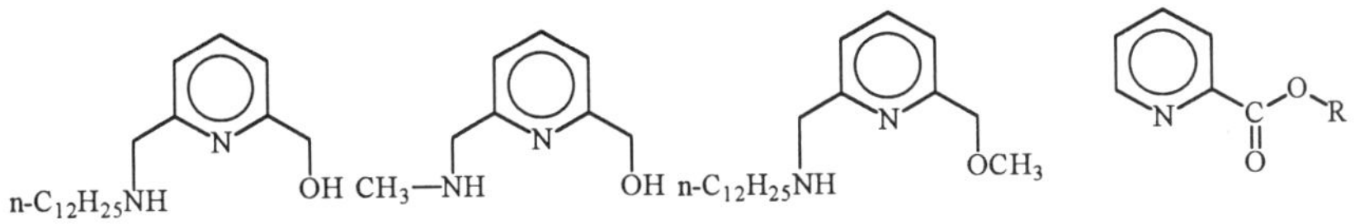

Hydrolytic Catalysts as Metalloenzyme Models

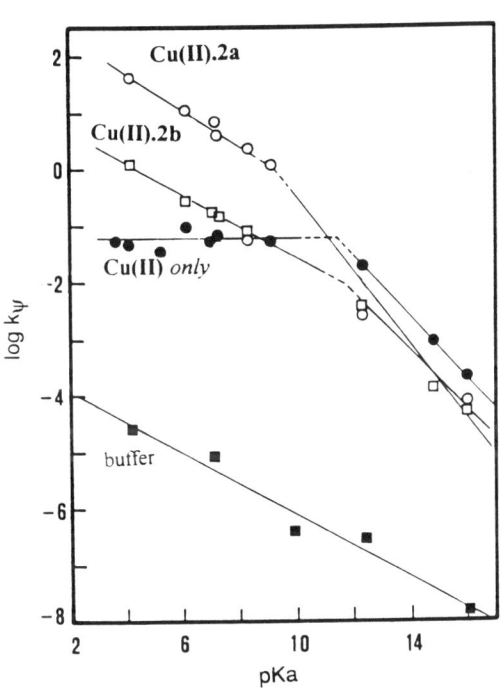

R (in 4): a: 2-chloro,4-nitrophenyl (3.5); **b:** 2,4-dinitrophenyl (4.1); **c:** pentachlorophenyl (5.2); **d:** 2-nitro,5-fluorophenyl (6.1); **e:** 4-nitrophenyl (7.1); **f:** 2-nitrophenyl (7.2); **g:** 3-nitrophenyl (8.3); **h:** 3-chlorophenyl (9.1); **i:** phenyl (9.9); **j:** tri-fluoroethyl (12.4); **k:** 2-methoxyethyl (14.8); **l:** ethyl (16.0)

Chart 2

Figure 1: Brønsted plots (log K_ψ vs pK_a of the conjugate acid of the leaving group) of the cleavage of picolinates **4a - 4j** induced by Cu(II) in the presence and absence of ligands **2a** and **2b**.

the nucleophilic species may be identified as the hydroxy function of the ligand (Scheme 2, B) with an apparent pK_a = *ca* 9 (**2a**) or 11 (**2b**), at least in the case of substrates with good leaving groups. In the case of esters with poorer leaving groups, the nucleophile may become (again) a metal ion coordinated water molecule within the tertiary complex (Scheme 2, C). Mode C is supported by the fact that the plots for $pK_a > 12$ are very close with similar slopes and the O-methylated micellar complex of **2c** is not a better catalyst than the metal ion alone.

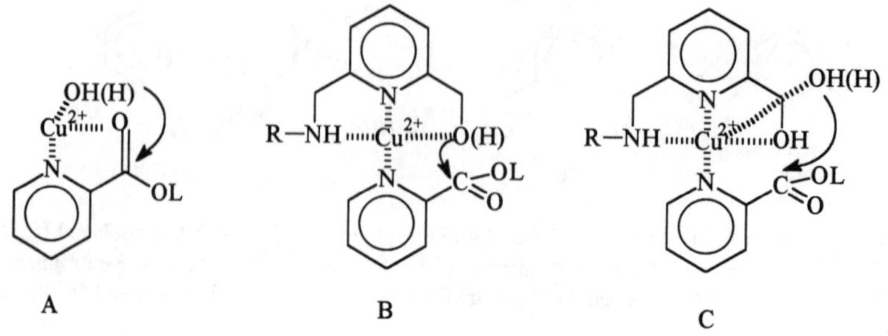

Scheme 2

In the case of *activated* picolinates, all the kinetic benefits of the metal ion are in full display so that the rate becomes insensitive to the leaving group effect. When the Cu(II) is associated with ligand molecules, it activates its hydroxy function with kinetic benefits which become more and more relevant as the pK_a of the leaving group decreases; moreover, all the catalytic effects are remarkably enhanced in the cationic micellar aggregate. In the case of *unactivated picolinates*, the metal ion alone promotes the hydrolysis more (albeit slightly) effectively than when it is complexed with any ligand used.[15] Thus the complexes with **2a** (or **2b**) are very effective for the less needy substrates and somehow fail in the case of unactivated ones; they still remarkably enhance the hydrolysis rate over that in buffer but add nothing to the effect of the metal ion alone. From the point of view of enzyme modelling, as pointed out by Menger,[10] the hydrolysis of *authentic* substrates present a challenge that nature, but not organic chemistry, has already met.

4 ENANTIOSELECTIVITY IN THE CLEAVAGE OF α-AMINO ACID ESTERS

The tight interaction between ligand and substrate within the ternary complex involved in the cleavage of activated α-amino acid stimulated the investigation of the enantioselective effects due to chiral ligands of the general structure **3**. Early investigations[9] using micelles of ligand **3a** indicated that relevant enantioselections may be achieved with a variety of activated (*p*-nitrophenyl, PNP) esters. In the search for the source of the enantioselectivity, we have recently undertaken a kinetic study employing structurally different ligands **3-3'** (see Chart 3) under a variety of conditions (different pH, temperature, cosurfactants). The Table shows some of the significant results obtained using ligands **3** and different PNP esters (Phe = phenylalanine, Phg = phenylglycine, Leu = leucine, CBZPhe = carbobenzoxyphenylalanine) under the standard conditions: [Cu(II)] = 8.3 × 10^{-5}M, [ligand] = 1.5 × 10^{-4}M (unless otherwise indicated), in MES buffer, pH = 5.5, 25 °C.

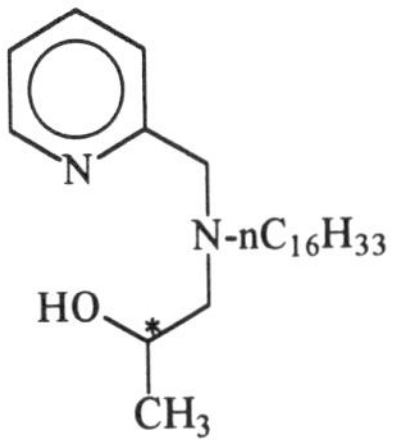

Chart 3

Ligands **3a**, 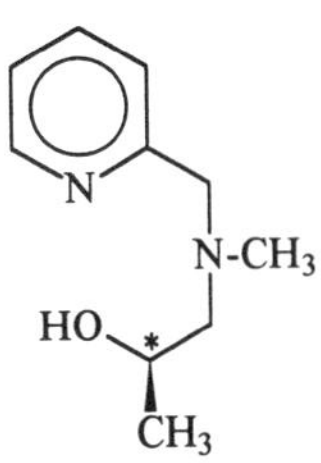 **3b**, 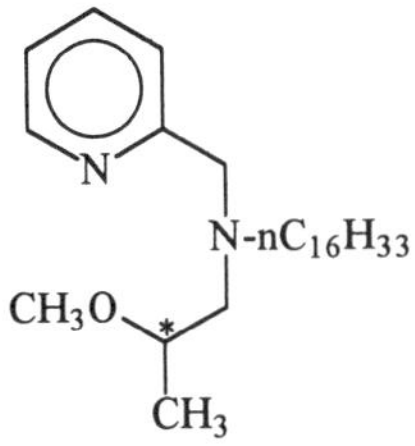 **3c**, 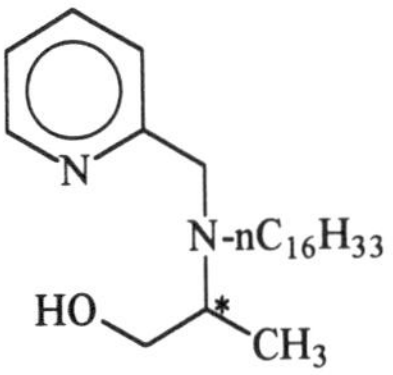 **3'm**, 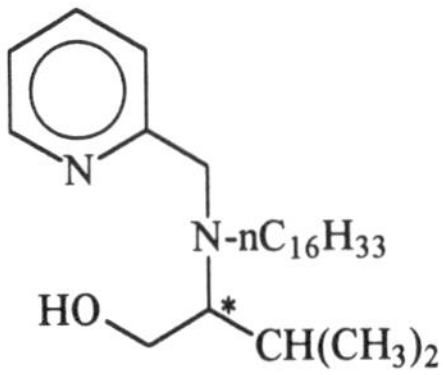 **3'i**, 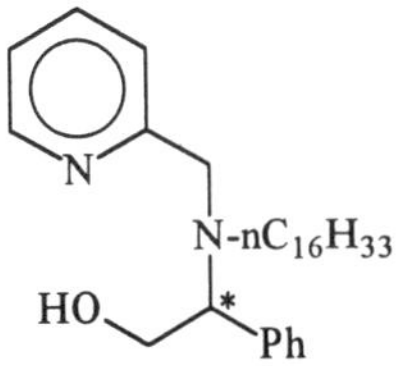 **3'ph** (pyridyl-methyl amino alcohol structures with varying N-substituents and side chains).

Table: Enantioselectivity factors (ef)[a] for the esterolytic cleavage of PNP esters of α-amino acids employing ligands **3a-c** and **3'm-ph**.

Ligand	3a	3a	3a	3a	3b	3c	3'm	3'm	3'i	3'ph
Substrate	Phe	Phg	Leu	CBZPhe	Phe	Phe	Phe	Phe	Phe	Phe
ef	7.4	10.5	6.0	1.5	1.6	1.1	9.5	7.8[b]	6.3[b]	8.0[b]

[a] ef = the ratio of the pseudo-first-order rate constants observed for the two enantiomers of the ester: k_R/k_S or k_S/k_R; [b] conditions: [ligand] = 0.2 mM; [CTABr] = 2 mM.

The most relevant observations are as follows: (a) in each case explored, the S-ligand reacts faster with the R-ester and *vice versa* (the same *efs*, R/S or S/R, were obtained using both enantiomers of **3a** and of the substrates); (b) much of the enantioselectivity depends on the formation of a ternary complex [ligand/Cu(II)/ester]. The kinetic version of a Job plot,[15] shown in Figure 2, clearly indicates that the maximum acceleration (and enantioselection) is obtained for a 1:1 [ligand/Cu(II)] complex. Within the ternary complex, the ligands' hydroxy function (with an apparent pK_a of 7.4 in the case of **3a**, and of 6.6 in that of **3'm**, as evaluated from measurements at different pHs) as well as the amino group of the ester are clearly involved. Somewhat surprisingly, the proximity of the hydroxy function with the chiral centre of the ligand is not a

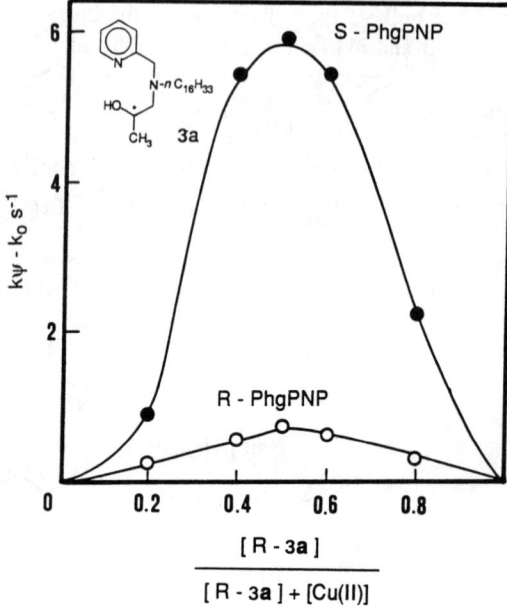

Figure 2: Job plots for the cleavages of (R)- and (S)- amino acid esters induced by **3a**.Cu(II).

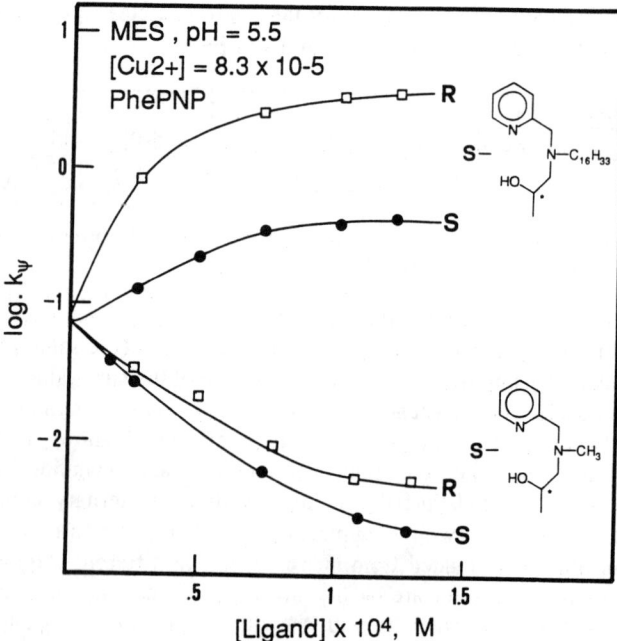

Figure 3: Enantioselectivity factors illustrated by rate profiles for hydrolyses of PhePNP induced by Cu(II) with either a micellar or non-micellar ligand.

relevant factor nor is the bulk of the substituents at the chiral carbon of either ligand or ester (compare **3** and **3'**, Table); (c) the enantioselectivity factors (*efs*) are much larger in the case of micellar than in that of non-micellar ligands (compare **3a** and **3b**, Table) and, as clearly shown in the rate profile of Figure 3, the overall rate effects are dramatically different. By addition of (non micellar) **3b** the apparent inhibition is due to complexation of the metal ion to give Cu(II)·**3b** which is a worse catalyst than the free metal ion and only slightly enantioselective; (d) the enantioselectivity of **3a** may be remarkably influenced when embedded in a matrix of aggregated inert surfactants. The effect of comicellization on both rate and *ef* is small in the case of cationic surfactants (such as CTABr) and is quite large (the *ef* decreases or vanishes) in the case of neutral and anionic surfactants (such as BRIJ35 or SDS). Quite interestingly, preliminary experiments indicate that the *ef* increases remarkably (up to 27, in the case of PhgPNP using **3'i**) when the chiral ligand is in a vesicular matrix of cationic double-tailed surfactants (such as the dioctadecyldimethylammonium bromide).

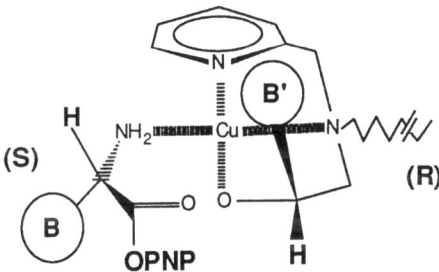

Figure 4: Explanation of enantioselection in terms of a productive ternary complex.

Our earlier explanation[9] was based on the idea that the enantioselection is the result of steric interactions between the bulkiest group of the ligand and of the ester within the productive ternary complex (see Figure 4) which would favour the R/S or S/R interaction. The present results rule out the primary role of steric interactions as very little effect is observed on changing the chiral carbon of the ligand and the bulk and orientation of the substituent bound to it. A different explanation would point to a change in the 'solvation' requirements of the two different diastereomeric ternary complexes; the less hydrated one would be either more tightly bound to the aggregate or inserted in a more hydrophobic region of the micelle or vesicle. Further study is in progress to provide a rationale for the enantioselectivity effects observed and to improve the efficacy of the system.

REFERENCES

1. J A Hartsuck and W N Lipscomb in 'The Enzymes', 3rd ed P Boyer Ed, vol 3, New York, 1971; L Stryer, 'Biochemistry', Freeman and Co, New York, 1986; A L Lehninger, 'Principles of Biochemistry', Worth, New York, 1982
2. M L Bender and B W Turnquist, *J Amer Chem Soc*, 1957, **79**, 1889; M A Wells and T C Bruice, *J Amer Chem Soc* 1977, **99**, 5341; P Nanjappan and T Czarnik, *J Amer Chem Soc*, 1987, **109**, 1826; J Chin and M Banaszczyk, *J Amer Chem Soc*, 1989, **111**, 2724; J Suh, *Bioorg Chem*, 1990, **18**, 345.
3. D W Christianson and W N Lipscomb, *Acc Chem Res*, 1989, **22**, 62 and references therein.
4. J H Fendler, 'Membrane Mimetic Chemistry', Academic, New York, 1975; C A Bunton and G Savelli, *Adv Org Chem*, 1986, **22**, 231; M K Jaim, 'Introduction to Biological Membranes', Wiley, New York, 1988; U Tonellato, *Colloids Surf*, 1989, **35**, 121.
5. F M Menger, *Angew Chem Int Ed Engl*, 1991, **30**, 1086.
6. W A Tagaki and K Ogino, *Top Curr Chem*, 1985, **128**, 144, and references therein.
7. F M Menger, L H Gan, E Johnson, and D H Durst, *J Amer Chem Soc*, 1987, **109**, 2800; S H Gellman, R Petter, and R Breslow, *J Amer Chem Soc*, 1986, **108**, 2388; J G J Weiningen, A Koudijs, and J F J Engbersen, *J Chem Soc Perkin Trans 2*, 1991, 1121.
8. (a) R Fornasier, D Milani, P Scrimin, and U Tonellato, *Gazz Chim Ital*, 1986, **116**, 55; (b) R Fornasier, P Scrimin, P Tecilla, and U Tonellato, *J Amer Chem Soc*, 1989, **111**, 224; (c) P Scrimin, P Tecilla, and U Tonellato, *J Org Chem 1991*, 56, 161; (d) G De Santi, P Scrimin, and U Tonellato, *Tetrahedron Lett*, 1990, **31**, 4791; (e) P Scrimin, P Tecilla, and U Tonellato, *J Phys Org Chem*, 1992, **5**, 619.
9. R Fornasier, P Scrimin, and U Tonellato, *J Chem Soc Chem Comm*, 1988, 716.
10. F Menger and M Ladika, *J Amer Chem Soc*, 1987, **109**, 3145.
11. J Chin, *Acc Chem Res*, 1991, **24**, 145.
12. T H Fife and Przystas, *J Amer Chem Soc*, 1985, **107**, 1041.
13. J N Brønsted and K Pederson, *Z Phys Chem*, 1924, **108**, 185; M Eigen, *Angew Chem Int Ed Engl*, 1964, **3**, 1.
14. A Williams, *Acc Chem Res*, 1989, **22**, 387.
15. P Scrimm, P Tecilla, and U Tonellato, *J Org Chem*, 1994, **59**, 18.
16. P Job, *Ann Chim (Rome)*, 1928, **9**, 113.

SECTION B

Physical Aspects of Organic Reactivity

GAS PHASE ION CHEMISTRY: KINETIC AND MECHANISTIC ASPECTS

Fulvio Cacace

Dipartimento di Studi di Chimica e Tecnologia delle Sostanze Biologicamente Attive Università di Roma "La Sapienza" - 00185 Rome, Italy

1 INTRODUCTION

Much of the intellectual challenge to the theory of organic reactivity is related to the influence of the environment, which affects the intrinsic kinetic and mechanistic features of the reacting system, preventing, *inter alia*, direct comparison of the results obtained in solution with those from theoretical approaches.

The influence of the reaction medium is particularly profound when charged reactants are involved, which accounts for the great interest and potential utility of extending the study of ion chemistry to the gas phase, where the complicating effects of a bulk-phase solvated environment are largely absent.

This expectation is reinforced by the unqualified success of thermodynamically oriented mass spectrometric studies, that have provided a wealth of data on free gaseous ions.[1,2] The results have allowed certain important trends of equilibrium data to be corrected, *eg* the basicity order of primary, secondary and tertiary alkylamines, the acidity order of alcohols, *etc*, distorted in solution by differential solvation effects.

Kinetic and mechanistic correlations between gas-phase and condensed-phase reactions have proved much more difficult, and the diverging, frequently opposite trends observed in many important classes of reactions, several of which are discussed in the following sections, have delineated a sharp dichotomy between the two areas of ion chemistry, preventing their constructive comparison.

Such a dichotomy, which is both unsatisfactory and artificial, can be traced to undue generalization of the kinetic data obtained under the very special conditions prevailing in low-pressure mass spectrometric experiments, discussed in the following section.

1.1. Mass Spectrometric Approaches

The principal mass spectrometric techniques used in kinetically oriented studies, and their maximum operating pressure, are the following ones:[3]

- Fourier-transform ion cyclotron resonance (FT-ICR) mass spectrometry, ca 10^{-6} torr.
- Ion cyclotron resonance (ICR) mass spectrometry, ca 10^{-5} torr.
- Electron-bombardment (EB) flow reactor, 10^{-3} torr.
- Chemical ionization (CI) mass spectrometry, 1-5 torr.
- Flowing afterglow (FA) mass spectrometry, 10 torr (He).
- Selected-ion flow tube (SIFT), 10 torr (He).
- Pulsed high-pressure (PHP) mass spectrometry, 5 torr.

It is apparent from the above, that most widely employed mass spectrometric techniques are characterized by low operating pressures. This has important kinetic consequences, especially regarding meaningful comparison with solution kinetics characterized by complete thermal equilibration of the reacting system. Consider a bimolecular ion-molecule reaction

$$A^+ + B \rightarrow C^+ + D \qquad (1)$$

All procedures for generating the charged species A^+ from an isolated neutral precursor, either by electron impact, photoionization, multiphoton absorption, *etc*, or by reaction with another charged species (charge transfer, proton addition, hydride or halide transfer, *etc*) impart excess internal energy to the A^+ ions formed, whose population is characterized, in general, by a non-Boltzmann energy distribution. In order to remove the excess energy it is necessary to ensure that the ionic reactants undergo a sufficient number of thermalizing collisions with the unreactive molecules of a bath gas. This condition, although a necessary one, is not sufficient *per se* to ensure compliance of reaction (1) with thermal kinetics.

In fact, the collision of an isolated ion with a neutral molecule gives an ion-molecule adduct, also denoted as 'electrostatic', or simply 'early' complex, stabilized by electrostatic ion-dipole or ion-induced dipole interactions, which can undergo back dissociation or evolve into the products (Figure 1).[4,5] In the low-pressure range typical of mass spectrometric experiments, where collisional thermalization is inefficient, the electrostatic energy $E°$ released by the ion-molecule association and ranging from a few kcal mol^{-1} up to over 25 kcal mol^{-1}, remains stored in the internal degrees of freedom of the complex, available to overcome to intrinsic ('chemical') energy barrier $E^{\ddagger}$ to the products. In the frequent case where $E° > E^{\ddagger}$, there is enough energy to cross the energy barrier, and any limitations of the reaction rate can only arise from entropic factors.

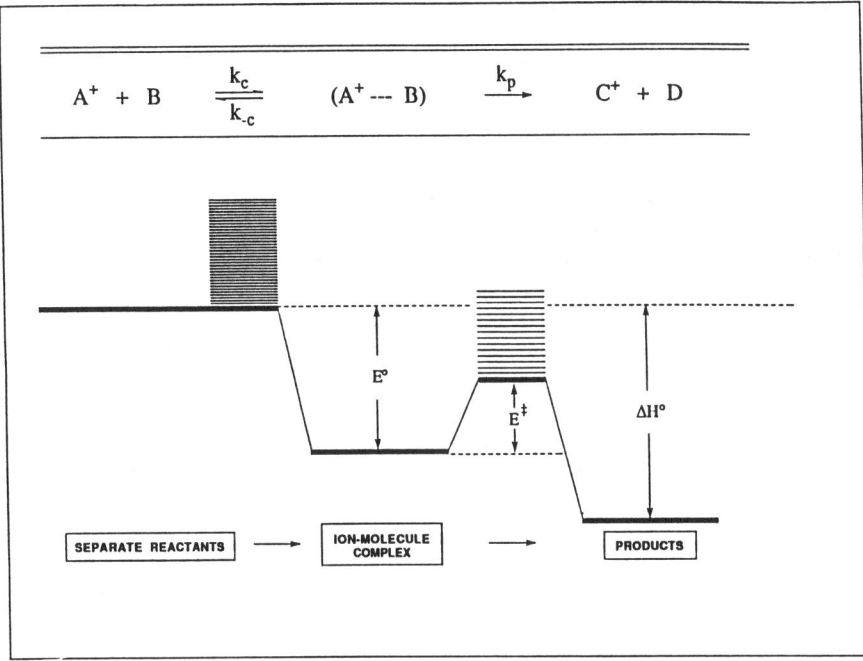

Figure 1: Schematic energy diagram of a gas-phase ion-molecule reaction.

As a consequence, most exothermic ion-molecule reactions proceed in low-pressure gases at collision rate, with *negative* temperature coefficients, which has led to the widespread, misconception these processes have no activation barriers. Even in those cases where $E^\circ < E^\ddagger$, the quantity experimentally accessible, *eg* from temperature-dependence studies, is not $E^\ddagger$, but rather the $E^\ddagger - E^\circ$ difference. In any case, the rate constants from low-pressure studies are not thermally averaged, but represent purely phenomenological constants averaged over a non-Boltzmann energy distribution, and measured at a high, if poorly defined, 'temperature' of the isolated ion/molecule pair undergoing conversion to the products.

Another limitation is related to the structural characterization of the charged species by mass spectrometric methods, whose inadequate structural discrimination is compounded by the long delay (10 ms) before structural assay, typical of techniques such as collisionally activated dissociation (CAD) and mass analyzed ion kinetic energy (MIKE) spectrometry. Long lifetimes and inefficient collisional stabilization account for the extensive fragmentation and/or isomerization processes that almost invariably affect the results of 'structurally diagnostic' mass spectrometric techniques.

1.2. The High-Pressure Radiolytic Approach

The above outlined limitations of purely mass spectrometric techniques have prompted the development of an integrated approach, based on the joint application of mass spectrometric and high-pressure radiolytic methods.[6,7]

The latter involves γ irradiation of the substrate of interest, S, highly diluted in a bulk gas, M, whose ionization produces well-defined cations, R^+, that undergo many thermalizing collisions before reacting with S, much in the same way as in CI mass spectrometry. In the unreactive bulk gas the cations R^+ are sufficiently long-lived to interact with as little as 10^{-5}-10^{-4} mol % of a reactive substrate prior to neutralization.[8] The charged intermediates formed are rapidly (10^{-8}-10^{-6} s) trapped by suitable nucleophiles and converted into neutral end products, amenable to analysis by GC and GC/MS. The most valuable features of the radiolytic approach, the wide pressure range (from a few torr to several atmospheres), the meaningful definition and the wide range of the reaction temperature, coupled with the structural and stereochemical characterization of the products, pertain exactly to those areas where mass spectrometry suffers from recognized limitations, which makes the two techniques highly complementary. Such features allow extension to the gas phase of many classical mechanistic tools, including competition kinetics, pressure- and temperature-dependence studies, isotopic labelling, *etc*, which, coupled with appropriate mass spectrometric techniques, make the integrated approach particularly powerful. It compares favourably with other approaches based on the isolation of the neutral products from gas-phase ionic reactions, *ie*, the electron-bombardment (EB) flow technique, limited, as mass spectrometry, to a restricted pressure range,[9] and the powerful β-decay technique, which requires multiply tritiated precursors and preliminary study of their decay-induced fragmentation.[10]

An inherent complication of the radiolytic technique arises from the incursion of reactive neutral species, *ie*, free radicals, cogenerated with the R^+ ions, which makes it necessary to ascertain the ionic origin of the products of interest. While the problem requires detailed case-by-case consideration, several criteria have proved generally useful. First, the ionic reaction of interest must be shown by preliminary application of CIMS to occur in the specific system investigated. Second, the yields of neutral products of ionic origin are unaffected by the addition of radical scavengers, *eg*, O_2 in gaseous alkanes. Third, formation of neutral products from ionic reactions is depressed by additives capable of intercepting their charged precursors, *eg*, gaseous nucleophiles that efficiently trap the R^+ ions. In many systems, application of the above criteria and of additional, more specific tests has ruled out significant radical contributions to the formation of the products of interest.

Most of the radiolytic studies have been performed in gaseous systems at atmospheric pressure, where collisions of the ionic reactants, and of the ion-molecule complex, with the molecules of the bath gas occur at a frequency exceeding 10^{11} s^{-1}.

Nevertheless, even such a high rate of thermalizing collisions can be insufficient to ensure that the ion-molecule reaction of interest obeys fully thermal kinetics, a condition which must be verified by showing that the rate constants measured are significantly lower at any given temperature than the collision rate of the ions with the molecules of the bath gas and, in addition, are unaffected by further increasing the pressure of the latter.

2 PROTON MIGRATION IN AROMATIC SYSTEMS

Because of their inherent simplicity and general importance, proton-transfer reactions are appealing models for the study of other reactions and occupy a pre-eminent position in the development of theoretical organic chemistry.[11,12] In particular, intramolecular proton migrations have received a great deal of attention since the wide application of powerful physical methods such as dynamic NMR has allowed protonated organic molecules to be studied under 'long-life' conditions. The results concerning hydrogen migrations in arenium ions[13] show that in general these processes are characterized by an appreciable activation energy which depends in most cases on the nature of the medium. As a typical example, Olah and coworkers have reported an activation energy of 10 ± 1 kcal mol^{-1} for degenerate 1,2-H$^+$ shifts in the model benzenium ion, $C_6H_7^+$, dissolved in a HF/SbF$_5$/SO$_2$ClF/SO$_2$F$_2$ mixture,[14] and comparable activation energies have been measured for proton migrations in a large number of other arenium ions.[13]

Energy barriers for 1,2-H$^+$ shifts in arenium ions have been also evaluated by theoretical methods. Referring to $C_6H_7^+$, the height of the barrier calculated at the 4-31 G level of theory is *ca* 21 kcal mol^{-1},[15] reduced to 8-10 kcal mol^{-1} at the MP2/6-31G** level,[16] a result much closer to the value measured in solution.

2.1. Mass Spectrometric Results

Intraannular hydrogen migration in gaseous arenium ions, the so-called 'hydrogen-ring walk' is a ubiquitous process long detected in the mass spectrometric study of arenium ions, whose rate is generally too fast to allow kinetic characterization.[17] The extensive evidence for fast intraannular H$^+$(D$^+$) migration in arenium ions has led to the view that the activation energy for the process is much lower in the gas phase than in condensed media. This position is well exemplified by the 'dynamic' model of arenium ions suggested to account for the unexpectedly high entropy of protonation of certain substituted benzenes. The model embodies the concept of a stable, yet 'dynamic' structure characterized by a low activation barrier, < 5 kcal mol^{-1} for the internal translation of the extra proton across a broad potential well within the molecule.[18] Although such a speculative model has not survived reassessment of its experimental foundations,[19] it aptly illustrates the impact of the copious mass spectrometric evidence pointing to fast and extensive hydrogen migration in gaseous arenium ions.

Interannular $H^+(D^+)$ transfer has also been observed in bicyclic arenium ions from selectively deuterated molecules, eg from α,ω-diphenylalkanes having a fully deuterated ring. These ions, obtained from highly exothermic processes, undergo metastable loss of benzene, giving charged fragments whose isotopic composition reveals extensive H/D interannular mixing.[17] It has been shown that at least 20 consecutive ring-to-ring proton-transfer steps are required to achieve the observed near random H/D distribution over the two rings.[20,21]

These results should not be taken as an indication that interannular proton migration is characterized by a much lower activation barrier in the gas phase than in solution. In fact, in the mass spectrometric experiments, the extent of the $H^+(D^+)$ transfer is inferred from the composition of fragments formed when the polycyclic arenium ions undergo metastable dissociation. This is an endothermic process accessible only to the ions with a large excess of internal energy, equivalent, from the kinetic standpoint, to a high, if unknown, 'temperature' of the reacting species, which allows even processes characterized by large activation energies to occur at high rates.

2.2. High-pressure Studies. Monocyclic Ions

The first attempt at a quantitative kinetic evaluation of thermal hydrogen migration in gaseous arenium ions has involved alkylation of phenyltrimethylsilane by radiolytically formed $i\text{-}C_3H_7^+$ ions in C_3H_8.[22] Preliminary mass spectrometric, radiolytic and theoretical studies[23-26] had shown that, owing to its high basicity, the ring carbon bearing the $SiMe_3$ group is a proton sink, to which protons bound to other ring positions tend to migrate.

In addition, it was known that the arenium ion protonated at the carbon bearing the $SiMe_3$ substituent can be deprotonated by strong, hindered nitrogen bases (B_1) such as NEt_3, whereas oxygenated bases (B_2) such as $c\text{-}C_6H_{10}O$ promote instead selective desilylation, eg

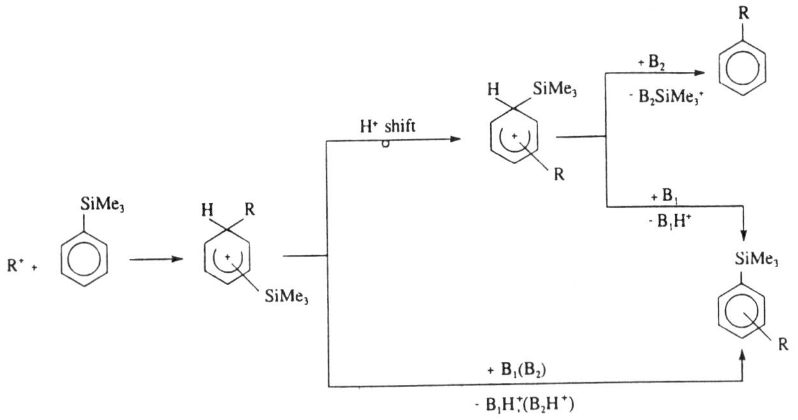

Scheme 1.

R = i-Pr, B_1 = NEt_3, B_2 = c-$C_6H_{10}O$

Following the above preliminary studies, radiolytic experiments were performed in propane gas at pressures up to 3040 torr in the temperature range from 37.5 to 100°C, in order to measure the ratio of alkylated to alkyldesilylated products formed from the reaction of phenyltrimethylsilane and tolyltrimethylsilanes with radiolytically formed i-$C_3H_7^+$ ions. Attention was focused, in particular, on the dependence of the above ratio on the pressure, the temperature and the concentration of the B_1 and B_2 bases.

The results fit a simple kinetic model, summarily illustrated in Scheme 1, and lend themselves to a quantitative treatment which allows one to evaluate the rate constant of the (1) → (2) isomerization, k_i ~ 1.6 · 10^9 s^{-1} at 310 K. The constant derived in this way does not refer to a specific 1,2-H^+ shift in a single arenium ion, since the alkylation yields a mixture of isomeric ions (1), whose conversion to (2) can, in principle, require several consecutive 1,2-H^+ shifts.

The k_i constant is rather the weighted average of individual rate constants of the slowest step in each of the sequences of H^+ shifts necessary to convert the various ions (1) into the corresponding isomers (2). Despite its composite nature, the k_i value derived from the above experiments is significant, in that in C_3H_8 at 3040 torr the rate of (1) → (2) isomerization is lower by a factor of 30 than the rate of collision of (1) with the molecules of the bath gas, and one can therefore assume that the isomerizing arenium ions are equilibrated with the gaseous environment *before* undergoing H^+ shifts, and, therefore, obey thermal kinetics.

In a subsequent study, the gas-phase t-butylation of selectively deuterated toluenes has been exploited for a more quantitative kinetic characterization of thermal hydron migration in arenium ions.[27] This study will be illustrated in some detail, since it provides a typical example of the application of the radiolytic technique to kinetic problems. The irradiated systems consist of isobutane gas at atmospheric pressure, containing traces of p- or m-D-toluene, a radical scavenger (O_2), and variable amounts of a gaseous base. In addition to the desired electrophile, the t-butyl cation, the radiolysis produces other reactive species, ie free radicals, scavenged by O_2, and fragment ions

$$i\text{-}C_4H_{10} \rightsquigarrow \begin{cases} t\text{-}C_4H_9^+ + H^\cdot \\ i\text{-}C_3H_7^+ + CH_3^\cdot \\ i\text{-}C_3H_7^\cdot + CH_3^+, \text{etc.} \end{cases} \quad (4)$$

that are converted into t-butyl ions via hydride-ion abstraction from the bulk gas.[28]

$$i\text{-}C_4H_{10} + i\text{-}C_3H_7^+ \rightarrow t\text{-}C_4H_9^+ + C_3H_8 \quad (5)$$

The base, typically pyridine (P) or triethylamine, capable of deprotonating at collision rate gaseous arenium ions, serves two purposes. First, it provides a criterion to verify the ionic origin of a given product, since the yields of ionic products are expected to decline, and eventually to vanish, at increasing concentrations of the base which intercepts the ionic reactants. Furthermore, the lifetime of the ions depends, in a quantitatively predictable way, on the concentration of the base. This provides the means to define the time allowed for the hydron shifts in the arenium ion, a key kinetic parameter.

The structure, the thermochemical properties and the reactivity of $t\text{-}C_4H_9^+$ are well documented by extensive theoretical, mass spectrometric and radiolytic studies,[28-31] which have established, in particular, the general outline of gas-phase aromatic t-butylation, adapted in Scheme 2 to the specific case of interest, ie alkylation of p-D-toluene.

The analysis of the products shows that deuterium is partially retained, even when substitution occurs at the *para* position, pointing to intraannular D^+ shifts from the *ipso* position, which increase with the temperature and decrease at higher base concentrations.

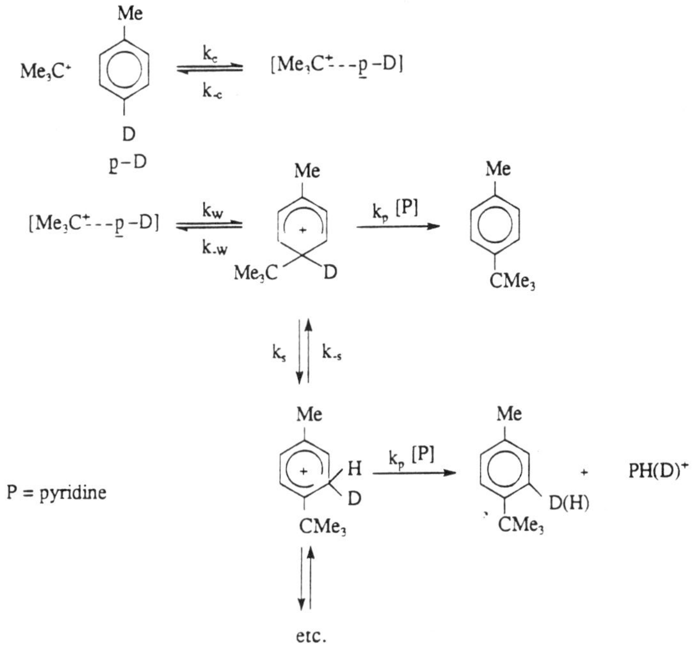

Gas Phase Ion Chemistry: Kinetic and Mechanistic Aspects

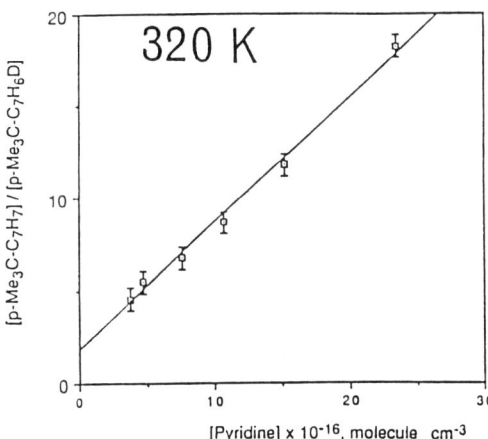

Scheme 2.

Figure 2: Dependence of the ratio of unlabelled to deuterated products from the gas-phase alkylation of *p*-D-toluene on the concentration of pyridine.

Quantitative treatment of the data has been based on two assumptions, both independently verified: (*i*) only the first D$^+$ shift occurs to any significant extent, and, (*ii*) there are no isotope effects in the deprotonation of the arenium ions by very strong gaseous bases. Under the steady-state approximation, the ratio of the unlabeled to the deuterated end product is given by the expression which accomodates well the experimental results (Figure 2):

$$\frac{[\underline{p}\cdot Me_3C-C_7H_8]}{[\underline{p}-Me_3C-C_7H_7D]} = \frac{k_{-s}+k_p[P]}{k_s}$$

Analogous plots are obtained for reactions carried out at different temperatures, from 47 to 120°C, all characterized by an intercept very close to unity, showing that k_{-s} is invariably negligible with respect to k_s. From the slope, equal to $k_p[P]/k_s$, one can calculate k_s under the assumption, supported by mass spectrometric and radiolytic evidence, that pyridine deprotonates the arenium ions at collision frequency. The latter can be calculated with good accuracy using the ADO or the AADO theory,[32] which in turn allows one to derive the k_s rate constant at any given temperature. From experiments performed at different temperatures, one can construct an Arrhenius plot, which is linear over the entire temperature range investigated (47 to 120°C) as shown in Figure 3, and derive

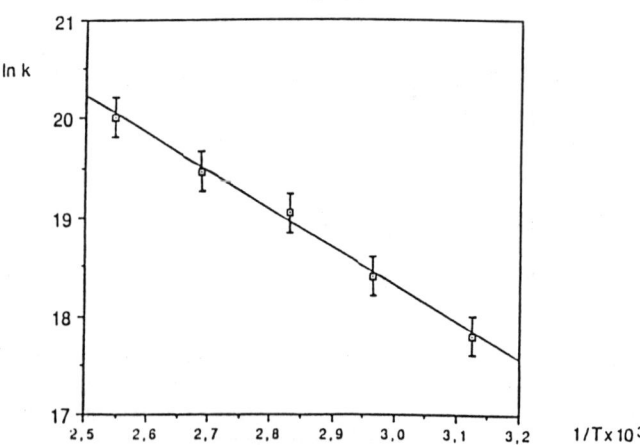

Figure 3: Arrhenius plot showing the dependence of k_s on the temperature in the range from 47 to 120°C.

the activation parameters of the D$^+$ shift studied (Table 1). The kinetics of the corresponding H$^+$ shift have been studied alkylating m-D-toluene and measuring the decrease of the D content in the p-t-butylated product at different temperatures and concentrations of bases (Table 1).

Overall, the results demonstrate the possibility of investigating *thermal* kinetics of hydron shifts in gaseous arenium ions. As apparent from Table 1, the rates of H$^+$(D$^+$) migration are at least 2 orders of magnitude lower than the rate of collision of the arenium ions with the isobutane molecules, showing that the isomerizing ions are thermally equilibrated with the bath gas.

Table 1: Kinetic Parameters of Thermal Hydron Shifts in Arenium Ions.

PROCESS	METHOD	logA	E kcal mol^{-1}	Ref.
H$^+$ SHIFTS IN C$_6$H$_7^+$	SOLUTION, NMR	15.9 ± 2	10 ± 1	14
	4-31G*, THEORY	—	20.6	15
	MP2/6-31G**, THEORY	—	9.3 ± 3	16
D (H) SHIFTS IN Me—⟨•⟩—CMe$_3$ with D(H)	RADIOLYSIS, ISOBUTANE, 720 torr $k_D = 5.2 \pm 0.2\ 10^7\ s^{-1}$ (320 K) H/D KIE = 8 ± 4 (320 K)	12.9 ± 0.2	7.6 ± 0.2 (6.2 ± 0.2)	27
INTERANNULAR D SHIFT IN D$_3$⟨•⟩—D—⟨⟩—H$_5$ with CMe$_3$	RADIOLYSIS, ISOBUTANE, 720 torr $k_D = 2.9 \pm 0.6\ 10^6\ s^{-1}$ (320 K)	11.9 ± 0.3	8.0 ± 0.2	33
INTERANNULAR H SHIFT IN H$_3$⟨•⟩—H—⟨⟩—D$_5$ with CMe$_3$	RADIOLYSIS, ISOBUTANE, 720 torr $k_H = 1.3 \pm 0.4\ 10^7\ s^{-1}$ (320 K)	11.4 ± 0.4	6.3 ± 0.2	33

Furthermore, *thermal* hydron shifts are characterized by activation energies close to those measured in solution and calculated by the most recent theoretical methods, and display an appreciable H/D kinetic isotope effect.

2.3. Interannular Hydron Migration

As discussed in a previous section, extensive ring-to-ring hydron transfer has been observed in gaseous protonated α,ω-diphenylalkanes, and it is too fast to allow kinetic evaluation by mass spectrometric techniques. The problem has recently been addressed using the same high-pressure technique employed in the study of intraannular hydron shifts, by alkylating 1,2-diphenylethane having a fully deuterated ring with radiolytically formed t-butyl cations in isobutane at atmospheric pressure, over a wide temperature range and in the presence of variable amounts of bases.[33]

Substitution takes place at the *meta* and *para* positions of both the deuterated and the unlabeled ring, and analysis of the neutral products reveals significant interannular isotopic mixing, whose extent increases at the higher temperatures, but is much less extensive than in mass spectrometric experiments, being limited to the partial exchange of a single hydron. The reactivity pattern is outlined in Scheme 3, that refers to *para* alkylation of the deuterated ring.

The relevant measurable quantities are the yields of the 'singly-scrambled', and the 'unscrambled' *p*-t-butylated products, whose ratio can be deduced from the relative intensity of the corresponding benzyl ion in the mass spectra of the neutral products.

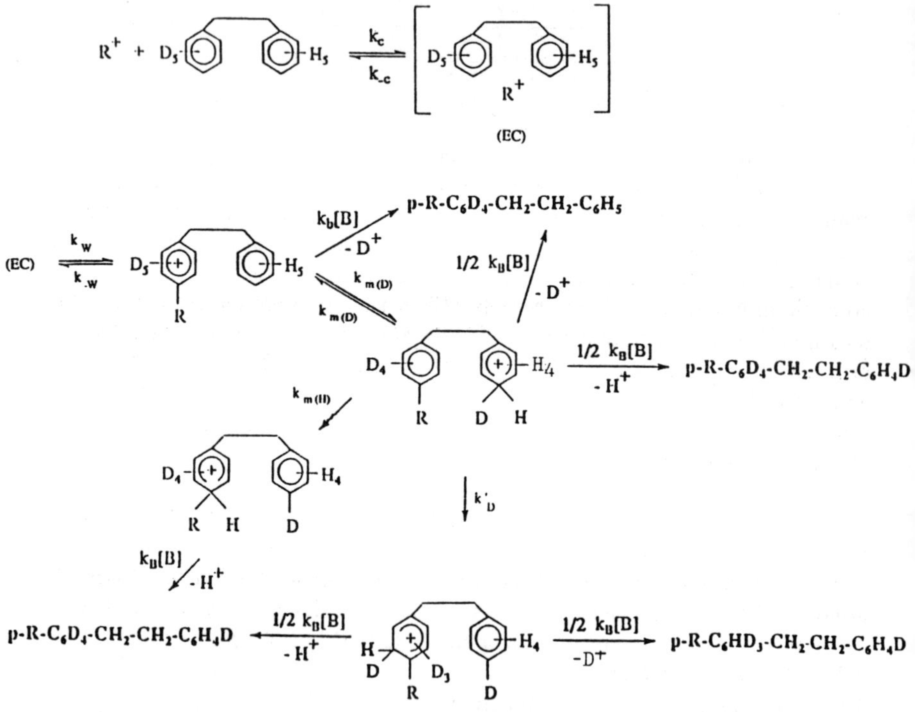

Scheme 3.

Under reasonable simplifying assumptions one obtains the expression

$$\frac{[Me_3CC_6D_4CH_2CH_2C_6H_5]}{[Me_3CC_6D_4CH_2CH_2C_6H_4D]} = \frac{i\,(C_6H_5CH_2^+)}{i\,(C_6H_4DCH_2^+)} = \frac{2\,k_b\,[B]}{k_{m(D)}}$$

that accommodates well the experimental results of systematic base-dependence studies. The temperature dependence, studied in the range from 47 to 150°C, gives a linear Arrhenius plot and hence the activation parameters of *thermal* deuteron ring-to-ring transfer. Analogous results have been obtained considering the substitution at the unlabeled ring, which has allowed evaluation of the activation parameters of the interannular H^+ transfer.

The results, illustrated in Table 1, show that the activation barrier for *thermal* ring-to-ring hydron transfer is similar to that for 1,2-hydron shifts in monocyclic arenium ions. Nevertheless, the preexponential factor for interannular transfer is lower, probably reflecting the geometrical constraints posed on the arrangement of the two rings in the transition state of hydron transfer, and hence the operation of conformational factors.

3   AROMATIC NITRATION

Aromatic nitration is probably the organic reaction most thoroughly investigated for more than a century and its mechanistic and kinetic study continues to play a central role in the development of theoretical organic chemistry.[34,35] Yet, in many respects, its mechanism is not established and it has been suggested that it would be preferable to think of a variety of mechanisms,[36] especially since aromatic nitration is a reaction heavily affected by environmental factors, *eg* its measured second-order rate constant in H_2SO_4 is known to decrease by four orders of magnitude for every 10% decrease of the weight % acidity of H_2SO_4,[37] and the role of a solvent cage in stabilizing the $[ArH-NO_2]^+$ 'encounter pair' is today a widely accepted kinetic feature of the nitration of activated aromatics in mixed-acid solutions.[38]

In view of the severe complicating effects of condensed reaction media, it is understandable that as soon as ICR and other suitable mass spectrometric techniques became available, numerous studies were performed to investigate aromatic nitration in the gas phase.

3.1.   Mass Spectrometric Results

The studies of Bursey and coworkers, performed by ICR mass spectrometry at a pressure of *ca* 10^{-5} torr,[39,40] mark the first striking discrepancy with the reactivity pattern long established in solution. Nitronium ion was found to undergo oxygen transfer

$$C_6D_6 + NO_2^+ \rightarrow C_6D_6O^+ + NO \qquad (6)$$

as well as charge transfer to benzene without detectable formation of nitrated adducts.

$$C_6D_6 + NO_2^+ \rightarrow C_6D_6^+ + NO_2 \qquad (7)$$

The latter could be obtained by the process

$$C_6D_6^+ + NO_2 \rightarrow [C_6D_6 NO_2]^+ \qquad (8)$$

Working in a high-pressure ion source in He up to 10 torr, Schmitt et al confirmed the lack of nitrating ability of NO_2^+, and the formation of nitrated adducts, $C_6H_6NO_2^+$, from reaction (8).[41] As remarked by Olah,[42] it is surprising that the latter process is still efficient in going from high-pressure mass spectrometry to ICR experiments, where the pressure is almost six orders of magnitude lower and true addition processes such as (8) can hardly be detected. In fact, the lack of deactivating collisions causes the association energy to remain stored in the adduct, which is prone to back dissociation or fragmentation, except in the unlikely event of IR-photon emission.

Table 2: Rate of Nitration of Aromatic Substrates by Gaseous $CH_2=O-NO_2^+$ Relative to that of Benzene.

Neutral reactant	k_{ArX}/k_{ArH} rate[43]
Benzene	1.0
Benzene-d_6	1.0
Aniline	0
Anisole	0
Nitrobenzene	10
Pyrrole	0
Ethylbenzene	0
Toluene	0.3
Cyclopropylbenzene	0
Chlorobenzene	0.25
α,α,α,-Trifluorotoluene	0.3
Pyridine	0.3
Fluorobenzene	0.55
Xylene (o, m, or p)	0
o-Difluorobenzene	0.25
m-Difluorobenzene	0.3
p-Difluorobenzene	0
Pentafluorobenzene	0
p-Bromoanisole	0

The above difficulties have prompted the use of 'solvated' forms of nitronium ion, such as $CH_3CHONO_2^+$ and $CH_2=O-NO_2^+$. Dunbar, Shen and Olah[43] carried out extensive ICR studies, including competition experiments, using $CH_2=NO-O_2^+$ from the ionization of $C_2H_5ONO_2$, and were able to detect formation of nitrated adducts from many aromatic substrates, as shown in Table 2. However, the effects of the substituents ran contrary to the order of reactivity prevailing in solution, eg nitrobenzene was found to react *faster* than benzene, whose reactivity exceeded that of toluene, *etc*.

A subsequent study by Ausloos and Lias,[44] who employed other 'solvated' forms of NO_2^+, including $C_2H_5O(NO_2)_2^+$, was successful in demonstrating the formation of nitrated adducts from various aromatic substrates, but again a paradoxical reactivity order was observed, highly deactivating substrates reacting *faster* than the activated ones.

Table 3: Rate of Gas-Phase Nitration *via* Aromatic Radical Cations.

ArH	Nitration via aromatic radical cation	k, cm^3 molecule^{-1} s^{-1} [46]
C_6H_6	yes	2.4 x 10^{-11}
$C_6H_5CH_3$	yes	1.2 x 10^{-11}
p-$C_6H_4(CH_3)_2$	yes	1.7 x 10^{-11}
Mesitylene	yes	
1,2,4-Trimethylbenzene	yes	
C_6H_5OH	yes	
C_6H_5F	yes	3.7 x 10^{-11}
C_6H_5Cl	yes	
o-$C_6H_4F_2$	yes	
m-$C_6H_4F_2$	yes	1.2 x 10^{-12}
p-$C_6H_4F_2$	yes	2.9 x 10^{-12}
1,2,4-$C_6H_3F_3$	yes	
1,2,3,4-$C_6H_2F_4$	no	
1,2,3,4-$C_6H_2F_4$	no	
Furan	no	
Pyridine	no	
m-$FC_6H_4CF_3$	very slow	

The above mass spectrometric experiments provide no direct information on the structure of the $[ArH \cdot NO_2]^+$ ions. Whereas the observed occurrence of the proton transfer

$$[ArH \cdot NO_2]^+ + B \rightarrow ArNO_2 + BH^+ \qquad (9)$$

where B= pyridine or tetrahydrofuran favours a σ-complex structure of the nitrated adduct,[41] other studies cast doubts on the conclusiveness of this kind of evidence.[45]

Reaction (8) is also characterized by substituent effects that contrast with those prevailing in solution, as shown by the rate constants summarized in Table 3, eg the reactivity of benzene exceeds that of toluene, but is lower than that of fluorobenzene.[46]

In conclusion, the picture delineated by the mass spectrometric studies of aromatic nitration is scarcely consistent with the reactivity pattern long established in solution, in that the nitronium ion, the well-established nitrating reagent in condensed phases, fails to nitrate aromatic substrates in the gas phase, and other cations that yield nitrated adducts display a selectivity that runs contrary to that observed in solution.

Clearly, at the low pressures typical of mass spectrometric experiments the nitration reactions not only fail to obey thermal kinetics, which alone would prevent meaningful comparison with the corresponding thermal reactions occurring in solution, but suffer as well from the incursion of competitive processes such as charge transfer and fragmentation which affect to an unpredictable and variable extent the observed nitration rate of different substrates.

3.2. Atmospheric-pressure Nitration

The reaction has been studied with the radiolytic technique in CH_4 at 760 torr using a nitrating cation obtained from the exothermic proton transfer

$$CH_3ONO_2 + AH^+ \rightarrow [CH_3ONO_2]H^+ + A \qquad (10)$$

promoted by radiolytically formed methonium and ethyl ions (A= CH_4 and C_2H_4). Preliminary CI experiments showed that protonated methyl nitrate exists in two isomeric forms, distinguishable by their MIKE and CAD spectra:[47]

$$CH_3\overset{+}{O}(H)-NO_2 \qquad CH_3O-\overset{+}{N}\underset{OH}{\overset{O}{\diagup}}$$

(3) (4)

Mass spectrometric evidence, supported by high-level *ab initio* results,[48] identifies protomer (3) as the most stable, and undoubtedly it represents the predominant species from reaction (10) in the radiolytic nitration experiments. Further CI[49] and ICR[50] studies demonstrated that protonated methyl nitrate reacts with aromatic substrates, *eg* benzene, according to the nitronium-ion transfer process

$$[ROH-NO_2]^+ + C_6H_6 \rightarrow ROH + [C_6H_6NO_2]^+ \qquad (11)$$

The MIKE and CID spectra of the adduct, recorded *ca* 10 ms after its formation in the CI ion source, are indistinguishable from those of a $C_6H_5NO_2H^+$ model ion prepared by protonation of nitrobenzene, which suggests a σ-complex structure of the primary nitrated adduct.[51]

Following these preliminary mass spectrometric experiments, a systematic study of atmospheric-pressure aromatic nitration has been undertaken, based on the radiolysis of systems containing CH_4, CH_3ONO_2 and PhX substrates in the typical molar ratios 1,000 : 20 : 1. The results of competition experiments, and the isomeric composition of the products, summarized in Table 4, allow evaluation of both the substrate and the positional selectivity of the reaction, which conforms to solution-chemistry trends, characterizing $[CH_3OH-NO_2]^+$ as the first well behaved nitrating cation in the gas phase.[51]

The partial rate factors plotted against the σ^+ constants of the X substituents fit a Hammett-type diagram which is linear over an extended range of σ^+ values, with a ρ^+ constant of -3.87. The latter is appreciably less negative than in condensed-phase reactions, *eg* ρ^+ = -6.53 and -9.7 in nitration performed in Ac_2O and in H_2SO_4, respectively, a result not unexpected for a gaseous cation lacking a solvation shell and a counterion (Figure 4).

Reaction (11) is viewed as the nucleophilic displacement of the ROH group by the aromatic substrate, an interpretation consistent with the influence of the leaving group on the reactivity of $[ROH-NO_2]^+$ nitrating cations.

Inductively stabilizing groups such as those containing R = i-C_3H_7 and i-C_5H_{11} depress the nitrating ability, enhanced instead by the presence of electron-withdrawing groups which make nitration more indiscriminate, *eg* the k_{PhMe}/k_{PhH} reactivity ratio decreases from 7.6 to 2.1 down to 1.2 in passing from R = CH_3 to R = CF_3CH_2 and to R = $(CF_3)_2CH$.[52]

Table 4: Selectivity and Orientation of Aromatic Nitration by $[CH_3ONO_2]H^+$ in CH_4 at 37.5 °C, 760 torr.

C_6H_5X (X)	$k_{C_6H_5X}/k_{C_6H_6}$	Orientation (%)		
		ortho	meta	para
H	1	–	–	–
CH_3	5.1	59	7	34
C_2H_5	5.6	47	4	49
C_3H_7	7.0	50	4	46
$CH(CH_3)_2$	6.0	31	5	64
c-C_3H_5	10.6	72	6	22
$C(CH_3)_3$	8.4	17	8	75
C_6H_5	1.5	40	4	56
OCH_3	7.6	41	?	59
F	0.15	14	13	73
Cl	0.19	36	10	54
CF_3	0.0037	–	100	–
Mesitylene	8.1	–	–	–

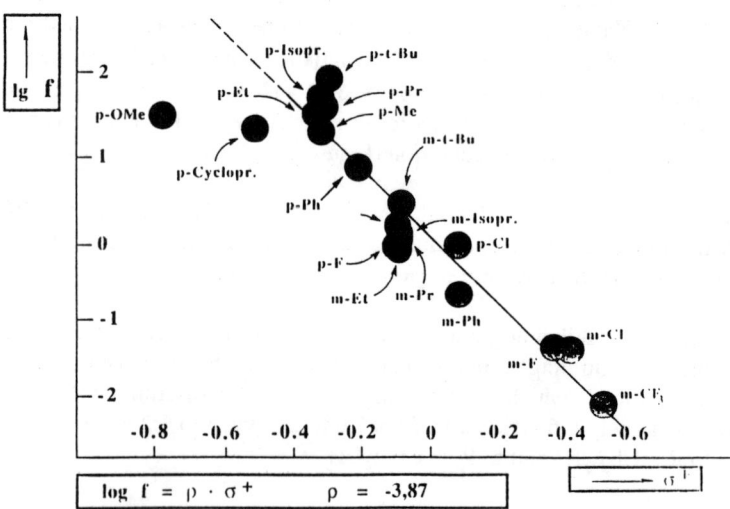

Figure 4: Hammett plot of gas-phase aromatic nitration by MeOH-NO_2^+ in CH_4 at 37 °C.

Another kinetic feature of gas-phase nitration presents intriguing analogies with condensed-phase nitration. From Figure 4 it is apparent that the most activated substrates, eg PhOMe and Ph-c-C$_3$H$_5$, display a negative deviation from the linear Hammett correlation, indicating that the rate of the gas-phase nitration tends toward a limiting value which cannot be further increased by the presence of more activating substituents. Such a trend has been explained with the formation of an 'early' electrostatic complex:

$$[ROH-NO_2]^+ + PhX \underset{k_{-1}}{\overset{k_1}{\rightleftharpoons}} \underset{\text{Early complex}}{[ROH-NO_2^+]\cdots PhX} \xrightarrow{k_2} \sigma\text{-complex}$$

(11a)

Whenever the presence of a strongly activating substituent makes $k_{-1} \ll k_2$, formation of the early complex becomes rate determining and the nitration proceeds at the limiting rate, ie with unit collision efficiency. This hypothesis is supported by the results obtained from the gas-phase nitration of benzylmesitylene, whose two rings are characterized by a widely different activation, roughly comparable to that of toluene (T) and of isodurene (I).

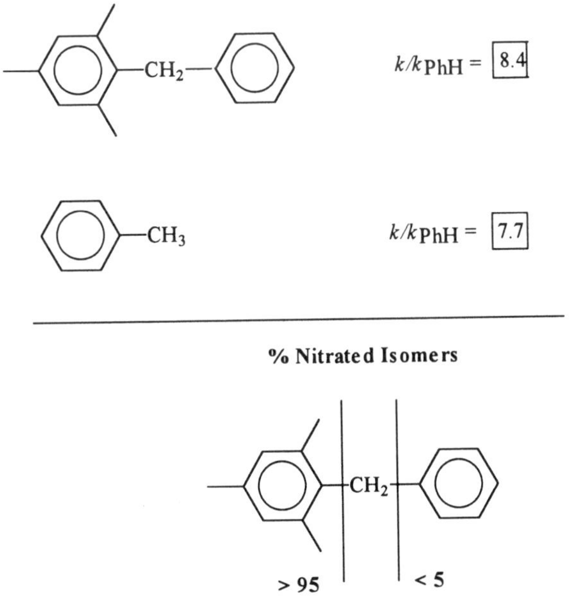

[structure: mesityl-CH$_2$-phenyl] k/kPhH = 8.4

[structure: Ph-CH$_3$] k/kPhH = 7.7

% Nitrated Isomers

[structure: mesityl-CH$_2$-phenyl]

> 95   < 5

Scheme 4

The results of competition experiments involving toluene and isodurene, the separate moieties linked together in benzylmesitylene, show that k_I/k_T is low, 1.5 at 37.5°C, whereas the interannular selectivity observed in the nitration of benzylmesitylene is as high as 20 in favour of the activated ring, despite its unfavourable (2/5) statistical ratio (Scheme 4).[53]

Formation of the 'early' complex, which is stabilized by electrostatic ion-molecule interactions, is the gas-phase counterpart of the rate-determining formation of 'early' complexes in condensed-phase nitration of activated aromatics, suggested by Olah,[54] based on Dewar's π-complex concept,[55] and by Schofield,[56] who envisaged a solvent cage that confines the reacting pair by a viscosity barrier.

Other significant features of condensed-phase aromatic nitration, including the irreversible character of the electrophilic attack, the lack of an appreciable H/D kinetic isotope effect, *etc*, have been observed in the gas phase as well.[51]

In conclusion, when studied under conditions that allow meaningful comparison with solution, in particular inside the thermal-kinetics domain, gas-phase aromatic nitration displays kinetic and mechanistic features fully consistent with those of condensed-phase nitration, reproducing not only its general reactivity trends but also its kinetic peculiarities, *eg* the 'encounter-rate' nitration of activated aromatics.

4 AROMATIC SILYLATION

In the previous sections it has been shown that meaningful kinetic and mechanistic correlation is possible between reactions long known in solution and the corresponding gas-phase processes when the latter are studied with appropriate high-pressure techniques.

The opposite case of ionic reactions first demonstrated in the gas-phase and only subsequently reproduced in solution is by far less common. One such rare example is direct electrophilic aromatic substitution by $SiMe_3^+$ ion. It had long been established that even the most activated aromatics do not undergo C-silylation in solution[57] a failure explained by the facile $SiMe_3^+$ loss from any silylated arenium ions initially formed.[58] Silylated adducts were observed by Wojtyniak and Stone from the reaction

$$SiMe_3^+ \; + \; C_6H_6 \; \rightarrow \; [C_6H_6-SiMe_3]^+ \quad (12)$$

in a mass spectrometric study performed at 3-5 torr. However, based on the failure to detect deprotonation (13) of the adduct by gaseous bases the authors

$[C_6H_6-SiMe_3]^+ + B \rightarrow C_6H_5SiMe_3 + BH^+$ (13)

excluded the occurence of aromatic substitution by $SiMe_3^+$, assigning the observed adduct the π-complex structure (5), instead of the σ-complex structure (6).[59]

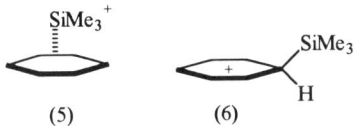

(5) (6)

A combination of radiolytic and mass spectrometric techniques allowed Fornarini to reverse the above structural assignment, reaching the conclusion that aromatic substitution by $SiMe_3^+$ does actually occur in the gas phase, via the intermediacy of the the σ-complex (6).

It was conjectured that the inefficiency of deprotonation (13) could reflect, rather than the π-complex structure of the silylated adduct, the incursion of a faster competitive process, ie $SiMe_3^+$ transfer to the base.

$\xrightarrow{k_{13}} C_6H_5SiMe_3 + BH^+$ (14)

$\xrightarrow{k_{14}} C_6H_6 + BSiMe_3^+$ (15)

It was correctly surmised that the k_{13}/k_{14} branching ratio, very low for the bases used by Wojtyniak and Stone,[59] could be shifted in favour of deprotonation (13) by a judicious choice of the base, which should combine a high proton affinity with a low nucleophilicity. This turned out to be correct, since ICR experiments showed that strong, hindered bases such as NEt_3 successfully accomplish deprotonation of silylated adducts. Particularly cogent is the evidence from the actual isolation of $C_6H_5SiMe_3$ from the radiolysis of $CH_4/SiMe_4/RC_6H_5$ mixtures containing O_2 as a radical scavenger and NEt_3 as the gaseous base. The irradiation promotes the ionic sequence yielding trimethylsilylated aromatics as the neutral end products. The results, confirmed

$CH_4 \rightsquigarrow C_nH_5^+$ (n = 1,2) (16)

$C_nH_5^+ + SiMe_4 \rightarrow C_nH_4 + CH_4 + SiMe_3^+$ (17)

$SiMe_3^+ + RC_6H_5 \rightarrow [RC_6H_5SiMe_3]^+ \xrightarrow[-NEt_3H^+]{+NEt_3} RC_6H_4SiMe_3$ (18)

and extended by additional theoretical, ICR and radiolytic studies,[24,25] conclusively show that the silylated adduct can be deprotonated by suitable bases of low nucleophilicity and therefore must be assigned the σ-complex structure (6), which in turn proves the occurence of direct aromatic substitution by $SiMe_3^+$ ions. The positional and substrate selectivity, deduced from atmospheric-pressure competition experiments and from the isomeric composition of the products summarized in Table 5, characterize $SiMe_3^+$ as a typical, moderately selective electrophile with large steric requirements. Typically, no substitution occurs *ortho* to *two* methyl groups, irrespective of the activation level of the substrate, as shown by the lack of reactivity of mesitylene and of position 2 of *m*-xylene. Substitution *ortho* to *one* methyl group occurs only at strongly activated sites, such as position *4* of *m*-xylene.

Table 5: Selectivity of Gas-Phase Aromatic Silylation[a].

Substrate	Toluene	o-Xylene	m-Xylene	p-Xylene	Mesitylene
Apparent k_S/k_T ratio:	1.0 (1.0)[b]	3.3 (2.0)	1.5 (1.6)	0.6 (< 0.05)	<0.05(<0.05)
Orientation[c]:	*ortho* 0 (0)%	1,2,<u>3</u>:0 (0)%	1,2,<u>3</u>:0 (0)%	1,<u>2</u>,4 100 (100)%	
	meta 14 (6)%	1,2,<u>4</u>:100 (100)%	1,3,<u>4</u>:70 (87)%		
	para 86 (94)%		1,3,<u>5</u>:30 (13)%		

[a] Values in parentheses refer to gas-phase substitution by t-Bu^+. [b] k_T/k_B = 3.8 ± 0.4.
[c] Position of the $SiMe_3$ group underlined.

There exists a close analogy with the reactivity of t-butyl cation, although $SiMe_3^+$ is less selective and characterized by less demanding steric requirements, consistent with the longer C–Si bond, 1.88 Å *vs* 1.54 Å of the C–C bond, which appreciably reduces its effective bulk.

Competition between desilylation and deprotonation of the *ipso*-silylated intermediates (6) suggests their scarce tendency to undergo isomerization by 1,2-H shifts before interception by the base. This can be the result either of unfavourable thermochemical factors or of exceedingly high activation barriers accompanying the intramolecular H transfer. The same factors may be also responsible for the extensive reversibility accompanying t-butylation of arenes in the gas phase, as well as the tendency of t-butylated aromatics to undergo pronounced de-t-butylation by attack of gaseous cations.[26] A proper perspective on these points, which are strictly connected with the long-known electrophilic cleavage of aryl-silicon bonds in solution involving the formation of the *ipso* σ-complex in the rate-determining step,[60] was obtained by an integrated mass spectrometric (0.5 torr, 40-185°C), radiolytic (95-700 torr, 37°C), and computational (SCF STO-3G) investigation of gas-phase electrophilic aromatic desilylation and de-t-butylation.[25]

The salient result of the study is the higher efficiency of desilylation than of de-t-butylation of aromatic substrates following the attack of charged electrophiles, such as CH_5^+, $C_2H_5^+$, i-$C_3H_7^+$, and t-$C_4H_9^+$. Several lines of evidence identify the key difference between the outcoming trimethylsilyl and t-butyl groups with their different effects on the basicity of the *ipso* carbon, hence on the stability of the corresponding *ipso*-substituted arenium ions (6) relative to that of their isomers protonated at other positions. Such a difference is in part due to electronic factors since a silyl substituent, relative to a t-butyl group, is known to destabilize an adjacent, and stabilize a vicinal, positively charged center, hence favouring the mesomeric structures of the *ipso* arenium ions (6) over their isomeric forms. Steric factors also play a role in stabilizing the *ipso*-silylated complexes (6) relative to their t-butylated analogs. Theoretical calculations have pointed out that the aromatic rings in the latter intermediates are not rigorously planar as a result of the significant steric repulsion between the bulky t-butyl substituent and the aromatic π system that causes a destabilizing deformation of the ring. The greater length of the C-Si bond relative to the C-C bond reduces the effective bulk of the trimethylsilyl group and hence the steric destabilization in the ipso-silylated intermediates (6). On the whole, favorable electronic and steric factors contribute to increase the basicity of the *ipso* position of trimethylsilylbenzene to an extent that it is comparable with that of its *para* position. Relatively unfavorable factors make the basicity of the *ipso* position of t-butylbenzene signficantly lower (*ca* 11 kcal mol^{-1}) than that of its *para* position.

Since the aromatic silylation *via* normal electrophilic substitution was unprecedented in the gas phase as well in solution, its selectivity, orientation and steric requirements are of interest to mechanistic organic chemistry.

However, the most striking consequence of the above gas-phase studies has been the development of a preparative method for the synthesis of phenyltrimethylsilanes based on direct aromatic substitution of arenes by a Me$_3$SiCl: AlCl$_3$ reagent under Friedel-Crafts conditions in the presence of a hindered amine,[61] which reproduces in solution the conditions established by Fornarini in her gas-phase study.

This is one of the preciously rare cases where a gas-phase, mechanistically oriented study of ionic reactions paved the way to the development of a synthetic method in condensed phases.

5   THE CYCLOHEXYL CATION

The solvated cyclohexyl cation has long enjoyed a respectable position as a legitimate ionic intermediate in a variety of reactions occurring in solution, such as deamination, solvolysis, acid-induced isomerization, *etc.*[62] The situation, however, is entirely different for the free cyclohexyl cation in media of very low nucleophilicity, or in the gas phase. Indeed, all attempts to generate c-C$_6$H$_{11}$$^+$ ions in superacidic solutions invariably fail, the 1-methylcyclopentyl ion being the only species detectable by NMR spectroscopy even at temperatures as low as -100°C.[63] Understandably, this failure led to the suggestion that the free cyclohexyl cation is inherently unstable, rearranging instantly,[64] or even before its actual formation, *ie* in its 'incipient' state,[65] into the 1-methylcyclopentyl ion, and this view was extended to the gas state.

5.1. Mass Spectrometric Evidence

The results of careful CAD and MIKE mass spectrometric[66] and EBFlow studies[67] have long corroborated the above hypothesis, since no experimental evidence for the existence of the gaseous cyclohexyl cation could be obtained. Three possible reasons for such a failure can be envisaged: (i) free cyclohexyl cation cannot be generated being inherently unstable; (ii) free cyclohexyl cation can be generated, but the activation barrier for its conversion to the thermodynamically more stable 1-methylcyclopentyl isomer is inherently low, so that its isomerization takes place in a time too short relative to that required for mass spectrometric structural assay (> 10^{-5} s); (iii) free cyclohexyl cation is generated by mass spectrometry and EBFlow techniques with an excess of internal energy sufficient to overcome a significant activation barrier for its fast conversion to the 1-methylcyclopentyl isomer.

5.2. High-Pressure Radiolytic Studies

Discrimination among the above possibilities has been provided by radiolytic experiments, which allowed generation of a free cyclohexyl cation from suitable precursors and kinetic investigation of its isomerization under carefully selected conditions. Thus, gaseous $C_6H_{11}^+$ ions have been produced at room temperature, at pressures ranging from 40 to 740 torr, by hydride-ion abstraction from c-C_6H_{12} by radiolytically formed H_3^+ and i-$C_3H_7^+$ ions and trapping with water.[68] Isolation of cyclohexanol and cyclohexanone, accompanied by 1-methylcyclopentanol, in proportions increasing with decreasing exothermicity of the $C_6H_{11}^+$ formation processes and with increasing total pressure of the gaseous mixture, provided evidence for the existence of the cyclohexyl cation in the dilute gas state, with a lifetime in excess of 10^{-7} s, and confirmed its facile rearrangement to the more stable 1-methylcyclopentyl ion. This conclusion, while excluding hypothesis (i), fails to discriminate between the two remaining hypotheses (ii) and (iii).

To answer these questions the gas-phase isomerization of cyclohexyl ions, obtained *via* protonation of cyclohexene and of bicyclo(3.1.0)hexane and *via* hydride-ion abstraction from cyclohexane by $C_nH_5^+$ (n = 1,2) and i-$C_3H_7^+$ ions, has been investigated at pressures up to 1480 torr, and in the temperature range 310 to 353 K.[69] Linear Arrhenius plots for the cyclohexyl cation → 1-methylcyclopentyl cation isomerization in propane at 1480 torr and in methane at 750 torr were obtained. The complete analogy of the two sets of data, despite the different exothermicity of the formation process and the different deactivating efficiency of the bath gas, indicates complete thermal equilibration of the cyclohexyl cation before its isomerization to 1-methylcyclopentyl ion.

Regression analysis of the two mutually consistent sets of data leads to a pre-exponential factor of $10^{12\pm1.3}$ s^{-1} and to an activation energy of 7.4 ± 1 kcal mol^{-1}. The results are in favor of an isomerization process characterized by a relatively low activation barrier (hypothesis (ii)), which, under the low-pressure conditions typical of mass spectrometric techniques, causes complete rearrangement of cyclohexyl to 1-methylcyclopentyl cation prior to structural assay, irrespective of the actual internal energy excess of the rearranging ion.

From the example illustrated, the very short sampling time emerges as a most valuable feature of the radiolytic technique, which allows rapid trapping and structural characterization of a variety of short-lived free ions, such as the cyclohexyl cation. Moreover, the possibility of evaluating the Arrhenius parameters of a given process over an extended bath-gas pressure range allows one to ascertain whether the short-lived ion attains thermal equilibrium and, therefore, whether its conversion to the products lends itself to standard kinetic treatment and to meaningful kinetic comparison with analogous ionic reactions in solution.

6 CONCLUSION

The previous examples demonstrate the fundamental kinetic and mechanistic similarity of condensed-phase and gas-phase ionic reactions that can emerge only when the latter are studied in a pressure domain where thermal kinetics are obeyed. This condition is hardly met by current mass spectrometric techniques,[70] whose kinetic results are in general not directly comparable to those obtained in solution. The gap is bridged by an integrated approach based on the coordinated application of mass spectrometric and radiolytic techniques, characterized by a sufficiently wide pressure range and by a high structural discrimination.

Application of the radiolytic technique to a steadily growing number of organic reactions is being systematically pursued,[6] and has already provided the general outline of a gas-phase ion chemistry that combines the advantage of a uniquely simple reaction environment with the possibility of a direct comparison with condensed-phase ion chemistry.

REFERENCES

1 D H Aue and M T Bower, 'Gas Phase Ion Chemistry', M T Bowers, Ed, Academic Press, New York, 1979, Vol 2, chapter 9.
2 S G Lias, J E Bartmess, J F Liebman, J L Holmes, R D Levin, W G Mallard, and *J Phys Chem Ref Data*, 1988, 17, suppl 1.
3 M Speranza, *Int J Mass Spectrom Ion Processes*, 1992, **118/119**, 395.

4   T F Magnera and P Kebarle, 'Ionic Processes in the Gas Phase' A Ferreira, Ed D Reidel, Dordrecht, Holland, 1983, p 15 and references therein.
5   W N Olmstead and J L Brauman, *J Am Chem Soc*, 1979, **101**, 3715.
6   F Cacace, *Radiat Phys Chem*, 1982, **20**, 99.
7   F Cacace, *Accounts Chem Res*, 1988, **21**, 215.
8   S G Lias, 'Interactions between Ions and Molecules' P Ausloos, Ed Plenum Press, New York 1975, p 541 and references therein.
9   T H Morton, *Radiat Phys Chem*, 1982, *20*, 29.
10   F Cacace, *Science*, 1990, **250**, 392.
11   G A Olah and Y K Mo, 'Carbonium Ions', G A Olah, P von R Schleyer, Eds Wiley-Interscience New York 1976, V 5, p 2135.
12   V A Koptyug, *Izv Akad Nak SSSR*, 1974, 1081.
13   V A Koptyug, 'Contemporary Problems in Carbonium Ion Chemistry III' *Topics in Current Chemistry*, F L Boshcke Ed, Springer-Verlag Berlin 1984.
14   G A Olah, R H Schlosberg, R D Porter, Y K Mo, D P Kelly, and G D Mantescu, *J Am Chem Soc*, 1972, **94**, 2034.
15   W J Hehre and J A. Pople, *J Am Chem Soc*, 1972, **94**, 6901.
16   P v R Schleyer, private communication quoted in ref 18.
17   D Kuck, *Mass Spectrometric Reviews*, 1990, **5**, 583 and references therein.
18   R S Mason, M T Fernandez, K Jennings, *J Chem Soc Faraday Trans 2*, 1987, **83**, 89.
19   A Porry, M T Fernandez, M Garley, and R S Mason, *J Chem Soc Faraday Trans*, 1992, **88**, 3331.
20   D Kuck, J Schneider, H F Grützmacher, *J Am Chem Soc*, 1979, **101**, 7154.
21   D Kuck, W Bäther, and H F Grützmacher, *Int J Mass Spectrom Ion Processes*, 1985, **67**, 75.
22   M Attinà, F Cacace, and A Ricci, *J Am Chem Soc*, 1991, **113**, 5937.
23   S Fornarini, *J Org Chem*, 1988, **53**, 1314.
24   F Cacace, M E Crestoni, S Fornarini, and R Gabrielli, *Int J Mass Spectrom Ion Processes*, 1988, **84**, 17.
25   F Cacace, M E Crestoni, G de Petris, S Fornarini, and F Grandinetti, *Can J Chem*, 1988, **66**, 3099.
26   L Xiaoping and J A Stone, *Int J Mass Spectrom Ion Processes*, 1990, **101**, 149.
27   F Cacace, M E Crestoni, and S Fornarini, *J Am Chem Soc*, 1992, **114**, 6776.
28   F Cacace and P Giacomello, *J Am Chem Soc*, 1973, **95**, 5851 and references therein.
29   D K Sen Sharma, S Ikuta, and P Kebarle, *Can J Chem*, 1982, **60**, 2325.
30   J M Stone and J A Stone, *Int J Mass Spectrom Ion Processes*, 1991, **109**, 247.
31   F Cacace and G Ciranni, *J Am Chem Soc*, 1986, **108**, 887.

32 T Su, M T Bowers, 'Gas Phase Ion Chemistry', M T Bowers Ed Academic Press, New York 1979, Vol 1, p 83.
33 F Cacace, M E Crestoni, S Fornarini, and D Kuck, *J Am Chem Soc*, 1993, **115**, 1024.
34 C K Ingold 'Structure and Mechanism in Organic Chemistry', Bell, London 1953.
35 R Taylor, 'Electrophilic Aromatic Substitution', Wiley & Sons, New York 1990.
36 K Schofield, 'Aromatic Nitration', Cambridge University Press, 1980 p 5.
37 J W Barnett, R N Moodie, K Schofield, and J B Watson, *J Chem Soc Perkin II*, 1975, 648.
38 J H Ridd, *Adv Phys Org Chem*, 1978, **16**, 1.
39 S A Benezra, M K Hoffmann, and M M Bursey, *J Am Chem Soc*, 1970, **92**, 7501.
40 M K Hoffmann and M M Bursey, *Tetrahedron Lett*, 1971, 2539.
41 R J Schmitt, S E Buttrill, Jr, and D S Ross, *J Am Chem Soc*, 1981, **103**, 5265.
42 G A Olah, R Malhotra, and S C Narong, 'Nitration', VCH Publishers New York 1989.
43 R C Dunbar, J Shen, and G A Olah, *J Am Chem Soc*, 1972, **94**, 6862.
44 P Ausloos and S G Lias, *Int J Chem Kinet*, 1978, **10**, 657.
45 W D Reents, Jr, and B S Freiser, *J Am Chem Soc*, 1980, **102**, 271.
46 R J Schmitt, S E Buttrill, Jr, and D S Ross, *J Am Chem Soc*, 1984, **106**, 926.
47 G De Petris, *Org Mass Spectrom*, 1990, **25**, 83.
48 T J Lee and J E Rice, *J Am Chem Soc*, 1992, **114**, 8247.
49 M Attinà and F Cacace, *J Am Chem Soc*, 1986, **108**, 318.
50 Early ICR studies demonstrated the occurrence of process (11), see ref 40. Recent FT-ICR measurements give $k_{11} \sim 5 \cdot 10^{-11}$ cm^3 molecule^{-1} s^{-1} (Cacace *et al*, unpublished results).
51 M. Attinà, F Cacace, and M Yañez, *J Am Chem Soc*, 1987, **109**, 5092.
52 M. Attinà, F Cacace, and A Ricci, *Tetrahedron*, 1988, **44**, 2015.
53 M Attinà, F Cacace, and G de Petris, *Angew Chem, Int Ed Engl*, 1987, **99**, 1174.
54 G A Olah, S Kuhn, and S H Flood, *J Am Chem Soc*, 1961, **83**, 4571.
55 M J S Dewar, 'The Electronic Theory of Organic Chemistry', Oxford University Press, London 1946.
56 J G Hoggett, R B Moodie, J R Penton, and K Schofield, 'Nitration and Aromatic Reactivity' Cambridge University Press, London 1971.
57 D Hähich and F Effenberger, *Synthesis*, 1979, 884.
58 G A Russel, *J Am Chem Soc*, 1959, **81**, 4831.
59 A C M Wojtyniak and J A Stone, *Int J Mass Spectrom Ion Processes*, 1986, **74**, 59.
60 C Eaborn, *J Organomet Chem*, 1975, **100**, 43.

61 G A Olah, T Bach, and G K Surya Prakash, *J Org Chem*, 1989, **54**, 3770.
62 'Carbonium Ions', G A Olah, and P v R Schleyer Ed, Wiley, New York 1970, Vol 2, chapters 14,15.
63 G A Olah, J M Bellinger, C A Caps, and J Lukas, *J Am Chem Soc*, 1967, **89**, 2692.
64 E M Arnett, and C Petro, *J Am Chem Soc*, 1978, **100**, 5408.
65 G A Olah, and J Lukas, *J Am Chem Soc*, 1968, **90**, 933.
66 C Wesdemiotis, R Wolfschutz, and H Schwarz, *Tetrahedron*, 1979, **36**, 275.
67 W J Marinelli, and T H Morton, *J Am Chem Soc*, 1979, **101**, 1908.
68 M Attinà, F Cacace, and P Giacomello, *J Am Chem Soc*, 1981, **103**, 4771.
69 M Attinà, F Cacace, and A di Marzio, *J Am Chem Soc*, 1989, **111**, 6004.
70 However, a kinetic study performed by very high pressure (610 torr) ion mobility mass spectrometry has recently been reported, *cfr* W B Knighton, J A Bognar, P M O'Connor, and E P Grimsud, *J Am Chem Soc* 1993, submitted.

ACKNOWLEDGEMENTS

The author expresses his appreciation to the talented and motivated members of his group, in particular to M Attinà and S Fornarini, senior Authors of the contributions illustrated, together with M E Crestoni and A Ricci. The financial support of Ministero dell'Università e della Ricerca Scientifica e Tecnologica (MURST) and of Consiglio Nazionale delle Ricerche (CNR) is acknowledged.

HYDROGEN-ENE AND METALLO-ENE REACTIONS

Alwyn G Davies

Chemistry Department, University College London, 20 Gordon Street, London WC1H OAJ, UK.

1   METALS AS HYDROGEN EQUIVALENTS

Reactions which involve the hydrogen centre H in organic compounds RH, frequently have their counterparts in reactions involving the metal centres M in organometallic compounds RML_n (L = ligand).[1] This is much more than a superficial analogy. It extends across all the main mechanistic classes of reactions - homolytic, heterolytic, pericyclic, and photolytic - and may include even details of the transition state.

This principle is important in designing organic syntheses. If a reaction occurs at H in RH, there is little that can be done at the monovalent hydrogen centre, apart perhaps from hydrogen bonding, to modulate the reactivity. In a compound RML_n, on the other hand, the identity of the metal M can be varied, and an unlimited variety of ligands L can be assembled around the metal. It is thus possible to vary the reaction site to achieve the optimum reactivity, and chemo-, regio-, and stereo-selectivity.

The ene reaction should benefit from this approach. In its usual form it takes place at an allylic hydrogen centre as illustrated in equation 1.[2,3]

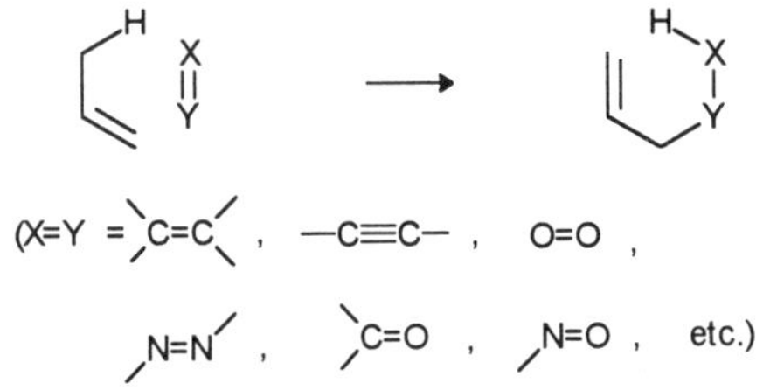

$$(X=Y = \:\!C=C\:\!, \quad -C\equiv C-, \quad O=O, \quad N=N, \quad C=O, \quad N=O, \quad etc.)$$

(1)

The enophiles X=Y belong to the same group of reagents as the dienophiles in the diene (Diels-Alder) reaction. The reactivity of alkenes and alkynes is rather low, but can be enhanced (as in the Diels-Alder reaction) by the presence of electron attracting substituents, *eg* (1), particularly in the presence of Lewis acids. Singlet oxygen, which may be represented as O=O (and derived from dye-sensitized photoexcitation of triplet oxygen) is an important enophile, leading to the formation of allylic hydroperoxides.[4,5,6]

$$HC{\equiv}CCO_2Et \qquad EtO_2CN{=}NCO_2Et$$

1   **2**   **3**

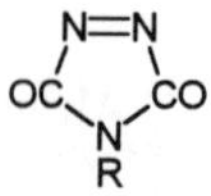

Azo compounds similarly have their reactivity enhanced by electron attracting substituents as in diethyl azodicarboxylate (**2**) or the triazolinediones (**3**). Aldehydes, unless they carry electronegative groups (*eg* as in chloral), react only sluggishly unless a Lewis acid is present.

The ene reaction thus provides a context in which the principle of metals acting as hydrogen equivalents might be applied with advantage. Will allylmetallic compounds show migration of the metal (the metallo-ene reaction, equation 2) similar to the migration of hydrogen in the hydrogen-ene reaction ? Can we then vary the metal and its ligands to achieve a more useful reaction ? This review considers first some relevant aspects of the hydrogen-ene reactions, then surveys what progress we have made in developing the metallic counterparts.

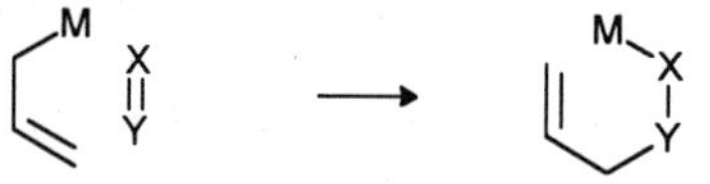

(2)

2   THE MECHANISM OF THE H-ENE REACTION

The diene reaction provides the classic example of a cycloaddition reaction proceeding through a concerted, cyclic transition state (**4**).

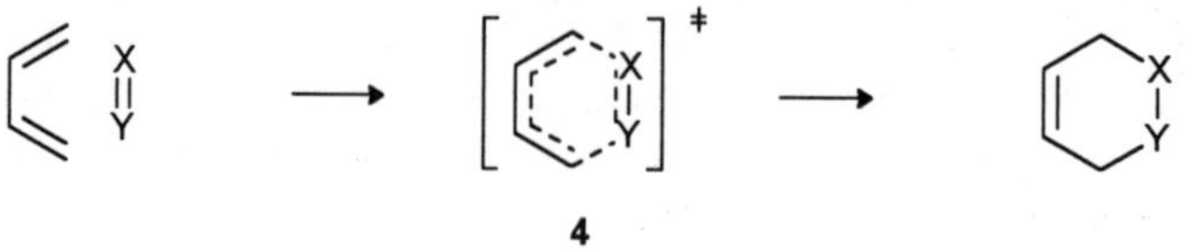

(3)

As the diene and ene reactions have much in common, it would be reasonable to expect that the M-ene reactions involve a similar pericyclic process (equation 4).

This would appear not to be correct: evidence from hydrogen kinetic isotope effects, backed up by stereochemical studies, shows that electrophilic

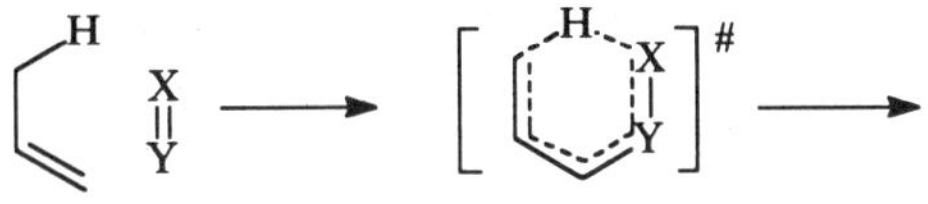

$$\overset{H}{\underset{Y}{\overset{X}{\|}}}\diagdown \longrightarrow \left[\overset{H\cdots X}{\underset{Y}{\bigcirc}} \right]^{\#} \longrightarrow \overset{H_{\diagdown}X}{\underset{Y}{\bigcirc}}$$

(4)

attack of the enophile upon the ene gives first the intermediate adduct (**5**); if the enophile is singlet oxygen, this intermediate is referred to as a perepoxide. The hydrogen is then transferred in a second step.

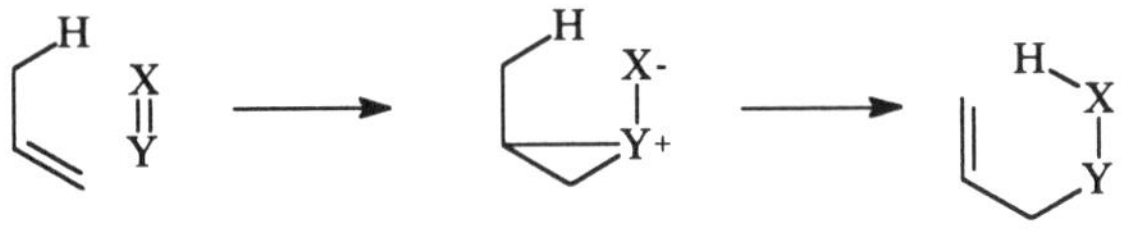

$$\overset{H}{\underset{Y}{\overset{X}{\|}}}\diagdown \longrightarrow \overset{H}{\underset{Y+}{\overset{X-}{\bigcirc}}} \longrightarrow \overset{H_{\diagdown}X}{\underset{Y}{\bigcirc}}$$

                         5

(5)

The kinetic hydrogen isotope effect on ene reactions involving a variety of enophiles is consistently greater for the *trans*- or *gem*-dimethyldi(trideuteriomethyl)ethene than it is in the *cis*-isomer; an example is shown in Scheme 1.[7]

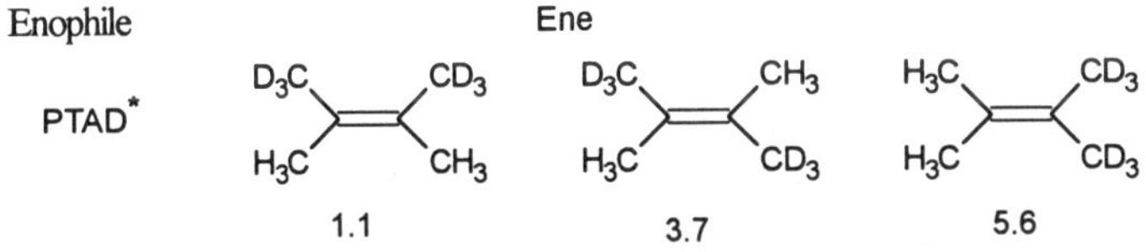

Enophile	Ene		
PTAD*	D_3C—CD_3 / H_3C—CH_3	D_3C—CH_3 / H_3C—CD_3	H_3C—CD_3 / H_3C—CD_3
	1.1	3.7	5.6

Scheme 1. Values of k_H / k_D *PTAD = N-phenyltriazolinedione

If a single pericyclic transition state (**4**) were involved, the kinetic isotope effect would be expected to be similar for *cis*-, *trans*-, and *gem*-isomers. The negligible isotope effect which is observed for the *cis*-isomer, and the significant one for the *trans*- and *gem*-isomers, can best be explained by the formation of an intermediate adduct (**7**) or (**8**).

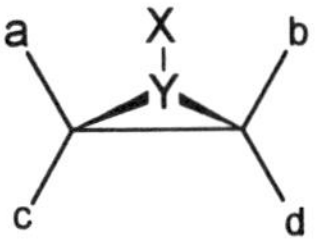

           7 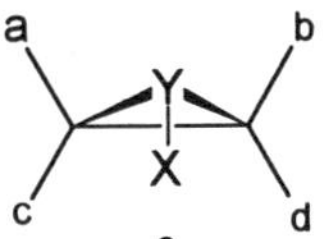   **8**

If these two isomeric adducts are formed with a negligible isotope effect, as would be expected, and irreversibly, the value of k_H/k_D in the *cis* derivative (a = b = R^H, c = d = R^D) is preordained as unity. In the *trans*-isomer (a = d = R^H, b = c = R^D), or the *gem*-isomer (a = c = R^H, b = d = R^D), each isomeric intermediate can still select between R^H and R^D in proceeding to product, and a significant isotope effect would be expected, as is observed.

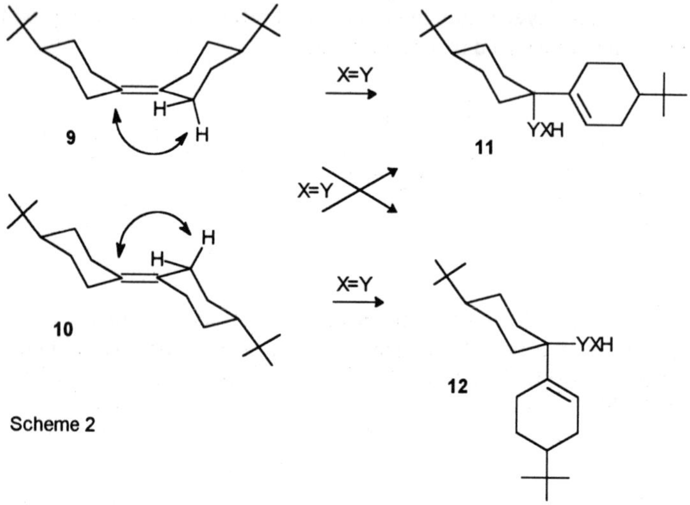

Scheme 2

Stereochemical evidence in favour of a stepwise mechanism is illustrated in Scheme 2.[8,9] If the t-butyl substituents in *syn-* and *anti-*1-(4-t-butylcyclohexylidene)-4-t-butylcyclohexane [(9) and (10) respectively] are equatorially oriented during the attack of the enophile, the steric requirements of a pericyclic transition state would permit transfer of only axially directed hydrogen at the 2-position, giving only one product [(11) or (12) respectively]. In fact, both enes react with a variety of enophiles to give the two products (11) and (12), suggesting the existence of an intermediate which can undergo conformational change before proceeding to products.

3   THE SCHENCK AND SMITH REARRANGEMENTS

Allylic hydroperoxides formed by the ene reactions of singlet oxygen can rearrange in a manner which has not been identified for the products of the reaction of any other enophile. The two types of reaction are illustrated in Scheme 3.

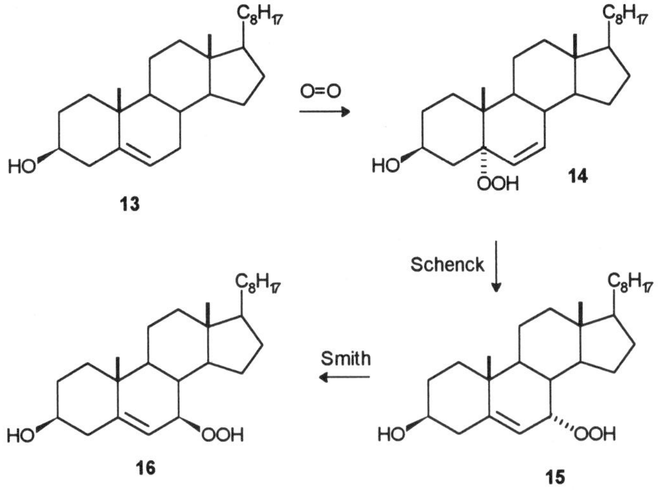

Scheme 3

Cholesterol (**13**) reacts with singlet oxygen to give the 5α-hydroperoxide (**14**);[10] in a non-polar solvent such as chloroform, this rearranges during about 1 day to give the 7α-hydroperoxide (**15**) (the Schenck rearrangement),[10] then, further, during about 1 week, to give the 7ß-hydroperoxide (**16**) (the Smith rearrangement).[11] It is well established that both these rearrangements are chain processes involving allylperoxyl radicals (equations 6 and 7, and 8 and 9, respectively).[10,12,13,14]

If the Schenck rearrangement is carried out under an atmosphere of $^{18}O_2$, there is no incorporation of labelled oxygen into the product, but under the same conditions the Smith rearrangement gives substantial incorporation of the label.[15,16] On the other hand, the hydroperoxide derived from methyl oleate undergoes rearrangement to incorporate a small amount of oxygen from the atmosphere, the amount (hexane > decane > octadecane) decreasing with increasing viscosity of the solvent.[17,18] It seems likely that both the Schenck and Smith rearrangements involve dissociation of the allylperoxyl radical; this may give at one extreme a charge transfer complex between the allyl radical and oxygen, and at the other two kinetically independent fragments, and the degree to which the separation occurs depends on the structure of the allylic fragment, and on the viscosity of the medium.

Here then is another context in which we might hope to observe metals acting as hydrogen equivalents. Will the allylperoxymetallic compounds, which we hope to be able to obtain by metalloene reactions, give the metallic equivalent of the Schenck and Smith rearrangements?

4 THE REACTION OF METALLOENES WITH ENOPHILES

A number of examples of metalloene reactions are to be found in the literature, and some important examples are listed at the end of this paper. For example, the reaction of an allylic Grignard reagent with a carbonyl compound may be included in this category, and the word metalloene was coined by Oppolzer[19] in the context of the reaction of allylpalladium compounds with alkenes. However there appeared to be no previous examples of reactions with singlet oxygen as the enophile; this is largely because allylic derivatives of many metals (*eg* boron, aluminium, magnesium, zinc, cadmium) react very rapidly with triplet oxygen by a radical chain process, and singlet oxygen, as it is usually (photolytically) generated, is in mixture with triplet oxygen. The allyl derivatives of the Group 14 metals (Si, Ge, Sn, and Pb), in contrast, are stable towards triplet oxygen, and our first experiments were aimed at identifying a metalloene reaction between singlet oxygen and allyltin compounds which we had readily available.

When a solution of allyltrimethyltin containing Rose Bengal as a photosensitizer and saturated with oxygen was irradiated with sodium light, a reaction occurred which is much faster than that of an allylic hydrocarbon. The products are shown in equation 10.[20]

(10)

17 50%   **18** 25%   **19** 25%

The principal product was the 3-stannylallyl hydroperoxide (**17**) resulting from a conventional H-ene reaction. Also present was the allylperoxytin compound (**18**) which we were looking for, formed by a metalloene process, and to our surprise, the 1,2-dioxolan (**19**).

When the enophile was N-phenyltriazolinedione (PTAD), the reaction was again rapid; no H-ene reaction occurred, and the principal product (**20**) was that of the metalloene reaction, with again some cyclisation to give (**21**) (equation 11).

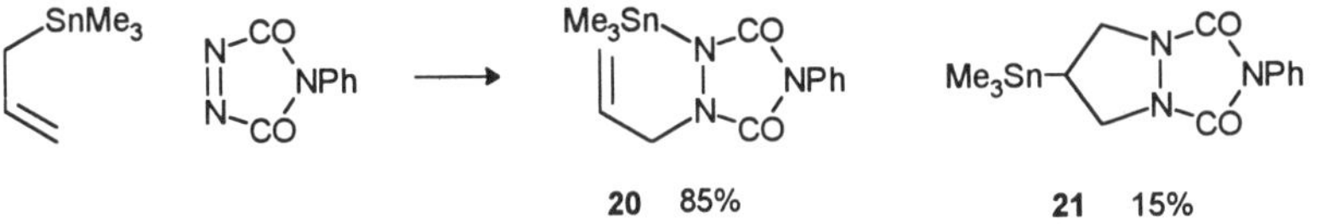

$$\text{SnMe}_3\diagup\diagdown + \underset{\text{N-CO}}{\overset{\text{N-CO}}{\|}}\text{NPh} \longrightarrow \underset{\mathbf{20}\ 85\%}{\text{Me}_3\text{Sn}\diagdown\diagup\diagdown\diagup\text{NPh}} + \underset{\mathbf{21}\ 15\%}{\text{Me}_3\text{Sn}\diagup\diagdown\text{NPh}} \quad (11)$$

4.1 The Reaction Mechanism

A reasonable mechanistic model for these reactions, based on what is known of the hydrogen-ene reactions, is shown in Scheme 4.

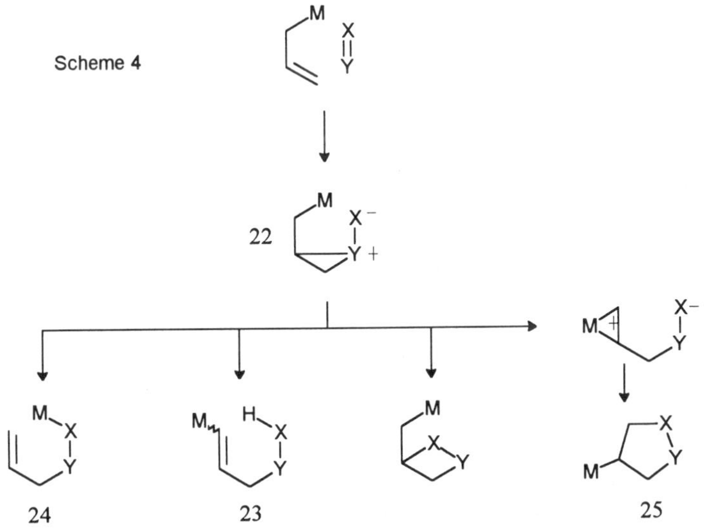

Scheme 4

22

24   **23**   **25**

The product of the H-ene reaction (**23**), of the M-ene reaction (**24**) and of cyclisation (**25**) are probably formed through the common intermediate (**22**). The enhanced rate results from the hyperconjugation of the carbon-metal bond with the π-system which makes it more reactive towards the electrophilic enophile in forming this intermediate. This hyperconjugation has the further effect of placing the metal in the axial position in the transition state (**26**) through which the hydrogen migration occurs, leaving the metal in a *cis* position in the alkene which is formed, whereas an alkyl substituent gives mainly a *trans* product through the transition state (**27**).

26

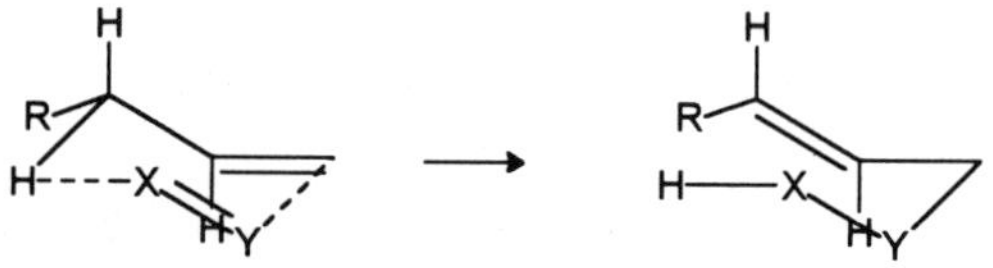

27

The dioxolane product (**19**) results from the interaction of the carbon-metal bond with the partial positive charge which exists on the C-2 of the allylic system in the intermediate; nucleophilic attack upon C-3 then closes the ring and completes the migration of the metal.

4.2 The Effect of the Ligands

The effect of varying the ligands about the tin in the reaction with singlet oxygen is shown in Table 1.[20,21] The 13-C NMR chemical shift of the (C-1) allylic methylene group provides a measure of the electronegativity of the metal. As δ_C increases, the yield of the product from the M-ene reaction increases, and if the tin carries one or more electronegative ligands such as halide or carboxylate and δ_C is > *ca* 23, this is the only product.

Tetraallyl tin is an interesting exception to this rule in that it gives only the metalloene process with singlet oxygen although δ_C is only 16.1. This reaction is important in that it gives an easy and relatively safe route to a useful derivative of simple allylic hydroperoxides. We believe that carbon-metal hyperconjugation with the π-systems enhances the electronegativity of the tin without affecting the chemical shift of the methylene carbon, then after the reaction of the first molecule of oxygen, the peroxyl group which is formed has the usual effect of an electronegative substituent.

Table 1. Reactions of allyltin compounds with singlet oxygen

Reactant	δ(CH₂)		Percentage yields	
⬈SnMe₃	18.1	25	50	25
⬈SnBu₃	16.2	18	40	42
⬈Sn(C₆H₁₁)₃	14.1	5	37	58
⬈SnPh₃	17.8	42	53	5
⬈SnBu₂Cl	23.7	100	0	0
⬈SnBu₂OAc	24.9	100	0	0
[⬈SnCl₂]₂	30.2	100	0	0
[⬈SnCl₂ bipy]₂	48.6	100	0	0

4.3 The Effect of the Metal

The effect of varying the metal is shown in Table 2;[22] PTAD is now the enophile, to provide a convenient rate of reaction with the range of enes. Allylsilanes usually show the H-ene reaction, together with sometimes a small proportion of the cyclisation process. The highest yield of cyclised product is usually obtained with allylgermanes. The tin compounds usually give a substantial amount of the metalloene reaction, and allyllead compounds show predominantly or exclusively the metalloene process. Again then a more electropositive metal favours migration of the metal.

Table 2. Reactions of allyl-silicon, -germanium, -tin, and -lead compounds with PTAD

Reactant		Percentage yields	
allyl–SiPh$_3$	0	100	0
allyl–GePh$_3$	20	30	50
allyl–SnPh$_3$	55	0	45
allyl–PbPh$_3$	63	0	37

As silicon appeared capable of the cyclisation and not the migration process, it seemed possible that a pure cyclisation reaction could be observed if the hydrogen-ene reaction were blocked. This indeed can be done: trimethyl(1,1-dimethylallyl)silane reacts slowly with PTAD to give only the cyclised product (equation 12).[22]

(12)

Mercurinium ions involving interaction of a positive charge with mercury on a ß-carbon atom, are familiar in mercuration reactions of alkenes, and it seemed probable that the reaction of allylmercury compounds with enophiles might give predominantly cyclised compounds with migration of the mercury. The reactions with singlet oxygen are complicated by the fact that some photolytic decomposition of the allylmercury compounds occurs, but both here, and in the reaction with PTAD, which is clean, the metalloene reaction predominates (equation 13).[21,22]

4.4 The Effect of Lithium Perchlorate

A number of reactions, for example Diels-Alder cycloadditions, have recently been shown to be surprisingly susceptible to catalysis by lithium perchlorate (typically 5 M in ether). As Diels-Alder reactions and ene reactions have many common characteristics, it seemed possible that H-ene and M-ene reactions might similarly be usefully accelerated.

The effect of lithium perchlorate on the rates of some H-ene and M-ene reactions involving diethyl azodicarboxylate or triazolinediones is shown in Table 3.[23]

Both the H-ene and M-ene reactions with these azo-enophiles are accelerated, some by a factor of up to 250, and this will extend the scope of the metalloene reaction to less reactive combinations of reagents. There appears to be some acceleration when singlet oxygen is the enophile, but there the catalysis does not seem to be sufficient to be experimentally useful.

The way in which the catalysis operates is not clear, but alternatives which have been suggested include the operation of the lithium as a Lewis acid on the enophile, the stabilisation of a polar transition state by the polar salt, or the internal pressure which is exerted on the hydrophobic reactants by the hydrophilic medium. It is not possible at the present time to select between these alternatives: a Lewis acid interaction with the enophiles can be envisaged, the transition state through which the intermediate (19) is formed is indeed polar, and at least one metalloene reaction (that between an allyltin compound and an aldehyde) has been shown to have a negative volume of activation and to be accelerated by high pressures.

Table 3 LiClO₄ catalysis of H-ene and Sn-ene reactions

Ene	Enophile	Product	Solvent	[LiClO₄]	Time
cyclopentene	2	cyclopentenyl-triazolidinedione-NMe	Et₂O	0	10 days
				5 M	1 h
cyclohexene	2	cyclohexenyl-triazolidinedione-NMe	Et₂O	0	4 days
				5 M	4 h
dicyclopentylidene	1	cyclopentenyl-cyclopentyl-N(CO₂Et)-NH-CO₂Et	Et₂O	0	4 days
				5 M	4 h
			EtOAc	0	4 days
				2.35 M	1 day
	2	cyclopentenyl-cyclopentyl-triazolidinedione-NMe	Et₂O	0	<5 min
				5 M	<5 min
Bu₃Sn-allyl	1	allyl-N(SnBu₃)-N(CO₂Et)-CO₂Et	Et₂O	0	28 h
				5 M	12 min
	3	allyl-triazolidinedione(SnBu₃)-NPh	Et₂O	0	1 h
				5 M	<5 min
Ph₃Sn-allyl	1	allyl-N(SnPh₃)-N(CO₂Et)-CO₂Et	Et₂O	0	80 min
				2 M	<5 min
				5 M	<5 min

1 EtOCON=NCO₂Et   2 4-methyl-1,2,4-triazoline-3,5-dione   3 4-phenyl-1,2,4-triazoline-3,5-dione

Table 4

Mikhailov (M = B) Sakurai (M = Si) Tagliavini (M = Sn)

Lehmkuhl

Oppolzer

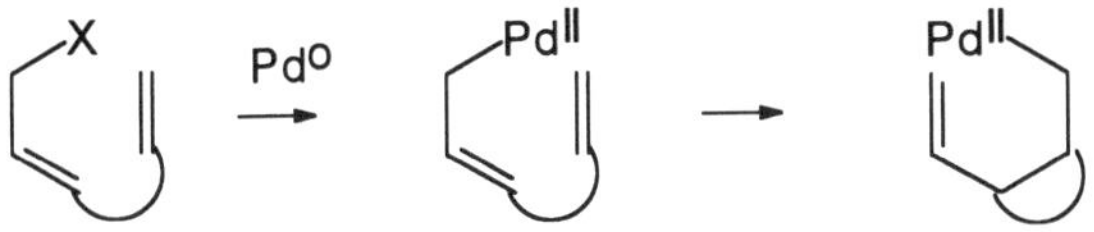

Herndon

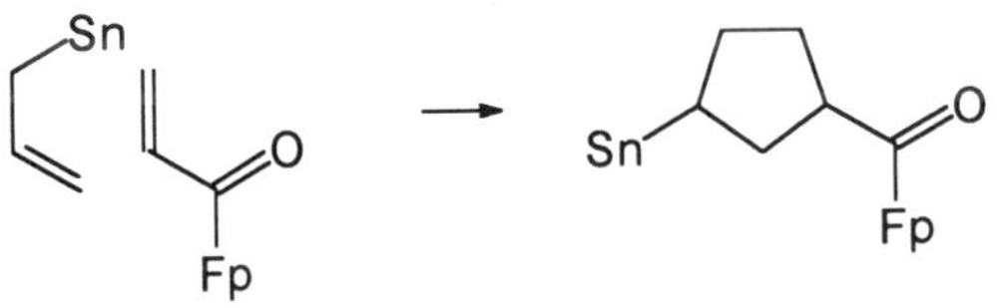

Wojcicki M = Mn, Fe, Co

Jefford

4.5 The Metallo-Schenck and Metallo-Smith Rearrangements

Our attempts to prepare 3ß-hydroxy-5α-tributylstannylperoxycholest-6-ene by the metalloene reaction of singlet oxygen with 3ß-hydroxy-7α-tributylstannylcholest-5-ene were unsuccessful because the reaction occurs only on the ß-face of the molecule, probably because the α-side is too sterically hindered. However the 5α-hydroperoxide can be stannylated, and then tested for the Schenck and Smith rearrangements. These do occur (equation 14), at about the same rates as for the hydroperoxides.[24] Presumably the mechanisms are the same as those as shown in equation 6-9 except that reactions 7 and 9 are replaced by the equivalents involving homolytic substitution at tin rather than at hydrogen.

(14)

5 CONCLUSION

The ene reaction, and its related Schenck and Smith rearrangements illustrate well the principle that metals can be used to advantage as hydrogen equivalents in organic synthesis. We have been concerned here mainly with the enophiles O=O and RN=NR, but there are already a number of other reactions in the literature which can be regarded as belonging to this group of reactions, and these, with the names of the workers associated with them, are listed in Table 4.

REFERENCES

1. A G Davies, *J Organomet Chem*, 1982, **239**, 87.
2. H M R Hoffmann, *Angew Chem, Int Ed Engl*, 1969, **8**, 556.
3. G V Boyd, in *The Chemistry of Double Bonded Functional Groups*, ed S Patai, Wiley, Chichester, 1989, Chap 8, p 4774. R W Denney and N Nickon, *Org React*, 1973, **20**, 133.
5. A A Frimer, *Chem Rev*, 1979, **79**, 359.
6. A A Frimer, and L M Stephenson in *Singlet Oxygen Vol II* ed A A Frimer, C R C Press, Boca Raton, 1985, pp 67-91.
7. C-C Cheng, C A Seymour, M A Petti, and F D Greene, *J Org Chem*, 1984, **49**, 2910.
8. H-S Dang and A G Davies, *J Chem Soc, Perkin Trans 2*, 1991, 721.
9. E W H Asveld and R M Kellog, *J Org Chem*, 1982, **47**, 1250.
10. G O Schenck, O A Neumuller, and W Eisfeld, *Annalen*, 1958, **618**, 202.

11   J I Teng, M J Kulig, L L Smith, G Kan, and J E V Lier, *J Org Chem,* 1973, **38**, 119.
12   W F Brill, *J Am Chem Soc,* 1965, **87**, 3286.
13   W F Brill, *J Chem Soc, Perkin Trans 2,* 1984, 621.
14   N Porter and P Zuraw, *J Chem Soc, Chem Commun,* 1985, 1472.
15   A L J Beckwith, A G Davies, I G E Davison, A Maccoll, and M H Mruzek, *J Chem Soc, Perkin Trans 2,* 1989, 815.
16   N Porter and J S Wujek, *J Chem Soc, Chem Commun,* 1987, 621.
17   K A Mills, S E Caldwell, G R Dubay, and N A Porter, *J Am Chem Soc,* 1992, **114**, 9689.
18   S L Boyd, R J Boyd, Z Shi, L R C Barclay, and N A Porter, *J Am Chem Soc,* 1993, **115**, 687.
19   W Oppolzer, *Angew Chem, Int Ed Engl,* 1989, **28**, 38.
20   H-S Dang and A G Davies, *Tetrahedron Lett,* 1991, **32**, 1745.
21   H-S Dang and AG Davies, *J Organomet Chem,* 1992, **430**, 287.
22   J Cai and A G Davies, *J Chem Soc, Perkin Trans 2,* 1992, 1743.
23   A G Davies and W J Kinart, *J Chem Soc, Perkin Trans 2,* 1993, 2281
24   H-S Dang and A G Davies, *J Chem Soc, Perkin Trans 2,* 1992, 1095.

CHEMISTRY BY COMPUTER: A THEORETICAL APPROACH TO STRUCTURE AND MECHANISM

Leo Radom

Research School of Chemistry, Australian National University, Canberra, ACT 0200, Australia.

1 INTRODUCTION

Chemistry is traditionally an experimental science. However, recent advances in computer technology and the development of highly efficient computer algorithms have opened the way for a viable alternative approach to chemistry: chemistry by computer. The computer calculations that I will be describing in this article are based on *ab initio* molecular orbital theory,[1] a mathematical procedure that uses no experimental information in its implementation other than the values of the fundamental physical constants (such as the speed of light). Such calculations may be used to predict properties of primary interest to chemists, such as the structures of molecules and the mechanisms of reactions in which they are involved. The aim of the present article is to show some of the ways in which this might be achieved. The examples used to illustrate the approach are taken largely from recent research carried out in Canberra.

2 METHODS

It is useful to begin by discussing some of the principal capabilities of *ab initio* molecular orbital theory.[1] Most of these stem from the ability to calculate the energy of a system as a function of geometry. This enables the construction of energy profiles of the type displayed in Figure 1, showing schematically how the energy changes as one proceeds from one minimum on the potential energy surface (representing the structure AB) to another minimum (representing AB'). We can calculate equilibrium structures, the structures that correspond to minima on the surface. We can calculate transition structures (denoted TS), structures that represent saddle points separating minimum energy structures. And we can calculate vibrational frequencies both for equilibrium structures and transition structures. These can be used to rigorously characterize equilibrium structures (– all the frequencies are real) and transition structures (– one frequency is imaginary). They are also useful in calculating zero-point vibrational energies, temperature corrections to reaction enthalpies, and reaction rates via transition-state theory. We can also calculate reaction energies ΔE, dissociation energies D, and reaction barriers $\Delta E^{\#}$.

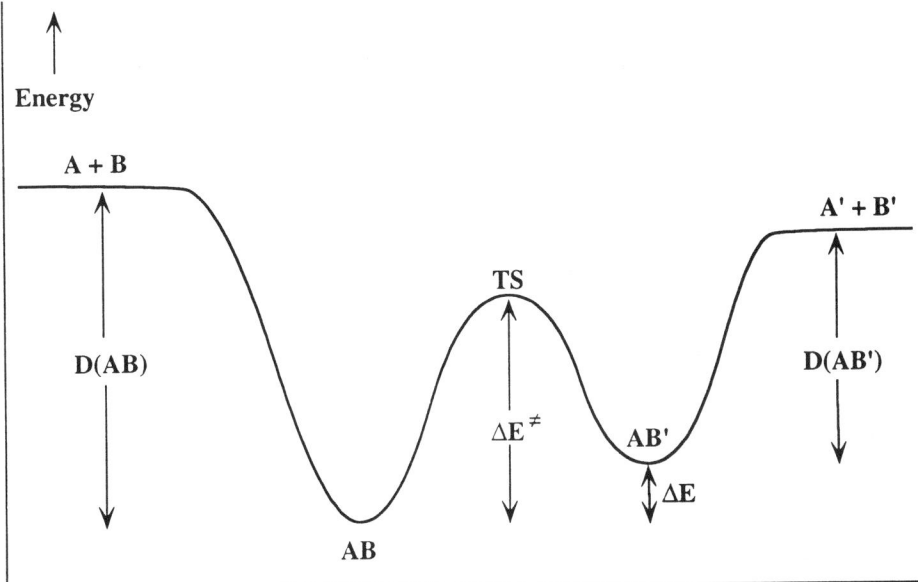

Figure 1: Information associated with potential energy profiles accessible from *ab initio* calculations.

It really goes without saying that the computer calculations are best used in those situations for which theory has an advantage over experiment. The advantage might be one of cost: for example, it is much less expensive to determine accurately the structure of a small molecule using theory than to carry out a microwave spectral analysis. The advantage might be that theory provides direct information whereas an uncertain path needs to be traversed in going from experimental data to the desired information. Or it might be that we need to deal with short-lived or dangerous molecules. In respect of this last point, it is important to note that theory can be applied as readily to reactive species as to normal, stable molecules. On the other hand, the detailed experimental characterization of reactive species is, as we know, a much more difficult task.

The quality of an *ab initio* calculation depends on the size of the basis set used and on the extent of incorporation of electron correlation. Better calculations are, however, computationally more expensive. We are therefore faced in our calculations with striking a compromise between the accuracy we desire and the expense we can afford.

Figure 2, which may be referred to as a Pople diagram, shows this more explicitly. In carrying out an *ab initio* calculation, it is necessary to choose a basis set and a level of incorporation of electron correlation. Basis sets have names such as STO-3G (a small basis set), 6-31G* (a medium-sized basis set)

Basis sets	Electron correlation procedures HF MP2 MP3 MP4 QCISD(T)
STO-3G 3-21G 6-31G* 6-311G** 6-311+G(3df,2pd)	

Figure 2: Pople diagram showing frequently used levels of *ab initio* molecular orbital theory.

and 6-311+G(3df,2pd) (a large basis set).[1,2] The quality of the basis set improves as we move down the table. Electron correlation procedures have names such as MP2, MP3, MP4 and QCISD(T),[1,3] corresponding to better and better calculations as we move to the right. We could choose HF/3-21G, which is a low level of theory because it involves a small basis set and no electron correlation. On the other hand, QCISD(T)/6-311+G(3df,2pd) would be a high level of theory – a large basis set and a good correlation procedure. The ultimate level of theory would involve an infinite basis set and complete incorporation of electron correlation, and then we would have the exact solution of the non-relativistic Schrödinger equation and the exact answers to many of our problems. Unfortunately, this is not a practical proposition and we are forced to compromise, the ultimate choice depending on the problem at hand and the computing resources that are available.

One specific level of theory that I will have reason to discuss is G2, recently introduced by Pople and co-workers[4,5] with the aim of predicting thermochemical data for a wide variety of systems to so-called chemical accuracy – roughly 10 kJ mol^{-1}. G2 corresponds effectively to calculations at the QCISD(T)/6-311+G(3df,2p) level on MP2/6-31G* optimized geometries, together with zero-point vibrational energy and isogyric corrections. In the broad context of Figure 2, it corresponds to a high level of theory.

Chemistry by Computer: A Theoretical Approach to Structure and Mechanism

3 APPLICATIONS

Up to this point, I have been discussing in general terms the sorts of things that theory can do for us. The remainder of this article will be devoted to specific snap-shot examples, taken from our research in Canberra. Some of the examples represent very recent work. Others involve older work. The latter are specially included to exemplify situations in which the theoretical predictions have subsequently been verified by experiment.

3.1 Determination of Molecular Structure

The first example deals with predictions of accurate molecular structures. In a recent study,[6] we explored the possible existence of trithia[1.1.1]propellane, the sulfur analogue of [1.1.1]propellane, first synthesized by Wiberg about 10 years ago.[7] As a preliminary to calculations on the unknown thiapropellane molecule, we carried out calculations of the structures of related systems at the same level of theory. We can see (Figure 3) that the agreement between theory and experiment is very good: bond lengths agree to within a few thousandths of an Ångstrom and bond angles to a few tenths of a degree.

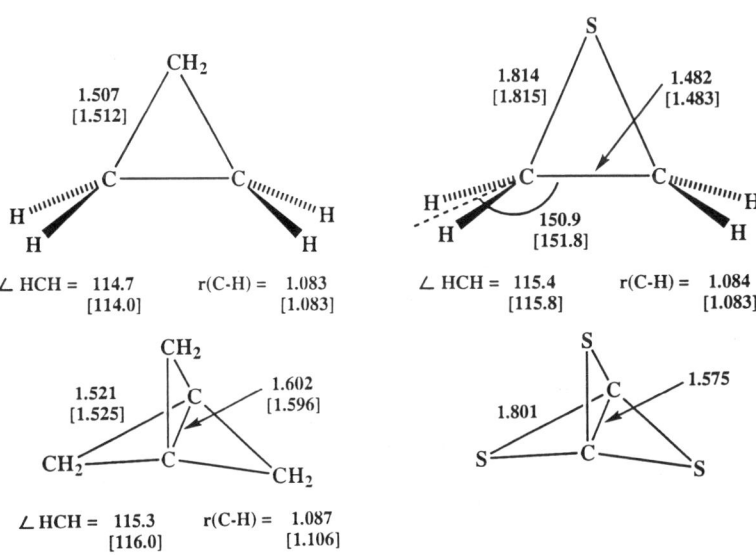

Figure 3: Calculated (MP2/6-311G(MC)*) and experimental (in square brackets) structures for cyclopropane, thiirane, [1.1.1]propellane and trithia[1.1.1]propellane (from ref 6). Bond lengths in Å, bond angles in degrees.

In particular, the length of the interesting interbridgehead bond comes out well: 1.602 Å versus 1.596 Å experimentally, indicating a bond somewhat longer than normal C–C single bonds. The results for the three known systems lend confidence to our predictions for the unknown trithiapropellane which is predicted to have shorter bonds than the parent propellane – 1.575 versus 1.602 Å – or thiirane – 1.801 versus 1.814 Å

3.2   Determination of Microwave Transition Frequencies

Next, we move on to the prediction of rotational transition frequencies and microwave spectra. The example I have chosen is the isoformyl cation, COH$^+$, which had not been experimentally observed at the time we reported the results of our calculations back in 1981.[8] Our predicted J = 0 →1 rotational transition frequency for COH$^+$ was ν = 89.0 ± 0.8 GHz. But why do we want to know this frequency?

The reason is that the predicted microwave frequencies are in turn useful in providing an interstellar fingerprint. The normal procedure for identifying an interstellar molecule is through comparison of frequencies observed in radioastronomical experiments with reference laboratory spectra. In the case of a molecule such as COH$^+$ which had not been previously observed in the laboratory, the theoretical predictions can be useful. Indeed, Gudeman and Woods[9] in laboratory experiments in Wisconsin searched in exactly the range suggested by our calculations – 89.0 ± 0.8 GHz – and found ν = 89.487 GHz. Based on these results, Woods, Irvine and others[10] searched in the interstellar gas cloud Sagittarius B2 and found a signal at 89.487 GHz. A new interstellar molecule had been added to the list in this manner.

3.3   Determination of Vibrational Spectra

The calculations allow us to predict vibrational frequencies and also infrared intensities, hence full infrared spectra. Let me exemplify this with results[6] for the [1.1.1]propellane molecule, which we met earlier. Wiberg's experimental infrared spectrum[11] is compared with our calculated spectrum in Figure 4. The agreement is quite remarkable. It is difficult not to be impressed that a purely mathematical theory can produce something as intricate as an infrared spectrum to this degree of accuracy. Remember that the theoretical procedure involves no experimental information other than the values of the fundamental constants. Again, the good agreement for the known molecule lends confidence to the prediction of the infrared spectrum of the unknown trithiapropellane.

3.4   Determination of Stable Isomers

Another useful capability of the calculations is the prediction of possible stable isomers of any particular composition. For example, suppose we wanted to know the structures and energies of all possible isomers of composition $C_2H_3O^+$, with a view to identifying which of them might be experimentally accessible. We can do this systematically and the results of a study of this type, carried out several

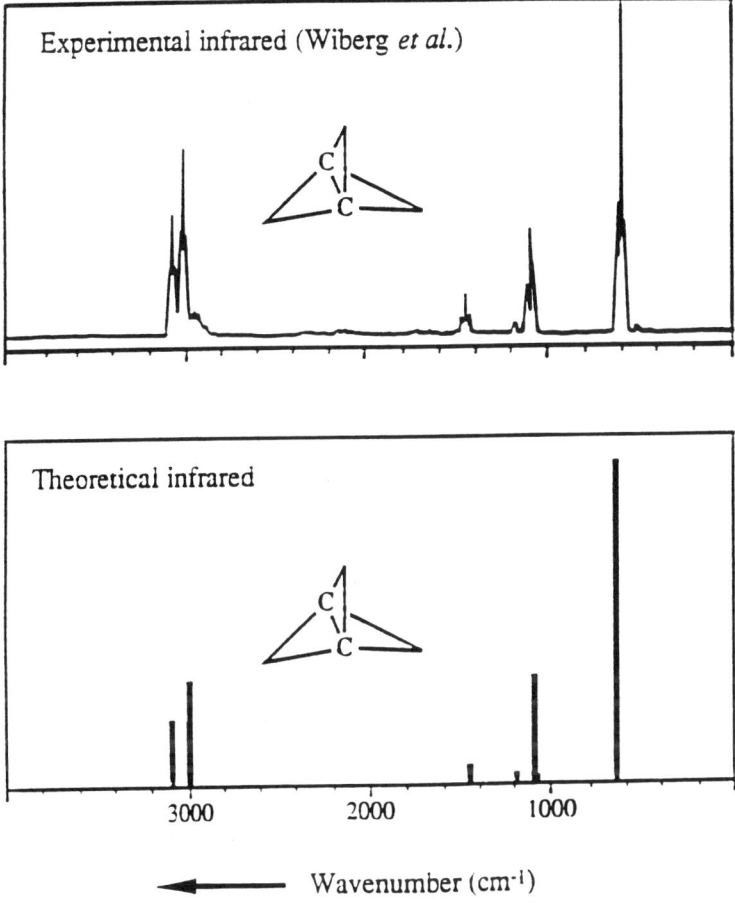

Figure 4: Experimental and theoretical (MP2/6-31G*) infrared spectra for [1.1.1]propellane (from ref 6).

years ago,[12] are shown in Figure 5. This gives the various structural isomers with relative energies shown in kJ mol^{-1} below each structure. The first three – acetyl cation, 1-hydroxyvinyl cation and oxiranyl cation – were the only experimentally observed isomers at the time. Our calculations suggested that of the remaining isomers only one, CH_3OC^+, has an energy comparable to those of the observed isomers. The others all have relative energies above 300 kJ mol^{-1}. We suggested that CH_3OC^+ should be experimentally observable.

A couple of years later a multinational paper appeared[13] with the intriguing title 'CH_3OC^+: A Long-sought Molecule, Predicted to Exist by Theory, Identified' and provided experimental confirmation of the theoretical prediction. Again, we have useful interplay between theory and experiment.

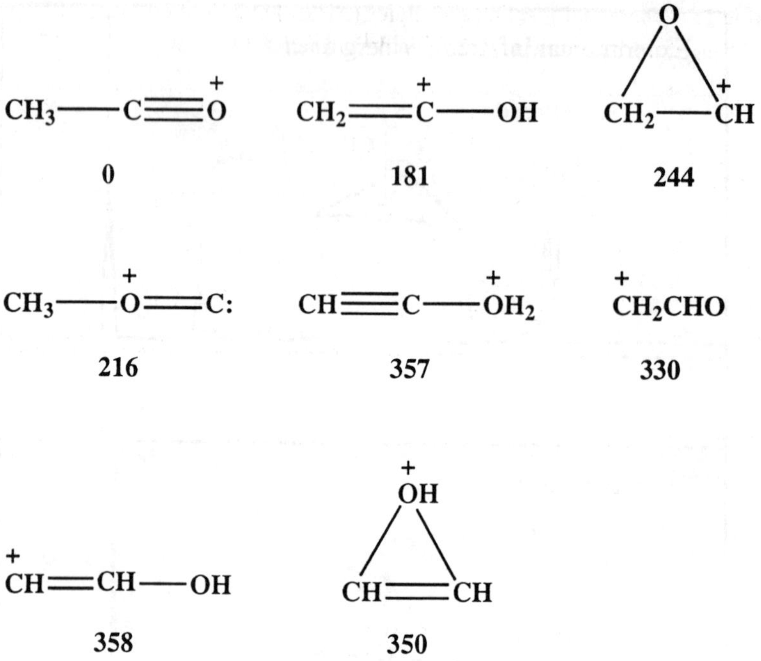

$$CH_3\text{—}\overset{+}{C}\text{≡}O \qquad CH_2\text{=}\overset{+}{C}\text{—}OH \qquad \underset{CH_2\text{—}\overset{+}{CH}}{\overset{\overset{\displaystyle O}{\diagup\diagdown}}{}}$$

0   181   244

$$CH_3\text{—}\overset{+}{O}\text{=}C\text{:} \qquad CH\text{≡}\overset{+}{C}\text{—}OH_2 \qquad \overset{+}{C}H_2CHO$$

216   357   330

$$\overset{+}{C}H\text{=}CH\text{—}OH \qquad \underset{CH\text{=}CH}{\overset{\overset{\displaystyle \overset{+}{O}H}{\diagup\diagdown}}{}}$$

358   350

Figure 5: Calculated relative energies (MP3/6-31G**//4-31G, kJ mol^{-1}) of possible stable isomers of $C_2H_3O^+$ (from ref 12).

3.5 Determination of Heats of Formation

Theory can also be used to obtain thermochemical information such as heats of formation. Let me present results from a recent study of the formaldimine (CH_2=NH) molecule as an example.[14] Despite it being a very simple molecule, there is presently no agreement on an experimental value of the heat of formation of formaldimine: the experimental values[15-20] span from 69 to 135 kJ mol^{-1}. The theoretical approach is firstly to construct a formal reaction involving formaldimine in which the heats of formation are known for all the remaining species, for example,

$$CH_2\text{=}NH + CH_3OH \rightarrow CH_2\text{=}O + CH_3NH_2 \qquad (1)$$

The energy of this reaction (ΔH) is then calculated at a suitably high level of theory and combined with the experimental heats of formation for the other species – in this case, formaldehyde, methylamine and methanol – to obtain the heat of formation of formaldimine using the formula:

$$\Delta H_f(CH_2\text{=}NH) = \Delta H_f(CH_2\text{=}O) + \Delta H_f(CH_3NH_2) - \Delta H_f(CH_3OH) - \Delta H \qquad (2)$$

Table 1: Theoretical reaction enthalpies (ΔH_{298}) and derived heats of formation ($\Delta H_{f}^{o}{}_{298}$) for formaldimine (G2, kJ mol^{-1}) from ref 14.

Reaction	ΔH_{298}	$\Delta H_{f}^{o}{}_{298}$
$CH_2=NH + CH_3\text{-}CH_3 \rightarrow CH_3\text{-}NH_2 + CH_2=CH_2$	29.9	83.3
$CH_2=NH + CH_3\text{-}OH \rightarrow CH_3\text{-}NH_2 + CH_2=O$	-19.9	89.8
$CH_2=NH + CH_4 \rightarrow NH_3 + CH_2=CH_2$	-0.8	81.6
$CH_2=NH + H_2 \rightarrow CH_3\text{-}NH_2$	-105.2	82.2
$CH_2=NH \rightarrow \frac{1}{2}CH_3\text{-}NH_2 + \frac{1}{2}HC\equiv N$	-33.0	89.1
$CH_2=NH \rightarrow HC\equiv N + H_2$	39.2	95.9
$CH_2=NH + NH_4^+ \rightarrow CH_2\text{-}NH_2^+ + NH_3$	-15.0	84.3
$CH_2=NH + H^+ \rightarrow CH_2\text{-}NH_2^+$	-868.6	83.6
$CH_2=NH \rightarrow HC=NH^+ + H^-$	1007.5	84.7
$CH_2=NH \rightarrow C + 3H + N$	1756.2	87.2

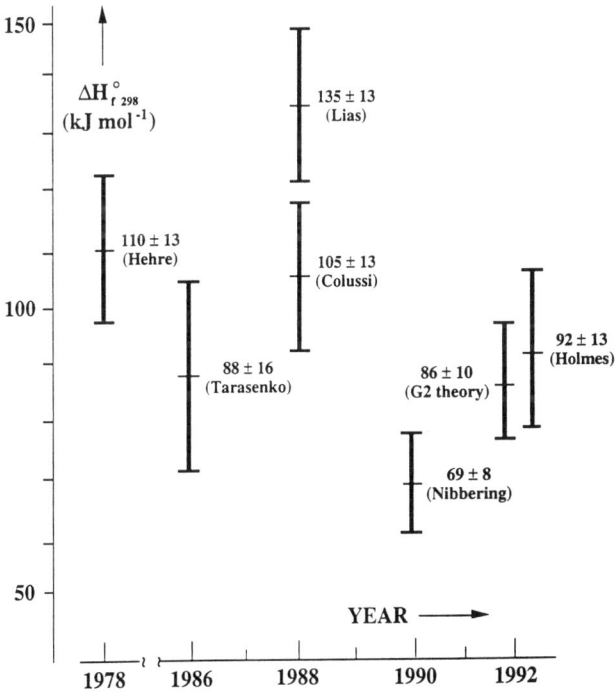

Figure 6: Comparison of experimental and theoretical values of the heat of formation of formaldimine ($CH_2=NH$) (from ref 14).

This can be repeated for a number of different reactions to obtain a collection of independent estimates of the heat of formation of formaldimine. The reactions in Table 1 include, for example, protonation, deprotonation and atomization. We can see that the ΔH_f values for formaldimine obtained in this manner lie in quite a narrow range and this lends confidence to our prediction, based on these results, of a heat of formation of 86 ± 10 kJ mol^{-1}.

Figure 6 shows the great variation in the experimental estimates from the literature, spanning a range from 69 to 135 kJ mol^{-1}. The 135 kJ mol^{-1} entry is the recommended value in a widely used 1988 compendium.[18] We would question this assignment. There has been one experimental paper[20] on this topic since we published our theoretical result and it gives 92 ± 13 kJ mol^{-1}.

3.6 Determination of Gas-Phase Acidities and Basicities

Theory can also make predictions about gas-phase acidities and basicities. The particular example that I wish to present concerns the remarkably high acidities observed in solution for ynols compared with enols by Kresge, Wirz and coworkers.[21] We were interested to see if this behaviour carried over to the gas phase.[22,23]

Table 2 presents a comparison of our calculated gas-phase acidities with experimental values. The first point to note is that the theoretical results are in very good agreement with experiment. The largest difference between theory and experiment is 12 kJ mol^{-1} and that is in a case where the experimental acidity is quoted with an uncertainty of ±11 kJ mol^{-1}. The second point is that ethynol is indeed very acidic. It is about 100 kJ mol^{-1} more acidic than vinyl alcohol, with an acidity comparable to that of HCl. The high acidity of ynols is thus predicted to carry over from solution to the gas phase.

Table 2: Theoretical (G2) and experimental gas-phase acidities (ΔH_{acid}, kJ mol^{-1}) (298 K).

Molecule	Theory[a]	Experiment[b]
CH_3OH	1601	1595
CH_3CHO	1537	1533
CH_2CO	1539	1527
$CH_2=CHOH$	1491	1492
HCl	1398	1395
$HC\equiv COH$	1392	-

[a] From ref 23; [b] from ref 18.

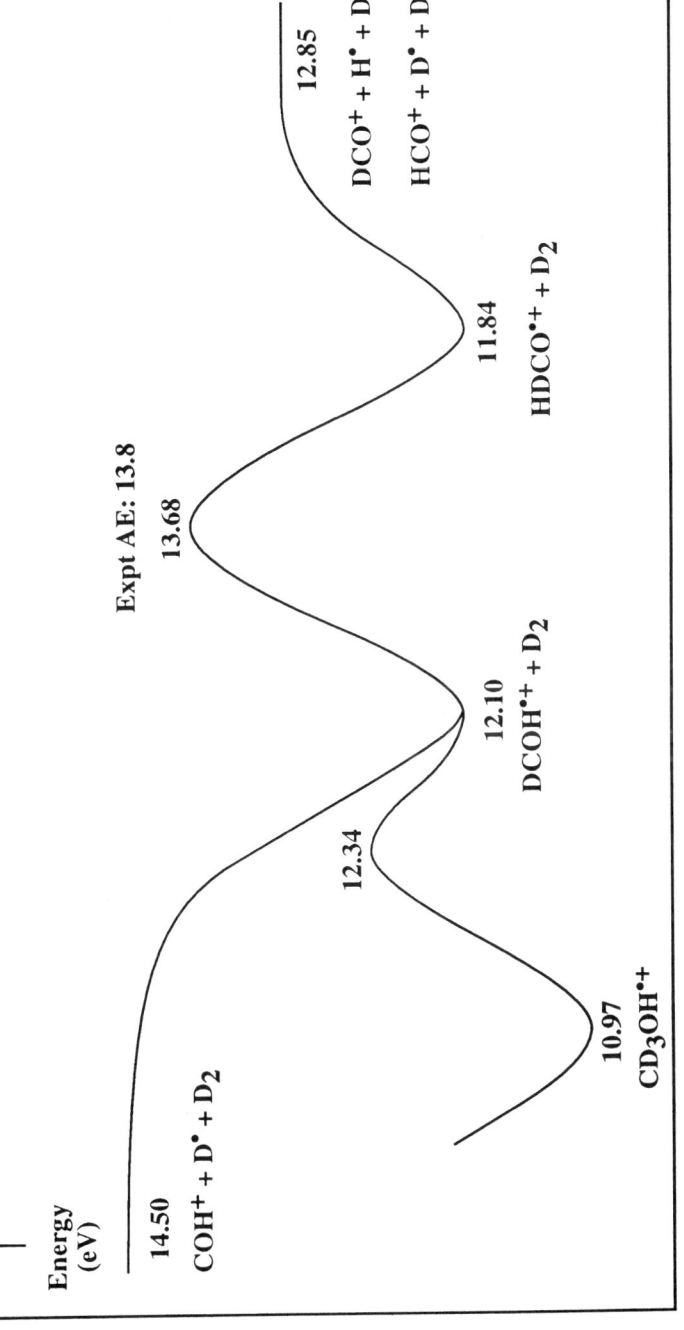

Figure 7: Schematic energy profile (G2) showing the production of COH$^+$ and HCO$^+$ from CD$_3$OH *via* DCOH$^{\bullet+}$ (from ref 26).

3.7 Determination of Reaction Mechanism

Finally, I want to say something about reaction mechanism and the prediction of preferred reaction products. This example involves the production and appearance energy of the isoformyl cation, COH^+. Several years back, John Holmes and coworkers[24] reasoned that if you start with $DCOH^{·+}$ (produced from CD_3OH), loss of $H^·$ should lead to a mass 30 ion with the formyl cation structure (DCO^+) and its appearance energy (other than an isotope effect) would represent the appearance energy of HCO^+, while loss of $D^·$ would give a mass 29 ion with the isoformyl cation structure and this would give the appearance energy of COH^+. However, the appearance energy obtained for COH^+ in this manner (13.8 eV) is much lower than expected on the basis of the theoretical energy for COH^+ (14.5 eV). Why is this the case?

The potential energy diagram of Figure 7 immediately gives the answer.[25,26] COH^+ *can* be produced by direct loss of deuterium from $DCOH^{·+}$ but this would cost 14.5 eV. However, a much lower energy pathway involves initial rearrangement of $DCOH^{·+}$ to the deuterium isotopomer of formaldehyde cation ($HDCO^{·+}$) which can then lose either $H^·$ or $D^·$, the latter corresponding to formation of HCO^+. Theory thus predicts that the experimental appearance energy for mass 29 corresponds not to the formation of COH^+ but rather to formation of HCO^+ via the second pathway, and on this basis there is good agreement between theory and experiment: 13.68 eV vs 13.8 eV.

The theoretical prediction that the 'HOC^+' ion from $DCOH^{·+}$ is really formyl cation and not isoformyl cation was subsequently confirmed by collisional activation experiments carried out in John Holmes' laboratory in Ottawa.[25]

4 RECENT CASE STUDIES

In the the remainder of this article, two very recent pieces of work, one dealing with structure and the other dealing with mechanism, are discussed in greater detail as case studies.

4.1 The Quest for a Neutral Hydrocarbon Containing a Planar Tetracoordinate Carbon Atom

The quest for a neutral hydrocarbon containing a planar tetracoordinate carbon atom is a subject that has attracted a great deal of attention over the past 20 years.[27-29] The first major paper on the topic was that of Roald Hoffmann and co-workers in 1970.[30] They wrote that 'the tetracoordinate tetrahedral carbon has magnificently served biological systems for millions of years and our imaginations for but a century. We here open the problem of stabilizing tetracoordinate planar carbon'. However, they caution that 'it would seem too much to hope for a simple carbon compound to prefer a planar to a tetrahedral structure'.

In relation to this last comment, I should point out that there is a cost of about 550 kJ mol^{-1} involved in making methane square-planar, a non-trivial amount of energy! A striking characteristic of the ground singlet state of planar methane is that the highest occupied molecular orbital is a lone pair centred on carbon. There are just six electrons left in the σ system to form the four C–H bonds which makes the σ system electron deficient. Thus planar methane is π-electron rich and σ-electron deficient.

However, in a landmark 1976 paper, Paul Schleyer and co-workers[31] examined theoretically a number of lithio- and bora- substituted systems and found several for which a planar structure was predicted to actually lie lower in energy than the tetrahedral-type structure, including 1,1-dilithiocyclopropane (**1**) and 3,3-dilithio-1,2-diboracyclopropane (**2**).

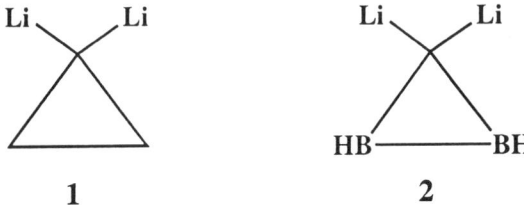

There are in fact two discrete approaches that have been employed to achieving the goal of planar tetracoordinate carbon. The first, which may be described as electronic, involves selecting substituents which will preferentially stabilize a planar disposition of bonds at carbon over the normal tetrahedral arrangement. Withdrawal of electrons from the lone pair orbital at the planar carbon by π-electron-accepting substituents, σ-electron donation to the carbon, and incorporation of the target carbon atom into a small ring have all been found to be particularly effective in this regard. As I noted above, this approach has met with some success, several molecules containing a planar tetracoordinate carbon atom having been theoretically characterized including, for example, 1,1-dilithiocyclopropane (**1**).

The alternative approach may be described as mechanical, the aim in this case being to achieve planarity at the target carbon atom by constraining the bonds through appropriate linkages. An example is [4.4.4.4]fenestrane (**3**), sometimes referred to as windowpane. The mechanical approach would seem the only viable approach for obtaining a *neutral saturated hydrocarbon* containing a planar tetracoordinate carbon atom. Although molecules with substantial deviations from the normal tetrahedral arrangement at carbon have been constructed in this manner, a completely planar disposition of bonds at tetracoordinate carbon has not yet been attained.

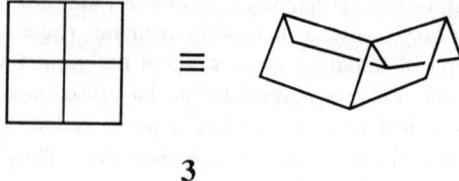

3

Several of the hydrocarbons examined previously that show bonding arrangements approaching planarity at one or more tetracoordinate carbons (*eg* [2.2.2.2]paddlane (**4**) and bowlane (**5**)[32,33]) are based on the skeleton, shown in **6**, in which the target carbon atom is attached to four ethyl-type groups. The problem with this skeleton, however, is that its symmetry does not even allow the *possibility* of planarity at the quaternary carbon atom. This would require a plane of symmetry containing the bonds at the quaternary carbon. Because the regions above and below the quaternary carbon in these molecules are different, planarity could only arise by accident.

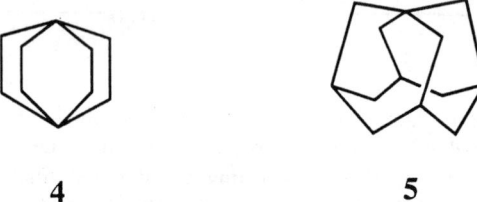

4 **5**

The same argument applies to the various fenestranes (*eg* **3**), the *cis* forms of which are described by the skeleton shown in **7**. Again, symmetry does not even allow the *possibility* of planarity at the quaternary carbon atom. Planarity could only arise by accident, or if the four carbon atoms attached to the quaternary carbon were also planar tetracoordinate. The latter would, of course, be prohibitively costly from an energy point of view.

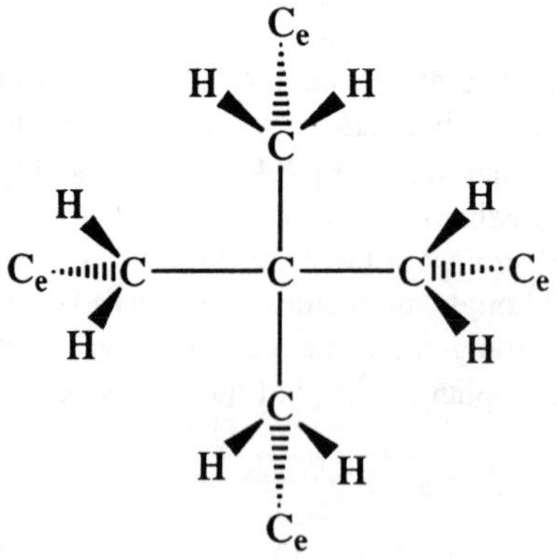

6 (C_4) 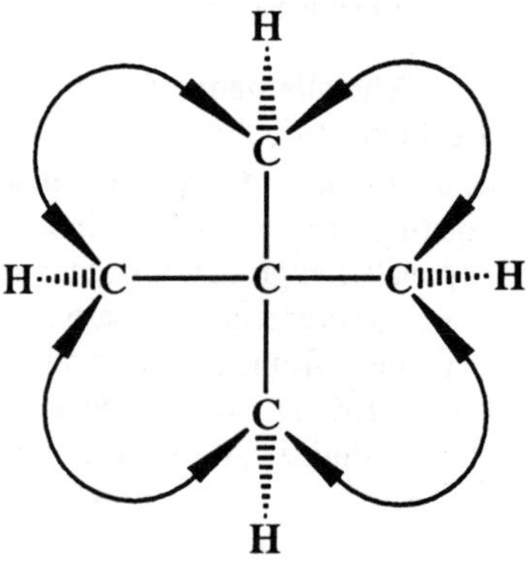   **7 (C_4)**

Our design strategy[34] has been based on new skeletons, such as **8** and **9**, that *do* have the appropriate symmetry to allow a planar disposition of bonds. The potentially planar tetracoordinate carbon is bonded in these cases to four isopropyl-type groups. Of course, the planar arrangement is not *necessarily* preferred. For example, if the four substituents are *simply* isopropyl groups, the molecule would be free to distort to a tetrahedral arrangement and undoubtedly does so. Our task therefore is to apply suitable constraints to the eight carbon atoms designated C_i.

8 (C_{4h})   **9** (D_{2h})

In order to maintain the possibility of a planar tetracoordinate carbon in structures related to our basic skeleton, we require the groups capping the top and the bottom of **8** or **9** to be identical. We have examined a number of structures of this type (*eg* **10–12**). Again, I should stress that although a planar tetracoordinate carbon is *allowed* in all of these structures, it is not *required*.

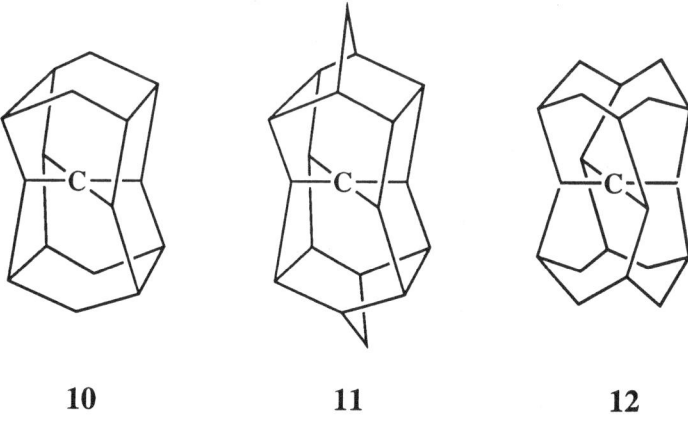

10   **11**   **12**

The molecule (**12**) with the closest approach to planarity that we have achieved to date[34] has molecular formula $C_{21}H_{28}$ and is derived from our basic skeleton **8** by incorporating the C_i carbon atoms on the top and the bottom in crown cyclooctane rings. We have called the molecule *octaplane* reflecting the cyclooctane caps and the near-planar carbon. The systematic name is octacyclo[9.7.1.1^{5,17}1.^{7,13}0.^{3,14}0.^{6,15}0.^{9,16}0.^{15,19}]heneicosane.

The equilibrium structure of octaplane (**12**, Figure 8) has S_4 symmetry with two bonds coming up and two going down and CCC angles of about 169 degrees. In an exactly planar form, the angle would be 180° while for a tetrahedral carbon the angles are 109.5°, so we have moved a large part of the way to planarity. Some of the more interesting bond lengths in octaplane are also included in Figure 8. The C–C lengths at the quaternary carbon atom – 1.583 Å – and the C–C lengths elsewhere in the molecule – up to 1.615 Å – are only marginally longer than normal, and comparable to bond lengths in X-ray crystal structures of some of the smaller fenestrane derivatives (*eg* 1.56, 1.57 and 1.60 Å).[29] The fact that these bond lengths are not excessively long is very important and certainly an encouraging first sign as to the stability of this molecule.

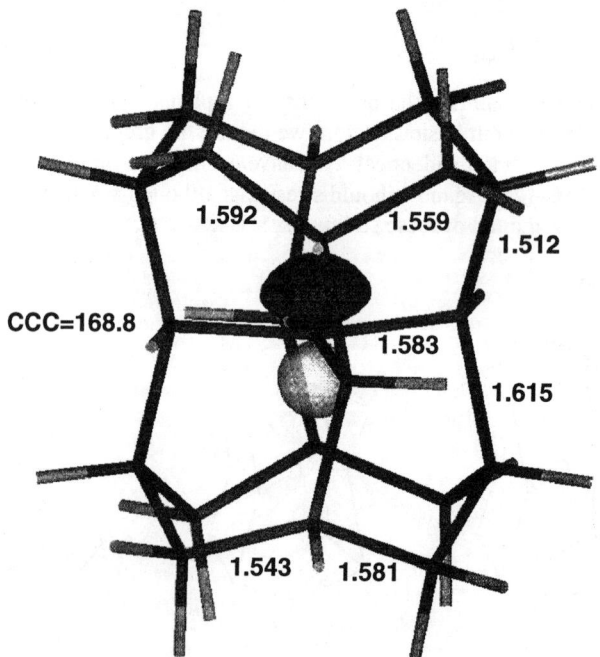

Figure 8: Calculated (HF/6-31G*) structure and highest-occupied molecular orbital (HOMO) of octaplane (from ref 34). Bond lengths in Å, bond angles in degrees.

A striking electronic feature of octaplane (**12**) is the highest occupied molecular orbital (HOMO) which is dominated by the lone pair on the quaternary carbon, also shown in Figure 8. Intriguingly, the lone pair lies within the cage formed by the remaining carbon atoms, a feature which may have interesting consequences on some of the chemical properties of octaplane.

One immediate consequence is that the ionization energy might be expected to be particularly low. We have carried out calculations on octaplane cation to test this prediction.[35] The first point of interest from our calculations is that octaplane cation is predicted to contain a *fully planar tetracoordinate carbon atom*, ie it has C_{4h} symmetry (Figure 9).

The ionization energy of octaplane is calculated as 4.62 eV (MP2/6-31G*). Test calculations for a selection of molecules for which experimental data are available and/or for which accurate (G2) calculations can be performed (eg NH_3, CH_2, CH_4) show that ionization energies at the MP2/6-31G* level are generally too low by about 0.3–0.7 eV. On this basis, we estimate that the ionization energy for octaplane is approximately 5 eV.

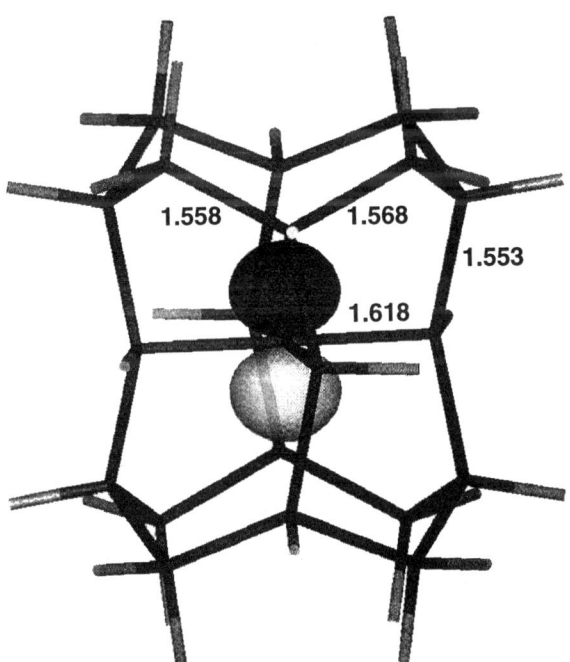

Figure 9: Calculated (HF/6-31G*) structure and singly-occupied molecular orbital (SOMO) of octaplane cation (from ref 35). Bond lengths in Å, bond angles in degrees.

The ionization energy of 5 eV is strikingly low. For comparison, the lowest value for a neutral saturated hydrocarbon recorded in a recent compendium[18] is 7.1 eV (for tetra(*t*-butyl)tetrahedrane)). Indeed, the predicted ionization energy of octaplane is comparable to the experimental values for the alkali metals lithium and sodium (5.39 and 5.14 eV, respectively).

Octaplane is a member of the family of molecules called *alkaplanes* in which the near-planar carbon is capped by cycloalkane moieties.[34] Thus, when the capping groups are cyclohexanes, we have hexaplane (**10**) and when they are bicycloheptanes, we have biheptaplane (**11**). We are carrying out additional calculations to see if we can find capping groups that will result in a fully planar carbon.

In the meantime, we note that octaplane (**12**) shows the closest approach to date to a planar tetracoordinate carbon atom in a neutral hydrocarbon. The lone pair in the cage and the particularly low ionization energy are intriguing features. We are encouraged by various results from our calculations – the reasonable bond lengths and strain energies – to believe that octaplane may well be synthesizable. We believe that it is an attractive synthetic target.

4.2 Addition of Methyl Radical to Alkenes

The mechanistic problem with which I wish to conclude concerns the addition of methyl radical to alkenes.[36,37] This is of fundamental interest and also is of some practical relevance as the propagation step in polymerization reactions. Because the reaction is accelerated by electron-withdrawing groups on the alkene substrate, it had been concluded that polar effects are important and, in particular, that the methyl radical is nucleophilic in such reactions.[38–40] We wanted to critically assess these conclusions.

In order to do so, we have studied[36,37] methyl radical addition to a series of alkenes:

$$CH_3^{\bullet} + CH_2=CHX \rightarrow CH_3CH_2CHX^{\bullet} \quad (3)$$

with X = H, CH_3, NH_2, OH, F, SiH_3, Cl, CN, CHO, and NO_2 at a moderately high level of theory (QCISD(T)/6-311G**//HF/6-31G* + ZPVE), and obtained energy profiles of the type shown in Figure 10. The calculations allow us to characterize the reactants, the product, and the transition structure in such reactions. We can calculate, among other things, the reaction barrier, the reaction exothermicity, and the direction and extent of charge transfer in the transition structure. The questions we seek to answer include: do polar effects have a significant influence, is the methyl radical nucleophilic, and are enthalpy effects important?

Let me address the third question first – the possible influence of reaction enthalpy on the reactivity of methyl radical with alkenes. We can tackle this quite straightforwardly by plotting reaction barrier against exothermicity for a range of substituents X (Figure 11). When we do this, we observe a nice straight line correlation with an R^2 value of 0.96. This excellent correlation suggests that reaction thermodynamics has a very strong, if not dominant, influence on the rate of methyl radical addition to alkenes, consistent with classical rate-equilibrium considerations.

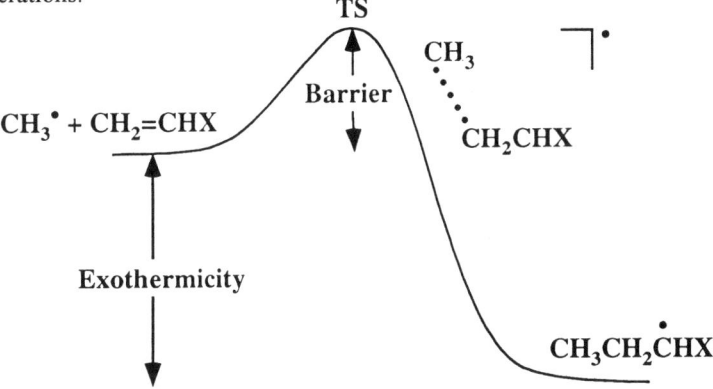

Figure 10: Schematic reaction profile for the addition of methyl radical ($CH_3\bullet$) to alkenes (CH_2=CHX).

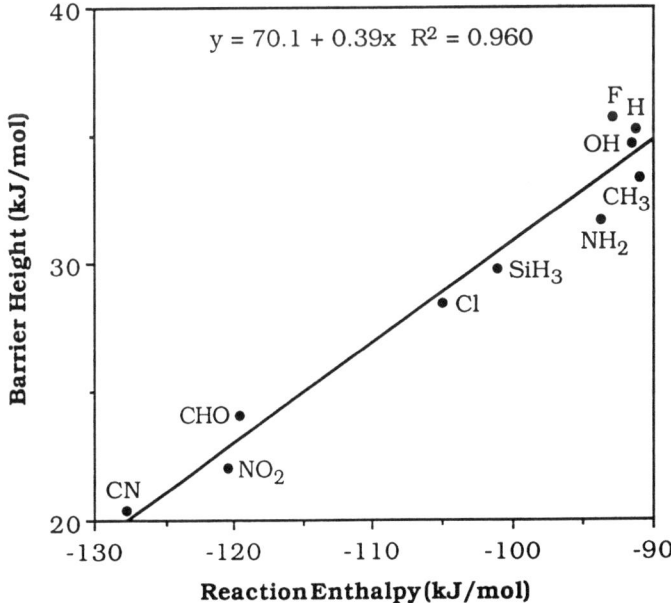

Figure 11: Plot of barrier height versus reaction enthalpy for the addition of methyl radical to alkenes (CH_2–CHX, X–H, CH_3, NH_2, OH, F, SiH_3, Cl, CN, CHO, NO_2) (QCISD(T)/6-311G** + ZPVE, kJ mol^{-1}) (from ref 36).

What about the importance of polar effects? We have looked at two measures of the direction and extent of charge transfer in the transition structure. First, we calculate the relative energies of the charge-transfer states $CH_3^+/CH_2=CHX^-$ and $CH_3^-/CH_2=CHX^+$, based on the energies of separated reactants. We observe (Table 3) that in most cases, including the unsubstituted case, the energy of the charge-transfer state $CH_3^-/CH_2=CHX^+$ is lower than the charge-transfer state $CH_3^+/CH_2=CHX^-$, ie methyl prefers to be an electron acceptor than an electron donor. The second measure comes from the calculated charges in the transition structures, which predict that methyl normally carries a negative charge; it is positive only for X = CHO, NO_2, and CN. Thus the charge analysis also says that methyl is generally an electron acceptor. So our calculations on two counts suggest that the answer to the second question that I had posed is that methyl radical does not display general nucleophilic behavior, in contrast to the current conventional wisdom.

The other question we want to address is whether polar effects have an important influence on reactivity in methyl radical addition reactions. The charges (Table 3) tell us that that there is some charge transfer in the transition structure but they don't say anything about the energetic consequences. We want to know whether polar effects are important energetically – do they influence barrier heights?

Table 3: Calculated charge-transfer energies (eV) and charges related to methyl radical addition reactions to $CH_2=CHX$.[a]

X	Charge Transfer Energy[b]		Charge[c]
	$CH_3^+/CH_2=CHX^-$	$CH_3^-/CH_2=CHX^+$	
F	11.39	10.33	-0.012
H	11.63	10.54	-0.017
OH	11.52	9.22	-0.029
CH_3	11.59	9.78	-0.024
NH_2	11.69	8.14	-0.039
SiH_3	10.69	10.13	-0.009
Cl		9.94	0.000
CHO	9.74	10.17	0.006
NO_2	8.98	11.83	0.030
CN	10.00	10.94	0.012

[a] From refs 36 and 37.
[b] Charge-transfer energies of separated reactants, calculated at the G2 level.
[c] Amount of charge transferred (Bader analysis) from the methyl radical to the alkene in the transition structure (HF/6-31G*). A positive value indicates electron transfer from the radical to the alkene.

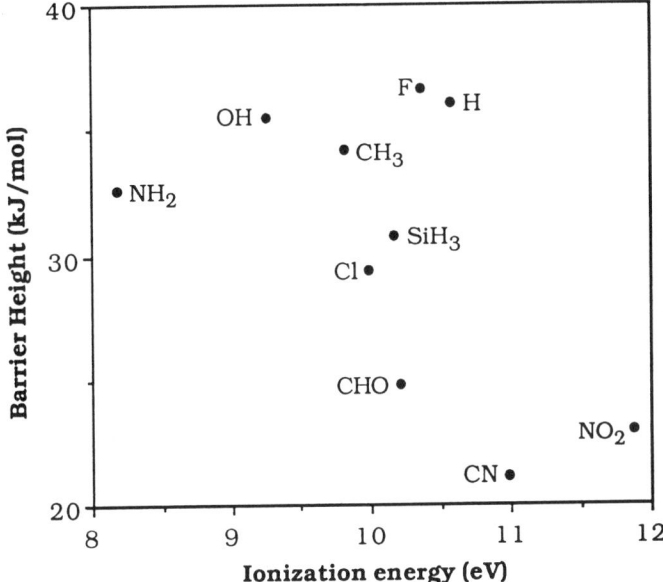

Figure 12: Plot of barrier height (kJ mol^{-1}) versus ionization energy (eV) for the addition of methyl radical to alkenes (CH$_2$=CHX, X=H, CH$_3$, NH$_2$, OH, F, SiH$_3$, Cl, CN, CHO, NO$_2$) (QCISD(T)/6-311G** + ZPVE (from ref 37).

Because the direction of charge transfer is generally from the alkene to the methyl radical, we might expect that if polar effects were important they would manifest themselves in a nice correlation of reaction barrier with the ionization energy of the alkene. However, when we plot our calculated barrier heights against the ionization energy of the alkene (Figure 12), the result shows random scatter.

On the other hand, when we plot reaction barrier against the electron affinity of the alkene (Figure 13), there *is* a reasonable correlation. This would not have been anticipated on the basis of the charge-transfer results. Only the three substituents CHO, NO$_2$, and CN might have been expected to show a correlation with electron affinities since they correspond to electron-accepting alkenes – but they in fact are not particularly well behaved in that respect in the correlation diagram. What is the significance then of the correlation of reaction barrier with electron affinity, mirrored also in the experimental observation that I mentioned earlier that electron-withdrawing substituents in the alkene generally lead to lower reaction barriers?

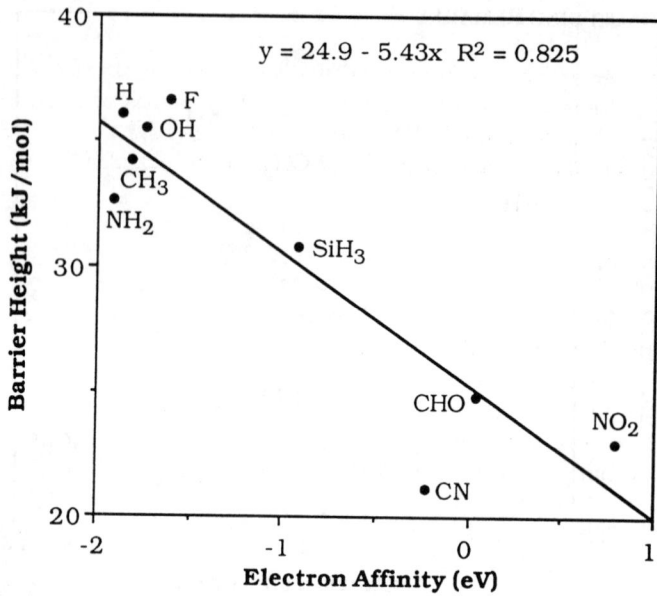

Figure 13: Plot of barrier height (kJ mol^{-1}) versus electron affinity (eV) for the addition of methyl radical to alkenes (CH$_2$=CHX, X=H, CH$_3$, NH$_2$, OH, F, SiH$_3$, Cl, CN, CHO, NO$_2$) (QCISD(T)/6-311G** + ZPVE (from ref 37).

The answer is that π-electron-accepting groups such as CHO, NO$_2$, and CN enhance reactivity because these groups tend to increase reaction exothermicity, and not because of induced polar character in the transition structure. These groups tend to increase reaction exothermicity, in part because they stabilise the product radical (CH$_3$CH$_2$CHX$^{•}$) formed from the addition of methyl radical to the alkene.

We have analysed our results also using the curve-crossing model of Pross and Shaik[41] and found that such an analysis supports our quantitative findings. In particular, it shows that, in the case of methyl radical plus ethylene, the charge-transfer states are at sufficiently high energy so as to preclude their mixing significantly into the ground-state wavefunction.

In summary, we find that polar contributions to the reactivity of methyl radical with alkenes are small and that reaction exothermicity is the main dominating factor. We find no evidence for the prevalent view that the methyl radical is generally nucleophilic towards alkenes. We are currently exploring other systems to find out when polar contributions *do* become significant energetically.

5 CONCLUDING REMARKS

Theory is quantitatively useful in predicting the properties of reactive molecules and the mechanisms of organic reactions. Many properties of interest are now accessible. The good results for known molecules lends confidence to the predictions in unknown cases (e.g. trithiapropellane and octaplane).

REFERENCES

1. W J Hehre, L Radom, P v R Schleyer, and J A Pople, 'Ab Initio Molecular Orbital Theory', Wiley, New York, 1986.
2. M J Frisch, J A Pople, and J S Binkley, *J Chem Phys*, 1984, **80**, 3265.
3. J A Pople, M Head-Gordon, D Fox, K Raghavachari, and L A Curtiss, *J Chem Phys*, 1989, **90**, 5622.
4. L A Curtiss, C Jones, G W Trucks, K Raghavachari, and J A Pople, *J Chem Phys*, 1990, **93**, 2537.
5. L A Curtiss, K Raghavachari, G W Trucks, and J A Pople, *J Chem Phys*, 1991, **94**, 7221.
6. N V Riggs, U Zoller, M T Nguyen, and L Radom, *J Am Chem Soc*, 1992, **114**, 4354.
7. K B Wiberg and F H Walker, *J Am Chem Soc*, 1982, **104**, 5239.
8. R H Nobes and L Radom, *Chem Phys*, 1981, **60**, 1.
9. C.S. Gudeman and R.C. Woods, *Phys Rev Lett*, 1982, **48**, 1344.
10. R C Woods, C S Gudeman, R L Dickman, P F Goldsmith, G R Huguenin, W M Irvine, A Hjalmarson, L A Nyman, and H Olofsson, *Astrophys J*, 1983, **270**, 583.
11. K B Wiberg, W P Dailey, F H Walker, S T Waddell, L S Crocker, and M D Newton, *J Am Chem Soc*, 1985, **107**, 7247.
12. R H Nobes, W J Bouma, and L Radom, *J Am Chem Soc*, 1983, **105**, 309.
13. B van Baar, P C Burgers, J K Terlouw, and H Schwarz, *J Chem Soc, Chem Commun*, 1986, 1607.
14. B J Smith, J A Pople, L A Curtiss, and L. Radom, *Aust J Chem*, 1992, **45**, 285.
15. D J DeFrees and W J Hehre, *J Phys Chem*, 1978, **82**, 391.
16. N A Tarasenko, A A Tishenkov, V G Zaiken, V V Volkova, and L E Gusel'nikov, *Izv Akad Nauk SSSR, Ser Khim*, 1986, 2397.
17. M A Grela and A J Colussi, *Int J Chem Kinet*, 1988, **20**, 713.
18. S G Lias, J E Bartmess, J F Liebman, J L Holmes, R D Levin, and W G Mallard, *J Phys Chem Ref Data*, 1988, **17**, Supplement 1.
19. R A Peerboom, S Ingemann, N M M Nibbering, and J F Liebman, *J Chem Soc, Perkin Trans 2*, 1990, 1825.
20. J L Holmes, F P Lossing, and P M Mayer, *Chem Phys Lett*, 1992, **198**, 211.

21 Y Chiang, A J Kresge, R Hochstrasser, and J Wirz, *J Am Chem Soc*, 1989, **111**, 2355.
22 B J Smith, L Radom, and A J Kresge, *J Am Chem Soc*, 1989, **111**, 8297.
23 B J Smith and L Radom, *J Phys Chem*, 1991, **95**, 10549.
24 P C Burgers, A A Mommers, and J L Holmes, *J Am Chem Soc*, 1983, **105**, 5976.
25 W J Bouma, P C Burgers, J L Holmes, and L Radom, *J Am Chem Soc*, 1986, **108**, 1767.
26 N L Ma, B J Smith, J A Pople, and L Radom, *J Am Chem Soc*, 1991, **113**, 7903.
27 A Greenberg and J F Liebman, 'Strained Organic Molecules', Academic Press, New York, 1978.
28 R Keese, *Nachr Chem Tech Lab*, 1982, **30**, 844.
29 W C Agosta, in 'The Chemistry Of Alkanes', S Patai and Z Rappoport, Eds, Chapter 20, Wiley, Chichester, 1992.
30 R Hoffmann, R W Alder and C F Wilcox, *J Am Chem Soc*, 1970, **92**, 4992.
31 J B Collins, J D Dill, E D Jemmis, Y Apeloig, P v R Schleyer, R Seeger, and J A Pople, *J Am Chem Soc*, 1976, **98**, 5419.
32 H Dodziuk, *J Mol Struct*, 1990, **239**, 167.
33 M P McGrath, H F Schaefer, and L Radom, *J Org Chem*, 1992, **57**, 4847.
34 M P McGrath and L Radom, *J Am Chem Soc*, 1993, **115**, 3320.
35 J E Lyons, D Rasmussen, M P McGrath, and L Radom, *Angew Chem*, in press.
36 M W Wong, A Pross, and L Radom, *J Am Chem Soc*, 1993, **115**, 11050.
37 M W Wong, A Pross, and L Radom, *Israel J Chem*, 1993, **33**, 415.
38 J M Tedder, *Angew Chem, Int Ed Engl*, 1982, **21**, 401.
39 B Giese, *Angew Chem, Int Ed Engl*, 1983, **22**, 753.
40 H Zipse, J He, K N Houk, and B Giese, *J Am Chem Soc*, 1991, **113**, 4324.
41 A Pross and S S Shaik, *Acc Chem Res*, 1983, **16**, 363.

ACKNOWLEDGEMENTS

I thank the very capable members of my research group and the colleagues elsewhere with whom I have collaborated whose efforts have made possible the work described herein. Their names are listed within the reference list. I gratefully acknowledge the continuing generous provision of computing resources by the Australian National University Supercomputing Facility.

DIRECT STUDIES OF THE REACTIVITIES OF SHORT-LIVED CARBOCATIONS

Robert A McClelland

*Department of Chemistry, University of Toronto,
Toronto, Ontario M5S 1A1, Canada.*

1   INTRODUCTION

Since the pioneering work of Olah, there have been extensive investigations of carbocations under 'stable ion' conditions. The non-nucleophilic solvents and counterions that are employed here are, however, quite different from ones where such cations are found as reactive intermediates. Thus such studies generally do not address questions of reactivity. Highly stabilized derivatives such as p-Me_2N-substituted triarylmethyl cations and tropylium ions had been examined in this regard, principally by Ritchie.[1] Short-lived cations, however, had been investigated only in terms of selectivities, that is, through ratios of rate constants as measured by the competition kinetics method or through kinetic analysis of common ion inhibition. Absolute rate constants had been estimated from such data, through a 'clock' approach based on the assumption that one of the nucleophiles reacted at the diffusion limit.[2,3]

The technique of flash photolysis provides a method for the actual observation of reactive intermediates, and in consequence, for the direct measurement of rate constants for their decay reactions. Prior to the mid-1980's, this method had been widely used to study neutral organic intermediates such as free radicals and carbenes. There had, however, been limited application to carbocations, although examples involving triarylmethyl cations[4] and vinyl cations[5] had appeared in the literature. This situation has changed over the last five years, with a number of cations of different types now being observed and their reactivity directly studied. This paper summarizes some general features of the use of flash photolysis for the study of cationic intermediates, and illustrates the approach with some recent examples, principally from the author's group.

2   FLASH PHOTOLYSIS STUDIES OF CARBOCATIONS - REQUIREMENTS

In the technique of flash photolysis a short pulse of light is used to generate the desired intermediate, and this species is detected and studied by some time-resolved method. For this technique to be applicable for the investigation of carbocations, the first requirement is that there are photochemical reactions in which such species form as intermediates. There is in fact good evidence from product studies that such systems do exist, in the form of photochemical analogs of S_N1 solvolysis and C=C protonations. In addition, several other routes have been uncovered during the course of the flash photolysis studies. A list of these various methods, with some key references, is given in Table 1. Examples of (a) - (d), (g) and (h) will be presented in the following text.

Table 1: Photochemical Reactions Producing Cations in Flash Photolysis Experiments.

(a) Photoheterolysis (photosolvolysis)	Refs 6, 7
(b) Alkene photoprotonation (photoaddition)	Refs 8, 9
(c) Aromatic photoprotonation (photoexchange)	Refs 10, 11
(d) C-C Fragmentation of photogenerated cation radicals	Refs 12, 13
(e) Ionization of excited radicals	Refs 14, 15
(f) Protonation of photogenerated carbenes	Refs 16 - 18
(g) Reaction of photogenerated cation with C=C to produce second cation	Ref 19
(h) Cleavage of photogenerated cation to produce second cation	Ref 13

An obvious further requirement is that the intermediate be detectable. Since the most common monitoring method involves optical spectroscopy, this requires the cation to have a UV-visible spectrum. In consequence, cations studied to date have been either 'benzylic' with one, two or three aryl groups attached to an adjacent C^+-center or 'cyclohexadienyl'. Limited studies involving conductivity detection have also been reported.[12,20] This method does not require an absorbing intermediate since it relies on the release of H^+ as the cation reacts. It appears, however, to be less generally useful.

A final requirement is that the cation survive at least as long as the pulse of light used for its generation. The most common current experimental set-up employs as the excitation source a laser with a 5 - 20 ns pulse width (ns flash photolysis), so that in general this is the lifetime limit. This does place restrictions on the structure of cations that can be investigated. As will be shown however, there are dramatic effects of solvent that can be exploited to observe inherently more reactive systems. There are other consequences of the ns limit, namely that the photochemical events leading to the cation are likely complete within this time, and moreover that the ion observed after the pulse is in the ground state, identical with the same intermediate when formed in a thermal reaction such as solvolysis. Evidence that the latter is the case has been presented in several publications.[6,7,21,22] Picosecond laser pulses have been employed in some cases,[17,23] although there are indications with very short-lived cations that there are differences when generated thermally and photochemically.[24] The reactivity of cations in the excited state has also seen recent investigation.[25]

3 IDENTIFICATION OF TRANSIENTS AS CARBOCATIONS

There are several criteria for identifying transients observed in flash photolysis experiments as carbocations. The most compelling is a match of the transient's UV-visible spectrum with one obtained under stable ion conditions. Solvent has little effect on the spectra of carbocations, with shifts in λ_{max} of at most 10 nm in moving from strong acids to alcohols and acetonitrile. In cases where this direct comparison is not possible, characteristics of the kinetics of decay can be used. Cations undergo exponential decay in the protic solvents normally employed in these experiments, and even in acetonitrile through the formation of nitrilium ions.[7] This decay is accelerated by added nucleophiles, while oxygen, an efficient quencher of radicals and triplets, has no effect. Obviously, photolysis products consistent with intermediate cations should also be observed. This is a particularly good indicator if a ratio of two rate constants directly measured in the flash photolysis experiments agrees with the same ratio measured through product analysis (the competition kinetics method).

4 THE AZIDE 'CLOCK' METHOD

An especially good 'quenching' agent is azide, a nucleophile with a long history in carbocation chemistry. Originally employed as a mechanistic probe in studies of solvolysis reactions, extensive data were generated showing a relation between the selectivity $k_{Az}:k_w$ in aqueous acetone and reactivity as measured by k_{ion}.[26]

$$RX \xrightarrow{k_{ion}} R^+ \begin{array}{c} \xrightarrow{+H_2O, -H^+;\ k_w} ROH \\ \xrightarrow{+N_3^-;\ k_{Az}} RN_3 \end{array} \qquad (1)$$

This has been a widely cited example of a reactivity-selectivity relation. More recently, it has been suggested to be an artificial one arising from the azide reaction being encounter controlled.[2,3,27] With this assumption, and taking k_{Az} as 5×10^9 M^{-1}s^{-1}, absolute rate constants for the solvent and other nucleophiles have been calculated from competition kinetic data[2,3] as well as common ion inhibition.[28] This approach essentially uses azide as a "clock" to obtain the rate constant for the other nucleophile.

Flash photolysis provides a test of this approach, since the individual rate constants can be directly measured. Such studies have been carried out, with tri- and di-arylmethyl cations in aqueous acetonitrile[29] and fluoro-substituted cations in aqueous trifluoroethanol.[30] Within each series, the more reactive derivatives do react with azide with a constant rate constant independent of structure. One example of this is shown in Figure 1. The limiting k_{Az} are in the vicinity of 10^{10} $M^{-1}s^{-1}$, the numbers being somewhat dependent on the solvent composition, as well as cation structure. These studies produce the conclusion that the 'azide clock' method with k_{Az} at 5×10^9 $M^{-1}s^{-1}$ does produce good estimates of rate constants for other nucleophiles, with the proviso that the cation be reasonably reactive. As a rough guide, a solvent rate constant greater than 10^5 s^{-1} (or a $k_{Az}:k_s$ ratio less than 10^5 M^{-1}) is probably sufficient to ensure this.

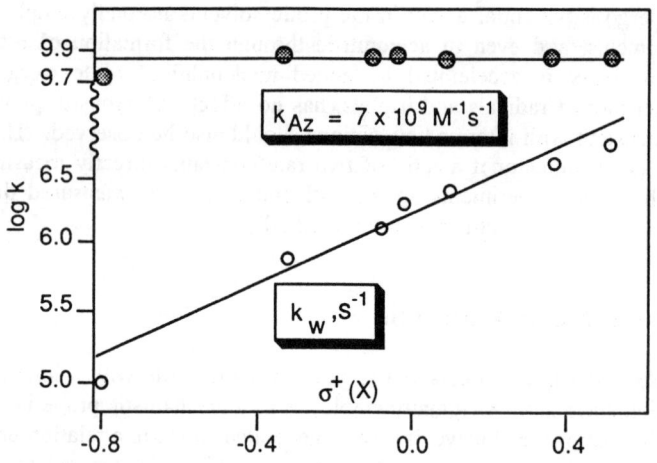

Figure 1: Rate constants at 20 °C for diarylmethyl cations $(4-MeOC_6H_4)(X-C_6H_4)CH^+$ in 1:4 acetonitrile:water.

5   EFFECT OF SOLVENT ON LIFETIMES OF CARBOCATIONS

Figure 2 combines data from a number of sources to illustrate the effect of solvent ROH on the 'kinetic' stability of benzylic-type cations. The solvent reactivity order, as expected, is methanol > water > trifluoroethanol (TFE) > 1,1,1,3,3,3-hexafluoroisopropyl alcohol (HFIP), and the differences can be large. The lifetime of the diphenylmethyl cation, for example, increases about nine orders in magnitude in proceeding from methanol to HFIP. The low reactivity of the latter obviously makes it a good solvent for observing reactive cations such as the cumyl and phenethyl ions.[9] The parent benzyl cation is probably just too short-lived in HFIP to be detected with nanosecond flash photolysis equipment.

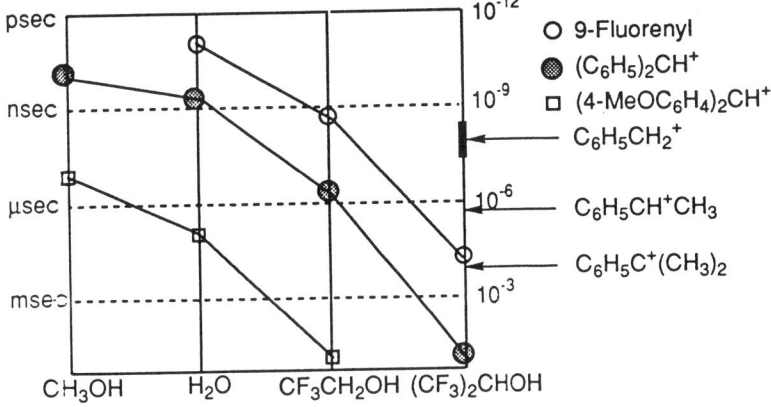

Figure 2: Lifetimes of carbocations. The solid line represents an estimated reactivity range for the benzyl cation.

6 DIRECT OBSERVATION OF CATIONIC INTERMEDIATES IN FRIEDEL-CRAFTS ALKYLATION

The low nucleophilicity of HFIP makes this a suitable solvent for investigating the reactions of carbocations with carbon nucleophiles. The 9-fluorenyl cation Fl$^+$, generated photochemically from 9-fluorenol in HFIP,[22] partitions between solvent capture and reaction with added aromatic compounds to give products of electrophilic aromatic substitution.[31] The latter reaction is quite rapid, particularly with aromatics such as anisole and mesitylene which quench Fl$^+$ with rate constants approaching the diffusion limit. Moreover, with these electron-rich compounds, the cyclohexadienyl cation that forms in the addition step can actually be observed to grow in as Fl$^+$ decays. Thus, in the same experiment both cationic intermediates of a Friedel-Crafts alkylation are directly observed. The situation with mesitylene is summarized below, the cyclohexadienyl cation (**1**) being observed at 360 nm.

7 PHOTOPROTONATION OF AROMATIC COMPOUNDS BY 1,1,1,3,3,3-HEXAFLUOROISOPROPYL ALCOHOL

Evidence that the 360 nm transient observed with the above system does represent the σ-complex (1) comes from experiments involving direct irradiation of mesitylene in HFIP. In this case the singlet excited state of the mesitylene is protonated by the solvent to give the cation (3).[10] This has both a similar spectrum and similar lifetime to the fluorenyl-adduct (1).

$$\text{Mesitylene} \xrightarrow{h\upsilon} S_1 \xrightarrow{H\text{-}OCH(CF_3)_2} \mathbf{3} \text{ (360 nm)} \qquad (3)$$

$$\xleftarrow{-H^+,\ 1 \times 10^5\ s^{-1}}$$

This is an example of photoprotonation, and illustrates a further feature of HFIP, an ability to protonate excited π systems. For this system, there is conclusive evidence that the 360 nm transient is (3), since this species, protonated mesitylene, has been characterized under stable ion conditions and its absorption spectrum reported.

8 LIFETIME OF THE PHENYL CATION IN 1,1,1,3,3,3-HEXAFLUOROISOPROPYL ALCOHOL

An approach similar to that described in equation 2 has been used to detect indirectly the phenyl cation generated upon photolysis of the benzenediazonium ion in HFIP.[32] The difference here is that the phenyl cation itself is not be observed. Irradiation of the diazonium ion in HFIP alone causes a bleaching of the absorbance of this species, and this is complete within the ns laser pulse. There is no transient absorbance, and in fact no further spectral change of any kind. When mesitylene is added in reasonable concentration however, a signal due to the cyclohexadienyl cation (4) is present, this species being characterized through comparison with the similar adducts (1) and (3). Despite being

$$PhN_2^+ \xrightarrow[\text{thermal}]{h\upsilon\ \text{or}} Ph^+ \xrightarrow{k_{Mes}[Mes.]} \mathbf{4}\ (350\ nm) \xrightarrow[1 \times 10^5\ s^{-1}]{-H^+} \mathbf{5} \qquad (4)$$

$$\xrightarrow{k_{HFIP}} ArOCH(CF_3)_2$$

generated in a bimolecular reaction, (4) is completely formed within the 20 ns excitation pulse. Thus, the flash photolysis experiments do not provide direct kinetic data for the reaction of the phenyl cation with either the solvent or mesitylene. The ratio k_{Mes}:k_{HFIP} can however be measured in the normal manner, through the analysis of the product ratio [5]:[ArOCH(CF$_3$)$_2$]. This has been done for both thermal and photochemical reactions, with the two in good agreement. The thermal studies also provide the information that, with mesitylene present, (5) forms with no change in the rate of decay of the diazonium ion from experiments in HFIP alone. This is classic evidence for an intermediate that partitions to products after the rate-determining step. The selectivity of this intermediate is however low, with k_{Mes}:k_{HFIP} = 4 M^{-1}. This number accounts for the failure to observe spectral changes associated with the phenyl cation in the flash photolysis experiments. This conclusion is reached on considering that the more stable Fl$^+$ reacts with mesitylene with a bimolecular rate constant of 2×10^9 M^{-1}s^{-1} (equation 2). The phenyl cation must react at least this fast, and probably slightly faster, at the diffusion limit of $3 - 5 \times 10^9$ M^{-1}s^{-1}. Thus with k_{Mes}:k_{HFIP} only equal to 4, k_{HFIP} must be of the order 10^9 s^{-1}, corresponding to a lifetime of about 1 ns. The conclusion is that the phenyl cation does have a finite existence in HFIP, sufficient for it to be trapped by added aromatics. However, this cation is too short-lived for detection with 20 ns excitation pulses.

9   OBSERVATION OF THE INITIAL STEP IN THE CATIONIC POLYMERIZATION OF STYRENE

Irradiation of styrene in HFIP provides a final example of a C-C bond forming reaction. In this case, the excited styrene undergoes protonation by the solvent at the β carbon, producing the phenethyl cation which is observed with flash photolysis at λ_{max} 315 nm.[9] At low ($< 10^{-4}$ M) initial concentrations of styrene, this cation has a lifetime of 2 ms and the major product is the Markovnikov solvent adduct (6). With initial styrene concentrations of 1 - 5 mM however, significant amounts of the styrene dimers (8) are observed, along with small amounts of trimers and higher oligomers.[33] In the flash photolysis experiments, the lifetime of the phenethyl cation is shortened, rate constants for its decay increasing in a linear fashion with styrene concentration with a quenching rate constant of 2×10^8 M^{-1}s^{-1}. Moreover, as the phenethyl cation decays at 315 nm, a second longer-lived transient with λ_{max} at 340 nm grows in with the same rate constant. A scheme consistent with these various observations is presented in equation 5. The acceleration of the decay of the phenethyl cation, coupled with the observation of dimer and oligomeric products, is obviously consistent with this cation being trapped by unreacted styrene at the higher concentrations of substrate. The new transient that appears at 340 nm can be assigned as the cation (7) formed by addition of the initial phenethyl cation to a second styrene. This cation reacts by an intramolecular Friedel-Crafts alkylation to give the dimers (8), or with additional styrene to give eventually the trimers, *etc*. There is no further spectral change associated with the formation of the cations that are the precursors of these higher oligomers. However, they may have identical spectral characteristics to dimer cation (7), so that the 340 nm transient could represent a collection of different cations. The growing cationic intermediate of the acid-initiated polymerization of styrene has been observed in stopped-flow experiments in dichloromethane, and it also has a λ_{max} at 340 nm.[34,35]

The flash photolysis experiments in HFIP represent the first observation of the initial phenethyl cation of this polymerization. One interesting finding is that this cation has a different λ_{max} than the dimer cation (7) and higher oligomeric cations. A possible explanation is that the phenyl group at the γ-carbon in these latter cations is interacting with the positive charge at the α-carbon, forming an internal charge-transfer complex. The phenethyl cation is also considerably more

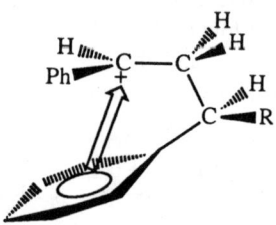

reactive. This is seen in the rate constants for the decay of the cations in HFIP (equation 5), with the difference being even more striking considering that the number for the 340 nm cation does not refer to reaction with solvent, but rather with the internal phenyl group. It is also seen in the rate constants for the addition to styrene, the propagation reaction in the polymerization. This number is 2×10^8 M^{-1}s^{-1} for the phenethyl cation, but is reduced to $\sim 10^5$ M^{-1}s^{-1} for 7,

Direct Studies of the Reactivities of Short-Lived Carbocations 309

the latter being calculated from the ratio of trimer:dimer products.[33] Propagation rate constants of 10^5 M^{-1}s^{-1} have also been obtained in the stopped-flow studies for the temperature (20 °C) of the flash photolysis experiments.[34,35] This difference in reactivity may be explained by a steric effect, the phenethyl cation being less crowded and thus more reactive. Stabilization of the dimer and oligomer cations by the adjacent phenyl group, as discussed above, could also contribute.

10 FRAGMENTATION OF BICUMENE

The systems of equations 1, 3, and 4 exemplify cases where a carbocation generated in some photochemical reaction reacts with a carbon nucleophile to give a second cation. The remaining systems will illustrate what is essentially the reverse, reactions where an initial cation undergoes a cleavage reaction.

The first example involves some diverse photoreactions of bicumene recently reported by Faria and Steenken.[13] In the non-polar solvent dichloromethane, excitation of this compound results in straightforward homolysis of the central C-C bond, resulting in the production of two cumyl radicals. In trifluoroethanol (TFE), photohomolysis is less efficient, and a two photon process is observed, resulting in the cumyl cation and cumyl radical. These species arise from the bicumene cation radical produced by biphotonic ionization.

$$\text{Ph}-\underset{\underset{\text{Me}}{|}}{\overset{\overset{\text{Me}}{|}}{\text{C}}}-\underset{\underset{\text{Me}}{|}}{\overset{\overset{\text{Me}}{|}}{\text{C}}}-\text{Ph} \xrightarrow[\text{TFE}]{h\upsilon} S_1 \xrightarrow[-e^-]{h\upsilon} \left[\text{Ph}-\underset{\underset{\text{Me}}{|}}{\overset{\overset{\text{Me}}{|}}{\text{C}}}-\underset{\underset{\text{Me}}{|}}{\overset{\overset{\text{Me}}{|}}{\text{C}}}-\text{Ph}\right]^{+\bullet} \xrightarrow{<10 \text{ ns}} \text{Ph}-\underset{\text{Me}}{\overset{\text{Me}}{\text{C}}}{}^+ \quad {}^{\bullet}\underset{\text{Me}}{\overset{\text{Me}}{\text{C}}}-\text{Ph} \quad (6)$$

This is an example of route (d) of Table 1. The cation radical in this sequence is not observed in the flash photolysis experiments, indicating that the lifetime for the C-C fragmentation is less than 10 ns. The cumyl cation is also not observed, since its lifetime in TFE is too short. The presence of the cation can be inferred by the observation of the ether PhC(Me)$_2$OCH$_2$CF$_3$ in the product mixture. In HFIP the cumyl cation is longer-lived, being observed for example in the photoprotonation of α-methylstyrene.[9] Indeed, this same cation is produced on photolysis of bicumene. Now, however, there is no signal for the cumyl radical, and moreover, the photolysis has become monophotonic. The observation of cumene as a product of this reaction, along with the ether (9) and α-methylstyrene, leads to the following mechanism.

$$\text{Ph}-\underset{\underset{\text{Me}}{|}}{\overset{\overset{\text{Me}}{|}}{C}}-\underset{\underset{\text{Me}}{|}}{\overset{\overset{\text{Me}}{|}}{C}}-\text{Ph} \xrightarrow[\text{HFIP}]{h\nu} \begin{array}{c} \text{[cyclohexadienyl cation]} \\ \text{Me-C-Me} \\ | \\ \text{Me-C-Me} \\ | \\ \text{Ph} \end{array} \xrightarrow{<10 \text{ ns}} \begin{array}{c} \text{[diene]} \\ \text{Me}\diagdown \diagup \text{Me} \\ \text{C} \\ \text{Me}\diagdown \diagup \text{Me} \\ \underset{|}{\overset{+}{C}} \\ \text{Ph} \end{array} \begin{array}{c} \longrightarrow \text{PhCH(Me)}_2 \\ \\ \text{PhC(Me)=CH}_2 \\ \longrightarrow \\ \text{PhC(Me)}_2\text{OCH(CF}_3)_2 \\ \mathbf{9} \end{array} \quad (7)$$

This has an initial photoprotonation of one of the rings of the bicumene, as in equation 2. This produces a cyclohexadienyl cation that rapidly undergoes side chain C-C fragmentation to give isocumene, the precursor of cumene, and the cumyl cation.

11   C-C Fragmentation of a Cyclohexadienyl Cation - A Reverse Friedel-Crafts Reaction

The second example also involves fragmentation of a cyclohexadienyl cation produced by photoprotonation, but here the departing cation is directly attached to the aromatic ring, so that the cleavage constitutes the reverse of a Friedel-Crafts alkylation. The background here is the observation that 1,3-dimethoxybenzene and some simple derivatives undergo photoprotonation in HFIP with a high degree of selectivity for the 2-position.[11] This produces 2,6-dimethoxybenzenium ions (**10**) that are relatively long-lived in HFIP and can be detected as transients with λ_{max} 400 - 440 nm. This situation contrasts with that

$$\underset{\underset{\text{H H } \mathbf{11}}{}}{[\text{MeO, R, OMe cation}]} \xrightleftharpoons[]{\text{conc. H}_2\text{SO}_4} \underset{R = H, \text{Me}}{[\text{MeO, R, OMe}]} \xrightarrow[\substack{4 \times 10^3 \text{ s}^{-1} (R = H) \\ 1 \times 10^3 \text{ s}^{-1} (R = \text{Me})}]{h\nu, \text{ H-OCH(CF}_3)_2} \underset{\mathbf{10}}{[\text{MeO, H, R, OMe cation}]} \quad (8)$$

in the ground state, where C4 is protonated producing 2,4-dimethoxybenzenium ions that can be observed with λ_{max} 320 - 340 nm in concentrated acids. In terms of the products of photolysis in HFIP, there is no net photoreaction other than exchange of the hydrogen at C2. This can be observed with the parent 1,3-dimethoxybenzene upon irradiation in $(CF_3)_2CHOD$. With 2-methyl-1,3-dimethoxybenzene, where the hydrogen introduced from the solvent must be removed, there is no new product even on prolonged irradiation.

This situation changes with the 1,3-dimethoxybenzene derivative (12) where the substituent at C2 is diphenylmethyl, a group forming a relatively stable cation.[36] Now irradiation in HFIP produces the cleavage products 1,3-dimethoxybenzene and the ether $Ph_2CHOC(CF_3)_2$. This is obviously consistent with protonation at C2 to give the cation (13) followed by ejection of the diphenylmethyl cation. Interestingly some rearranged 4-diphenylmethyl-1,3-dimethoxybenzene (15) is also formed. The relative amount of this product increases with increasing irradiation time, a result that can be explained by combination of the diphenylmethyl cation with 1,3-dimethoxybenzene as the latter accumulates. However, even on extrapolation to zero irradiation time, (15) is formed in 17% yield, and this material must arise from an intramolecular rearrangement. Flash photolysis sheds further light on the situation. With a nanosecond apparatus, the transient spectrum observed immediately after the laser pulse consists of a band in the region 320 - 340 nm, with a second band centred at 440 nm that is superimposable on the spectrum of the diphenylmethyl cation. In other words, there is a prompt (within 20 ns) formation of the diphenylmethyl cation, plus some other species. Spectral changes are then observed with a rate constant of 5×10^5 s^{-1}, and these represent a growth of additional diphenylmethyl cation while the transient at lower wavelength decays with the same rate. There is approximately twice as much diphenylmethyl cation formed in the delayed process as in the prompt process. The diphenylmethyl cation eventually decays, but this is a slow process (0.1 s) in HFIP. The initial formation of the diphenylmethyl cation can be observed with a picosecond apparatus, in the form of a rise at 440 nm that occurs with a rate constant of 10^9 s^{-1}.[37]

These results can be explained in terms of the mechanism of equation 9, the partitioning percentages given in this scheme arising from a considering of the products at zero time and the ratio of the initial and delayed processes observed with flash photolysis. The initial species that is formed in the photoprotonation is the 2,6-dimethoxybenzenium ion (13), as in the case of 1,3-dimethoxybenzene itself. The cation (13) is however short-lived, with a lifetime of 1 ns. It reacts 28% by cleavage to give 1,3-dimethoxybenzene and diphenylmethyl cation, this process representing the prompt formation of the latter observed in the ns flash photolysis experiments. The remaining 72% of (13) reacts by rearrangement to the thermodynamically more stable 2,4-

dimethoxybenzenium ion (14). This species is observed in the region 320 - 340 nm, as is characteristic for this type of benzenium ion. This cation reacts partly (23%) by deprotonation to give the rearranged product (14), although its predominant reaction is cleavage. This represents the delayed formation of the diphenylmethyl cation that is observed in the nanosecond experiments.

The surprising aspect of these results is not so much that the cation (13) loses the diphenylmethyl cation, but that its predominant reaction is a very rapid rearrangement to (14). The observation that the analog (10) with R = methyl does not undergo this chemistry suggests the rearrangement is not concerted, so that a π-complex (16) is a likely intermediate in the rearrangement. This would also serve as a common intermediate for the loss of the diphenylmethyl cation from the two cyclohexadienyl cations.

12 SILYLSUBSTITUTED CYCLOHEXADIENYL CATIONS - A PHOTOCHEMICAL PROTODESILYLATION

The final example is related to the one above, with the difference that this involves a silicon substituent at C2. Superficially the chemistry appears similar. The compound (**17**) (2-SiMe$_2$Ph), for example, undergoes desilylation upon irradiation in HFIP, giving equivalent amounts of 1,3-dimethoxybenzene and the silyl ether PhMe$_2$SiOCH(CF$_3$)$_2$ as the only observable products.[38]

This raises two questions - can the cyclohexadienyl cation (**18**), presumably the intermediate in this reaction, be observed, and what is the mechanism of the desilylation? The species (**18**) is an example of a β-silylsubstituted carbocation. Although the β-silicon is well established to exert a considerable stabilizing effect,[39] observations of such cations, even under stable ion conditions, are limited to the recent report of a 2-trimethylsilylallyl cation[40] and some β-silylsubstituted vinyl cations.[41] In general, attempts to prepare such ions results in the products of desilylation, even under non-nucleophilic stable ion conditions.[42] As for the desilylation, two mechanisms can be envisaged - a one-step associative reaction in which there is nucleophilic assistance, and a two-step dissociative reaction proceeding by way of a silyl cation. The latter in a sense would be analogous to the situation with the 2-Ph$_2$CH system of equation 9.

Flash photolysis experiments in HFIP with several examples of (**17**) reveal a transient with λ_{max} at 380 nm, the position independent of the nature of the substituents on silicon. The assignment of this species as the cation (**18**) is based upon several arguments - similarity to the spectrum of the parent 2,6-dimethoxybenzenium ion (λ_{max} = 410 nm), kinetic behaviour characteristic of cations - exponential decay unaffected by oxygen but accelerated by methanol, and consistency with the products, including quantitative agreement in a rate constant ratio (see later). That the λ_{max} of (**18**) is not identical with that of the parent probably reflects a perturbation of the p-system of the cyclohexadienyl cation. β-Silicon stabilizes adjacent C$^+$ through a hyperconjugative or 'vertical'

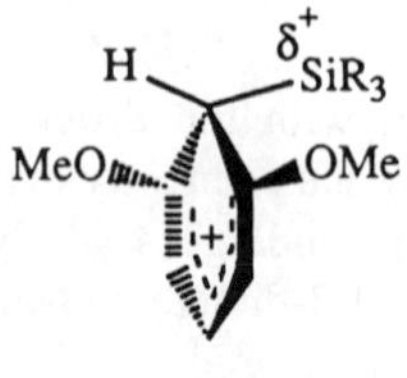

              H   δ⁺SiR₃   δ⁺SiMe₃
                                              H....C....H
         MeO⋯⌬⋯OMe   ⌬⌬⌬
              ⌬+⌬   (fluorenyl⁺)

               18 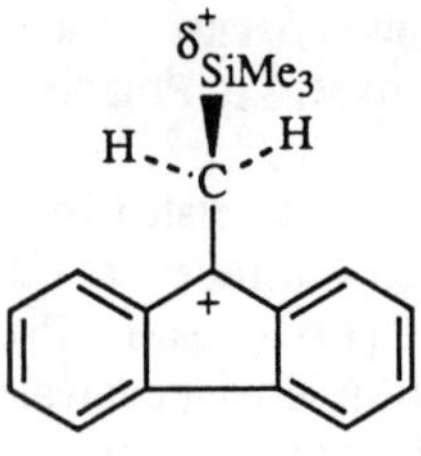   **19**

interaction,[39] and, as shown by the structure below, the appropriate geometry is achievable with **(18)**. The effect of this interaction on the absorption spectrum is difficult to predict. The experimental result is that there is a hypsochromic shift. A similar, albeit more pronounced, effect is seen with the cation **(19)**, the 9-(trimethylsilylmethyl)-9-fluorenyl cation, which has recently been observed upon flash photolysis of the appropriate fluorenol in TFE.[43] This cation has its principal λ_{max} at 370 nm, with a shoulder at ~410 nm. 9-Alkyl-9-fluorenyl cations have their λ_{max} at 485 nm, with a shoulder near 450 nm.

Having identified the transient as **(18)**, an important consideration is whether this decays only by desilylation or whether deprotonation to return to the starting material also contributes. That the latter is not the case is best illustrated

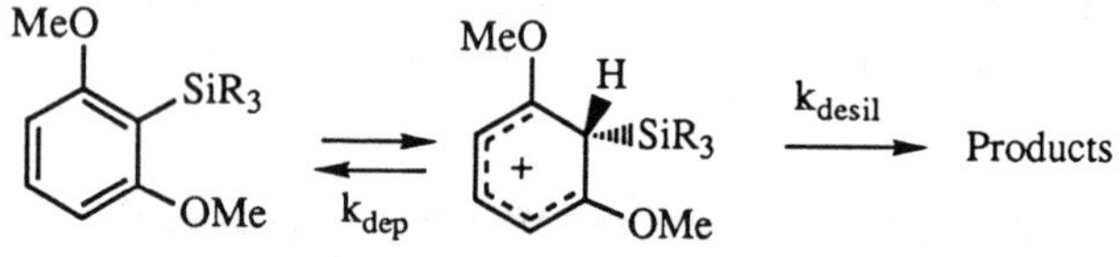

$$\underset{\textbf{17}}{\begin{array}{c}\text{MeO}\\ \ce{C6H4(SiR3)(OMe)}\end{array}} \underset{k_{dep}}{\overset{}{\rightleftarrows}} \underset{\textbf{18}}{\begin{array}{c}\text{MeO, H, SiR3, OMe}\\ \text{arenium}^+\end{array}} \xrightarrow{k_{desil}} \text{Products} \qquad (12)$$

by an observation that methanol accelerates the decay of the transient, but has no effect on the quantum yield for disappearance of the starting material. If the added alcohol were accelerating decay by acting as a base, then it would inhibit product formation. A second piece of evidence is that the rate constants for the decay of **(18)** are significantly larger than rate constants for 2,6-dimethoxybenzenium ions that react by deprotonation. For example, **(18)**

(PhMe$_2$Si) decays in HFIP with a rate constant of 7×10^4 s^{-1}, 70-times faster than the decay of the analog (10) with a 2-methyl group. The 2-silyl group is expected to stabilize the cation and thus decrease the rate constant for deprotonation. That the silylsubstituted (18) actually reacts more rapidly indicates that another pathway, desilylation, is available. It can also be noted that a situation with desilylation faster then deprotonation has been suggested previously in studies of thermal protodesilylation. The evidence with these systems generally points to the protonation step as being rate-limiting,[44] so that once formed, the cyclohexadienyl cation goes on to products and $k_{desil} > k_{dep}$.

The ability to study the kinetics of the desilylation of (18) provides an opportunity for directly probing the mechanism of the reaction, something that is not possible in the thermal studies. The most revealing observation is that nucleophiles such as methanol added to the HFIP accelerate the decay of the transient cation. This is obviously consistent with nucleophilic participation, that is, the associative pathway. This conclusion is reinforced through a quantitative comparison of kinetic and product results obtained with the PhMe$_2$Si compound. The experiments with added methanol result in the observation of two ethers, PhMe$_2$SiOCH(CF$_3$)$_2$ and PhMe$_2$SiOCH$_3$, in a total concentration equal to that of 1,3-dimethoxybenzene. From the methanol dependence of the relative yields of the ethers, the trapping ratio, k_{MeOH}:k_{HFIP}, is calculated as 27 M^{-1}. These same

two rate constants are individually determined in the flash photolysis experiments. Their ratio is 24 M^{-1}, within experimental error the same as the number from the product analyses. This, of course, is classic evidence of a bimolecular mechanism in a solvolysis system, rate accelerations and product ratios in quantitative agreement. There is a possibility that the solvent reaction is unimolecular, but this is ruled out by the entropies of activation, -24 cal mol^{-1} deg^{-1} for k_{HFIP} and -10 cal mol^{-1} deg^{-1} for k_{MeOH} (with PhMe$_2$Si). The large negative number for the solvent is clearly of the magnitude consistent with a bimolecular reaction. The solvent $\Delta S^{\neq}$ is in fact more negative than the value for MeOH. This may be related to release of a solvent molecule from a hydrogen-bonded complex {Me(H)O:·····H-OCH(CF$_3$)$_2$} in the transition state with the latter. As a further comparison, an example of a dissociative decomposition of a benzenonium ion exists, the diphenylmethyl-substituted **(14)** where Ph_2CH^+ is actually observed to form as the cyclohexadienyl cation decays in HFIP (equation 9). This system has $\Delta S^{\neq} = +13$ cal mol^{-1} deg^{-1}.

The kinetics, therefore, offer unambiguous proof that the desilylation involves nucleophilic participation, and that a free silyl cation does not form. Hammett ρ values for a series of **(18)** ($ArMe_2Si$), however, offer evidence for some R_3Si^+-character in the transition state, particularly for the reaction with the solvent. The silicon atom in this cation bears a partial positive charge due to its interaction with the neighboring p system. In an associative transition state this charge can either decrease or increase depending on the relative amounts of C-Si bond breaking and Si-O bond making. For the methanol reaction, the ρ is 0.0,

$$\text{(14)}$$

indicating no change in charge at the transition state. For the solvent reaction, however, ρ is -1.3, pointing to an increase in positive charge at silicon as the reaction proceeds. In other words, with HFIP as the nucleophile, C-Si bond cleavage is more advanced than Si-O bond formation at the transition state, and the silicon has partial Si^+-character. For comparison here, hydride transfer from $ArMe_2SiH$ to a diaryl-methylcarbenium ion, a reaction where an intermediate $ArMe_2Si^+$ is proposed, has ρ = -2.7.[45] The cations **(18)** also exhibit a reactivity order $Ph_2MeSi > PhMe_2Si > Me_3Si$, with the bulkier systems reacting more quickly, and this is also consistent with a relatively loose transition state. This order is in fact observed with both methanol and HFIP, with the differences being smaller with the more nucleophilic methanol, where the ρ value indicates greater nucleophilic participation in the transition state.

13   SUMMARY

In summary, these examples illustrate the application of the flash photolysis method to the study of carbocations. This technique, whose emphasis is reactivity, provides a complement to investigations carried out under stable ion conditions, where the general objective is information about structure. The latter, by providing a non-nucleophilic medium, allows for the leisurely examination of cations by techniques such as NMR and IR spectroscopy. Flash photolysis however provides the opportunity to generate cations in the presence of nucleophiles, that is under the conditions where they are normally found as reactive intermediates. As illustrated here, cations of S_N1 solvolyses can be directly observed reacting with the alcohol solvent and added nucleophiles such as azide. Using the weakly nucleophilic solvent HFIP, C-C bond forming reactions with carbon nucleophiles can be directly studied, with both the initial and product carbocations in many cases observable. Friedel Crafts alkylations have been studied in this way, as has the initiation in the cationic polymerization of styrene. As shown by the final two examples, flash photolysis can also be employed to study cations that are not accessible under stable ion conditions because of the presence of rapid rearrangements or other fragmentation reactions.

REFERENCES

1   C D Ritchie, *Acc Chem Res*, 1972, **5**, 348.
2   J P Richard and W P Jencks, *J Am Chem Soc*, 1982, **104**, 4689.
3   J P Richard, M E Rothenberg, and W P Jencks, *J Am Chem Soc*, 1984, **106**, 1361.
4   V B Ivanov, V L Ivanov, and M G Kuzmin, *J Org Chem, USSR*, 1972, **8**, 626; *Mol Photochem* 1974, **6**, 125; V L Ivanov, V B Ivanov, and M G Kuzmin, *J Org Chem, USSR*, 1972, **8**, 1263.
5   W Schnabel, I Naito, T Kitamura, S Kobayashi, and H Taniguchi, *Tetrahedron*, 1980, **36**, 3229. S Kobayashi, T Kitamura, H Taniguchi, and W Schnabel, *Chem Lett*, 1983, 1117. F I M Van Ginkel, R J Visser, C A G O Varma, and G Lodder, *J Photochem*, 1985, **30**, 453.
6   R A McClelland, V M Kanagasabapathy, N Banait, and S Steenken, *J Am Chem Soc*, 1989, **111**, 3966.
7   J Bartl, S Steenken, H Mayr, and R A McClelland, *J Am Chem Soc*, 1990, **112**, 6918.
8   R A McClelland, F Cozens, and S Steenken, *Tetrahedron Lett*, 1990, **31**, 2821.
9   R A McClelland, C Chan, F Cozens, A Modro, and S Steenken, *Angew Chem, Int Ed Engl*, 1991, **30**, 1337.
10 S Steenken and R A McClelland, *J Am Chem Soc*, 1990, **112**, 9648.
11 N Mathivanan, F Cozens, R A McClelland, and S Steenken, *J Am Chem Soc*, 1992, **114**, 2198.
12 S Steenken and R A McClelland, *J Am Chem Soc*, 1989, **111**, 4967.
13 J L Faria and S Steenken, *J Phys Chem*, 1992, **97**, 10869.
14 J L Faria and S Steenken, *J Am Chem Soc*, 1990, **95**, 1277.
15 B R Arnold, J C Scaiano, and W G McGimpsey, *J Am Chem Soc*, 1992, **114**, 9978.
16 W Kirmse, J Kilian, and S Steenken, *J Am Chem Soc*, 1990, **112**, 6399.

17   J E Chateauneuf, *J Chem Soc, Chem Commun*, 1437, 1991.
18   S T Belt, C Bohne, G Charette, S E Sugamori, and J C Scaiano, *J Am Chem Soc*, 1993, **115**, 2200.
19   F Cozens, J Li, R A McClelland, and S Steenken, *Ang Chem, Int Ed Engl*, 1992, **31**, 743.
20   See also S Steenken, J Buschek, and R A McClelland, *J Am Chem Soc*, 1986, **108**, 2808; R A McClelland, and S Steenken, *J Am Chem Soc*, 1988, **110**, 5860.
21   R A McClelland, V M Kanagasabapathy, and S Steenken, *J Am Chem Soc*, 1988, **110**, 6913.
22   R A McClelland, N Mathivanan, and S Steenken, *J Am Chem Soc*, 1990, **112**, 4857.
23   S L Mecklenburg and E F Hilinski, *J Am Chem Soc* 1989, **111**, 5471.
24   S Jaarinen, J Nuranen, and J Koskikallio, *Int J Chem Kin*, 1985, **17**, 925; S Anderson and K Yates, *Can J Chem*, 1988, **66**, 2412.
25   R E Minto and P K Das, *J Am Chem Soc*, 1989, 111, 8858.
26   R J Sneen, V J Carter, and P S Kay, *J Am Chem Soc*, 1966, **88**, 2594; D A Raber, J M Harris, R E Hall, and P R Schleyer, *J Am Chem Soc* 1971, **93**, 4821.
27   R Ta-Shma and Z Rappoport, *J Am Chem Soc*, 1983, **105**, 6082.
28   T L Amyes and W P Jencks, *J Am Chem Soc*, 1989, **111**, 7882.
29   R A McClelland, V M Kanagasabapathy, N Banait, and S Steenken, *J Am Chem Soc*, 1991, **113**, 1009.
30   R A McClelland, F L Cozens, S Steenken, T L Amyes, and J P Richard, *J Chem Soc, Perkin Trans 2*, 1993, 1717.
31   F Cozens, J Li, R A McClelland, and S Steenken, *Ang Chem, Int Ed Engl*, 1992, **31**, 743.
32   S Steenken, C S Q Lew, and R A McClelland, unpublished.
33   R A McClelland, A Modro, P A Davidse, and R A McClelland, unpublished.
34   M Villesange, A Rives, C Bunel, J-P Vairon, M Froeyen, M Van Beylen, and A Persoons, *Makromol Chem, Macromol Symp*, 1991, **47**, 271.
35   T Kunitake and K Takarabe, *Macromolecules*, 1979, **12**, 1061.
36   E MacKnight, J L Faria, R A McClelland, and S Steenken, unpublished observations.
37   C S Q Lew, L Johnston, N Schepp, and R A McClelland, unpublished observations.
38   C S Q Lew and R A McClelland, *J Am Chem Soc*, 1993, **115**, 11516.
39   J B Lambert, *Tetrahedron*, 1990, **46**, 2677.
40   G K S Prakash, V P Reddy, G Rasul, J Casanova, and G A Olah, *J Am Chem Soc*, 1992, **114**, 3076.

41 H-U Siehl, F-P Kaufmann, Y Apeloig, V Braude, D Danovich, A Berndt, and N Stamitis, *Ang Chem, Int Ed Engl*, 1991, **30**, 1479; H-U Siehl and F-P Kaufmann, *J Am Chem Soc*, 1992, **114**, 4937; H-U Siehl, F-P Kaufmann, and K Hori, *J Am Chem Soc*, 1992, **114**, 9343.
42 G A Olah, A L Berrier, L D Field, and G K S Prakash, *J Am Chem Soc*, 1982, **104**, 1349.
43 C S Q Lew, R A McClelland, L J Johnston, and N P Schepp, *J Chem Soc, Perkin Trans 2*, 1994, 395.
44 C Eaborn, *J Organomet Chem*, 1975, **100**, 43.
45 H Mayr, N Basso, and G Hagen, *J Am Chem Soc*, 1992, **114**, 3060.

NITROSATION AND NITRIC OXIDE CHEMISTRY

D Lyn H Williams

Department of Chemistry, University of Durham, Durham DH1 3LE, UK.

1 INTRODUCTION

It is an honour to be invited to speak at the Ingold Symposium within this Conference. As an undergraduate and postgraduate in Sir Christopher's Department at UCL I benefited from his lectures and general presence in the Department. Ingold was also indirectly responsible for arousing my interest in nitrosation chemistry in the following way. In the 1950's and 1960's there was quite a lot of interest in the determination of the reaction mechanisms for the rearrangement reactions of a family of *N*-substituted aromatic systems (equation 1). At that time I spent some time with the late Professor E D Hughes working

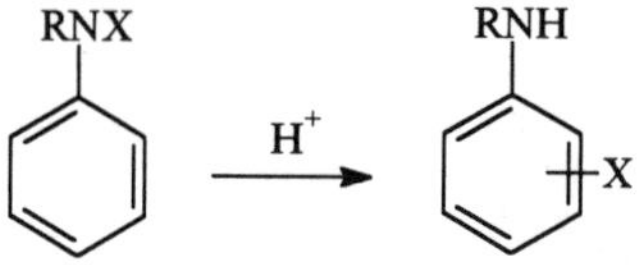

$$\text{RNX} \xrightarrow{H^+} \text{RNH} \cdot X \tag{1}$$

X = NO_2, OH, Cl, NO, Alk, NHAr *etc.*

on the mechanism of the nitramine rearrangement, *ie* X=NO_2. During the course of this work I noted the following extract (referring to the apparently similar reaction for X=NO - the Fischer-Hepp rearrangement) in Ingold's book[1]:- 'What is now needed in order to establish the mechanism firmly is a kinetic study of the rearrangement, and, as far as possible, of the separate reactions (1), (2) and (3) (see Scheme 1) just as in the example of the chloramine rearrangement.' When

$$C_6H_5\text{NRNO} + HX \underset{2}{\overset{1}{\rightleftharpoons}} C_6H_5\text{NHR} + XNO$$

$$\downarrow 3$$

$$4\text{-NOC}_6H_4\text{NHR} + HX$$

Scheme 1: Early suggestion for the mechanism of the Fischer-Hepp rearrangement.

the same quotation appeared in the second edition of the book 16 years later,[2] I felt that it was time to undertake such a study. The results of our experiments[3] finally showed clearly that the reaction took a different pathway, involving intramolecular rearrangement which was concurrent with reversible denitrosation (Scheme 2). All of the evidence for the original Scheme 1 was based on product studies particularly involving the ability to trap out XNO (for

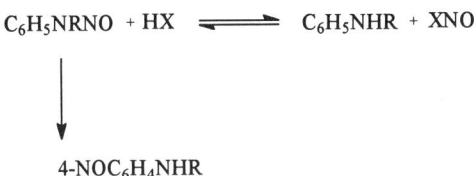

Scheme 2: Correct intramolecular mechanism for the Fischer-Hepp rearrangement.

X=Cl). This work provides an example which shows that the trapping of an intermediate from a reaction does not *necessarily* mean that the intermediate is along the reaction pathway leading to the product of the reaction studied. All of the early trapping and other experiments are equally well accounted for by Scheme 2. The experiments which did distinguish between the two mechanisms involved measurement of the product ratio (rearrangement:denitrosation *ie* 4-nitroso product:amine) and also the rate constant for the reaction as a function of the concentration of trap added to remove XNO. We used sulfamic acid, hydrazine, hydrazoic acid *etc.* The product ratio decreased with increasing trap concentration to a *constant value*. This can only be accounted for by Scheme 2 whereas for Scheme 1 the product ratio should decrease towards zero as the trap concentration increased. Similarly, the observed rate constant increased to a constant value, again an observation which can only be accommodated by the outline mechanism in Scheme 2.

Historically, nitrosation reactions were much less studied from the mechanistic point of view than the formally related nitration reactions, even though the range of reactions is greater for the former. However in recent years the balance has been redressed to a considerable degree and the study of nitrosation reactions has proved to be a fruitful one for mechanistic study.

2 REAGENTS FOR EFFECTING NITROSATION

By far and away the most convenient and useful reagent for effecting nitrosation is nitrous acid, generated *in situ* in aqueous solution from sodium nitrite and mineral acid. This system is now well understood (see Scheme 3) mainly due to

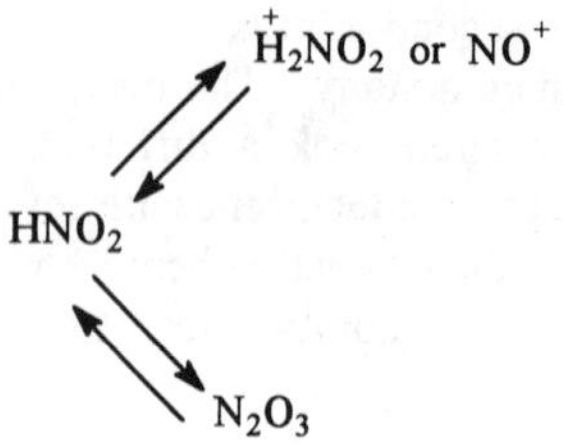

$$HNO_2 \rightleftharpoons H_2NO_2^+ \text{ or } NO^+$$
$$HNO_2 \rightarrow N_2O_3$$

$$HNO_2 + H^+ + X^- \rightleftharpoons XNO + H_2O$$

$$(X = Cl, Br, I, SCN, SC(NH_2)_2 \text{ etc})$$

Scheme 3: Nitrosating agents from nitrous acid.

the work of Ridd[4] who showed that the effective electrophile is either the nitrous acidium ion $H_2NO_2^+$ (or free NO^+) or dinitrogen trioxide N_2O_3, depending on the reaction conditions. The former predominates at higher acidities and low [HNO_2] whereas the latter takes over at low acidities particularly when [HNO_2] is high. The two pathways can clearly be distinguished by the kinetic order of the reaction with respect to [HNO_2]. In nitration reactions the formation of NO_2^+ as the reagent was detected kinetically by the zero order dependence upon [substrate] for very reactive substrates.[5] In aqueous media the analogous situation has not been experimentally detected for nitrosation but in acetonitrile solvent,[6] nitrosation (using alkyl nitrites and also nitrous acid) of alcohols and thiols is kinetically zero order with respect to the substrate, a result which has been taken to mean that rate-limiting formation of NO^+ occurs under these conditions.

In the presence of non-basic nucleophiles (X^-) equilibrium formation of XNO occurs which then can act as the nitrosating species. Catalysis by X^- (which is easily measured kinetically) depends both on the equilibrium constant for XNO formation and the rate constant for XNO reaction.[7] The former dominates which makes thiourea the most effective catalyst (*eg* in the nitrosation of morpholine).[8]

Other nitrosating agents, many of which can be used in non-aqueous solvents, include nitrosonium salts, such as $NO^+BF_4^-$ and $NO^+HSO_4^-$, alkyl nitrites RONO, nitrosyl acetate NOOAc (which is probably the effective agent when sodium nitrite is dissolved in acetic acid), the nitrogen oxides N_2O_3 and N_2O_4, some nitrosamines R_2NNO (and nitrososulfonamides and nitrosamides) and some metal nitrosyl complexes *eg* $[Fe(CN)_5NO]^{2-}$.

2.1 Reaction at Nitrogen
Reaction at nitrogen centres is very well-known, yielding (Scheme 4), diazonium

$$RNH_2 \rightarrow RN_2^+$$

$$R_2NH \rightarrow R_2NNO$$

$$R_3N \rightarrow R_2NNO + KETONE + N_2O$$

$$RCON(H)R \rightarrow RCON(NO)R$$

Scheme 4: Amine and amide nitrosation.

ions (and often deamination products), nitrosamines, nitrosamides *etc*. With tertiary amines an interesting reaction occurs[9] which involves N-C bond fission and results in the formation of a nitrosamine, together with a carbonyl compound (often a ketone) and nitrous oxide. All of these reactions involve N-N bond formation initially by electrophilic attack by a carrier of NO^+ (XNO).

2.2 Reaction at Oxygen and Sulphur
Alcohols and thiols yield alkyl nitrites and thionitrites (or nitroso thiols) respectively, on reaction with nitrous acid or any other nitrosating agent (Scheme 5). For alkyl nitrites, this is the reaction used in their synthesis and the procedure

$$ROH + HNO_2 \rightleftharpoons RONO + H_2O$$

$$RSH + HNO_2 \rightarrow RSNO + H_2O$$

Scheme 5: Alcohol and thiol nitrosation.

depends on the fact that the alkyl nitrite has a lower boiling point than the alcohol and can be distilled out from the equilibrium mixture. Thionitrites are much less well known, partly because of their instability generally, but are recently being studied more thoroughly because of their possible role *in vivo*, connected with the now well established role of nitric oxide in a range of human physiological processes including vasodilation. This aspect will be discussed later in this paper.

3 ALIPHATIC C-NITROSATION
Oxime formation from ketones, keto esters and related compounds, is a well-known synthetic procedure,[10] particularly useful for the introduction of a C=N

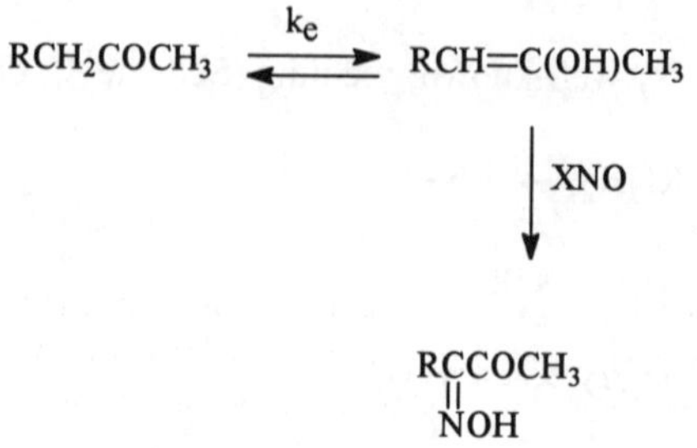

k_e
RCH_2COCH_3
$RCH{=}C(OH)CH_3$
XNO
$RCCOCH_3$
NOH

Scheme 6: Nitrosation of ketones.

bond. It is only relatively recently[11] that the reaction (for simple ketones) has been shown to proceed via the enol tautomers, just as in halogenation. Either the enolisation or the reaction of the enol (Scheme 6) can be made rate limiting depending on the relative rates of ketonisation and nitrosation of the enol. Values of k_e (the rate constant for enolisation) agree with those obtained in halogenation and in hydrogen isotope exchange experiments. For some of the ketones, *eg* acetone and methyl ethyl ketone, the reaction of ClNO and BrNO (generated *in situ*) are at the diffusion controlled limit. For the more acidic enols[12] (*eg* those derived from Meldrum's Acid, trifluoroacetylacetone and hexafluoroacetylacetone) it has been shown that a reaction pathway is also possible via the enolate anion (Scheme 7), which may in certain circumstances

Ketone
Enol
Enolate
XNO
XNO
Oxime
Oxime

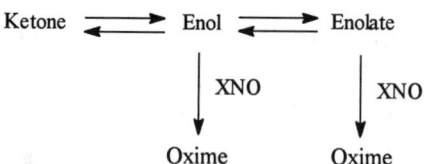

Scheme 7: Enol and enolate reactions.

dominate. More recently, interest has centred around enols derived from other carbonyl-containing compounds, notably carboxylic acids, esters and amides. Nitrosation (and halogenation) of malonic acids proceeds *via* their enol tautomers.[13] An interesting feature here is that acid catalysis of enolisation occurs intramolecularly (and not intermolecularly) via a cyclic six membered ring transition state. More recently[14] we have identified kinetically the enol form of malonamide both in nitrosation and halogenation (Scheme 8).

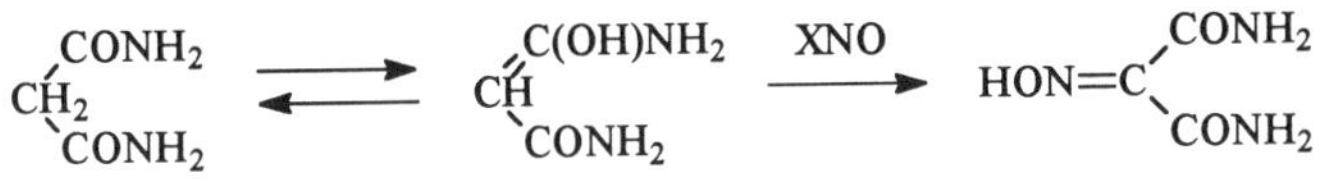

CH₂(CONH₂)₂ ⇌ CH(C(OH)NH₂)(CONH₂) —XNO→ HON=C(CONH₂)₂

Scheme 8: Reaction *via* the enol form of malonamide.

This is believed to be the first example of the kinetic identification of the enol form of an amide. Not unexpectedly, XNO species, I_2 and Br_2 all appear to react at the same rate (*ie* the diffusion rate). This leads to an estimated value of $4(\pm2) \times 10^{-10}$ for K_E the equilibrium contant for enol formation.

A formally similar situation occurs in the nitrosation of aliphatic nitro compounds,[15] a reaction which was studied in the last century by Victor Meyer.[16] The reactive intermediate is the corresponding nitronic acid (Scheme 9). However, formation of the nitronic acid from the nitro compound by an acid

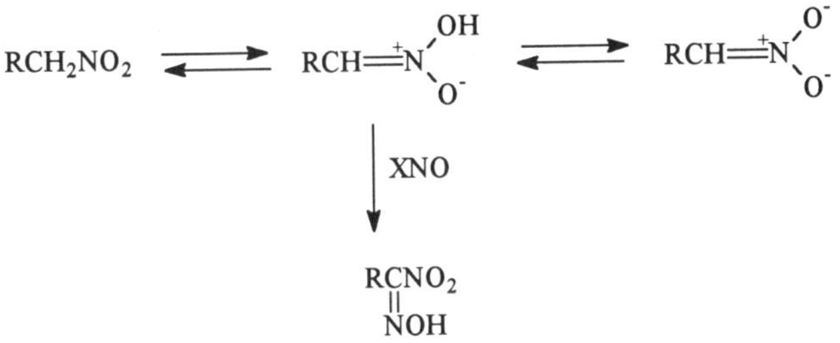

$$RCH_2NO_2 \rightleftharpoons RCH=\overset{+}{N}(OH)(O^-) \rightleftharpoons RCH=\overset{+}{N}(O^-)(O^-)$$

↓ XNO

$$RCNO_2 \\ \| \\ NOH$$

Scheme 9: Nitronic acid intermediates in the nitrosation of nitro compounds.

catalysed pathway is too slow to allow nitrosation to take place, so it has to be developed from the nitronate ion (formed by dissolving the nitro compound in base) by acidification. The nitronic acid formed in this way is sufficiently stable (with reference to tautomerisation to the nitro compound) to allow reaction with XNO.

Reaction can also occur *via* a carbanion intermediate even in mildly acid solution in the reaction of malononitrile with nitrous acid in the presence of X⁻ at *ca* pH 3.[17] Analysis of the data reveals, not surprisingly, that the carbanion is the most reactive of all substrates which have been examined in nitrosation chemistry. A general reactivity trend ClNO > BrNO > ONSCN > ONS⁺C(NH₂)₂ has been established for a range of substrates. Here, however, the last three react with the carbanion with rate constants near to the diffusion limit.

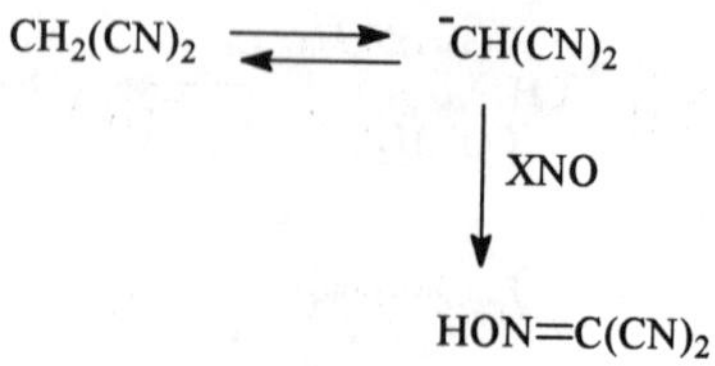

$$CH_2(CN)_2 \rightleftharpoons {}^-CH(CN)_2$$

$$\downarrow XNO$$

$$HON=C(CN)_2$$

Scheme 10: Reaction of the carbanion from malononitrile.

4   NITRIC OXIDE CHEMISTRY

Since 1987 there have been many spectacular discoveries implicating nitric oxide NO in a range of physiological processes. Initially it was shown that NO controls vasodilation and blood pressure; later discoveries identified it as a retrograde messenger in the central nervous system. It is involved in some of the processes in the brain and attacks tumour cells. *In vivo*, NO is synthesised enzymatically from L-arginine. It is believed that failure to synthesise NO or to deliver it is a prime cause of impotence in some men. Because of the range and importance of these discoveries, NO has been much in the news. It was voted 'Molecule of the Year' in 1992 by Science and it has hit the headlines on a number of occasions, and a recent Horizon television programme told the story of its fame. There are numerous review articles,[18,19] each one relating new findings.

We have recently been investigating some of the chemistry of NO which is likely to have some bearing on the understanding of the mechanisms of its actions *in vivo*, and in particular have been looking at the question of NO release from some nitroso compounds particularly from thionitrites.

Nitric oxide as a stable free radical, does not act (and is not expected to act) itself as a conventional electrophilic nitrosating agent. However there has been much confusion in the literature because even small quantities of oxygen which might be present allow nitrosation to occur, *via* it is thought (Scheme 11) oxidation to NO_2 and combination with more NO to give dinitrogen trioxide, a well known electrophilic nitrosating agent. When oxygen is rigorously excluded, nitric oxide will not give nitrosamines from secondary amines,[20] nor thionitrites from thiols[21] (except at very high pH) as outlined in Scheme 12.

$$2NO + O_2 = 2NO_2$$

$$NO_2 + NO = N_2O_3$$

Scheme 11: N_2O_3 Formation from NO.

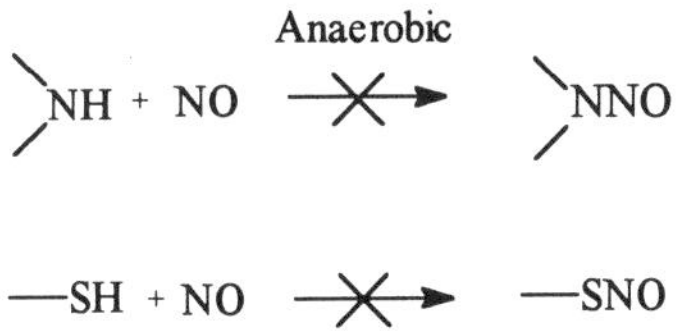

$$\text{\textbackslash NH + NO} \xrightarrow{\text{Anaerobic}} \text{\textbackslash NNO}$$

$$-\text{SH + NO} \xrightarrow{\hspace{1cm}} -\text{SNO}$$

Scheme 12: Failure of NO as an electrophile.

The reaction of NO with O_2 in the gas phase is a classic example of a third order reaction and has been much studied. The same reaction in aqueous solution has recently been investigated kinetically[22] (Scheme 13) and the same rate law applies. Using the measured value of the third order rate constant, one

$$2NO + O_2 = 2NO_2 \text{ in water}$$

$$\text{Rate} = k[NO]^2[O_2]$$

$$k = 5 \times 10^6 \text{ dm}^6 \text{ mol}^{-2} \text{ s}^{-1}$$

Scheme 13: Rate equation for the reaction of NO with O_2.

can estimate that in water saturated with oxygen and with nitric oxide concentration at say 1×10^{-8} mol dm^{-3}, the half-life is approximately 3 hours. This is typically the *in vivo* concentration of NO, so that oxidation under these conditions is not important.

5 NITROVASODILATORS

Drugs which have been used medically to treat angina and other circulatory problems and also to induce lowering of blood pressure during surgical operations include alkyl nitrites (used nitrite of amyl was used in a Sherlock Holmes story 'The case of the reluctant patient'!), sodium nitroprusside and most importantly glyceryl trinitrate GTN. It is now known from the work mainly in the Wellcome Research Laboratories that each of these generates NO *in vivo*. It is relatively easy to see simple non-enzymic pathways for NO release from RONO and [Fe(CN)$_5$NO]$^{2-}$ but it is less easy in the case of GTN. There have been some studies made[23] which reveal that thiols are required for this purpose, and the tolerance of patients after some time to GTN may result from thiol depletion; however, the mechanism is not yet completely understood.

5.1 Thionitrites (or Nitrosothiols)

We have recently been concerned with possible NO release from thionitrites since they are also known to be vasodilators and inhibit platelet aggregation, and

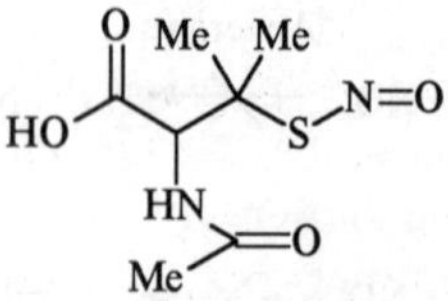

S-NITROSO-N-ACETYLPENICILLAMINE (SNAP)

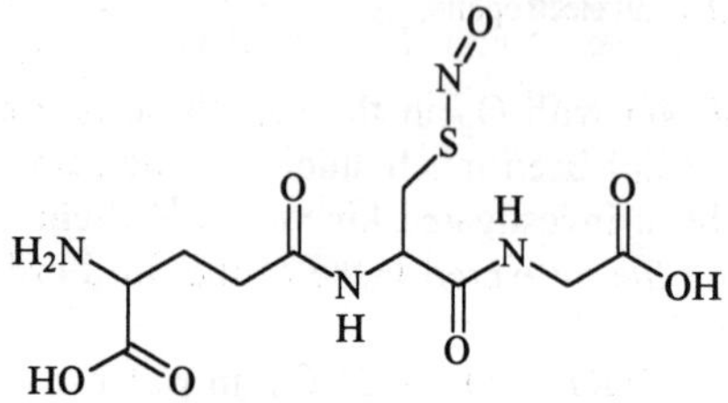

S-NITROSOGLUTATHIONE (SNOG)

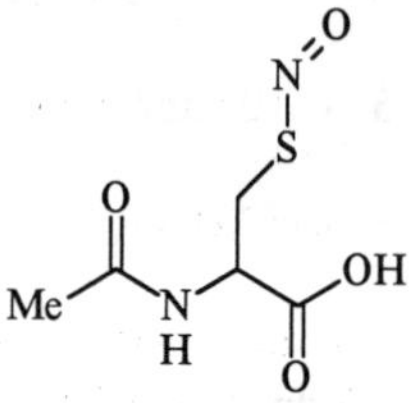

S-NITROSO-N-ACETYLCYSTEINE (SNAC)

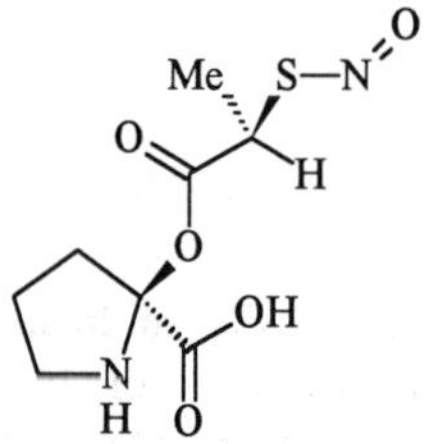

S-NITROSOCAPTOPRIL (SNOCAP)

Scheme 14: Some thionitrites (or nitrosothiols).

have synthesised the four compounds shown in Scheme 14. To date we have only examined the decomposition reactions of SNAP and to a more limited extent SNOG, in water and in aqueous buffers. For SNAP we find quantitative release

$$2RSNO \xrightarrow{H_2O} RSSR + 2NO_2^-$$

Scheme 15: Decomposition of RSNO.

of nitrite ion (as measured by diazotisation and azo coupling), and also recovered the disulfide (Scheme 15). We have followed the kinetics of the decomposition both by disappearance of the absorbance at 340 nm due to the thionitrite and also by sampling and nitrite ion analysis. Both methods show the same characteristics. In general the results were very erratic giving no clear cut order of reaction - sometimes we found reasonable fits for first order reactions and at other time half-order reactions. The half-life also varied wildly and we were unable to get consistent results with our collaborators in St Andrews University using the same samples of SNAP. Further, the rate profiles were not independent of the nature of the buffer used. Quantitative NO_2^- formation can readily be explained if NO is first formed which yields N_2O_3 (Scheme 11), which in acid solution gives nitrous acid, and in solutions of pH > ~ 4 nitrite ion (Scheme 16). If the decomposition is carried out in the presence of N-methylaniline, then

$$N_2O_3 + H_2O \rightarrow 2HNO_2$$

$$N_2O_3 + 2OH^- \rightarrow 2NO_2^- + H_2O$$

$$N_2O_3 + Amine \rightarrow Nitrosamine$$

Scheme 16: Reactions of N_2O_3.

quantitative formation of the corresponding nitrosamine results (even at pH 7). If, however, oxygen is rigorously excluded, then the very little nitrosamine is formed, implying very strongly that the whole process depends upon NO release.

Following suggestions made at an informal seminar at the Wellcome Research Laboratories, we examined the effect of metal ions upon the decomposition of SNAP. With added EDTA, even at concentrations as low as 1×10^{-5} mol dm^{-3} we find that decomposition is halted implying strongly that we have eliminated the metal ion catalysed reaction. Further, addition of aliquots of a range of metal ions in excess of the [EDTA] have enabled us to show that catalysis occurs most strongly with Cu^{2+} and Fe^{2+}, with some activity from Ag^+ but no measurable catalysis from Zn^{2+}, Ca^{2+}, Mg^{2+}, Ni^{2+}, Co^{2+}, Mn^{2+}, Cr^{3+} or Fe^{3+}. The extent of Cu^{2+} catalysis is shown in Figure 1. Further, we were able to correlate semi-quantitatively our earlier catalytic results with the [Cu^{2+}] measured by atomic absorption occurring in the various water supplies used (see Figure 2).

Catalysis by Cu(II) species is known in quite a number of organic reactions and indeed there is a report[24] that deamination of arylamines can be effected by thionitrites in acetonitrile using Cu(II) halides. Catalysis of biologically important reactions by copper species is also well-known[25] and is thought to involve in most cases Cu(I)⇌Cu(II) interconversions.

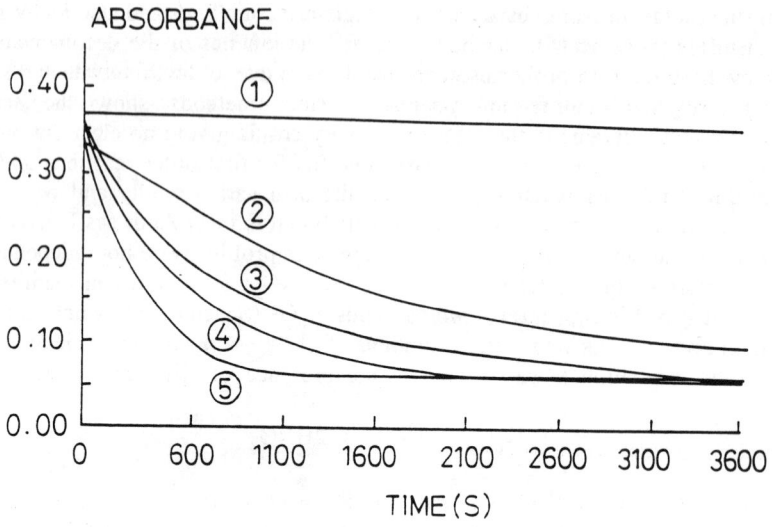

Figure 1: Absorbance time plots for the decomposition of SNAP (5×10^{-4} mol dm^{-3}), (1) no added Cu^{2+}; (2) [Cu^{2+}] 1×10^{-5} mol dm^{-3}; (3) [Cu^{2+}] 5×10^{-5} mol dm^{-3}; (4) [Cu^{2+}] 1×10^{-4} mol dm^{-3}; and (5) [Cu^{2+}] 5×10^{-4} mol dm^{-3}.

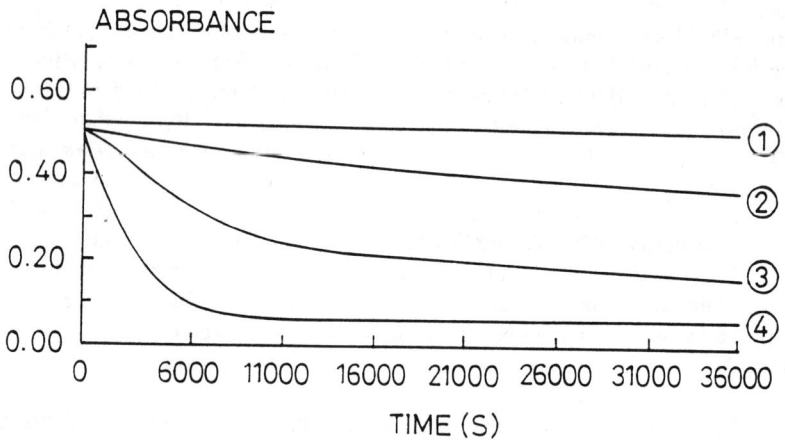

Figure 2: Absorbance time plots for the decomposition of SNAP (5×10^{-4} mol dm^{-3}), (1) with added EDTA (3×10^{-5} mol dm^{-3}); (2) in 'new' distilled water ([Cu^{2+}] 0.006 ppm); (3) in 'old' distilled water ([Cu^{2+}] 0.059 ppm); and (4) in tap water ([Cu^{2+}] 0.198 ppm).

The rate of Cu^{2+}-catalysed decomposition of SNAP is markedly pH dependent. The reaction is much faster at pH 7 than it is at lower pH values. At this stage we have no further evidence as to the nature of the mechanism of the catalytic reaction, but suggest a possible scenario in Scheme 17. Here Cu^+ is developed by reaction with RS^- (a well-known reaction) present in small

$$RS^- + Cu^{2+} \rightarrow RS^\bullet + Cu^+$$

$$Cu^+ + RSNO \rightarrow [RSNO.Cu]^+$$

$$[RSNO.Cu]^+ \rightarrow RS^- + NO + Cu^{2+}$$

$$2RS^\bullet \rightarrow RSSR$$

Scheme 17: Possible outline mechanism for Cu^{2+} catalysis.

quantities from RSH present in the RSNO sample. Complexation occurs at nitrogen or possibly sulfur in SNAP which then breaks up releasing RS^-, NO and Cu^{2+}. At this stage, this must be regarded as speculation, but more work is in progress. The involvement of RS^- would account for the pH dependence, since $[RS^-]$ would be very much higher at pH 7 than pH 4.

In addition to Cu^{2+} and Fe^{2+} catalysis, we find reaction with Hg^{2+}, which brings about decomposition of SNAP very rapidly - much more rapidly than the Cu^{2+} catalysed reaction. However we find that this is not a true catalytic process, in that the extent of SNAP decomposition is governed by the $[Hg^{2+}]$ until it is in stoichiometric excess when decomposition becomes quantitative. The final product is also HNO_2 or NO_2^-. This is in fact a known reaction and has been used as the basis of an analytical procedure for thiol determination.[26] All of our work so far suggests that the Hg^{2+} reaction releases NO^+ rather than NO in agreement with the suggestion by Saville[26] and is given in Scheme 18.

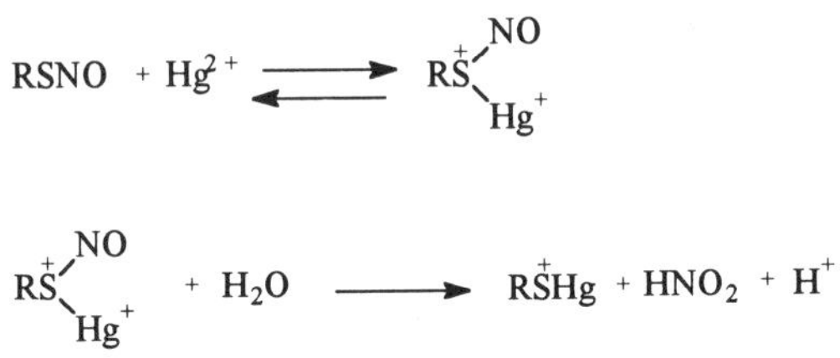

$$RSNO + Hg^{2+} \rightleftharpoons R\overset{+}{S}\!\!\begin{array}{c}NO\\ Hg^+\end{array}$$

$$R\overset{+}{S}\!\!\begin{array}{c}NO\\ Hg^+\end{array} + H_2O \longrightarrow R\overset{+}{S}Hg + HNO_2 + H^+$$

Scheme 18: Mechanism for the interaction of RSNO with Hg^{2+}.

As yet, we have not examined the Cu^{2+} catalysis widely. The reaction of SNOG however is significantly slower than that of SNAP, but can be made to occur reasonably at pH 7. The decomposition of SNOG with Hg^{2+} takes place rapidly; we have not yet carried out the stopped-flow kinetic studies on the Hg^{2+} reaction to obtain relative rate constants. Clearly much more work needs to be done.

Finally, we report that thionitrites can readily transfer the NO group to another thiol group *ie* transnitrosation can be brought about. Reaction is very fast - much faster than the Cu^{2+} catalysed reactions - and has been followed kinetically by stopped-flow spectrophotometry. To date we have worked only with the transfer from SNAP to the thiol group in thioglycolic acid. The second order rate constants are very dependent upon the pH of the solution and show a characteristic S-shaped curve indicating that the reactive species is the thiolate anion. The results are consistent with the published pK_a value for thiol ionisation in thioglycolic acid. All of the evidence points to a direct reaction, *ie* a nucleophilic attack by the thiolate anion RS^- at the nitroso nitrogen atom of SNAP as in Scheme 19. An analogous reaction occurs whereby the nitroso group

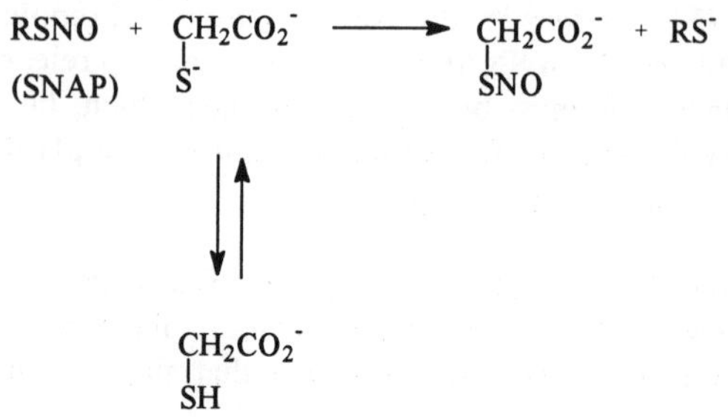

$$RSNO + \underset{S^-}{CH_2CO_2^-} \longrightarrow \underset{SNO}{CH_2CO_2^-} + RS^-$$
(SNAP)

$$\updownarrow$$

$$\underset{SH}{CH_2CO_2^-}$$

Scheme 19: Transnitrosation from RSNO to a thiol.

from a range of alkyl nitrites is transferred to a series of thiolate anions,[27] and has been interpreted mechanistically in the same fashion. As yet we have insufficient results to establish whether thionitrites are more reactive than the corresponding alkyl nitrites.

REFERENCES

1   C K Ingold, *Structure and Mechanism in Organic Chemistry*, Bell and Sons Ltd, 1st edition, 1953, 615.
2   C K Ingold, *Structure and Mechanism in Organic Chemistry*, Bell and Sons Ltd, 2nd edition, 1969, 902.
3   D L H Williams, *Tetrahedron*, 1975, 1343; *J Chem Soc, Perkin Trans 2*, 1982, 801.
4   J H Ridd, *Quart Rev*, 1961, **15**, 418.

5   K Schofield, *Aromatic Nitration*, Cambridge University Press, 1980, p 8-14.
6   M J Crookes and D L H Williams, *J Chem Soc, Perkin Trans 2*, 1988, 571.
7   D L H Williams, *Nitrosation*, Cambridge University Press, 1988, p 10-24,
8   T A Meyer and D L H Williams, *J Chem Soc, Perkin Trans 2*, 1981, 361.
9   P A S Smith and R N Loeppky, *J Am Chem Soc*, 1967, **89**, 1147.
10 O Touster, *Organic Reactions*, ed R Adams, Wiley, New York, 1953, **7**, Ch 6, p 327.
11 J R Leis, M E Peña, D L H Williams, and S D Mawson, *J Chem Soc, Perkin Trans 2*, 1988, 197.
12 M J Crookes, P Roy, and D L H Williams, *J Chem Soc, Perkin Trans 2*, 1989, 1015; P H Beloso, P Roy and D L H Williams, *ibid*, 1991, 17.
13 D L H Williams and A Graham, *Tetrahedron*, 1992, **48**, 7973.
14 D L H Williams and L Xia, *J Chem Soc, Perkin Trans 2*, 1993, 1429.
15 E Iglesias and D L H Williams, *J Chem Soc, Perkin Trans 2*, 1988, 1035.
16 V Meyer, *Ber Dtsch Chem Ges*, 1873, **6**, 1492; V Meyer and J Locher, *ibid*, 1874, 7, p 788 and 1506.
17 E Iglesias and D L H Williams, *J Chem Soc, Perkin Trans 2*, 1989, 343.
18 S Moncada, R M J Palmer, and E A Higgs, *Pharmacological Reviews*, 1991, **43**, 109.
19 A R Butler and D L H Williams, *Chem Soc Rev*, 1993, **22**, 233.
20 B C Challis and S A Kyrtopoulos, *J Chem Soc, Perkin Trans 1*, 1979, 299; *J Chem Soc, Perkin Trans 2*, 1978, 1296.
21 W A Pryor, DF Church, C K Govindan, and G Crank, *J Org Chem*, 1982, 47, 156.
22 D A Wink, J F Darbyshire, R W Nims, J E Saavedra, and P.C. Ford, *Chem Res Toxicol*, 1993, **6**, 23; H H Awad and D M Stanbury, *Int J Chem Kinet*, 1993, **25**, 375.
23 M Feelisch, *J Cardiovascular Pharmacology*, 1991, S25.
24 S Oae and K Shinhama, *Org Prog Proced Int*, 1983, **15**, 165.
25 B Hathaway in 'Comprehensive Coordination Chemistry', Ed G Wilkinson, Vol 5, p 720, Pergamon, 1987.
26 B Saville, *Analyst*, 1958, **83**, 670.
27 H Patel and D L H Williams, *J Chem Soc, Perkin Trans 2*, 1990, 37.

ACKNOWLEDGEMENTS

I wish to thank various members of my research group who have been involved in aspects of this work. I also thank Dr Elena Peña and Dr Ramon Leis (University of Santiago de Compostela, Spain), Dr Emilia Iglesias (University of La Coruña, Spain) and Dr A R Butler (University of St Andrews) for fruitful collaboration, and also various funding agencies including the SERC, the Royal Society, and the Wellcome Trust for financial support.

FORMATION AND STABILITY OF REACTIVE INTERMEDIATES OF ORGANIC REACTIONS IN AQUEOUS SOLUTION

Tina L Amyes,*[≠] John P Richard,[≠] and Vandanapu Jagannadham[≠]

*Department of Chemistry, University of Kentucky,
Lexington, KY 40506-0055, USA.*

[≠] Present address: Department of Chemistry, FNSM Complex, University of Buffalo, NY 14260, USA.

1   FORMATION AND STABILITY OF SIMPLE OXYGEN ESTER AND SIMPLE THIOL ESTER ENOLATES

The enolates of carboxylic acids are putative intermediates in the biologically important racemisation and elimination reactions catalysed by numerous enzymes (*eg*, mandelate racemase[1a] and enolase[1b]). Similarly, many enzyme-catalysed (*eg*, malate synthase[1c]) Claisen-type condensation reactions have been proposed to proceed through the enolates of simple[2] thiol esters. However, there are scant data pertaining to the formation and stability of such enolates in aqueous solution.[3] This has led to the question of whether these enzyme-catalysed reactions follow concerted pathways[1c,4,5] that avoid the formation of these species and which are *enforced* by their insignificant lifetime in the presence of acidic functional groups in an enzyme active site.[5] It is known that the enols and enolates of simple oxygen esters, which are models for carboxylic acids, and of simple thiol esters are highly unstable,[3] but there have been almost no determinations of rate or equilibrium constants for their formation in aqueous solution.[3,6] The goal of this work was to answer the following questions: (1) What are the pK_as of simple oxygen and thiol esters? (2) What are the lifetimes of simple oxygen and thiol ester enolates?

In principle, the pK_as of oxygen and thiol ester enolates in aqueous solution could be determined from rate constants for their hydroxide ion-catalysed formation, k_{OH}, and protonation by reaction with water, k_{HOH}, (Scheme 1). However, the determination of k_{OH} is severely hampered by the competing base-catalysed hydrolyses to give acetic acid (k_h). Bonhoeffer *et al* stated that 'the ionisation rate constants of the esters of acetic acid are not yet known, because hydrolysis proceeds much faster than exchange in alkaline deuterium oxide'.[7] Similarly, Lienhard found that the hydroxide ion-catalysed exchange of tritium from ^3H_2O into ethyl thioacetate was accompanied by fast

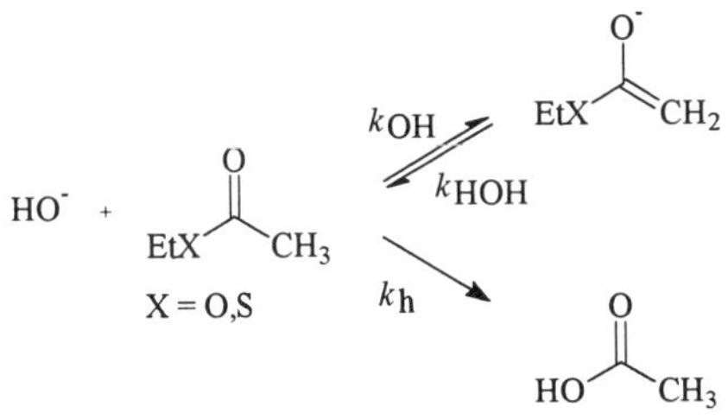

Scheme 1: Competing base-catalysed deprotonation and hydrolysis of oxygen and thiol esters.

competing hydrolysis to give acetate which had a specific radioactivity of only ca 0.5% that of the ^3H_2O.[8] We have developed the following experimental methods to determine rate and equilibrium constants for ionisation of the simple esters ethyl acetate and ethyl thioacetate[9] in aqueous solution: (1) We study deprotonation by tertiary amines which are more effective than lyoxide ion as catalysts of deprotonation. (2) We monitor deprotonation as exchange of the first α-proton for deuterium in 100% D_2O, which eliminates the discrimination isotope effect on the incorporation of the isotopically labelled hydrogen. (3) The incorporation of deuterium is monitored by the novel use of the ^2H perturbation of ^1H chemical shifts. This allows the detection of even very small (ca 1%) amounts of isotope exchange.

1.1 Exchange of the α-Protons of Ethyl Thioacetate and Acetone[9]

The exchange for deuterium of the first α-proton of each methyl group of ethyl thioacetate and of acetone in D_2O catalysed by 3-quinuclidinone was followed by ^1H NMR spectroscopy by monitoring the disappearance of the signal due to the α-CH_3 groups of the substrates (singlet) and the appearance of the cleanly resolved signal due to the α-CH_2D groups (triplet, 0.016 ppm upfield) of the products, during exchange of ca 30% and 37% of the first α-proton of each methyl group of ethyl thioacetate and acetone, respectively. During this time, there was no detectable formation of compounds containing α-CHD_2 groups, and there was negligible hydrolysis of ethyl thioacetate.

Pseudo-first-order rate constants, $k_{obsd} = k_B[B]$ (Scheme 2, B = 3-quinuclidinone), for exchange of the first α-proton of the individual methyl groups of ethyl thioacetate and acetone were obtained from the slopes of plots of reaction progress against time according to equation 1. In equation 1, $f(CH_3)$ is the fraction of unexchanged methyl groups remaining, calculated from the integrated areas (A) of the signals due to the α-CH_3 and the α-CH_2D groups using equation 2.

$$\ln f(CH_3) = -k_{obsd}t \qquad (1)$$

$$f(CH_3) = A(\alpha\text{-}CH_3)/\{A(\alpha\text{-}CH_3) + 1.5A(\alpha\text{-}CH_2D)\} \qquad (2)$$

The slopes of plots of k_{obsd} for the 3-quinuclidinone-catalysed exchange for deuterium of the first α-proton of ethyl thioacetate and acetone against [B] give $k_B = 2.2 \times 10^{-5}$ dm^3 mol^{-1} s^{-1} and 5.2×10^{-4} dm^3 mol^{-1} s^{-1}, respectively, for the deprotonation of ethyl thioacetate and acetone by 3-quinuclidinone.[10]

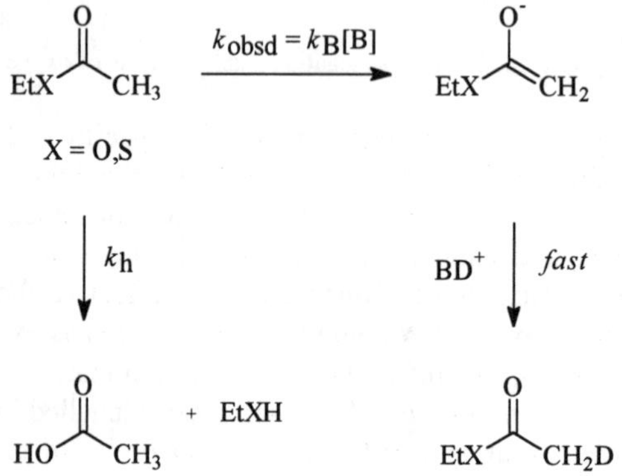

Scheme 2: Kinetic scheme for base-catalysed exchange for deuterium of the first α-proton of oxygen and thiol esters in D_2O.

1.2 Exchange of the α-Protons of Ethyl Acetate

Figure 1 shows partial 500 MHz ^{1}H NMR spectra of recovered ethyl acetate obtained during exchange of 1 - 10% of the first α-proton for deuterium in the presence of 3-substituted quinuclidine buffers in D_2O at 25 °C. The exchange reaction leads to replacement of the singlet at 2.053 ppm due to the α-CH$_3$ group of the substrate by an upfield triplet at 2.040 ppm ($J = 2.2$ Hz) due to the α-CH$_2$D group of the product. These exchange experiments are more complicated than those for ethyl thioacetate[9] for two reasons: (1) The competing hydrolysis of ethyl acetate to give acetic acid (Scheme 2) is > 10-fold faster than the exchange reaction, so that a high concentration of substrate (20 mmol dm^{-3}) was required in order to carry out the NMR analysis on the remaining ethyl acetate. However, in order to avoid medium effects, the rate constants for buffer-catalysed deprotonation must be obtained at [buffer] ≤ 0.50 mol dm^{-3}, and such concentrations do not adequately buffer the large amount of acetic acid produced by substrate hydrolysis. (2) The deuterium isotope shift of the α-protons of ethyl acetate (0.013 ppm) is considerably smaller than that for ethyl thioacetate (0.016 ppm/D), which results in only partial resolution of the triplet due to the α-CH$_2$D group of monodeuteriated ethyl acetate from the upfield ^{13}C satellite of the

Reactive Intermediates of Organic Reactions in Aqueous Solution 337

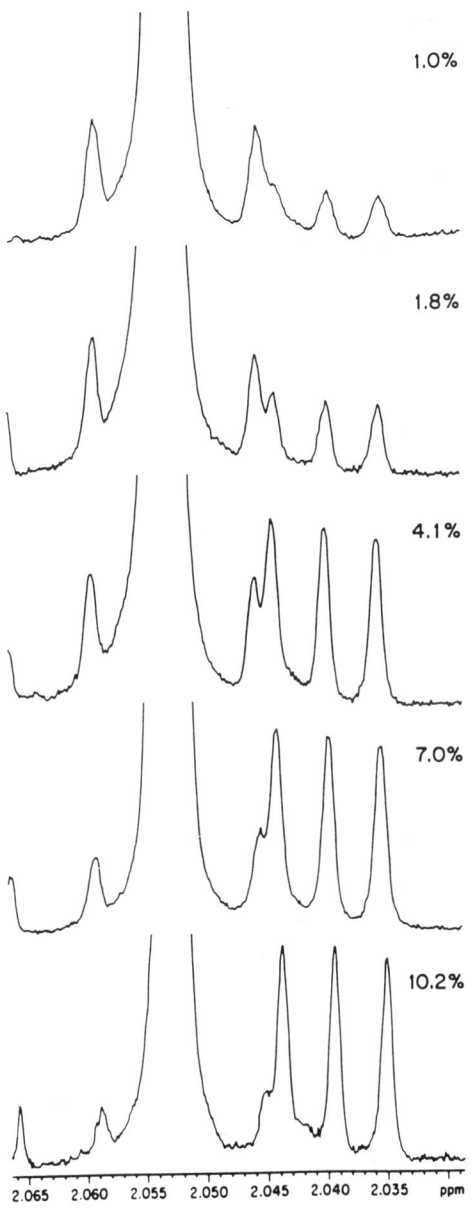

Figure 1: Partial 500 MHz ^{1}H NMR spectra (recorded in CDCl$_3$) of recovered ethyl acetate obtained during exchange of the first α-proton for deuterium in the presence of 3-substituted quinuclidine buffers in D$_2$O at 25 °C. The two peaks on either side of the large singlet at 2.053 ppm are ^{13}C satellites arising from coupling of the α-CH$_3$ protons to the neighboring carbonyl carbon. The fraction of exchange of the first α-proton is indicated at the top right of each spectrum.

singlet due to unreacted substrate (Figure 1). These difficulties were managed by the following protocols: (1) Rate constants were determined by following the exchange for deuterium of only 1 - 12% of the first α-proton of ethyl acetate, in order to minimise the extent of hydrolysis. (2) The pD of the reaction mixtures was maintained by the periodic addition of aliquots of KOD. (3) The area of the signal due to the α-CH_2D group of monodeuteriated ethyl acetate was determined from the integrated area of only the most upfield peak of this signal which was then multiplied by three to give $A(\alpha\text{-}CH_2D)$ (equation 2). This was then combined with the total integrated areas of the signal due to the α-CH_3 group of the substrate and the two most downfield peaks of the signal due to the α-CH_2D group of the product to give $A(\alpha\text{-}CH_3)$ (equation 2).

Pseudo-first-order rate constants, $k_{obsd} = k_B[B]$ (Scheme 2), for exchange of the first α-proton of ethyl acetate were obtained from the slopes of plots of reaction progress against time according to equation 1. The kinetic data are described by the rate law given in equation 3. Values of k_B (dm^3 mol^{-1} s^{-1}) for deprotonation of ethyl acetate by a series of 3-substituted quinuclidines and other tertiary amines (B) were obtained from the slopes of plots of $(k_{obsd} - k_o)$ against [B],[10] where k_o is the contribution of the deuteroxide ion-catalysed exchange reaction to the observed rate, calculated from $k_{OD} = 1.6 \times 10^{-3}$ dm^3 mol^{-1} s^{-1} and the observed pD of the solution.

$$k_{obsd} = k_{OD}[OD^-] + k_B[B] \qquad (3)$$

1.3 Brønsted Plot for Deprotonation of Ethyl Acetate

Figure 2 shows the Brønsted plot for deprotonation of ethyl acetate in D_2O by 3-substituted quinuclidines and other tertiary amines. The data for the structurally similar 3-substituted quinuclidines define a line of slope ß = 1.1 ± 0.1. Deuteroxide ion exhibits a 1200-fold negative deviation from the Brønsted plot. Such anomalously low reactivities of lyoxide ion are often observed for thermodynamically unfavorable proton transfer from carbon,[11] and they greatly facilitate the detection of general base catalysis of proton transfer.

The value of ß = 1.1 for deprotonation of ethyl acetate by 3-substituted quinuclidines shows that proton transfer is complete in the rate-limiting step. This strongly suggests that the thermodynamically highly unfavorable deprotonation of ethyl acetate to give the enolate is limited by the diffusional separation of the carbanion-ammonium cation ion pair (k_{-d}, Scheme 3). To the

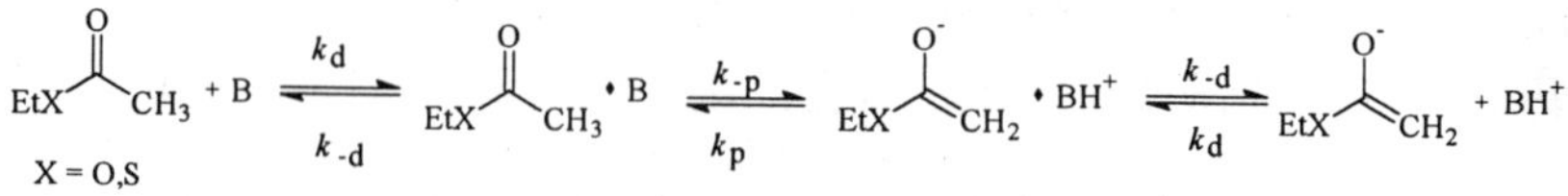

Scheme 3: Detailed kinetic scheme for proton transfer between oxygen and thiol esters and buffer bases.

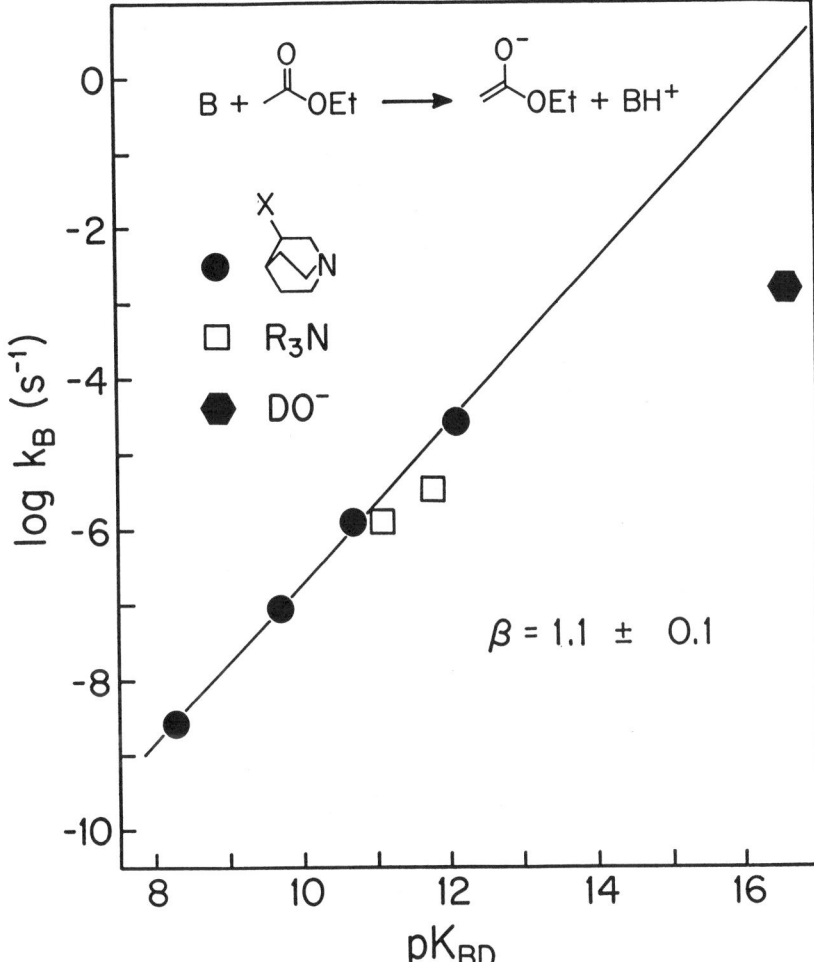

Figure 2: Brønsted plot for deprotonation of ethyl acetate by 3-substituted quinuclidines and other tertiary amines in D_2O at 25 °C and $I = 1.0$ (KCl). The values of pK_{BD} for the buffer bases in D_2O under the experimental conditions were determined from the measured solution pD.

best of our knowledge, ethyl acetate is the first example of a carbon acid in which the negative charge on carbon may be delocalised onto a neighboring carbonyl functional group and for which proton transfer is not rate-limiting.

1.4 Carbon Acid Acidities of Ethyl Acetate and Ethyl Thioacetate

The pK_as of ethyl acetate and ethyl thioacetate can be calculated from the rate constants for their deprotonation by a buffer base to give the free enolates (k_B, Scheme 4) and for protonation of these enolates by the conjugate acid of the base (k_{BH}, Scheme 4), according to equation 4.

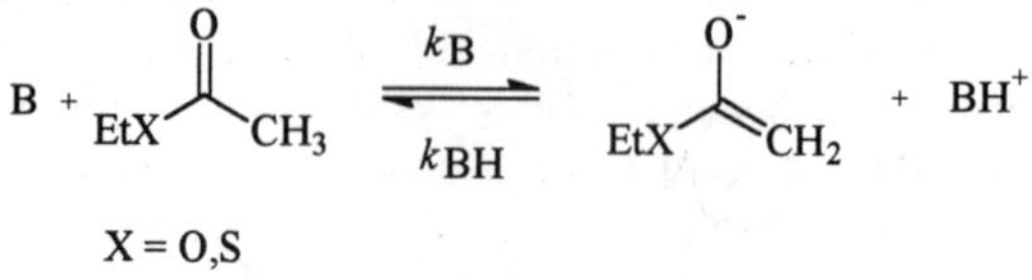

$$B + \underset{EtX}{\overset{O}{\|}}CH_3 \underset{k_{BH}}{\overset{k_B}{\rightleftharpoons}} \underset{EtX}{\overset{O^-}{\|}}CH_2 + BH^+$$

$$X = O, S$$

Scheme 4: Equilibrium proton transfer between oxygen and thiol esters and buffer bases.

$$pK_a = pK_{BH} + \log(k_{BH}/k_B) \qquad (4)$$

The value of $k_B = 5.2 \times 10^{-4}$ dm^3 mol^{-1} s^{-1} for deprotonation of acetone by 3-quinuclidinone and the values of $pK_a = 19$ for acetone[12] and $pK_{BH} = 7.5$ for 3-quinuclidinone[13] in water can be substituted into equation 4 to give $k_{BH} = 2 \times 10^8$ dm^3 mol^{-1} s^{-1} for reaction of the free enolate of acetone with protonated 3-quinuclidinone.[14] The deprotonation of ethyl thioacetate by 3-quinuclidinone is 12-fold slower than the deprotonation of a single methyl group of acetone by the same amine base. The substituent effect on k_{BH} for the protonation of the two enolates is expected to be no larger than that on the rates of their formation, so that the rate constant for reaction of the free enolate of ethyl thioacetate with protonated 3-quinuclidinone is estimated to be $k_{BH} = 2 \times 10^8 - 2 \times 10^9$ dm^3 mol^{-1} s^{-1}. The value of $k_B = 2.2 \times 10^{-5}$ dm^3 mol^{-1} s^{-1} and the limits on k_{BH} can be substituted into equation 4 to give $pK_a = 20.5 - 21.5$ for ethyl thioacetate.

The conclusion that the rate-limiting step for deprotonation of ethyl acetate by 3-substituted quinuclidines is the separation of the carbanion-ammonium cation ion pair suggests that the protonation of the enolate of ethyl acetate by protonated 3-substituted quinuclidines is essentially diffusion-limited with $k_{BH} = 5 \times 10^9 - 1 \times 10^{10}$ dm^3 mol^{-1} s^{-1}. The data for catalysis by 3-quinuclidinol, for which $pK_{BH} = 10.0$,[13] give $k_B = 1.35 \times 10^{-6}$ dm^3 mol^{-1} s^{-1}. These values of k_B, pK_{BH} and the limits on k_{BH} can be substituted into equation 4 to give $pK_a = 25.6 - 25.9$ for ethyl acetate.

These are the first experimental determinations of the carbon acid acidities of a simple oxygen ester and a simple thiol ester in aqueous solution. Ethyl acetate is 5 and 7 units less acidic than ethyl thioacetate and acetone, respectively. Using Marcus theory, Guthrie[6c] has calculated $pK_a = 25$ for acetic acid, and Kresge[6b] has determined an experimental value of $pK_a = 22$ for mandelic acid in which the enolate derives substantial additional stabilisation from a ß-phenyl group.

1.5 Lifetimes of Simple Oxygen Ester and Simple Thiol Ester Enolates

The reaction of the enolate of ethyl thioacetate with protonated 3-quinuclidinone is *slower* than the diffusional encounter of the two species, so that this reaction is limited by the *chemical* barrier to protonation of the enolate within the enolate-buffer acid ion pair (k_p, Scheme 3). Therefore, the simple thiol ester enolate has a significant lifetime in the presence of protonated 3-quinuclidinone. How long is this lifetime? The relationship $k_{BH} = K_{as}k_p$ ($K_{as} = k_d/k_{-d}$, Scheme 3), with an association constant for formation of a cation-anion ion pair of $K_{as} \approx 0.3$ mol^{-1} dm^3,[16] can be used to estimate $k_p = 6 \times 10^8 - 7 \times 10^9$ s^{-1} for collapse of BH$^+$•-CH$_2$COSEt by proton transfer to give B•CH$_3$COSEt, so that in the presence of a general acid of $pK_{BH} = 7.5$, the simple thiol ester enolate has an estimated lifetime ($1/k_p$) of $10^{-10} - 10^{-9}$ s. If an enzyme stabilises the thiol ester enolate relative to the thiol ester,[15a] then the lifetime of such carbanions in an enzyme active site may well be even longer than 10^{-9} s.

The diffusion-limited reaction of the enolate of ethyl acetate with protonated 3-substituted quinuclidines shows that for the simple oxygen ester enolate, $k_p > k_{-d}$ (Scheme 3), so that the barrier to protonation of the enolate is even smaller than that for diffusional separation of an ion pair in water, ie $k_p > 2 \times 10^{10}$ s^{-1}. Thus such enolates, if formed as intermediates of enzyme-catalysed reactions, are extremely short-lived. However, the enolate of ethyl acetate does have a significant lifetime in the presence of solvent water. The rate constant for reaction of the enolate of ethyl acetate with water can be calculated from equation 4, using $k_{OH} \approx 1 \times 10^{-3}$ dm^3 mol^{-1} s^{-1} for the hydroxide ion-catalysed deprotonation, to be $k_{HOH} = 1 \times 10^9$ s^{-1}. Thus, in aqueous solution, the simple oxygen ester enolate has a lifetime of ca 10^{-9} s.

2 KINETIC AND THERMODYNAMIC STABILITY OF α-OXYGEN AND α-SULFUR STABILIZED CARBOCATIONS

The effects of α-oxygen and α-sulfur substituents on carbocation stability have been studied extensively in both the gas phase[17a] and in solution,[17b] and by theory at several levels of calculation.[18,20] The gas phase data and theoretical calculations show that the relative stability of α-oxygen and α-sulfur carbocations depends on the nature of the substrate from which these species are generated. Thus the experimental energy change, ΔE, for the hydride transfer

3a X = O
3b X = S

CF$_3$CH$_2$X—⁺CH—C$_6$H$_4$—OMe (para)

$$HOCH_2^+ + CH_3SH \rightarrow CH_3OH + HSCH_2^+ \qquad (5)$$

$$ROCH_2^+ + ClCH_2SR \rightarrow ClCH_2OR + RSCH_2^+ \qquad (6)$$

reaction of equation 5 is *ca* -5 kcal/mol,[21] while the latest theoretical calculations for this reaction give ΔE = 1 - 2 kcal/mol,[18,20c,e] *ie*, the α-sulfur carbocation is of similar or greater thermodynamic stability than the corresponding α-oxygen species. The experimental energy change, ΔE = -2.4 kcal/mol, for the chloride ion transfer reaction of equation 6 (R = Me) favors the α-sulfur carbocation,[22] in excellent agreement with Apeloig's calculation at the MP3/6-31G*//3-21G level of ΔE = -3.0 kcal/mol (R = H).[18] However, in this case, the apparent greater stability of the α-sulfur carbocation has been shown by Apeloig[18] and Schleyer[20e] to result almost entirely from differences in the ground state energies of the neutral chloride adducts which favor ClCH$_2$OH over ClCH$_2$SH by 5 - 6 kcal/mol. By contrast, the solution studies show that α-oxygen stabilised carbocations generally form more rapidly in solvolytic and related reactions than the corresponding α-sulfur species, with rate constant ratios k_O/k_S ranging from 2 - 1500.[17b] We present kinetic and thermodynamic data for the formation of the α-oxygen and α-sulfur stabilised carbocations **3** in trifluoroethanol/water from the neutral precursors **1** and **2** that resolve this apparent contradiction between the greater ease of formation of α-oxygen than of α-sulfur stabilised carbocations in solution, and the results of the gas phase and theoretical studies which show that α-sulfur carbocations are of similar or greater thermodynamic stability than their α-oxygen counterparts.

2.1   Precursors to the Carbocations

The precursors to the α-oxygen and α-sulfur stabilised carbocations were prepared from 4-methoxybenzaldehyde by a divergent route *via* diazido-4-methoxyphenylmethane (Scheme 5). We have shown in previous work that the *gem*-diazide undergoes facile stepwise solvolysis via the diffusionally equilibrated α-azido-4-methoxybenzyl carbocation which may be trapped by nucleophiles.[23] Thus reaction of the *gem*-diazide with trifluoroethoxide ion in trifluoroethanol, or with trifluoroethanethiolate ion in DMSO/water, gives the monoazides **1a** and **1b**, respectively. The solvolyses of **1a** and **1b** are *ca* 100-fold slower than that of the *gem*-diazide, so that the former may be easily isolated before they undergo further reaction. Prolonged reaction of the *gem*-diazide with trifluoroethoxide ion, however, leads to the O,O-acetal **2a**. Similarly, the mixed O,S-acetal **2b** can be prepared by reaction of **1b** with trifluoroethoxide ion.

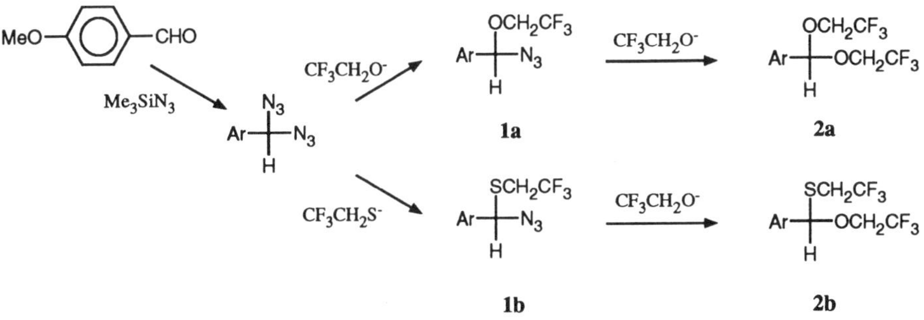

Scheme 5: Syntheses of precursors to α-oxygen and α-sulfur stabilised carbocations.

2.2. Lifetimes of the Carbocations

The monoazides **1a** and **1b** undergo solvolysis in 50:50 (v:v) trifluoroethanol/water at very similar rates, with $k_{solv} = 3.6 \times 10^{-4}$ and 4.2×10^{-4} s^{-1}, respectively (Table 1).

Figure 3 (top) shows that k_{obsd} for solvolysis of **1b** decreases 30-fold as [N$_3^-$] is increased from 0 to 0.4 mmol dm^{-3}. This strong common ion inhibition shows that **1b** reacts by a stepwise D$_N$ + A$_N$ mechanism[24] through the diffusionally-equilibrated α-sulfur stabilised carbocation **3b** which can be trapped by azide ion and by solvent (Scheme 6). The slope of the linear plot of the data according to equation 7, derived for Scheme 6, gives the rate constant ratio k_{az}/k_s = 71,000 mol^{-1} dm^3 for partitioning of the α-sulfur stabilised carbocation **3b** between reaction with azide ion and with solvent (Figure 3, bottom).

Similarly, k_{obsd} for the solvolysis of **1a** decreases 15-fold when [N$_3^-$] is increased from 0 to 0.25 mol dm^{-3}, and the fit of the data to equation 7 gives k_{az}/k_s = 60 mol^{-1} dm^3 for partitioning of the α-oxygen stabilised carbocation **3a** between reaction with azide ion and with solvent.

Table 1: Rate and equilibrium constants for formation and reaction of α-oxygen and α-sulfur stabilised 4-methoxybenzyl carbocations in 50:50 (v:v) trifluoroethanol/water.[a]

$$\text{MeO-}\underset{H}{\overset{XCH_2CF_3}{\text{C}_6H_4\text{-C-Y}}} \quad (+H^+) \underset{k_r}{\overset{k_f}{\rightleftharpoons}} \quad \text{MeO-}\underset{H}{\overset{XCH_2CF_3}{\text{C}_6H_4\text{-C}^+}} + Y^- (YH)$$

	X = O Y = N$_3$	X = S Y = N$_3$
k_{solv} (s^{-1}) [b]	3.6 x 10^{-4}	4.2 x 10^{-4}
k_{az} (dm^3 mol^{-1} s^{-1}) [c]	5 x 10^9	5 x 10^9
K_{az} (mol dm^{-3}) [d]	7.2 x 10^{-14}	8.4 x 10^{-14}
	X = O Y = OCH$_2$CF$_3$	X = S Y = OCH$_2$CF$_3$
k_H (dm^3 mol^{-1} s^{-1}) [e]	0.23	5.4 x 10^{-3}
k_{TFE} (dm^3 mol^{-1} s^{-1}) [f]	2.2 x 10^5	2.4 x 10^2
K_{TFE} [g]	1.1 x 10^{-6}	2.3 x 10^{-5}

[a] At 25 °C and $I = 0.50$ (NaClO$_4$). [b] First-order rate constant for solvolysis of **1**. [c] Diffusional rate constant.[25] [d] k_{solv}/k_{az}. [e] Second-order rate constant for HClO$_4$-catalysed C-O bond cleavage reaction of **2**. [f] Determined from k_{HOH}/k_{TFE} and k_s (s^{-1}) for **3**. [g] k_H/k_{TFE}.

$$k_{solv}/k_{obsd} = 1 + (k_{az}/k_s)[N_3^-] \quad (7)$$

The reaction of azide ion with a large range of unstable benzylic carbocations is diffusion-limited, with $k_{az} = 5 \times 10^9$ dm^3 mol^{-1} s^{-1}.[25] Therefore, the partitioning ratios k_{az}/k_s (mol^{-1} dm^3) can be combined with this value for k_{az} to give rate constants for reaction of the α-oxygen and α-sulfur stabilised carbocations **3a** and **3b** with a solvent of 50:50 (v:v) trifluoroethanol/water of $k_s = 8 \times 10^7$ and 7×10^4 s^{-1}, respectively. HPLC analysis of the products of the reactions of **1a** and **1b** in 50:50 (v:v) trifluoroethanol/water gave $k_{HOH}/k_{TFE} = 13.6$ and 10.1 for partitioning of **3a** and **3b**, respectively, between capture by water (which ultimately gives 4-methoxybenzaldehyde) and trifluoroethanol. These ratios were combined with the values of k_s (s^{-1}) to obtain second-order rate constants k_{TFE} (dm^3 mol^{-1} s^{-1}, Table 1) for reaction of **3a** and **3b** with trifluoroethanol.

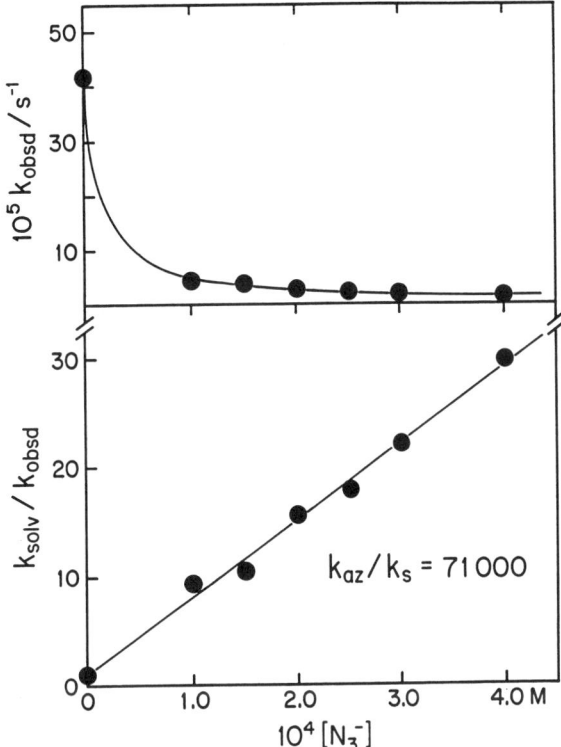

Figure 3: Top: Dependence of k_{obsd} on the concentration of added azide ion for the solvolysis of **1b** in 50:50 (v:v) trifluoroethanol/water at 25 °C and $I = 0.50$ (NaClO$_4$). Bottom: Linear plot of the data according to equation 7 of the text.

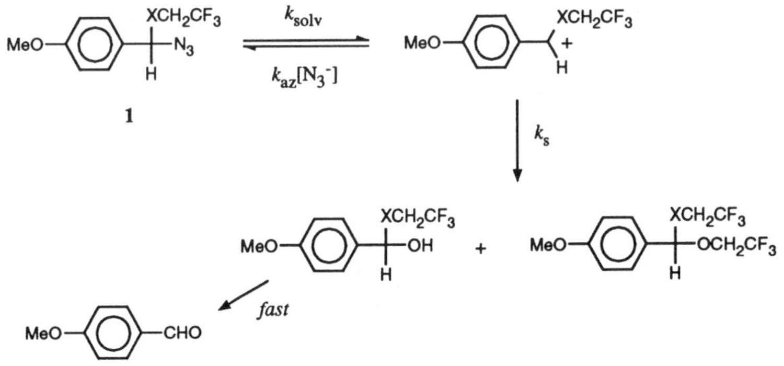

Scheme 6: Kinetic scheme for stepwise solvolyses of monoazides through α-oxygen and α-sulfur stabilised carbocation intermediates.

2.3 Equilibrium Constants

Table 1 gives rate and equilibrium constants for formation of the carbocations **3** from the azide ion adducts **1** and from the trifluoroethoxide ion adducts **2**. The equilibrium constants were calculated as the ratio of rate constants for formation and capture of the carbocations.

$$\underset{\text{MeO}}{\text{CF}_3\text{CH}_2\text{S}\diagdown\text{C}\diagup\text{SCH}_2\text{CH}_2\text{OH}}\text{ (4)} \qquad \underset{\text{MeO}}{\text{EtO}\diagdown\text{C}\diagup\text{SCH}_2\text{CO}_2\text{Me}}\text{ (5)}$$

The rate constants for formation of **3** from **2** are k_H (dm^3 mol^{-1} s^{-1}) for the perchloric acid-catalysed hydrolysis of the O,O and O,S acetals. The O,S acetal **2b** undergoes only C-O bond cleavage to give **3b**, as evidenced by the identical yields (76%) of **4** obtained from the reactions of **1b** and **2b** in the presence of 1% HO(CH$_2$)$_2$SH. Jensen and Jencks have reported a similar cleavage pattern for the acid-catalysed hydrolysis reaction of the closely related **5**.[26]

The equilibrium constants K_{az} (mol dm^{-3}) and K_{TFE} for the formation of **3a** and **3b** show that the α-oxygen and α-sulfur carbocations have about the same thermodynamic stability relative to the azide ion adducts, but that the α-sulfur carbocation is *ca* 1.8 kcal/mol more stable than its α-oxygen counterpart relative to the trifluoroethoxide ion adducts. The latter result is in agreement with the results of gas phase and theoretical studies which show that, relative to the chloride ion adducts, an α-SMe group provides slightly more stabilisation of the methyl carbocation than does an α-OMe group (equation 6).[18,22] The larger difference in K_{TFE} for formation of **3a** and **3b** from the trifluoroethoxide ion adducts **2** (20-fold) than in K_{az} (mol dm^{-3}) for formation of these species from the azide ion adducts **1** (1.2-fold) reflects a relatively large difference between strong ground-state stabilisation by oxygen-oxygen geminal interactions in **2a** and weaker stabilising oxygen-sulfur interactions in **2b**, and a smaller difference in such geminal interactions in **1a** and **1b**. The importance of these ground-state geminal interactions has been noted in earlier work.[18,20e,27]

6a X = O
6b X = S

7a X = O
7b X = S

2.4 Rate-Equilibrium Correlations

Figure 4 shows reaction coordinates for formation of the α-oxygen and α-sulfur stabilised carbocations from the trifluoroethoxide ion adducts, constructed using the data in Table 1. The rate constant for acid-catalysed formation of the *less stable* α-oxygen carbocation 3a from 2a is 40-fold larger than that for formation of 3b from 2b. The observation that the thermodynamically less stable α-oxygen carbocation is formed more rapidly shows that it is dangerous to infer the relative stabilities of α-oxygen and α-sulfur stabilised carbocations from their relative rates of formation in solvolyses and related reactions.[17b] Even more striking is the 900-fold greater reactivity of the α-oxygen than of the α-sulfur carbocation towards trifluoroethanol. A similar, although less extreme, result was noted by McClelland,[19b] who found that the rate constant for acid-catalysed cleavage of the hemiorthoester 6a to give 7a is 5-fold larger than that for cleavage of the sulfur analog 6b to give 7b. The two carbocations have very similar pK_R values, yet the reaction of 7a with water to regenerate 6a is 26-fold faster than that of 7b to give 6b.[19b] The observations may be summarised by the statement that there is a smaller intrinsic barrier to the formation and reaction of α-oxygen than of α-sulfur stabilised carbocations.

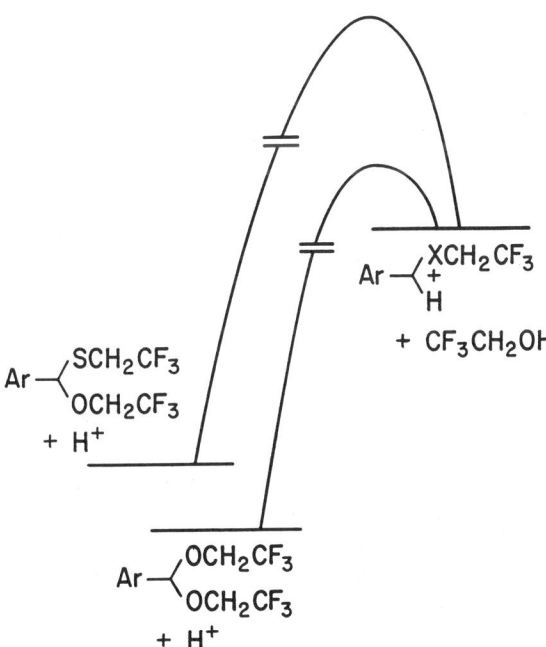

Figure 4: Reaction coordinates for acid-catalysed formation of 3a and 3b from the respective trifluoroethoxide ion adducts 2a and 2b, constructed using data from Table 1.

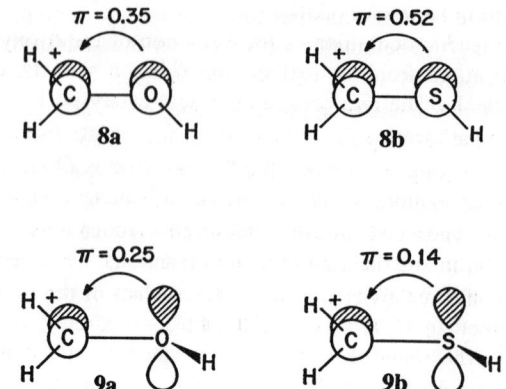

The intrinsic barrier for capture of resonance-stabilised carbocations by nucleophiles results largely from incomplete compensation of the energetically unfavorable loss of resonance interactions in the transition state by the energetically favorable bond formation to the nucleophile. The larger intrinsic barrier for capture of the α-sulfur than of the α-oxygen stabilised carbocation by solvent is therefore consistent with a more rapid fall-off in stabilising resonance electron donation from the α-sulfur in **3b** than from the α-oxygen in **3a** as the hybridisation at the benzylic carbon changes from sp^2 at the carbocation to more sp^3-like at the reaction transition state. Imbalance[28] or nonperfect synchronisation[29] between bond formation and the loss of resonance interactions leads to larger intrinsic barriers for reactions in which a reactant stabilising factor is lost relatively early along the reaction coordinate.[29,30] Thus the larger intrinsic barrier for capture of the α-sulfur carbocation suggests that overlap of the lone pair(s) of electrons on an α-oxygen atom with the p orbital on the benzylic carbon is effective at greater extents of bond formation and pyramidalisation of this centre than is overlap of the lone pair(s) of electrons on an α-sulfur atom.

Our notion that the lone pairs on oxygen are more effective than those on sulfur at providing resonance stabilisation by overlap with the p orbital on a partially pyramidalised adjacent carbon is given some credence by the calculations of Apeloig and Schleyer which show that an HO substituent provides substantial resonance stabilisation of the methyl carbocation in both the planar **8a** and perpendicular **9a** conformations.[18,20e] At the 6-31G* level, the rotational barriers for $HOCH_2^+$ and $HSCH_2^+$, which are a rough measure of the difference in the resonance stabilisation of the planar and the perpendicular conformers, are *ca* 26 and 42 kcal/mol, respectively.[18] At the same level of calculation, ΔE for equation 5 is 9 kcal/mol so that, in the perpendicular conformation, HO provides significantly more stabilisation (*ca* 25 kcal/mol) of the methyl carbocation than does HS.[18] Apeloig suggested this to be a consequence of the better π-donating ability of the 'sp^2 lone pair' on oxygen than of that on sulfur.[18] The smaller difference in the stabilisation of the methyl carbocation provided by oxygen in **8a** and **9a** than by sulfur in **8b** and **9b** is reflected in the larger increase in the C-X bond length and decrease in π electron donation from the heteroatom on moving from **8b** to **9b** than in the oxygen analogues.[18] We propose that pyramidalisation of the benzylic carbon in the transition state for capture of **3a** and **3b** by solvent is analogous to a 90° C-X bond rotation in **8a** and **8b** to give **9a** and **9b**, in that both events destroy the periplanarity of the overlapping orbitals on carbon and the heteroatom. In collaboration with Yitzhak Apeloig, the similarities between the two processes are being examined by *ab initio* calculations.

REFERENCES

1. (a) V M Powers, C W Koo, G L Kenyon, J A Gerlt, and J W Kozarich, *Biochemistry*, 1991, **30**, 9255. (b) J Stubbe and R H Abeles, *Biochemistry*, 1980, **19**, 5505. (c) J D Clark, S J O'Keefe, and J R Knowles, *Biochemistry*, 1988, **27**, 5961.
2. We define 'simple' oxygen and thiol esters as those in which the carbonyl group is adjacent to a carbon atom bearing no other electron-withdrawing groups.
3. A F Hegarty and P O'Neill in 'The Chemistry of Enols', Ed Z Rappoport, John Wiley and Sons, Chichester, 1990, Chapter 10.
4. J P Richard, Chapter 11 of ref 3. H F Gilbert, *Biochemistry*, 1981, **20**, 5643. J A Gerlt, J W Kozarich, G L Kenyon, and P G Gassman, *J Am Chem Soc*, 1991, **113**, 9667. J A Gerlt and P G Gassman, *J Am Chem Soc.*, 1992, **114**, 5928.
5. A Thibblin and W P Jencks, *J Am Chem Soc*, 1979, **101**, 4963.
6. (a) P O'Neill and A F Hegarty, *J Chem Soc, Chem Commun*, 1987, 744. (b) Y Chiang, A J Kresge, P Pruszynski, N P Schepp, and J Wirz, *Angew Chem, Int Ed Engl*, 1990, **29**, 792. (c) J P Guthrie, Can J Chem,1993, **71**, 2123.
7. K F Bonhoeffer, K H Geib, and O Reitz, *J Chem Phys*, 1939, **7**, 664.
8. G E Lienhard and T-C Wang, *J Am Chem Soc*, 1968, **90**, 3781.
9. T L Amyes and J P Richard, *J Am Chem Soc*, 1992, **114**, 10297.
10. The capture of the enolate by deuteriated buffer acid (BD$^+$) is fast and irreversible (Scheme 2) so that the rate constants for deuterium exchange are equal to those for formation of the enolate.
11. A J Kresge, *Chem Soc Rev*, 1973, **2**, 475.
12. Y Chiang, A J Kresge, Y S Tang, and J Wirz, *J Am Chem Soc*, 1984, **106**, 460. E Tapuhi and W P Jencks, *J Am Chem Soc*, 1982, **104**, 5758.
13. M J Gresser and W P Jencks, *J Am Chem Soc*, 1977, **99**, 6963.

14   This analysis neglects the secondary solvent isotope effect on k_B arising from its determination here in D_2O rather than in H_2O, but a small solvent isotope effect of $k_B(H_2O)/k_B(D_2O) = 1.1$ has been determined for similar systems.[15]
15   (a) J P Richard, *J Am Chem Soc*, 1984, **106**, 4926. (b) M W Washabaugh and W P Jencks, *J Am Chem Soc*, 1989, **111**, 674.
16   J P Richard and W P Jencks, *J Am Chem Soc*, 1984, **106**, 1373.
17   (a) Results of prominent gas phase studies are summarised in ref 18. (b) Results of prominent solution phase studies are summarised in ref 19a and in footnote 21 of ref 19b.
18   Y Apeloig and M Karni, *J Chem Soc, Perkin Trans 2*, 1988, 625.
19   (a) G Modena, G Scorrano, and P Venturello, *J Chem Soc, Perkin Trans 2*, 1979, 1. (b) L J Santry and R A McClelland, *J Am Chem Soc*, 1983, **105**, 3167.
20   (a) F Bernardi, I G Csizmadia, H B Schlegel, and S. Wolfe, *Can J Chem*, 1975, **53**, 1144. (b) F Bernardi, I G Csizmadia, and N D Epiotis, *Tetrahedron*, 1975, **31**, 3085. (c) F Bernardi, A Bottoni, and A Venturini, *J Am Chem Soc*, 1986, **108**, 5395. (d) T Clark and P von R Schleyer, *Tetrahedron Lett*, 1979, 4641. (e) P von R Schleyer, *Pure Appl Chem*, 1987, **59**, 1647.
21   R W Taft, R H Martin, and F W Lampe, *J Am Chem Soc*, 1965, **87**, 2490.
22   J K Pau, M B Ruggera, J K Kim, and M C Caserio, *J Am Chem Soc*, 1978, **100**, 4242.
23   T L Amyes and J P Richard, *J Am Chem Soc*, 1991, **113**, 1867.
24   IUPAC Commission on Physical Organic Chemistry *Pure Appl Chem*, 1989, **61**, 23. R D Guthrie and W P Jencks, *Acc Chem Res*, 1989, **22**, 343.
25   J P Richard, M E Rothenberg, and W P Jencks, *J Am Chem Soc*, 1984, **106**, 1361. T L Amyes and W P Jencks, *J Am Chem Soc*, 1989, **111**, 7888. J P Richard, *J Am Chem Soc*, 1989, **111**, 1455. R A McClelland, V M Kanagasabapathy, N S Banait, and S Steenken, *J Am Chem Soc*, 1991, **113**, 1009.
26   J L Jensen and W P Jencks, *J Am Chem Soc*, 1979, **101**, 1476.
27   J P Richard, T L Amyes, and D J Rice, *J Am Chem Soc*, 1993, **115**, 2523. Y Apeloig, R Biton, and A Abu-Freih, *J Am Chem Soc*, 1993, **115**, 2522.
28   D A Jencks and W P Jencks, *J Am Chem Soc*, 1977, **99**, 7948. A J Kresge, *Can J Chem*, 1974, **52**, 1897.
29   C F Bernasconi, *Adv Phys Org Chem*, 1992, **27**, 119.
30   T L Amyes, I W Stevens, and J P Richard, *J Org Chem*, 1993, **58**, 6057.

ACKNOWLEDGEMENTS

We thank the National Institutes of Health (Grant GM 39754 to JPR) for support of this work.

STRUCTURE REACTIVITY EFFECTS IN THE AQUEOUS CHEMISTRY OF ALKANE DIAZOATES

Jian Ho, Jari I Finneman, and James C Fishbein*

Wake Forest University, Winston-Salem, North Carolina, USA, 27109.

1   INTRODUCTION

A large number of compounds containing the *N*-alkyl-*N*-nitroso functionality have important biological activities. In most cases such compounds are highly mutagenic or carcinogenic, but in certain cases they are employed as cancer chemotherapeutic agents.[1] As indicated in Scheme 1, the powerful biological activity of *N*-alkyl-*N*-nitroso compounds is at least in part due to their decomposition to alkane diazoates that subsequently form carbocations or diazonium ions that can alkylate DNA.

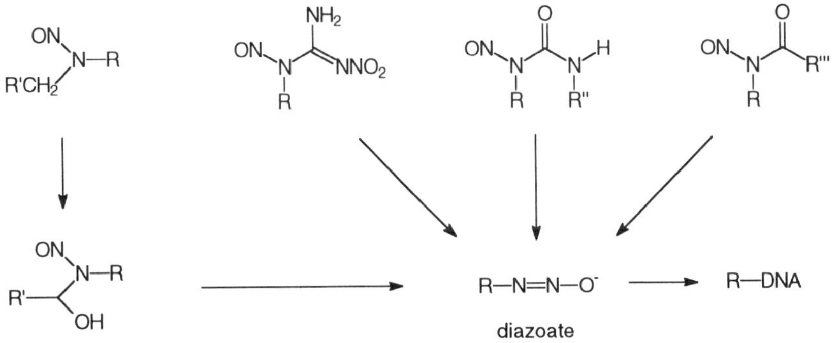

Scheme 1: Decomposition routes for nitroso compounds.

Alkane diazoates were synthesized nearly 100 years ago.[2] The *syn* and *anti* forms can be synthesized by separate routes and are not interconvertible, as indicated below:[3]

syn or Z   anti or E

The *syn* forms are reported to be qualitatively less stable than the corresponding *anti* isomers. Thus, the *anti*-benzyl compound is stable in cold water, whereas the *syn* form decomposes instantly.[4] A similar observation was made for the methyl isomers.[5] A number of experiments have been carried out to analyze the effects of structure on the products of decomposition. From these it was concluded that, in a number of cases, product determining steps involve the intermediacy of both diazonium ions and ion pairs in aqueous and non-aqueous solvents.[6]

There is a paucity of quantitative information concerning the chemistry of alkane diazoates and our own interest was prompted by the fundamental importance of these reactive intermediates in chemical carcinogenesis and cancer chemotherapy. From a biochemical perspective it is essential to know: the lifetime of these species in aqueous media; as precisely as possible the mechanisms of decomposition; the alkylating selectivities of the relevant intermediates; and lastly, how these parameters vary with diazoate structure.

The first quantitative analysis of the decay of one of the two simplest alkane diazoates, *anti* methyl diazoate, was recently reported.[7] It was concluded that the rate limiting step is the unimolecular decomposition of the diazoic acid to yield the diazonium and hydroxide ions as indicated in equation 1.

$$\text{Me-N=N-O}^- \underset{-H^+}{\overset{+H^+}{\rightleftharpoons}} \text{Me-N=N-OH} \xrightarrow{\text{rls}}_{--OH} \text{Me-N}\equiv\text{N}^+ \xrightarrow{H_2O} \text{MeOH} + H^+ + N_2 \quad (1)$$

A study on the kinetics of decay of a pair of secondary *anti*-alkane diazoates was interpreted as indicating that the rate limiting step for the 2-butyl compound similarly involved rate-limiting formation of the diazonium ion from the diazoic acid.[8] Direct formation of the diazonium ion in a concerted reaction, as in equation 1, was ruled out on the basis of the smaller rate constant for decomposition of the *anti*-1-phenylethyl system compared to the *anti*-2-butyl system.

It is reasonable to assume on the basis of the results with the secondary systems described above that *anti*-1-butyldiazoate decomposes with rate-limiting decomposition of the diazoic acid to give the diazonium ion. There is evidence for intermediacy of the primary diazonium ion from the work of Maskill and colleagues.[9] It was shown that in the nitrous acid catalyzed deamination of *N*-octylamine in acetic acid, which presumably occurs *via* a diazoic acid intermediate, the ratios of ester and alkene products is the same as those derived from *N*-octyltriazene or *N*-nitrosamide decomposition in acetic acid. Direct S_N2 substitution on the diazonium ion, as opposed to formation of the primary cation, was recently indicated by the 98% inversion in the alcohol product of a chiral 1-deuterio-1-butylamine decomposed in nitrous acid.[10]

A quantitative analysis of the reaction chemistry of primary alkane diazoates is of significant practical importance. The cancer chemotherapeutic agent bis-chloroethyl-nitrosourea and its fluoro analogue act at least in part through the intermediacy of the respective chloroethyl- and fluoroethyl-alkane diazoates.[11] In the case of primary diazoates with electron withdrawing groups attached, for example the chloroethyl- and fluoroethyl-diazoates, the nature of the rate limiting step is uncertain. For the more stable arene diazoates the rate limiting step for compounds with strong electron withdrawing substituents involves rate-limiting isomerization to the more reactive *syn* form.[12] This mechanism cannot be dismissed in the case of primary alkane diazoates with electron withdrawing groups attached. The two mechanisms - rate limiting diazonium ion formation and rate limiting isomerization, should have different dependencies on substituent electron-withdrawing ability. The rate constant should decrease with increasing strength of the electron withdrawing group in the case of rate-limiting diazonium ion formation, whereas the opposite is predicted for rate limiting isomerization, on the basis of what is observed in the case of arenediazoates. With these uncertainties in mind, we have set out to define clearly the mechanistic chemistry of the primary *anti* alkane diazoates.

2 PRIMARY ANTI-ALKANE DIAZOATES

A number of *anti*-primary alkane diazoates have been synthesized by the traditional reaction of the hydrazines with nitrite esters in the presence of base. The decomposition of several of these in aqueous media, followed by product analysis using a combination of gas chromatography and NMR techniques, gives the expected product types in yields that are quantitative or nearly so. This is illustrated in equations 2 - 4.

$$CF_3\text{-}CH_2\text{-}N{=}N\text{-}O^- \xrightarrow[H_2O,\ RT]{0.1\ M\ HClO_4} CF_3CH_2OH \qquad (4)$$

89%

The pH - rate profiles for the buffer-independent decomposition of a number of primary alkane diazoates, as well as the methyl and 2-butyl compound, are illustrated in Figure 1.

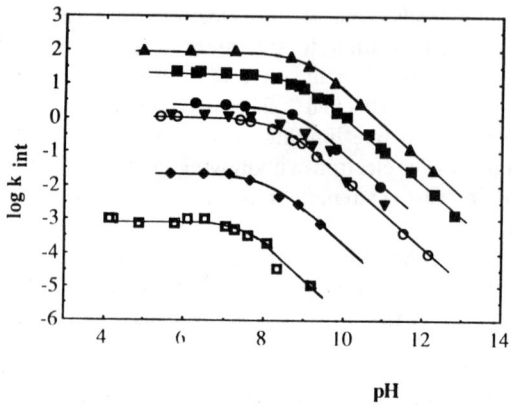

Figure 1: Plot of log k_{int}, the buffer independent rate constant for decomposition of *anti*-alkane diazoates against pH for reactions at 25 °C, ionic strength 1 M (NaClO$_4$), 4% 2-PrOH. (△), 2-butyl; (■), 1-butyl; (●), methyl; (▽), benzyl; (○), 2-methoxyethyl; (♦), 2-cyanoethyl; (□), 2,2,2-trifluoroethyl.

As indicated previously in the case of the methyl and 2-butyl compounds the results are consistent with the general mechanism of equation 5 in which the reaction involves the decomposition of a protonated form of the diazoate, DH.

$$R\text{-}N{=}N\text{-}O^- + H^+ \underset{}{\overset{K_a}{\rightleftharpoons}} D\text{-}H \xrightarrow{k_0} \text{products} \qquad (5)$$

Deuterium incorporation experiments, in which the diazoates are decomposed in D$_2$O followed by analysis of deuterium content in the products, rule out certain mechanisms that have been previously suggested. The result indicated in equation 6 rules out the possibility that the reaction occurs by elimination to give the diazoalkane that subsequently decomposes to products. This mechanism would require that all product molecules incorporate at least one deuterium atom from solvent. From the result in equation 7, the fact that most of the product of iodide ion substitution contains two deuterium atoms rules out a direct S_N2 substitution on the diazoic acid or direct decomposition of the diazoic acid to the carbocation in the case of the reaction of *anti*-trifluoroethyldiazoate. For both of these mechanisms no exchange would be expected. The observation of extensive exchange is consistent with what is predicted from the known chemistry of the trifluoroethyldiazonium ion/trifluoroethyldiazoethane system.[13]

$$\text{CH}_3\text{CH}_2\text{CH}_2\text{CH}_2-N=N-O^- \xrightarrow[\text{(D}_2\text{PO}_4^-, \text{D}_2\text{O)}]{pH_{obs} = 7} \text{CH}_3\text{CH}_2\text{CH}_2\text{CHD-OH} \quad > 95\% \; H_2 \quad (6)$$

$$CF_3\text{CH}_2-N=N-O^- \xrightarrow[\text{(D}_2\text{PO}_4^-, \text{1 M NaI)}]{pH_{obs} = 7} CF_3CD_2I \quad > 93\% \; D_2 \quad (7)$$

A common mechanism for all of the *anti*-primary alkane diazoates studied is indicated by the fit to a common line of the plot of the log of the pH independent limiting rate constant against σ^* as indicated in Figure 2. Also included are the values for the 2-butyl and methyl compounds. The slope of the line of (ρ^* =) -4.38 is similar or slightly more negative than the values of ρ^* observed for substituents in the remaining group in the acid catalyzed hydrolysis of acetals, and indicates that there is substantial positive charge accumulation in going from the ground state diazoic acid to the transition state. There is no evidence of an upward break in the plot with increasing electron withdrawal by the substituent, as would be expected for the onset of a new mechanism, such as rate limiting isomerization to the more reactive *syn* form.

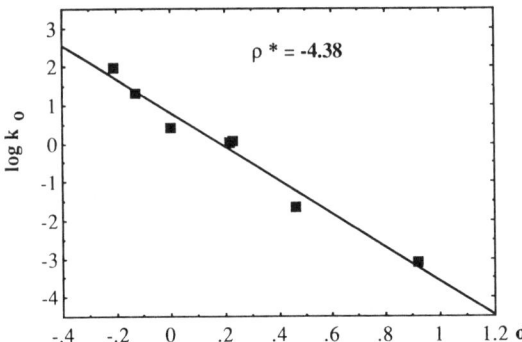

Figure 2: Plot of log k_o, the pH independent rate constant for decomposition of *anti*-alkane diazoates, against σ^*. Reaction rate constants at 25 °C, ionic strength 1 M (NaClO$_4$), 4% 2-PrOH.

On the basis of the deuterium isotope incorporation studies and the structure reactivity correlation in Figure 2, it is concluded that primary *anti* alkane diazoates decompose with rate limiting N-O bond fission of the diazoic acid to give the diazonium ion, as written for the methyl compound in equation 1. The small, slightly negative, values of $\Delta S^{\#} = -16 \pm 3$ and -7 ± 10 J K^{-1} mol^{-1} for the 1-butyl and 2,2,2-trifluoroethyldiazoates and the values for the pH independent reactions of k_{H_2O}/k_{D_2O} of 1.3 ± 0.1 and 1.8 ± 0.1 for the same two compounds, respectively, are consistent with some solvent electrostriction about the developing lyoxide ion.[7] The large decrease in the pH independent rate constant by factors of 800 and > 500 for the same two compounds, respectively, when the reaction is carried out in ethanol rather than water, are also consistent with a mechanism which involves ion pair formation from a neutral species.

3 *SYN* VERSUS *ANTI*: REACTIVITY AND CHEMISTRY

We have recently synthesized the *syn*-trifluoroethyldiazoate. The pH - rate profile for the buffer independent decomposition of the *syn* compound is qualitatively similar to that observed for the *anti* compound in Figure 1. The limiting rate constant for the pH independent reaction is 2800 × faster than that for the *anti*-trifluoroethyl compound.

In addition to this, the decomposition of the *syn*, but not the *anti* diazoic acid, is stimulated by increasing concentrations of carboxylic acid buffers as indicated in Figure 3. That the difference is not a function of reactivity is

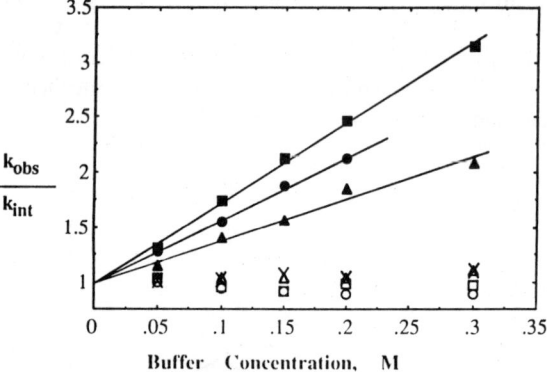

Figure 3: Plot of k_{obs}/k_{int} against buffer concentration for *syn*-2,2,2-trifluoroethyldiazoate (solid symbols), *anti*-2,2,2-trifluoroethyldiazoate (open symbols), and 1-butyldiazoate (X's). The constant k_{obs} is the observed first order rate constant at a given buffer concentration while the constant k_{int} is the value of the rate constant from extrapolation [buffer] = 0 of the plot of k_{obs} against [buffer]. The squares and X's are for experiments at 20% acetate anion buffer, the circles are for 50% acetate anion buffers and the triangles are for 20% cyanoacetate anion experiments.

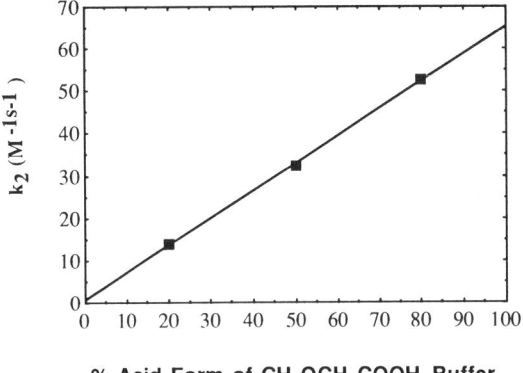

Figure 4: Plot of k_2, the second order rate constant for catalysis for a given buffer ratio, against '% Acid Form of Buffer' for methoxyacetic acid buffer at 25 °C, ionic strength = 1 M (NaClO$_4$).

indicated by the single experiment in Figure 3 with the *anti*-1-butyl compound, for which the buffer independent rate constant for decomposition is greater than that for the *syn*-trifluoroethyl compound. General acid catalysis is indicated by the data in Figure 4, which plots the observed second order rate constants for catalysis by methoxyacetic acid as a function of the 'Percent Acid Form' of the buffer. The intercept at 100% gives the catalytic constant, k_{HA}, for methoxyacetic acid. The catalytic constants for four acidic catalysts (acetic, formic, methoxyacetic and cyanoacetic) give a good straight line with slope α = 0.4 on Bronsted plot of log k_{HA}/P against pKa + p/q. We take the intermediate value of Bronsted α to indicate that a proton is 'in flight' in the rate limiting step for the general acid catalyzed reaction.

Two possible mechanisms for the observed general acid catalysis are indicated in equations 8 and 9 respectively. In theory, isotope incorporation experiments could distinguish between these possibilities except that in the present case, as indicated by the results above, the trifluoroethyldiazonium ion is in rapid protonic equilibrium with the diazoalkane, even in the presence of high concentrations of iodide ion. The mechanism of equation 8 involves rate limiting transfer of an exchangeable hydron, whereas the mechanism of equation 9 does

$$CF_3\text{-}N=N\text{-}OH \cdots H\text{-}A \longrightarrow CF_3CH_2N_2^+ \qquad (8)$$

$$CF_3\text{-}N=N\text{-}OH \xrightleftharpoons{H\text{-}A} \underset{CF_3}{\overset{H \quad H}{\text{C}}}\text{-}N=N\text{-}OH_2^+ \xrightarrow{A^-}_{k_2} CF_3CH=N=N \qquad (9)$$

not. The solvent deuterium isotope effect on the formic acid catalyzed decomposition reaction was determined to be 3.1 ± 0.2 (four trials). This value is clearly consistent with a primary isotope effect contribution expected for the mechanism of equation 8.

The mechanism of equation 9 is ruled out because the solvent isotope effect expected for such a mechanism can be shown to be too small to account for the experimentally observed effect. The application of fractionation factor theory to the mechanism of equation 9, which mechanism is only subject to secondary solvent isotope effects, can set an upper limit on the solvent isotope effect expected for the mechanism for equation 9. The expected solvent deuterium isotope effect can be expressed in terms of the relevant fraction factors as indicated in equation 10. We take ϕ_{AcOH} and ϕ_{AcO^-} as those determined for

$$\frac{k_{HA}}{k_{DA}} = \frac{1}{\phi_{AcO^-}} * \phi_{AcOH} * \left(\frac{1}{l}\right)^2 * \phi_{ROH} * \frac{k_2^H}{k_2^D} \qquad (10)$$

acetic acid and acetate, respectively, and 1 as that for a hydron attached to a hydronium ion, while $\phi_{ROH} = 1.$[14] These numerical values reduce the isotope effect expression to that indicated in equation 11, and an upper limit for the isotope effect can be determined if an upper limit for the isotope effect on the k_2 step can be determined. The maximum effect that could be expected on k_2

$$\frac{k_{HA}}{k_{DA}} = 2.26 * \frac{k_2^H}{k_2^D} \qquad (11)$$

would be for a transition state that is highly asymmetric with nearly complete proton transfer to the carboxylate anion to give a carbanion, with essentially no N-O bond cleavage. The maximum possible effect can then be written as the combination of the transfer fractionation factors for the acetate anion and the carbanion as in equation 12. The precise value for the ϕ_{C^-} is unknown, but a

$$k_2^H/k_2^D = \phi_{AcO^-}/\phi_{C^-} \qquad (12)$$

reasonable lower limit can be taken as $\phi_{C^-} > 0.8$. This lower limit is based on: (i) a survey of values that can be calculated from isotope effects on the ionization equilibrium of certain carbon acids; and (ii) the isotope effects on the rate constants for proton transfer involving carbon acids, for which the transition state for hydron transfer is very large or involves diffusional separation.[15] In all cases the lower limit for ϕ_{C^-} is between 0.8 and 1. Given the value $\phi_{C^-} = 0.8$, the upper limit for k_{HA}/k_{DA} can be computed as equal to 2.51. This is well below, outside experimental error, the observed value of $k_{HA}/k_{DA} = 3.1 \pm 0.2$ and therefore rules out the mechanism of equation 9.

4 SUMMARY

These experiments establish that the mechanism of decomposition of primary anti-alkane diazoates, with even quite strong electron withdrawing groups attached, involves rate limiting formation of the diazonium ion from the diazoic acid. These studies provide the first quantitative analysis of the difference in reactivity of *syn-* and *anti-*alkane diazoates, with the *syn-*2,2,2-trifluoroethyldiazoate being 2800 times as reactive as its *anti* isomer. Finally these studies establish a unique mode of decomposition for the *syn-*2,2,2-trifluoroethyldiazoate that involves the general acid catalyzed formation of the diazonium ion from the diazoic acid. The generality of this new aspect of diazoate chemistry is being studied presently.

REFERENCES

1. (a) P D Lawley, in *Chemical Carcinogens*; ed C D Searle; ACS Monograph 182, American Chemical Society, Washington DC, 1984.
  (b) W Lijinsky, *Chemistry and Biology of N-nitroso Compounds*, Cambridge University Press, Cambridge, 1992. (c) G M Blackburn, and B Kellard, *Chem Ind (Lond)*, 1986, 607.
2. A Hantzsch and M Lehmann, *Chem Ber*, 1902, **35**, 897.
3. Ref 2 and J Thiele, *Justus Leibigs Ann Chem*, 1910, **376**, 239.
4. J Thiele, *Chem Ber*, 1908, **41**, 2806.
5. J W Lown, S M S Cauhan, R R Koganty, and A-M Sapse, *J Am Chem Soc*, 1984, **106**, 6401.
6. R A Moss, *Acc Chem Res*, 1974, **7**, 421 and references within.
7. (a) J Hovinen and J C Fishbein, *J Am Chem Soc*, 1992, **114**, 366.
  (b) J Hovinen, J I Finneman, S N Satapathy, J Ho, and J C Fishbein, *ibid*, 1992, **114**, 10321.
8. J I Finneman, J Ho, and J C Fishbein, *J Am Chem Soc*, 1993, **115**, 3016.
9. H Maskill, R M Southam, and M C Whiting, *J Chem Soc, Chem Commun*, 1965, 496.
10. D Brosch and W Kirmse, *J Org Chem*, 1991, **56**, 907.
11. J W Lown, R R Koganty, U G Bhat, A-M Sapse, and E B Allen, *Drugs Exptl Clin Res*, 1986, **XII**, 463.
12. (a) E S Lewis and M P Hanson, *J Am Chem Soc*, 1967, **92**, 6268. (b) P D Goodman, T J Kemp, and P Pinot de Moira, *J Chem Soc Perkin Trans II*, 1981, 1221. (c) J Jahelka, O Machackova, V Sterba, and K Valter, *Coll Czech Chem Comm*, 1973, **38**, 706.

13 H Dahn and J H Lenoir, *Helv Chim Acta*, 1979, **62**, 2218. D T Loehr, D Armistead, J Roy, and H C Dorn, *J Fluorine Chem*, 1988, **39**, 283.
14 A J Kresge, R A More O'Ferrall, and M F Powell in *Isotopes in Org Chem*, eds E Buncel and C C Lee, 1987, **7**.
15 The transfer fractionation factor for cyanide ion has been calculated to be 0.84.[14] Values of between 0.80 and 0.90 can be calculated from the isotope effects on the equilibrium ionization of t-butylmalononitrile,[16] 2-acetylcyclohexane,[17] and methyl acetylacetone.[17] Finally, $0.8 < \phi_{C^-} < 1$ is calculated from isotope effects on rate constants for proton transfer involving carbon acids which have late transition states for proton transfer, or, for which the rate limiting steps involve diffusional separation of the carbanion-conjugate acid pair. These systems include the triethylamine catalyzed detritiation of 1,4-dicyano-2-butene,[17] the acetate catalyzed detritiation of t-butylmalononitrile,[18] the H_2O and acetate catalyzed detritiation of thiamine,[19] and the lyoxide ion catalyzed tritium incorporation into chloroform.[20] In the cases involving isotope effects on ionization equilibria, these values underestimate the value of ϕ_{C^-} needed here because the values may contain a contribution due to differences in the strength of solvent-carbanion hydrogen bond in H_2O compared to D_2O. In the presence case the asymmetric transition state contains an H in both H_2O and D_2O.
16 F Hibbert and F A Long, *J Am Chem Soc*, 1971, **93**, 2836.
17 E A Walters and F A Long, *J Am Chem Soc*, 1969, **91**, 3733.
18 M Hojatti, A J Kresge, and W H Wang, *J Am Chem Soc*, 1987, **109**, 4028.
19 M W Washabaugh and W P Jencks, *J Am Chem Soc*, 1989, **111**, 674.
20 A J Kresge and A C Lin, *J Am Chem Soc*, 1975, **97**, 6527.

RECENT ADVANCES IN THE APPLICATION OF CROSS-INTERACTION CONSTANTS

Ikchoon Lee

Department of Chemistry, Inha University, Inchon 402-751, Korea.

1 INTRODUCTION

For the last several years, we have been involved in developing cross-interaction constants as a mechanistic tool for organic reactions in solution.[1] In this talk, I wish to review some of the more recent applications of the cross-interaction constants to the elucidation of substitution reaction mechanisms.

2 NON-INTERACTIVE (OR ISOPARAMETRIC) PHENOMENON

In general, when two reactants interact, the effect of substituents in the two on the reactivity are not additive; the non-additive part consists of the cross-interaction term, ρ_{ij} in equation 1.

$$\log(k/k_o) = \rho_i{}^o\sigma_i + \rho_j{}^o\sigma_j + \rho_{ij}\sigma_i\sigma_j \tag{1}$$

where a degree symbol on ρ denotes that the other reactant is unsubstituted, *ie* σ_j or σ_i is zero (= H). This expression is obtained by a Taylor expansion of log k around $\sigma_i = \sigma_j = 0$ (for which $k = k_o$) and neglecting pure second order and higher order terms, $\rho_{ii} = \rho_{jj} = \cdots = \rho_{iij} = \cdots = \rho_{ijk} = 0$. The cross-interaction constant, ρ_{ij}, can alternatively be defined as the partial differential of ρ_i (or ρ_j) with respect to σ_j (or σ_i), equation 2.

$$\rho_{ij} = \frac{\partial^2 \log k}{\partial \sigma_i \partial \sigma_j} = \frac{\partial \rho_j}{\partial \sigma_i} = \frac{\partial \rho_i}{\partial \sigma_j} \tag{2}$$

Likewise, a Taylor expansion of log k around $pK_i = pK_j = 0$ gives us the Bronsted type cross-interaction constant, β_{ij}, equation 3.

$$\log (k/k_o) = \beta_i pK_i + \beta_j pK_j + \beta_{ij} pK_i pK_j \tag{3}$$

One advantage of using β_{ij} is that reactivities are not dependent explicitly on interaction between substituents, but on direct interaction between two reaction centres of the reactants, and an intervening group like CH_2 between a substituent and the reaction center in a reactant does not require any special consideration in interpreting the magnitude of β_{ij} in contrast to that of ρ_{ij} (*vide infra*).

Although two substituents of the reactants cross-interact normally, the two may become non-interacting under certain conditions. We can define a non-interactive parameter (NIP),[2] $\hat{\sigma}_j$, for any two substituents interacting with a sizable ρ_{ij} value, equation 4.

$$\log(k/k_o) = \rho_i^o \sigma_i + \rho_j^o \sigma_j + \rho_{ij}\sigma_i\sigma_j$$

$$= \rho_i^o \sigma_i + (\rho_j^o + \rho_{ij}\sigma_i)\sigma_j = \rho_i^o \sigma_i + \rho_j\sigma_j$$

$$= \rho_i^o \hat{\sigma}_i = -\frac{\rho_i^o \rho_j^o}{\rho_{ij}} = \text{const} \qquad (4)$$

where

$$\hat{\sigma}_i = -\frac{\rho_j^o}{\rho_{ij}}, \text{ at which } \rho_j \ (= \rho_j^o + \rho_{ij}\sigma_i) \text{ vanishes.}$$

Likewise, at $\hat{\sigma}_j = -\dfrac{\rho_i^o}{\rho_{ij}}$, $\rho_i = 0$. Thus, at a non-interactive point, $\hat{\sigma}_i$, the other substituent, σ_j, does not cause reactivity changes so all σ_j-substituted reactants have the same reactivity ($\rho_j = 0$), *ie* there is an isoparametric[2b] or isokinetic[2a] relationship. It is a characteristic of second order interactions that cross-interaction disappears at a non-interactive point and becomes isokinetic (equation 4). Thus ρ_j can change sign within a reaction series at the non-interactive point, $\hat{\sigma}_i$, and hence ρ_j is of only limited or relative mechanistic significance. Some examples are presented in Tables 1 ~ 2, where σ_X and σ_Y denote substituents in the nucleophile and substrate, respectively. For the reaction A - D in Table 1, the $\hat{\sigma}_Y$ values (not shown) are large (negative) and out of range of the σ values for substituents normally used in experimental studies. In contrast, $\hat{\sigma}_X$ values are well within the range so that $\hat{\sigma}_X$ values can be actually realized or observed. We see in the two examples, B and D, we actually encounter the cases of sign inversion for ρ_Y at $\hat{\sigma}_X$. This sign inversion can be interpreted as charge reversal at the reaction centre of the substrate indicating changes from a predominant bond making (BM) to a dominant bond breaking (BB) in the transition state (TS) at the NIP.

Table 1: Changes in ρ_Y with σ_X for reactions A - D showing the non-interactive point, $\hat{\sigma}_X$, at which $\rho_Y = 0$.

X	σ_X	ρ_Y			
		A	B	C	D
4-NH$_2$	-0.66			0.0	
H	0	1.03	0.63	-0.23	0.22
3-Cl	0.37	0.78	0.42	-0.46	0.0
3-NO$_2$	0.71	0.53	0.16	-0.62	-0.14
5-NO$_2$,3-CO$_2$Me	0.96	0.15	0.0		-0.35
3,5-(NO$_2$)$_2$	1.42	0.0	-0.37		
	$\hat{\sigma}_X$	1.41	0.94	-0.69	0.39

A $YC_6H_4COCl + XC_6H_4NH_2 \xrightarrow[25°C]{C_6H_5Cl - C_6H_{12}(1:1, v/v)}$

B $YC_6H_4COBr + XC_6H_4NH_2 \xrightarrow[25°C]{C_6H_5Cl - C_6H_{12}(1:1, v/v)}$

C $YC_6H_4CH_2Br + XC_6H_4NH_2 \xrightarrow[40°C]{CH_3NO_2}$

D $YC_6H_4CH_2Br + XC_6H_4NH_2 \xrightarrow[40°C]{1M \ Me_2SO \ in \ C_6H_5NO_2}$

Table 2: Calculated non-interactive parameter for nucleophilic substitution reaction of 1-phenylethyl chlorides with anilines in methanol.

$YC_6H_4CH(CH_3)Cl + XC_6H_4NH_2 \xrightarrow[65.0°C]{CH_3OH}$ [Desolvation barrier control is precluded].

Y	p-CH$_3$	p-CMe$_3$	$(\hat{\sigma}_Y^+)$	m-CH$_3$	H	p-Cl	m-Cl	m-NO$_2$
σ_Y^+	-0.31	-0.26	(-0.23)	-0.07	0.0	0.11	0.40	0.67
ρ_X	0.18	0.08	(0.0)	-0.29	-0.47	-0.69	-0.95	-1.60

$\rho_{XY} = -2.05$ (for Y = EDS) $\hat{\sigma}_Y^+ = -\dfrac{\rho_X^0}{\rho_{XY}} = -\dfrac{-0.47}{-2.05} = -0.23$

In the second example in Table 2, a NIP of $\hat{\sigma}_Y^+$ is obtained with a sign change of ρ_X at $\hat{\sigma}_Y^+$; ρ_X for the nucleophile, in contrast to ρ_Y for the substrate in the previous example, changes sign at the NIP, $\hat{\sigma}_Y^+ = -0.23$. This is very unusual in the sense that the reaction centre of the nucleophile can become more negatively charged in the TS for S_N reactions. Clearly $|\rho_X|$ cannot serve as a measure of the degree of bond making since a change from positive to negative degree of bond making is an absurd interpretation. This phenomenon can be rationalized with the TS imbalance[3] in an ion-pair (S_N2C^+) mechanism (*vide infra*).

This type of second order non-interactive or isoparametric phenomenon is not limited to the parameter $\hat{\sigma}$, *ie* isokinetic substituent. Any two of the rate variables, σ, pK_a, Y (ionizing power), n (nucleophilicity), T (temperature), *etc*, can be combined in the second-order expansion. We can thus generalize the isoparametric or non-interactive phenomenon and can define various other isokinetic parameters, *eg* isokinetic-basicity ($p\hat{K}_a$), -ionizing power, -medium ($\hat{Y}$), -nucleophile ($\hat{n}$), temperature ($\hat{T}$), *etc*, equations (5) and (6).

$$\log(k/k_o) = M_m^\circ m + N_n^\circ n + Q_{mn} mn \tag{5}$$

where $M_m^\circ = (\frac{\partial \log k}{\partial m})_{n=0}$, $N_n^\circ = (\frac{\partial \log k}{\partial n})_{m=0}$

and $Q_{mn} = (\frac{\partial^2 \log k}{\partial m \partial n})$

m,n = σ, pK_a, Y, n, T, *etc*, and M°, N° = ρ°, β, m, s, *etc*.

$$\log(k/k_o) = M_m^\circ m + (N_n^\circ + Q_{mn}m)n \tag{6}$$

$$= M_m^\circ m + N_n n \quad (N_n = N_n^\circ + Q_{mn}m)$$

$$= -\frac{M_m^\circ N_n^\circ}{Q_{mn}} \quad \text{(Isokinetic)}$$

for NIP, $\hat{m} = -\frac{N_n^\circ}{Q_{mn}}$ with $N_n = 0$

or

$$\hat{n} = -\frac{M_m^o}{Q_{mn}} \text{ with } M_m = 0.$$

Some possible combinations of the second-order expansion are:

$$\log(k/k_o) = \rho_i{}^o\sigma_i + \beta_j{}^o pK_j + \Theta_{ij}\sigma_i pK_j \qquad (7a)$$

$$= \rho_i{}^o\sigma_i + mY + \Theta_{iY}\sigma_i Y \qquad (7b)$$

$$= mY + K_T T + \Theta_{YT} YT \qquad (7c)$$

where $K_T = \dfrac{\partial \log k}{\partial T}$ and $\Theta_{ij} = -\dfrac{\partial^2 \log k}{\partial \sigma_i \partial pK_j}$ etc.

An example of the type given by equation (7b) is found in the solvolysis of $YC_6H_4SO_2Cl$ in $MeOH\text{-}MeNO_2$ mixtures are 45.0 °C.[4]

For this reaction at $\hat{Y} = -5.5$ ($\cong$ 50% MeOH) the solvolysis of substituted benzenesulfonyl chlorides becomes isokinetic or isoparametric ($\sigma_{Y1} = \sigma_{Y2} = \cdots = \sigma_{Yn}$).

3 SIGN OF ρ_{XZ} IN S_N2 TS

In the following, we adopt a convention of denoting three fragments in the S_N2 TS as X, Y and Z for nucleophile, substrate and leaving group (LG), respectively, scheme 1. Now let us consider the significance of the sign of ρ_{XZ} in the S_N2 TS. The ρ_{XZ} is defined as $\partial \rho_Z/\partial \sigma_X$ or $\partial \rho_X/\partial \rho_Z$, equation 2. This means that a stronger nucleophile with a more negative σ_X as well as a stronger nucleofuge (or better LG) with a more positive σ_Z leads to an earlier TS along the reaction coordinate when ρ_{XZ} is positive, or to a later TS when ρ_{XZ} is negative. For the

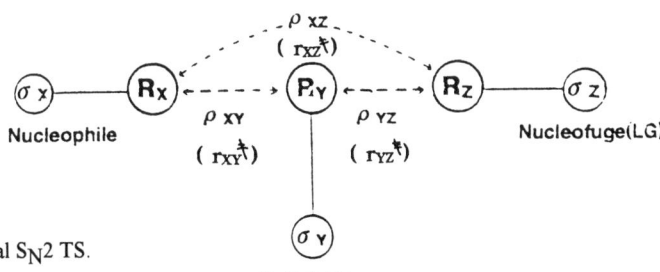

Scheme 1: Typical S_N2 TS.

former case, the TS variation is consistent with that predicted by the potential energy surface (PES) diagram (or More O'Ferrall-Jencks diagram),[3b,5] Figure 1, whereas for the latter case it is consistent with that predicted quantum mechanically. These TS variations are in fact substantiated by experimental trends in $|\rho_X|$, $|\rho_Z|$ and secondary α-deuterium kinetic isotope effects (SKIE) with deuterated substrates and nucleophiles.[1]

Let us look at the case of $\rho_{XZ} > 0$ more closely. Reactions with a positive ρ_{XZ} are characterized by thermodynamic barrier (TB) control, and a stronger nucleophile or nucleofuge invariably leads to an earlier TS in accordance with the predictions by a PES diagram. In a PES diagram (Figure 1) normally 'principal axis' (or reaction coordinate axis) from reactant to product is taken as 'thermodynamic' to represent the Hammond effect,[6] and 'auxiliary axis' perpendicular to this is taken as 'intrinsic' to represent anti-Hammond effect.[7] Substituent effects on TS variation can be classified into two types. Primary effect is given by the sign of ρ_{ii} (ρ_{XX} and ρ_{ZZ}) or vector direction on the axis of the PES diagram, indicating the effect of σ_X on ρ_X, ie on bond making (BM) and σ_Z on ρ_Z ie on bond breaking (BB). Secondary effect is obtained as a sign of ρ_{XZ} or direction of the vector sum of the two primary effects. In the PES diagram in Figure 1, two primary effects are given by vector directions on the two axes, $\overrightarrow{OA}$ and $\overrightarrow{OB}$ (or $\overrightarrow{OC}$), while secondary effects are represented by the vector sums, $\overrightarrow{OD}$ and $\overrightarrow{OE}$, which indicate a decrease in BM as well as BB (earlier TS) in agreement with a positive ρ_{XZ}.

4 MAGNITUDE OF ρ_{ij}

The magnitude of cross-interaction constants, ρ_{ij}, has been shown to be proportional to the change in the intensity of interaction ($\Delta I^{\neq}_{int}$) between the two substituents through the respective reaction centres in the activation process,[7] equation 8. Thus when there is no change in the intensity of interaction during the activation process, ρ_{ij} vanishes.

$$|\rho_{ij}| \propto |\Delta I^{\neq}_{int}(i,j)| \tag{8}$$

where $\Delta I^{\neq}_{int}(i,j) = I^{\neq}_{int}(i,j) - I^{o}_{int}(i,j)$

Recent Advances in the Application of Cross-Interaction Constants

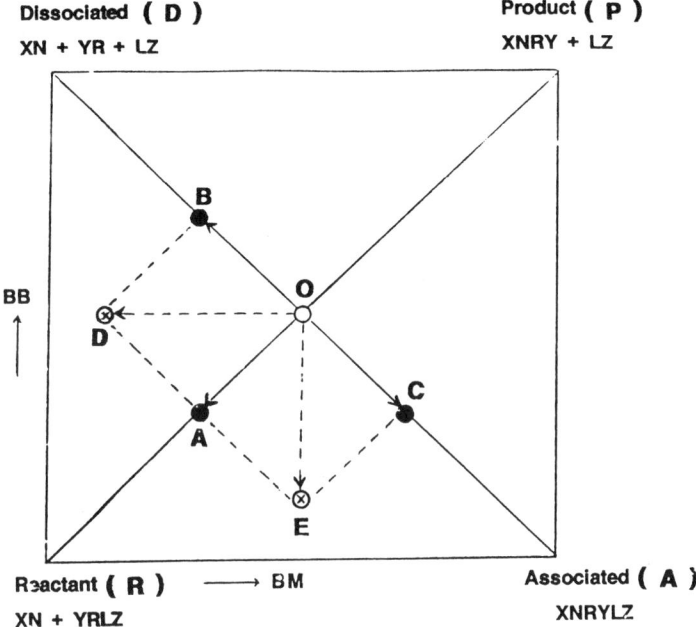

Figure 1: Potential energy surface (PES) diagram illustrating TS variation for typical S_N2 reaction. Vector sum of the two primary substituent effects, $\overrightarrow{OA}$ and $\overrightarrow{OB}$ (or $\overrightarrow{OC}$), results in a secondary effect $\overrightarrow{OD}$ (or $\overrightarrow{OE}$) which invariably leads to an earlier TS.

This means that σ_i has no effect on ρ_j or *vice versa*, equation 9.

$$\rho_{ij} = \frac{\partial \rho_j}{\partial \sigma_i} = \frac{\partial \rho_i}{\partial \sigma_j} = 0 \qquad (9)$$

Such cases are encountered in rate-limiting processes of BM and BB. For example, in S_N1 reactions bond cleavage is rate-limiting so that ρ_{XY} and ρ_{XZ} are zero (*eg* reactions of Y-α-t-butylbenzyl Z-arenesulfonates with X-aniline[9]), whereas in a rate-limiting BM process, ρ_{YZ} is zero (*eg* OH⁻ + $YC_6H_4COOC_6H_4Z$).[10] In general, when a nucleophile is involved in the rate-limiting step, the initial intensity of interaction is zero, $I°_{int}(X,j) = 0$ where $j = Y$ or Z, since initially the nucleophile is at an infinity distance and there is no interaction. As a result, equation 8 can be simplified into equation 10. Since the intensity of the interaction at a given state [TS or ground state (GS)] is related inversely to the distance r_{Xj} (scheme 1)[11] the magnuitude of ρ_{Xj} is inversely related to the distance $r_{Xj}^{\neq}$ in the TS, equation 11.

$$\Delta I_{int}^{\neq}(X,j) = I_{int}^{\neq}(X,j) \tag{10}$$

$$|\rho_{Xj}| \propto I_{int}^{\neq}(X,j) \propto \frac{1}{r_{Xj}^{\neq}} \tag{11}$$

A number of experimental results are reported on this type of inverse correlation.[1] A special case is that when the nucleophile is far away in the TS, $r_{Xj}^{\neq} = \infty$, ρ_{Xj} practically vanishes.[12]

When, however, i,j = Y,Z, the situation is different and equations 10 ~ 11 do not apply.[8] In this case bond breaking is rate-limiting and $|\rho_{YZ}|$ is now proportional to $r_{YZ}^{\neq}$, since $r^o{}_{YZ}$ (GS) is constant; for the BB process, $I_{int}^{\neq}(Y,Z) < I_{int}^o(Y,Z)$ and hence $\Delta I_{int}^{\neq}(Y,Z) < 0$. However, $I_{int}^o(Y,Z)$ and $r^o{}_{YZ}$ are constant quantities.

Thus $|\rho_{YZ}| \propto |\Delta I_{int}^{\neq}(Y,Z)| = |I_{int}^{\neq}(Y,Z) - I_{int}^o(Y,Z)|$

$$\propto |\Delta r_{YZ}^{\neq}| = |r_{YZ}^{\neq} - r_{YZ}^o|$$

$$\propto r_{YZ}^{\neq} \tag{12}$$

A relatively large magnitude of ρ_{YZ} was indeed observed (eg $|\rho_{YZ}| \cong 1.0$ in methanolysis of α-t-butylbenzyl arenesulfonates)[9] for an S_N1 reaction, in contrast to a smaller value (eg $|\rho_{YZ}| \cong 0.2$ in methanolysis of indan-2-yl arenesulfonates)[13] for an S_N2 process.

Since ρ_{Xj} is inversely dependent on distance, an intervening group between the substituent and reaction centre causes ρ_{Xj} to fall off by ca ½. However, if there are multiple routes of interaction in the TS, which are lacking in the GS, an exalted $|\rho_{ij}|$ due to additive effects through multiple routes is obtained.[1]

5 NON-INTERACTIVE AND TS IMBALANCE PHENOMENA

At a non-interactive point, $\hat{\sigma}_i$ ($\hat{\sigma}_i = -\rho_j^o/\rho_{ij}$), ρ_j is zero. In order that a non-interactive point, $\hat{\sigma}_i$, is an experimentally observable or realizable value, which is limited to within a narrow range of σ_i constants, $\sigma_i = -1.0$ to 1.0, $|\rho_{ij}|$ should be large compared to $\rho_j{}^o$. This means that two substituents σ_i and σ_j interact strongly in the TS, normally as a result of an extensive BM, but $\rho_j{}^o$ is small, *ie* weak susceptibility due to a small polar or resonance effect of substituent on the reaction centre within a reactant as a result of a small degree of reorganization in the TS. Often this results in a TS imbalance phenomenon in which development or reorganization of resonance (structure) lags behind the bond changes associated with proton transfer.[2a] It is therefore more likely that we observe the TS imbalance in proportion to $|\rho_{ij}|$. Such an example is the S_N reaction of 1-phenylethyl chloride (1-PEC) with anilines in methanol, which has been shown to proceed by an ion-pair, or S_N2C^+ mechanism.[2a] For the reactions of 1-PEC with relatively strong electron-donating groups (Y = EDS), ρ_{XY} was large and negative, $\rho_{XY} = -2.05$, but $\rho_X{}^o$ was relatively small, $\rho_X{}^o = -0.41$, leading to an observable non-interactive substituent at $\hat{\sigma}_Y{}^+ = -0.23$. For the substrates with $\sigma_Y{}^+ < \hat{\sigma}_Y{}^+$, negative charge accumulates on C_α as the $N-C_\alpha$ bond is being formed with concurrent proton transfer, due to reorganization of resonance structure lagging behind the BM. This results in an unusual, negatively charged nucleophilic centre, N, in the TS with positive ρ_X. In Figure 2, the lower right corner of Y = EDS represents the imbalanced species and TS1 has an imbalanced structure with a positive ρ_X. The upper left corner with electron-withdrawing group (Y = EWS) represents normal addition complex and TS3 is normal with a negative ρ_X. At TS2, the situation is intermediate with $\rho_X = 0$ (at $\sigma_Y{}^+ = -0.23$) having a constant reactivity, *ie* isokinetic, irrespective of the substituent in the nucleophile, σ_X.

6 KINETIC SOLVENT ISOTOPE EFFECT (KSIE)

Kinetic solvent isotope effect in the solvolysis of substituted (Y) substrates can be a useful mechanistic tool,[14] when the change of ρ_Y, $\Delta\rho_Y = \rho_{Y(SOH)} - \rho_{Y(SOD)}$, is considered in conjunction with ρ_{XY} (= $\Delta\rho_Y/\Delta\sigma_X$), the degree of bond formation. It has been shown that SOD (D_2O or CH_3OD) is a weaker nucleophile[15] (tantamount to having less negative or a more positive σ_X) than SOH (H_2O or CH_3OH), and hence the solvents, SOH and SOD, as a nucleophile may be considered to give a negative constant, $\Delta\sigma_X$ (= $\sigma_{X(SOH)} - \sigma_{X(SOD)}$). Thus $\Delta\rho_Y$ is proportional to ρ_{XY} with a sign reversal since $\Delta\sigma_X < 0$. It has been found that $\Delta\rho_Y$ is positive for S_N2 and S_AN reactions and zero for S_N1 reactions.[14] However, $\Delta\rho_Y$ becomes negative ($\rho_{XY} > 0$) for an ion-pair (S_N2C^+) mechanism (*eg* methanolysis of 1-PEC).[14] The three types of KSIE are graphically illustrated in Figure 3 a ~ c. Some examples are given in the Figures.

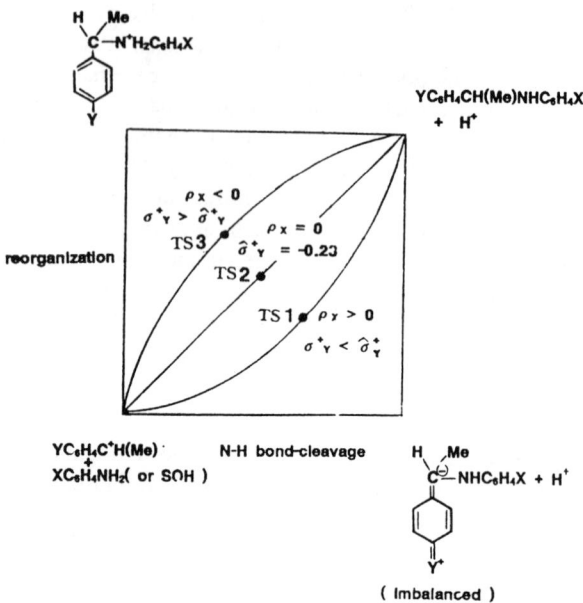

Figure 2: PES diagram for reactions of 1-phenylethyl chlorides with anilines in MeOH. Imbalanced TS, TS1 is characterized by a negative ρ_X and reorganization of delocalized structure, at lower right corner, lags behind the proton transfer.

7 TEMPERATURE AND SOLVENT EFFECTS ON ρ_{ij}

It is well known that numerical values of all first derivative susceptibility factors decrease with increasing temperature,[16] eg $\partial|\rho_i|/\partial T < 0$. However, the magnitude of second-order interaction constants ρ_{ij} does not change uniformly in a decreasing manner with temperature. The magnitude of ρ_{XY} was found to decrease with temperature indicating a decrease in the degree of BM, whereas an increase in the magnitude of ρ_{YZ} with temperature was observed reflecting an increase in BB with increase in temperature.[13] The size of ρ_{XZ}, on the other hand, decreases because higher temperatures cause BM to decrease but BB to increase. The magnitude of ρ_{YZ}, and hence the degree of BB, may increase with temperature if the numerator of the temperature coefficient of the normalized degree of bond fission, $a_{fi} = \dfrac{\rho_z}{\rho_e}$ is positive ie $\dfrac{\partial a_{fi}}{\partial T} = \dfrac{(\dfrac{\partial \rho_z}{\partial T}\rho_e - \dfrac{\partial \rho_e}{\partial T}\rho_z)}{\rho_e^2} > 0$ if the numerator is positive.

Recent Advances in the Application of Cross-Interaction Constants

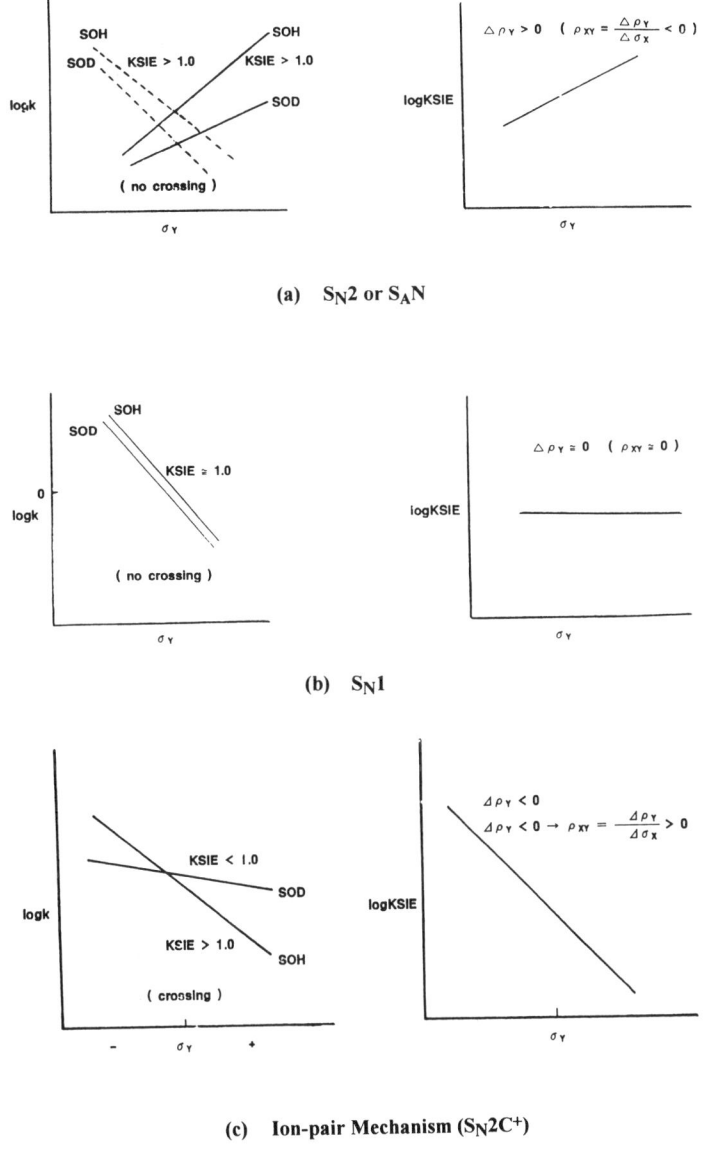

(a) S_N2 or S_AN

(b) S_N1

(c) Ion-pair Mechanism (S_N2C^+)

Figure 3: (a) Kinetic solvent isotope effects (KSIE = k_{SOH}/k_{SOD}) are always greater than unity and $\Delta\rho_X = \dfrac{\partial \text{KSIE}}{\partial \sigma_Y}$ is positive for S_N1 and S_AN processes, (eg solvolysis of benzyl, benzoyl, phenacyl and benzenesulfonyl halides in H_2O and MeOH.)
(b) For S_N1, $\Delta\rho_Y \cong 0$, (eg solvolysis of t-butylbenzyl arenesulfonates).
(c) For ion-pair mechanism (S_N1C^+), $\Delta\rho_Y$ becomes negative, (eg solvolysis of 1-phenylethyl chlorides in MeOH).

The magnitude of ρ_{YZ} and its solvent dependence are relatively large for S_N1 reactions due to extensive bond cleavage (*eg* for solvolysis of α-t-butylbenzyl arenesulfonates, $|\rho_{YZ}| \cong 0.6$ to 1.0). Ability of the LG to delocalize negative charge lowers[16] the Grunwald-Winstein coefficient m (equation 13)[17] greatly in the S_N1 TS, with a greater lowering for a better LG,

$(\frac{\partial m}{\partial \sigma_z} \cong 0.5$ for solvolysis of α-t-butylbenzyl arenesulfonates).[9]

$$\log(k/k_o) = mY \qquad (13)$$

The variation of ρ_{YZ} with solvent composition for S_N1 reactions does not depend on the ionizing power, Y, of the solvent but reflects the stability of the ion-pair intermediate; a more stable ion-pair intermediate leads to an earlier TS according to the BEP principle[18] (or Hammond postulate)[6] so that a lesser degree of BB, *ie* a smaller $|\rho_{YZ}|$ is obtained.

In contrast to S_N1 reactions, the magnitude of ρ_{YZ} and its solvent dependence are small for S_N2 reactions, $|\rho_{YZ}| \cong 0.2$ to 0.3. For the solvolysis of 5-substituted indan-2-yl arenesulfonates,[13] the magnitudes of both ρ_Z and ρ_Y decrease very small amounts with increasing ionizing power, Y, but that of ρ_{YZ} increases substantially with nucleophilicity, N_{OTS}, $(\frac{\partial |\rho_{YZ}|}{N_{OTS}} \cong +0.36)$ reflecting an increase in bond cleavage as the solvent nucleophilicity is increased.

In this review, I have presented the application of cross-interaction constants to the S_N type reactions only. However, the application can be extended readily to other reaction types.[1]

REFERENCES

1   (a) I Lee, *Chem Soc Rev*, 1990, **19**, 317; (b) I Lee, *Adv Phys Org Chem*, 1992, **27**, 57.
2   (a) I Lee, W H Lee, H W Lee, and T W Bentley, *J Chem Soc, Perkin Trans 2*, 1993, 141; (b) I V Shpan'ko, *Mendeleev Commun*, 1991, 119.
3   (a) C F Bernasconi, *Adv Phys Org Chem*, 1992, **27**, 119; (b) W P Jencks, *Chem Rev*, 1985, **85**, 511.
4   D D Sung, Y H Kim, Y M Park, Z H Ryu, and I Lee, *Bull Korean Chem Soc*, 1992, **13**, 599.
5   R A More O'Ferrall, *J Chem Soc* (B), 1970, 274.
6   G S Hammond, *J Am Chem Soc*, 1955, **77**, 334.

7   E R Thornton, *J Am Chem Soc*, 1967, **89**, 2915.
8   I Lee, *J Phys Org Chem*, 1992, **5**, 736.
9   I Lee, M S Choi, and H W Lee, *J Chem Res, (S)*, 1994, 92.
10 J F Kirsch, W Clewell, and A Simon, *J Org Chem*, 1968, **33**, 127.
11 (a) I Lee, J K Cho, and C H Song, *J Chem Soc Faraday Trans*, 1988, **84**, 1117; (b) I Lee, J K Cho, H S Kim, and K S Kim, *J Phys Chem*, 1990, **94**, 5190.
12 (a) C D Ritchie, *Acc Chem Res*, 1972, **5**, 348; (b) C D Ritchie, *Can J Chem*, 1986, **64**, 2239.
13 I Lee, Y S Lee, and H W Lee, *J Chem Soc Perkin Trans 2*, 1994, 1441.
14 (a) I S Koo, I Lee, J Oh, K Yang, and T W Bentley, *J Phys Org Chem*, 1993, **6**, 223; (b) I Lee, W H Lee, and H W Lee, *ibid*, 1993, **6**, 361.
15 (a) E M Arnett, and D R McKelvey, 'Solute-Solvent Interaction', ed J F Coetzee and C D Ritchie, Marcel Dekker, New York, p 353 (1969); (b) X G Zhao, S C Tucker, and D G Truhlar, *J Am Chem Soc*, 1991, **113**, 826.
16 G W Klumpp, 'Reactivity in Organic Chemistry' Wiley, New York, 1982, p 224.
17 E Grunwald, and S Winstein, *J Am Chem Soc*, 1948, **70**, 846.
18 M J S Dewar, and R C Dougherty, 'The PMO Theory of Organic Chemistry', Plenum, New York, 1975, Chapter 5.

ACKNOWLEDGEMENTS

I thank all the coworkers, who have contributed to this project of applying cross-interaction constants to organic reaction mechanism. Financial supports from the Ministry of Education, the Korea Science and Engineering Foundation, and the Korea Research Center for Theoretical Physics and Chemistry are gratefully acknowledged.

AMBIDENT REACTIVITY IN REACTIONS INVOLVING THE NITROSO GROUP

M Elena Peña[*]

Departamento de Química Física, Facultad de Química, Universidad de Santiago, 15706 Santiago de Compostela, Spain.

In the last few years, we have found that several apparently simple nitrosation reactions were mechanistically more complex than one would anticipate. This happened to be the case for the nitrosation of amides, some indoles and some S-containing amines and amino acids. All these nucleophiles share the property of being potentially ambident nucleophiles and the root cause for the detected 'anomalies' seems to lie in this ambident nature and a particular ability of the nitroso group to undergo rearrangement processes. More recently, we have extended the topic of ambident reactivity in nitrosation reactions to include nitrosating agents bearing two electrophilic centres. The study of such situations casts light on the factors determining reactivity in nucleophile-electrophile combinations. More complex situations can arise when both the nucleophile and electrophile have ambident natures, which broadens the scope of possibilities. In this paper, I shall make a survey of these different situations.

1 NITROSATION OF AMBIDENT NUCLEOPHILES

1.1. Nitrosation of Indoles and Amides

With indoles unsubstituted in position 3, kinetically 'normal' nitrosation takes place at this position to yield the corresponding *C*-nitroso compound which rapidly tautomerizes to the more stable oxime form.[1] However, when position 3 in the indolic ring is substituted, nitrosation leads to the formation of *N*-nitroso indoles.[2] A kinetic investigation[3] of the *N*-nitrosation of 3-substituted indoles revealed a number of 'anomalies' in the kinetic behaviour which were at odds with a mechanism involving direct *N*-nitrosation. In particular, the experimental rate equation (equation 1) only contained first order terms in nitrous acid and indole.

$$\text{rate (nitrosation)} = k\,[\text{HNO}_2]\,[\text{Indole}] \quad (1)$$

---

[*] Deceased 17th December, 1993.

$$HNO_2 + H^+ \rightleftharpoons NO^+ + H_2O$$

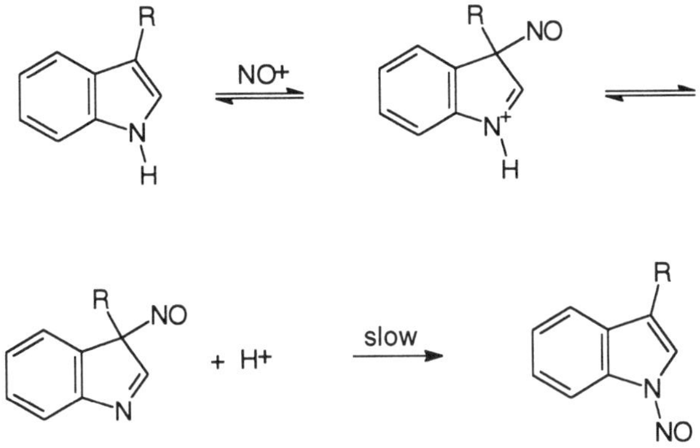

Scheme 1: Mechanism of the *N*-nitrosation of 3-substituted indoles.

The surprising feature in this equation is the lack of influence of acidity on the reaction rate which cannot be reconciled with any simple nitrosation mechanism. In particular, nitrosation in acidic media takes place *via* well-known nitrosating agents, which derive from nitrous acid protonation (NO^+ or $H_2NO_2^+$),[4] so that acid catalysis is a common characteristic of nitrosation reactions. Other unusual peculiarities were also detected. The usual catalysts of nitrosation reactions (Cl^-, Br^-, SCN^-), which catalyse the nitrosation of unsubstituted indoles,[1] did not affect the rate at which *N*-nitrosation took place. Finally, the solvent isotope effect for the reaction was *ca* 1, and not inverse as expected from the existence of a pre-equilibrium protonation of nitrous acid to yield the nitrosating species (*eg* NO^+).[5] All these observations led us to propose a mechanism (Scheme 1) in which reaction really started at position 3, followed by proton loss and by a rate-determining internal rearrangement (from C to N) to yield the thermodynamically more stable *N*-nitroso product.

The mechanism in Scheme 1 explains the lack of influence of acidity upon the reaction rate, the solvent isotope effect close to 1, and the absence of catalysis by the addition of halides, whose presence will not affect the *equilibrium* concentration of intermediate I.

A related situation had been found when studying the mechanism of *N*-nitrosation of amides, which is now recognized to be different[6] from that for *N*-nitrosation of amines. In particular, the reaction is strongly accelerated by buffers and shows primary solvent isotope effects,[7] both factors pointing to a proton transfer taking place in the rate determining step. Although early proposals[6,7] suggested the possibility that proton transfer occurred from the intermediate derived from direct N-nitrosation, that is, from an N-protonated N-nitrosamide (intermediate II, a very acidic species, pK_a well below -10),[8] a more detailed investigation revealed that this mechanism was not compatible with the magnitude of the isotope effects measured or the Brønsted slopes derived from buffer catalysis.[8] More detailed studies[8] on both the direct and reverse reaction revealed that the intermediate from which the proton was lost had indeed a pK_a not far from 0, an observation which led us to propose a new mechanism in which again the ambident nature of the amide played a key role (Scheme 2) and which keeps a parallelism with that observed for 3-substituted indoles.

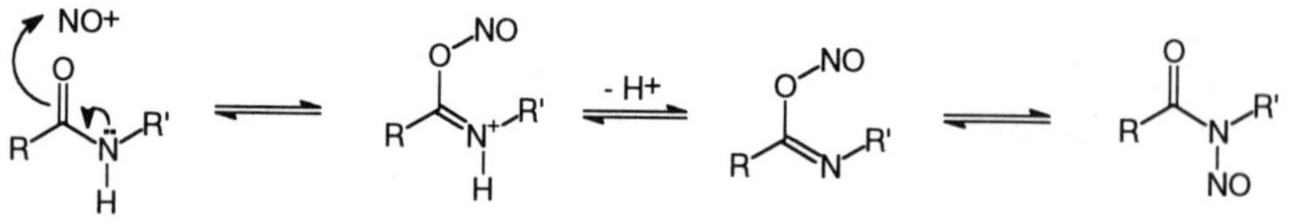

Scheme 2: Mechanism of the nitrosation of amides.

Some points are particularly appealing in both these mechanisms. The fact that reaction starts at oxygen seems to unify the nucleophilic behaviour of amides in many chemical reactions: protonation,[9] alkylation,[10] and nitrosation, and the same happens if nitrosation of indoles starts at position 3 of the ring where protonation also takes place.[11] However, the main limitation preventing general acceptance of the mechanism in Scheme 2 was the lack of conclusive experimental evidence. According to this mechanism, a limit should be reached at high acidity where the reaction rate should be limited by the internal O to N rearrangement and should therefore become acid-independent (as in the case of the indoles). Although this limiting condition could not be achieved in the nitrosation of aliphatic ureas or amides in water, recent work carried out using a mixture of acetonitrile-water (90:10) as the solvent allowed detection[12] of conditions under which the rate was in fact independent of acidity.

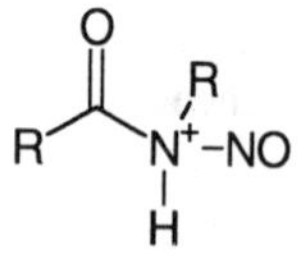

II

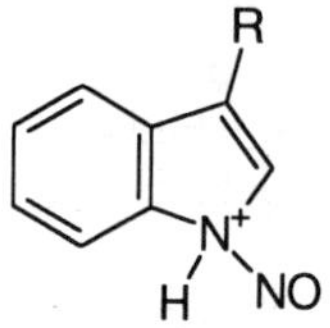

III

In the cases of the nitrosation of indoles, amides and other ambident compounds,[13] the advantage of the non-direct mechanisms seems to lie at least in part in the avoidance of the highly energetic intermediates (intermediates II and III) containing an extremely acidic $N(NO)H^+$ moiety. Let us remember that a pK_a of ca -12 has been recently proposed for a similar moiety derived from the much more basic aliphatic amines.[14]

1.2. Nitrosation of S-containing Amino Acids

In 1983 Williams and Meyer[15] observed that several sulfur-containing amino acids underwent N-nitrosation in acidic media much faster than the non sulfur-containing analogues, which pointed to the involvement of the nucleophilic S centre in the N-nitrosation reaction. MO studies,[16] using the frontier-orbital approach, suggested that the 'soft' electrophile NO^+ preferred to react at S rather than at N in these S-containing amino acids, which accounted for the enhanced reactivity observed. A later detailed analysis of the kinetic characteristics in the case of the biologically relevant N-nitrosation of thioproline[17] confirmed that reaction was several powers of ten faster than reaction of proline or other secondary amines without the sulfur atom. The experimental rate equation (equation 3) was again very unusual for amine nitrosation; it contained first order terms in protons, nitrous acid and *total* thioproline. Since thioproline in acid medium exists mainly in the N-protonated forms, this means that the reaction rate is proportional to the amount of N-protonated thioproline. This contrasts with the experimental rate equations for nitrosation of other amines, for which reaction rate is proportional to the concentration of *free* (unprotonated) amine (the reactive form).

$$\text{rate} = k\,[HNO_2]\,[H^+]\,[\text{Total thioproline}] \qquad (3)$$

Rate equation (3) indicates that N-protonated thioproline is reactive. This situation, which is impossible for 'normal' amines, becomes possible for thioproline whose additional nucleophilic centre - the S atom - can become a target for the NO group. A detailed kinetic analysis led us[17] to propose the mechanism in Scheme 3 for the reaction.

A familiar pattern of behaviour appears in Scheme 3: initial nitrosation on the sulfur, loss of a proton and a subsequent internal rearrangement of the nitroso group. The main advantage of this mechanism - derived from a quantitative analysis - is that it makes reactive a form (the protonated form) that would otherwise be unreactive. The question then arises as to what would happen if the molecule had the nitrogen free: would it still react at S - as predicted by the theoretical studies[16] - or would reaction at N be more favourable? This question can only be answered in a study of nitrosation in basic media, where the free base form dominates. Although conventional nitrosating agents do not exist in basic media, the use of alkyl nitrites[18] or N-methyl-N-nitroso-p-toluenesulfonamide (MNTS)[19] allows nitrosation under neutral or alkaline conditions.

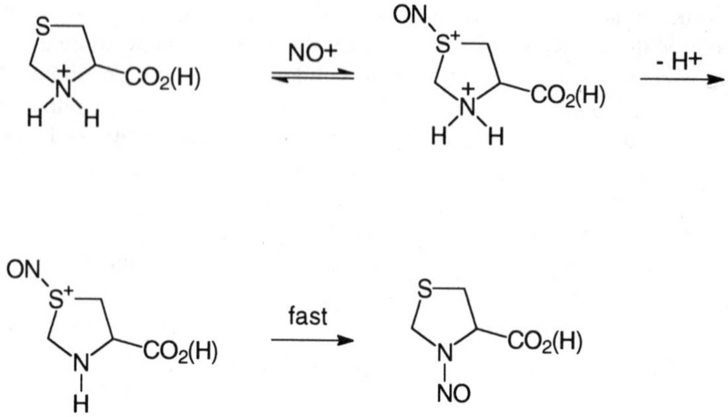

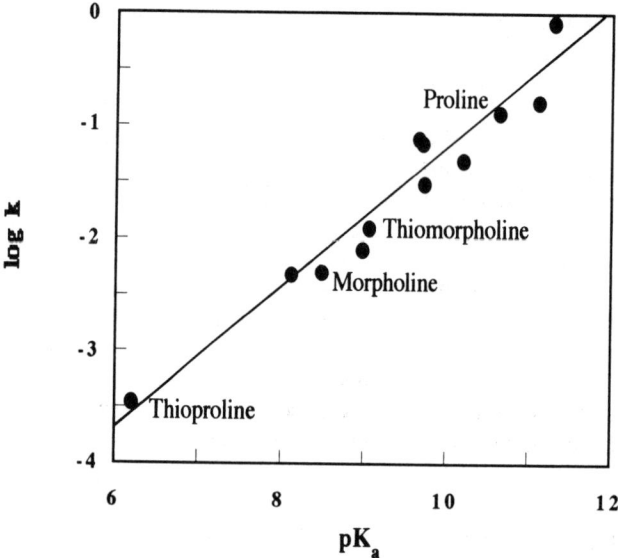

Scheme 3: Nitrosation of thioproline.

Our question could be answered when carrying out a kinetic study on the nitrosation of many different amines[19b] in neutral/basic media using MNTS as the direct nitrosating agent. Figure 1 shows the Brønsted plot relating the experimental rate constants for several secondary amines to their basicity.

Figure 1: Brønsted plot relating reactivity of several secondary amines towards the nitroso group of MNTS with their basicities (data taken from ref 19) [k is the second order rate constant in units $M^{-1}s^{-1}$].

The good relationship observed includes two members which are S-containing amines: thioproline and thiomorpholine, and undergo nitrosation unusually fast in acid medium.[17,20] Figure 1 shows that their reactivity in basic media towards MNTS is absolutely normal. They fall on the same line as non-S containing amines. The conclusion can be drawn that no advantage is obtained in these molecules from the presence of a S atom in the molecule, that is, a normal reaction - starting at N - takes place. We can conclude that the sulfur atom in thioproline is not more nucleophilic than the N atom towards nitrosating agents and that an initial S-nitrosation is only advantageous when the N atom is protonated and therefore unreactive. This contrasts with the predictions of the MO calculations,[16] the reasons for such a discrepancy remaining obscure.

2 REACTIONS WITH AMBIDENT ELECTROPHILES

Some of our recent work has concentrated on the reaction of nucleophiles towards ambident electrophilic nitroso compounds. Most of the work has referred to the study of the reactivity of MNTS, a molecule where a sulfonyl and a nitroso group can compete for the nucleophile.

I shall briefly summarize the site of reaction found for a wide range of nucleophiles. Nitrogen nucleophiles (primary, secondary, tertiary amines, hydrazine, hydroxylamine, azide ion, amino acids, semicarbazide),[19] I^-,[21] sulfur nucleophiles[21] (thiourea, SCN^-, thiosulfate, sulfite, *etc*) and carbon nucleophiles (enolates, nitronate ions)[22] all react at the nitroso group to yield the corresponding nitroso compounds (or their decomposition products). The situation is however more complex in the case of O nucleophiles. Simple O nucleophiles such as OH^- or alkoxides react at the sulfonyl group in a reaction leading to formation of alkylating species and presumably responsible for the potential carcinogenic properties of MNTS.[19] The sulfonyl group is also[21] the site of reaction of the α-effect nucleophile HOO^-; however, in the case of the α-effect nucleophile ClO^- a small percentage (*ca* 10%) of reaction takes place at the nitroso group, whereas the major reaction pathway continues to be reaction at the sulfonyl group. This pathway becomes more important with the anion of acetohydroxamic acid which clearly shows[21] an ambident pattern of reactivity (60% reacts at S, 40% reacts at the nitroso group). These results can be qualitatively rationalized in terms of the HSAB principle developed by Pearson.[23] The series of 'soft' nucleophiles (low charge density, delocalized, polarizable nucleophiles) react at the nitroso group, whereas 'harder' (less polarizable, high charge density) nucleophiles react at the sulfonyl group. Interestingly, the hardness diminishes for α-effect nucleophiles, which causes competition between the two reaction centres for some α-effect O nucleophiles. According to these ideas, the softer electrophilic centre in the molecule is the nitroso group.

Let us first explore the pattern of reactivity towards this 'soft' centre. Luckily, we observed that when nucleophiles reacted at the NO group of MNTS their reactivity was remarkably similar (usually within a factor < 3) to that exhibited towards 2-ethoxyethyl nitrite (EEN). This happened to be the case for amines, carbanions, sulfur nucleophiles and the anion of acetohydroxamic acid. Therefore, in the case of O nucleophiles reacting at the sulfonyl group of MNTS, a reliable estimation of their reactivity at the nitroso group could be made by studying the reaction using EEN as the electrophile.

There has been a tendency to correlate nucleophilicity towards the nitroso group (for example in the case of the denitrosation of nitrosamines in acid medium)[24] in terms of Pearson's nucleophilicity index[25] R_{MeI}. Such studies - carried out in acid media - only compared reactivity of a very restricted set of nucleophiles (Cl$^-$, Br$^-$, SCN$^-$ and thiourea). Our present work, carried out in neutral/alkaline medium allows comparison over a wider variety of nucleophiles. Figure 2(a) shows a plot of log k (the second order rate constant) *versus* R_{MeI} for several nucleophiles. The lack of correlation is absolute. In particular, the α nucleophile N_3^- (n = 5.8) is almost 100 times more reactive than I$^-$ (n = 7.4) and the strongest nucleophile according to this classification ($S_2O_3^{2-}$, n = 9) is, for example, less reactive than diethylamine (n = 6.7). In addition, our data show that N_3^- is a better nucleophile than OH$^-$. This order, as pointed out by Ritchie,[26] is the reverse of that expected on the basis of the Swain and Scott n scale, but is as expected from the nucleophilicity scale N_+. This scale, as Hoz[27] notes, seems to be followed by electrophiles characterized by a low-lying LUMO, mainly unsaturated compounds.

Figure 2 (b) shows the plot relating nucleophilic reactivity with N_+ values (taken from ref 28). The observed correlation, although not excellent, is reasonably good, especially if we take into account the variety of nucleophiles, both anionic and neutral, included and the fact that data obtained with MNTS and with EEN are shown in the same plot. In fact, deviations from the regression line are always lower than 1 log unit. Remarkably, the slope of the line is 1, as predicted by Ritchie's equation.

The scope of validity of the N_+ scale, which, using a single parameter N_+, has proven able to predict reactivities of nucleophiles with quite diverse electrophiles, is now enlarged to include nitroso compounds. In view of the variety of mechanisms and transition states involved in these reactions, even the different nature of the electrophilic atom (N in our case), the proposal that N_+ must be a measure of some intrinsic physical property of the nucleophiles is difficult to resist. The use of the curve-crossing model for nucleophile-electrophile combinations led Shaik[29] to propose a link between the value of N_+ and the vertical ionization potential (vIP) of the nucleophile, a magnitude which determines 'the energy gap' corresponding to a *single electron shift* between nucleophile and electrophile. It is tempting to follow Hoz[27] in suggesting that

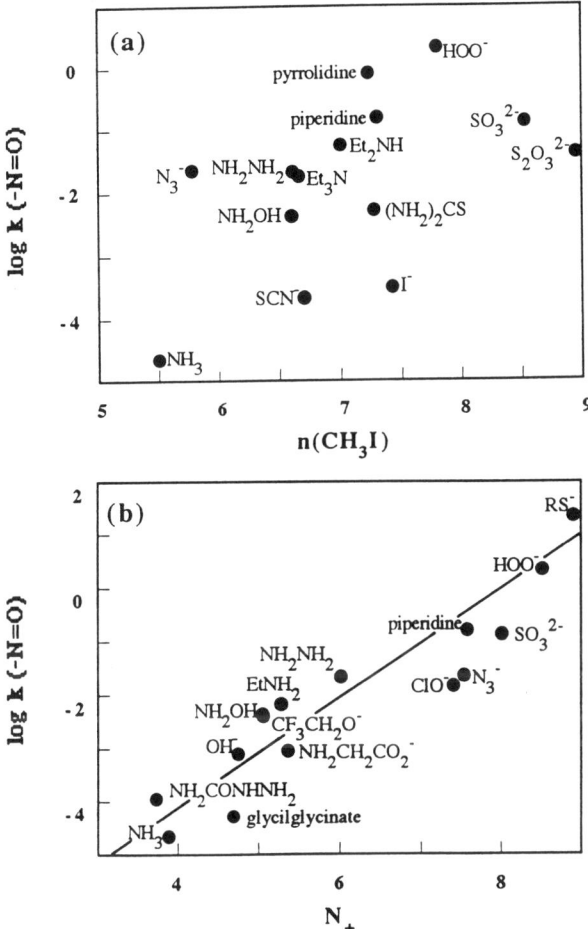

Figure 2: Correlation of nucleophilic reactivity k ($M^{-1}s^{-1}$) at 25 °C towards the nitroso group (of MNTS or EEN) in terms of (a) Pearson's n_{MeI} or (b) Ritchie's N_+ scales.

the reactivity of low-lying LUMO electrophiles with nucleophiles is governed by the formation of transition states with considerable radicaloid nature and adherence to the N_+ scale. This seems to be the case for reactions at the nitroso group.

Comparison between the trends in reactivity towards the nitroso group and towards the sulfonyl group is revealing. Although the number of nucleophiles for which rate constants towards the sulfonyl group of MNTS is very limited (O nucleophiles), our results indicate that the 2nd order rate constant for reaction of OH⁻ with the sulfonyl group (k = 0.12 M⁻¹ s⁻¹) is at least 100 times larger than the corresponding rate constant for N_3^- as the nucleophile (N_3^- reacts > 97% at the nitroso group with a rate constant k = 0.02 M⁻¹ s⁻¹). This indicates a pattern of reactivity absolutely different and suggests that nucleophilicity towards the sulfonyl group of MNTS is probably better described in terms of the Swain-Scott n scale.[30] This different pattern of reactivity can be attributed to a less radicaloid nature of the transition state through which reaction at this position takes place. It therefore seems that, whereas reaction at the softer electrophilic centre is governed by the N_+ scale, reaction at the harder sulfonyl group follows the n scale. According to Hoz,[27] reactions at the nitroso group will have a strong component of diradicaloid nature, whereas reaction at the sulfonyl group will not.

Further insight into the factors affecting reaction site or reactivity trends can be gained from the more classical concepts of charge-controlled and orbital-controlled reactions developed by Klopman.[31] The change in energy (ΔE) accompanying nucleophile-electrophile interaction is given by:

$$\Delta E = -\frac{q_n q_e}{\varepsilon R_{ne}} + \frac{2 C_n^2 C_e^2 \beta_{ne}^2}{E_{HOMO} - E_{LUMO}} \quad (4)$$

where the first term, containing charges of the nucleophile and electrophile, the distance between them, and the dielectric constant of the medium, is the Coulomb term, a measure of the electrostatic interaction. The second term or orbital term contains the energies of the HOMO and LUMO frontier orbitals, the MO coefficients of nucleophile and electrophile in these orbitals and the bonding integral β. When the HOMO-LUMO energy gap is very small, ie when single electron transfer is energetically 'easy', the orbital term dominates and the vIP of the nucleophile will be related to its overall reactivity. This seems to be the case for reactions at the nitroso group, which follow the N_+ scale, and are dominant for soft nucleophiles (carbanions, N nucleophiles, S nucleophiles, I⁻). On the contrary, reaction at the sulfonyl group is a result of a charge-controlled process and is therefore dominant for harder nucleophiles. This interpretation implies that competition between the two electrophilic centres in the molecule arises because S is the more electron-deficient centre in the molecule and coulombic interactions favour reaction at this position, but the LUMO frontier orbital of MNTS is located mainly on the nitroso group and therefore reaction at this position is orbitally favoured.

Further evidence for the key role played by the ease of electron transfer from the nucleophile in determining its reactivity towards the nitroso group was found very recently when working with ascorbate in basic media.[21] The reaction was studied at pH values in the range 10-13, and the influence of acidity upon the reaction rate is consistent with the reactive form of the nucleophile being the dianion of ascorbic acid ($pK_1 = 4$; $pK_2 = 12$).[32] There are two salient features in the reactivity of this O nucleophile. Although O nucleophiles (except for the α-effect nucleophiles ClO^-, HOO^- and acetohydroxamate anion) are usually quite poor nucleophiles (see Figure 2) towards the nitroso group, the second order rate constant for reaction between the dianion and EEN was measured as *ca* 300 M^{-1} s^{-1}, a value which makes the dianion of ascorbic acid the *strongest* nucleophile examined in the present work. If we compare its reactivity with that shown by the oxygen nucleophile of similar basicity, $CF_3CH_2O^-$, we find that the ascorbate dianion is some 10^5 times more reactive than trifluoroethoxide ion. In addition, analysis of the reaction products in the case of reaction with MNTS shows that nucleophilic attack takes place *exclusively* at the nitroso group of MNTS. (Nitrosation of ascorbic acid in acid medium has been kinetically studied and is a reaction of significant biological relevance.)[33] The extraordinary tendency of this O nucleophile to react with nitroso compounds can be linked to its reductive properties (ascorbic acid is a reductone). One is tempted to conclude that its great ability to undergo oxidation (electron transfer) is undoubtedly associated with a low vIP, which enhances its reactivity, makes the orbital term in equation (4) dominant and therefore leads to 100% reaction at the nitroso group.

3 REACTION OF AMBIDENT NUCLEOPHILES WITH AMBIDENT ELECTROPHILES

An interesting situation arose[21] during the study of the reaction of MNTS (an ambident electrophile) with the ambident nucleophile phenolate ion (C- and O-nucleophile). Analysis of the reaction products showed the existence of two reaction pathways: one leading to the formation of *p*-nitrosophenolate ion and the other leading to phenyl tosylate (see Scheme 4).

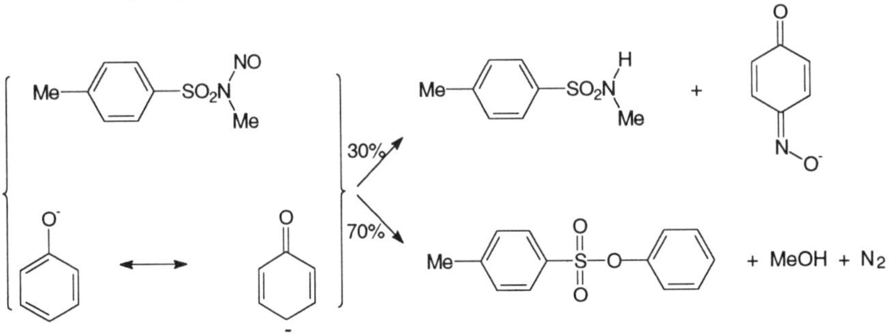

Scheme 4: Reaction of *N*-methyl-*N*-nitroso-*p*-toluenesulfonamide (MNTS) with phenolate.

We are in a situation of ambident reactivity: two products are formed, one of them corresponds to attachment of the O nucleophilic centre to the sulfonyl electrophilic centre, whereas the other results from attachment of the carbon nucleophilic centre to the nitroso group in MNTS. It then seemed obvious that alteration of the importance of the two competitive reaction pathways could be achieved by adequate substitution in the phenolate ion. The first choice was to work with the anion of *p*-cresol: methylation at the *para* position would prevent formation of the *p*-nitroso derivative, so we expected only the sulfonic ester would be formed during the course of the reaction. Surprisingly, analysis of the reaction products showed a decreased proportion of the sulfonic ester (which only accounted for *ca* 50% of the reaction); the remaining 50% still corresponded to nucleophilic attack on the nitroso group, as shown by formation of 50% *N*-methyl-*p*-toluenesulfonamide. The products of this attack at the nitroso group consisted of *o*-nitroso-*p*-methylphenolate ion and nitrite ion. These findings were at odds with an ambident pattern of reactivity such as that shown in Scheme 4. Furthermore, when the 2,4,6-trimethylphenolate ion was used the results were even more surprising. The observed reaction rates did not decrease upon methylation in the ring, and the sulfonic ester corresponding to attack at the sulfonyl group through the O atom was not present in the reaction products. Analysis of the reaction products showed quantitative formation of *N*-methyl-*p*-toluenesulfonamide and nitrite ion.

This means that methylation of the activated position of the ring does not result in reduced reactivity; on the contrary, the overall reaction rate increases upon methylation. Furthermore, C-methylation does not increase the products derived from nucleophilic attack *via* the O atom at the sulfonyl group, but increases the amount of reaction at the nitroso group.

These facts, together with the observed formation of nitrite ion in the reaction mixtures, which points to the formation at some stage of the reaction of intermediates bearing O-NO moieties, led us to propose that reaction of phenolate ions always occurred through the O atom, either at the nitroso or at the sulfonyl group, and that the final products observed depended upon the evolution of pathways of the unstable O-nitrosated intermediate. Apparently, this O-nitrosated intermediate can evolve in different ways: internal rearrangement of the NO group to yield stable C-nitroso compounds (in the oximate form) or, if substitution in the ring makes this rearrangement less likely or impossible, hydrolysis to yield nitrite ion.[34] The fact that the preferential reaction site of the phenolate oxygen moves from S to N upon substitution at the ring is in part a consequence of the sterically greater difficulty for reaction at the more crowded S atom, whereas the overall increase in reaction rate with methylation reflects the increased basicity (and nucleophilicity) of the O atom by the presence of electron-donating substituents in the ring. The kind of internal rearrangement we are proposing is somewhat similar to the Fischer-Hepp internal rearrangement undergone by *N*-methyl-*N*-nitrosoaniline.[4] The ability of the nitroso group to undergo intramolecular rearrangements seems a main factor governing its chemistry.

In conclusion, the versatile behaviour of the nitroso group in its ability to undergo internal rearrangements and of NO-carriers in their ability to take advantage of the existence of several nucleophilic centres in a substrate that can be nitrosated, result in a remarkable variety of reaction pathways in the chemistry of reactions involving the nitroso group.

REFERENCES

1. B C Challis and A J Lawson, *J Chem Soc, Perkin Trans 2*, 1973, 918.
2. P E Verkade, J Lieste, and E G G Werner, *Recl Trav Chim Pays-Bas*, 1945, **64**, 289.
3. C Bravo, P Hervés, J R Leis, and M E Peña, *J Chem Soc, Perkin Trans 2*, 1992, 185.
4. D L H Williams, 'Nitrosation', Cambridge University Press, 1988.
5. A Castro, M Mosquera, M F Rodríguez Prieto, J A Santaballa, and J Vázquez Tato, *J Chem Soc, Perkin Trans 2*, 1988, 1968.
6. G Hallett, and D L H Williams, *J Chem Soc, Perkin Trans 2*, 1980, 1372; J Casado, A Castro, M Mosquera, M F Rodríguez Prieto, and J Vázquez Tato, *Ber Bunsenges Phys Chem*, 1983, **87**, 1211.
7. J Casado, A Castro, J R Leis, M Mosquera, and M E Peña, *Monatsh Chem*, 1984, **115**, 1047.
8. A Castro, E Iglesias, J R Leis, M E Peña, and J Vázquez Tato, *J Chem Soc, Perkin Trans 2*, 1986, 1725.
9. R B Homer and C D Johnston, in 'The Chemistry of Amides', ed J Zabicky, Wiley, New York, 1970.
10. B C Challis, J N Iley, and H S Rzepa, *J Chem Soc, Perkin Trans 2*, 1983, 1037.
11. R L Hinman and E B Whipple, *J Am Chem Soc*, 1962, **84**, 2534; R L Hinman and J Land, *J Am Chem Soc*, 1964, **86**, 3796.
12. C Bravo, P Hervés, J R Leis, and M E Peña, *J Chem Soc, Perkin Trans 2*, 1991, 2091.
13. F Norberto, J A Moreira, E Rosa, J Iley, J R Leis, and M E Peña, *J Chem Soc, Perkin Trans 2*, in press.
14. L K Keefer, J A Hrabie, B D Hilton, and D Wilbur, *J Am Chem Soc*, 1988, **110**, 7459.
15. T A Meyer and D L H Williams, *J Chem Soc, Chem Commun*, 1983, 1067.
16. K A Jorgensen, *J Org Chem*, 1985, **50**, 4758.
17. A Castro, E Iglesias, J R Leis, J Vázquez Tato, F Meijide, and M E Peña, *J Chem Soc, Perkin Trans 2*, 1987, 651.

18   B C Challis, and D E G Shuker, *J Chem Soc, Chem Commun*, 1979, 315; S Oae, N Asai, and K Fujimori, *J Chem Soc, Perkin Trans 2*, 1978, 1124.
19   (a) A Castro, J R Leis, and M E Peña, *J Chem Soc, Perkin Trans 2*, 1989, 1861; (b) L García-Río, E Iglesias, J R Leis, M E Peña, and A Ríos, *J Chem Soc, Perkin Trans 2*, 1993, 29.
20   A Coello, A Gómez, F Meijide, and J. Vázquez Tato, *J Chem Soc, Perkin Trans 2*, 1989, 1677.
21   J R Leis, M E Peña, and A Ríos, manuscript in preparation.
22   J R Leis, M E Peña, and A Ríos, *J Chem Soc, Perkin Trans 2*, 1993, 1233.
23   R G Pearson, 'Hard and Soft Acids and Bases', Dowden, Hutchinson and Ross, Inc: Stroudsbourg, PA, 1973.
24   J T Thompson, and D L H Williams, *J Chem Soc Perkin Trans 2*, 1977, 1932.
25   R G Pearson, H Sobel, and J Songstad, *J Am Chem Soc*, 1968, **90**, 319.
26   C D Ritchie, *Acc Chem Res*, 1972, **5**, 34.
27   S Hoz, *Acc Chem Res*, 1993, **26**, 69.
28   C D Ritchie, *Can J Chem*, 1986, **64**, 2239.
29   S S Shaik, *J Org Chem*, 1987, **52**, 1563.
30   C G Swain and C B Scott, *J Am Chem Soc*, 1953, **75**, 141.
31   G Klopman, 'Chemical Reactivity and Reaction Paths', Wiley, New York, 1973.
32   M Kumler and T Daniels, *J Am Chem Soc*, 1935, **57**, 1929.
33   C A Bunton, H Dahn, and L Loewe, *Nature*, 1959, **183**, 163; S S Mirvish, *Toxicol Appl Pharmacol*, 1975, **31**, 325.
34   E Iglesias, L García-Río, J R Leis, M E Peña, and D L H Williams, *J Chem Soc, Perkin Trans 2*, 1992, 1673.

ACKNOWLEDGEMENTS

I would like to thank my co-workers whose names appear in the references, especially Dr E Iglesias for her continuous cooperation during the past few years and Miss Ana Ríos, who did much of the experimental work described in this communication. A special mention of course to my husband, Dr J R Leis, who shares with me the supervision of our research group. Financial support from the Dirección General de Investigación Científica y Técnica of Spain (project PB90-0767) is gratefully acknowledged.

SELF-ASSEMBLY OF [n]ROTAXANES

Martin Bělohradský, Douglas Philp,
Françisco M Raymo, and J Fraser Stoddart*

*School of Chemistry, University of Birmingham,
Edgbaston, Birmingham B15 2TT, UK.*

1 PREAMBLE

Self-assembly processes[1,2] are used by nature in several different guises to generate the wide diversity of structures observed at both the cellular and sub-cellular levels. In all the self-assembly pathways found in natural systems, we witness large structurally-complex assemblies constructed from small relatively simple sub-units, with or without some external influence being applied. The reason why such structures are formed so easily and precisely is a reflection of the mutual recognition between their sub-units. Usually, these arrays exhibit specific biological functions that could not be performed by any of their sub-units on their own. In the unnatural world created[3] by the chemist, the potential of wholly-synthetic systems,[4] displaying molecular recognition and mimicking nature in their self-assembly[5,6] and self-synthesis,[7,8] is now beginning to be realised. The ultimate goal for the contemporary chemist is to control the assembly, the form, and the function of man-made structures[9-15] with the same degree of precision that is displayed by nature.

2 INTRODUCTION

The idea of synthesising molecular compounds composed of separate entities that are mechanically linked to each other has aroused the interest of many investigators.[16-19] Rotaxanes are molecules containing a linear component (the rod) encircled by one or more macrocyclic components (the rings). In order to prevent the rings from slipping off the ends of the linear component, it must be terminated at both ends by large blocking groups (the stoppers), giving a so-called dumbbell component. An [n]rotaxane is formed by n components, namely the dumbbell plus n-1 rings. Thus, a [2]rotaxane incorporates one dumbbell and one ring, a [3]rotaxane — one dumbbell and two rings, and a [4]rotaxane — one dumbbell and three rings (Figure 1).

It has occurred to us that rotaxanes offer the possibility of constructing molecular devices by careful control of the interactions between the ring and

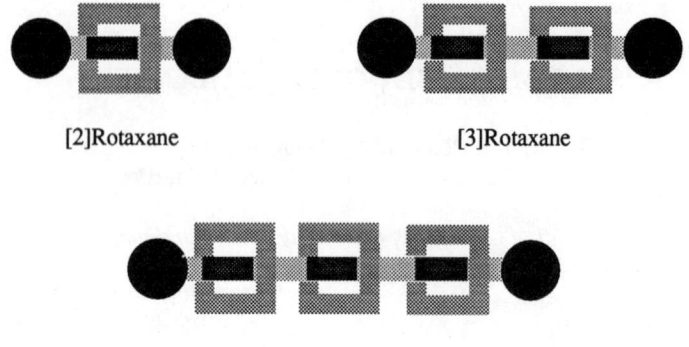

Figure 1: Schematic representation of a [2]rotaxane, a [3]rotaxane, and a [4]rotaxane.

dumbbell components. Therefore, our principal objective has been to develop synthetic strategies capable of generating molecular arrays with potential device-like character from appropriate molecular components by self-assembly processes. The observation that particular crown ethers, containing π-electron rich hydroquinol rings, bind to the well-known bipyridinium herbicide,[20] paraquat, prompted us to examine the ability of molecules incorporating such units[21] to form rotaxanes. The π-π stacking interactions between the π-electron deficient bipyridinium units and the π-electron rich hydroquinol rings of the crown ethers, together with the hydrogen bonding between the α-hydrogen atoms of the bipyridinium units and the oxygen atoms of the polyether chains, act as the main driving forces for the molecular self-assembly processes. A [2]rotaxane, incorporating two bipyridinium units both able to interact with the bisparaphenylene-34-crown-10 (**BPP34C10**) encircling the dumbbell component, is illustrated in Figure 2. This [2]rotaxane exhibits a shuttling of the ring along the rod of the dumbbell component between the two recognition sites. The rate of shuttling, which can be measured by variable temperature ^{1}H-NMR spectroscopy,[22] is *ca* 300,000 times a second at room temperature. By exploiting the stabilising noncovalent bonding interactions between bipyridinium units and **BPP34C10**, we have been able to devise three different synthetic approaches (Figure 3) to rotaxanes.[21] In the case of clipping,[21,23,24] a complete dumbbell-shaped component is preformed and the ring is then clipped around it. In the case of threading,[22,25,26] the complexation between a preformed ring and a rod-like component is followed by the covalent bonding of two stoppers which prevent the unthreading of the ring from the dumbbell component. In the case of slipping,[27] the ring and dumbbell-shaped components are preformed separately and then cajoled into associating one with the other under the influence of just the right amount of thermal energy.

Self-Assembly of [n]Rotaxanes

We have employed,[22,27,28] successfully both the threading and slipping procedures in order to synthesise rotaxanes formed by a π-electron deficient dumbbell component, incorporating from one to three bipyridinium units, encircled by one, two or three π-electron rich **BPP34C10** rings.

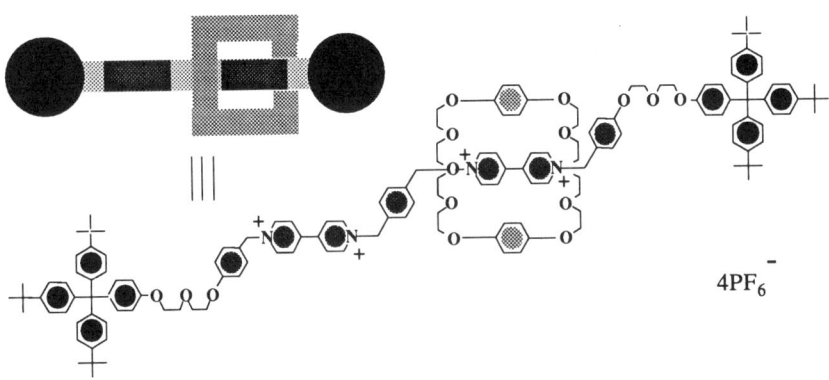

Figure 2: A [2]rotaxane incorporating two bipyridinium units in its dumbbell component.

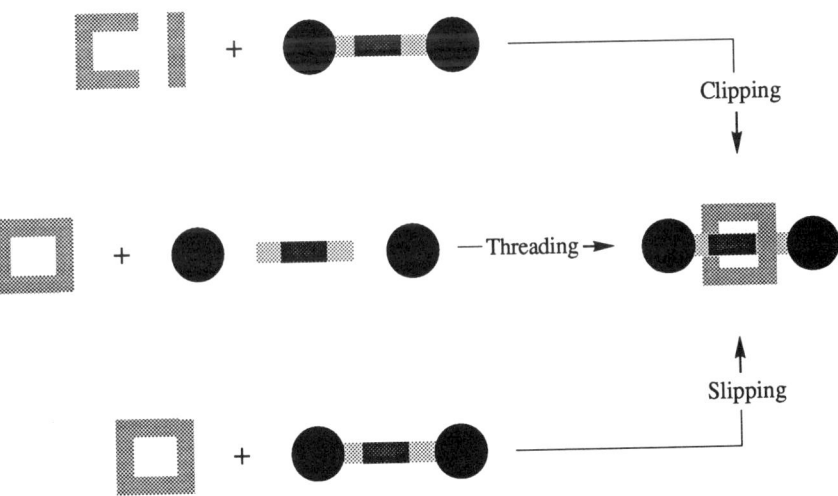

Figure 3: Three different synthetic approaches to rotaxanes.

We have yet to demonstrate the viability of the clipping approach to rotaxanes of the type described in Figure 2. Because of the instability of substituted bipyridinium units in basic media, a successful approach will require either acidic or neutral conditions for the cyclisation of a π-electron rich ring around a π-electron deficient dumbbell-shaped component. Esterification between a diol and a diacid dichloride might permit the clipping of macrocyclic lactones around the appropriate dumbbell molecule to afford the desired rotaxanes.

3 THE THREADING APPROACH

The self-assembly of rotaxanes, employing the threading methodology, is based on the mutual recognition between the π-electron deficient 4,4'-bipyridinium dications and the π-electron rich hydroquinol rings contained within the macrocyclic polyether. Preliminary results[22] showed high selectivity during this self-assembly route, requiring *two* bipyridinium dications to be present in the [2]rotaxane ultimately formed in order that the *one* **BPP34C10** ring could be incorporated into it. However, more detailed studies of the threading approach to the formation of rotaxanes has shown that the self-assembly process is strongly

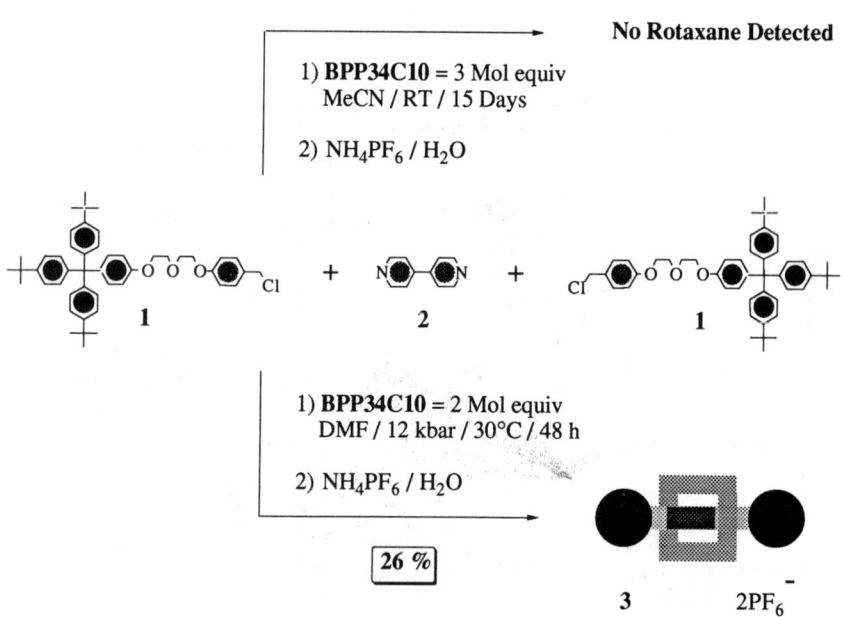

Scheme 1: Synthesis of a [2]rotaxane by threading under ultrahigh pressure conditions.

influenced by the ultrahigh pressures as well as by the molar ratio of the components forming the dumbbells and the ring component. The pronounced effect of the ultrahigh pressure on the threading process is illustrated by the fact that no rotaxanes have so far been isolated at ambient pressure with systems containing one or two bipyridinium units. Thus, when we reacted (Scheme 1) the benzylic chloride 1 with bipyridine 2, in the presence of the **BPP34C10** at ambient pressure, no rotaxane was detected. However, reaction between these same components under ultrahigh pressure conditions gave a 26 % yield of the [2]rotaxane 3. We have identified a similar marked influence of ultrahigh pressure in self-assembling systems containing two bipyridinium units (Scheme 2). Rotaxanes are formed only under ultrahigh pressure conditions. The proportions of the [2]rotaxane 5 and the [3]rotaxane 6 present in the reaction mixture depend upon the molar ratio of **BPP34C10** to the linear component 4. When a molar equivalent of 4 was reacted with 1.5 molar equivalents of the crown ether under ultrahigh pressure conditions in the presence of an excess of the benzylic chloride 1, the molar ratio of 5:6 was 6:1. However, when a fourfold molar excess of **BPP34C10** was employed, the molar ratio changed to 1:6 for **5:6**.

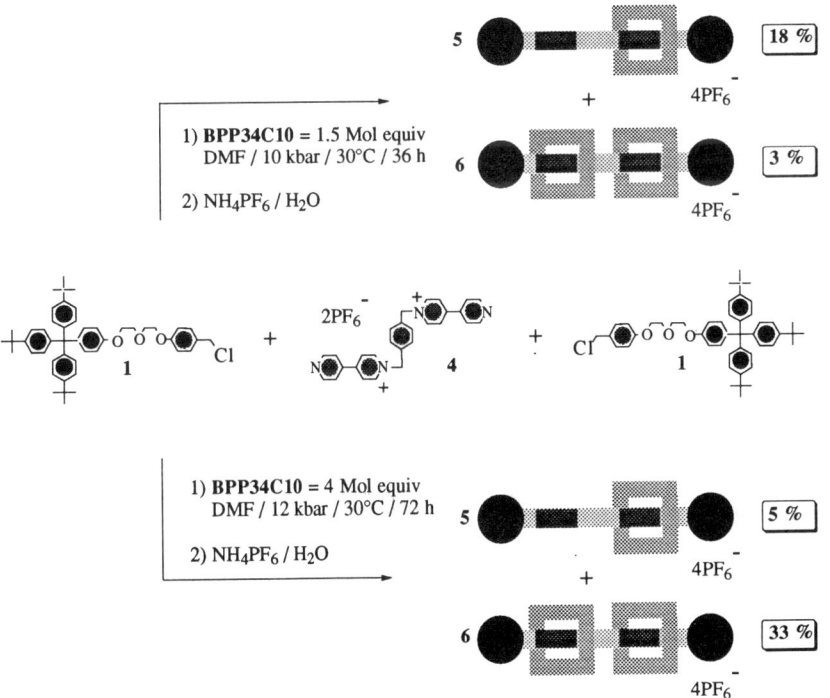

Scheme 2: Synthesis of a [2]- and [3]-rotaxane by threading under ultrahigh pressure conditions.

These results suggest that the key step in the threading processes — the formation of pseudorotaxane intermediates — is influenced strongly by pressure as well as by the relative concentrations of **BPP34C10** and the dumbbell component. Presently, we are unable to explain these results in terms of the relative stabilities of the complexes formed by the linear components and **BPP34C10** at ambient pressure. Starting from **7**, and by employing the threading approach (Scheme 3), we have been able to synthesise the [2]-, [3]-, and [4]-rotaxanes — **8**, **9**, and **10** respectively — incorporating three bipyridinium units. Surprisingly, even when 2.2 molar equivalents of **BPP34C10** were employed, the major product of the reaction was the [4]rotaxane **10**. The explanation may be related to the relative solubilities of the intermediate components. The solubility in DMF of the intermediate half-dumbbell component carrying five positive charges is undoubtedly low until it complexes with **BPP34C10**. This situation probably helps to determine the outcome of the self-assembly process. The more **BPP34C10** rings that are complexed, the greater will be the solubility of the intermediate pseudorotaxanes leading to the final products. In this manner, the most highly threaded [4]rotaxane **10** becomes a favoured product.

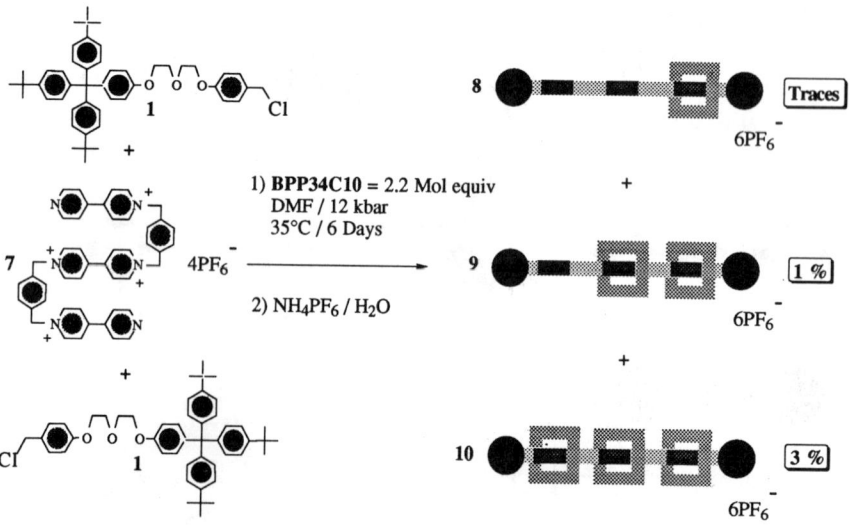

Scheme 3: Synthesis of a [2]-, a [3]-, and a [4]-rotaxane by threading under ultrahigh pressure conditions.

4 THE SLIPPING APPROACH

The stoppers, incorporating tris(4-*t*-butylphenyl)methyl groups, of the rotaxanes prepared by threading were designed to be large enough to prevent passage of the **BPP34C10** ring over them. We therefore reasoned that, by judicious adjustment of the size of the stoppers, we would arrive at a situation where the size complementarity between the **BPP34C10** ring and the stoppers would be such that the slipping of the macrocycle over them would become possible. The complexation of the bipyridinium dication within the dumbbell-shaped component by the macrocycle would then provide a *thermodynamic trap* for the crown ether, thereby increasing the activation energy (Figure 4) for the extrusion (dethreading) process with respect to the slipping (threading) process. In order to investigate the slipping approach as a synthetic route to rotaxanes, we have synthesised (Scheme 4) a range **12a-d** of dumbbells incorporating one bipyridinium unit, where the size of the stoppers was varied systematically. After heating **12a-c** with an excess of **BPP34C10** in acetonitrile at 55°C for 10 days, we were able to isolate the [2]rotaxanes **13a-c** in good yields.

By contrast, no rotaxane was isolated under otherwise identical conditions, starting from the dumbbell **12d** containing 4-isopropylphenyl-bis(4-*t*-butylphenyl)methyl groups as the stoppers. Thus, in practice, the barrier for the slipping process is surpassed on changing from the ethyl group to the branched isopropyl substituent.

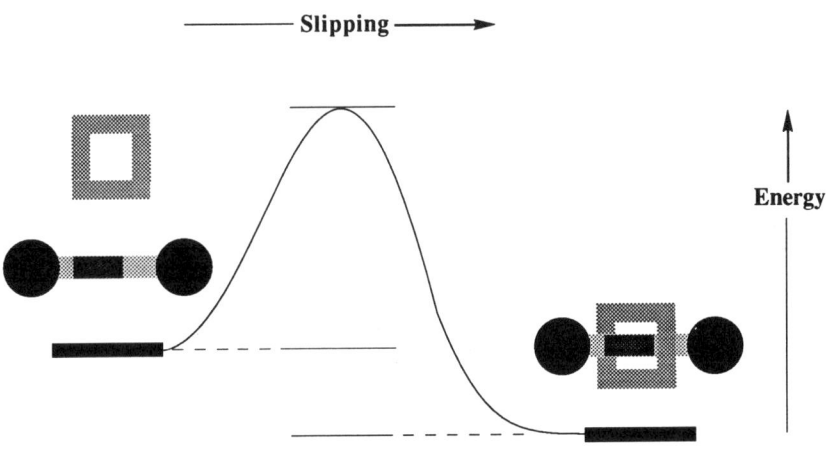

Figure 4: The formation of a [2]rotaxane by slipping driven thermodynamically.

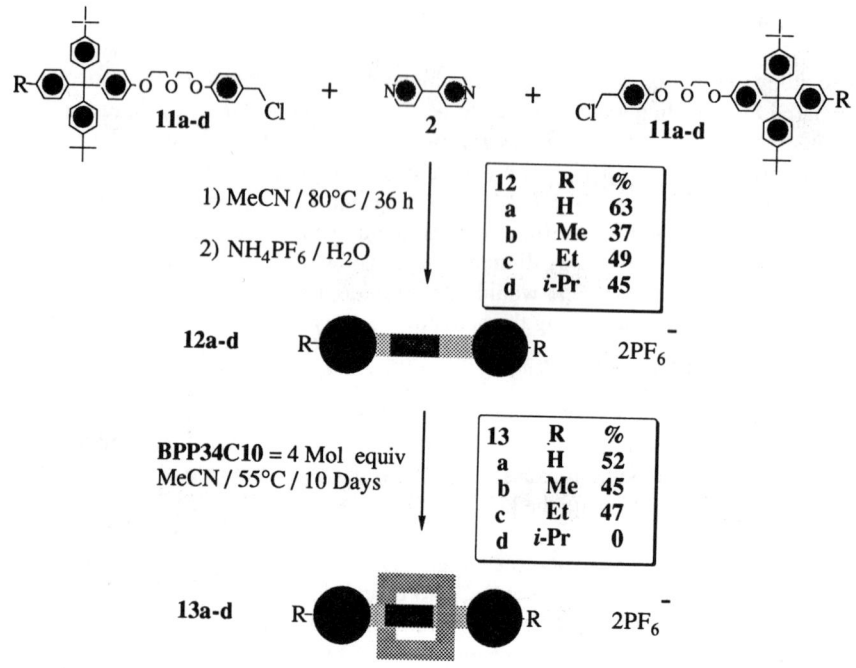

Scheme 4: Synthesis of some [2]rotaxanes by slipping.

The [2]rotaxanes **13a-c** are stable compounds at room temperature. They have been purified by silica gel chromatography and characterised by FABMS and by ^{1}H-NMR and ^{13}C-NMR spectroscopies. In contrast with the analogous [2]rotaxane **3**, prepared by the threading approach and incorporating the larger tris(4-*t*-butylphenyl)methyl stoppers, the extrusion of the **BPP34C10** can be observed by ^{1}H-NMR spectroscopy in d_6-DMSO at 100 °C. The successful syntheses in good yields of these [2]rotaxanes with one recognition site demonstrate the preparative utility of the slipping approach to rotaxane formation. Although size-complementarity has been exploited previously[29,30] in the statistical syntheses of [2]rotaxanes, the addition of a *thermodynamic trap*, in the form of noncovalent bonding interactions within the [2]rotaxane components, not only enhances the yields but also increases the inherent stabilities and information contents of the resulting molecular structures. These features and the synthetic simplicity of the slipping approach do much to recommend this method as an alternative synthetic procedure for the construction of larger oligorotaxanes and polyrotaxanes. We have applied the methodology to self-assembly rotaxanes containing two and three bipyridinium units.

Self-Assembly of [n]Rotaxanes   395

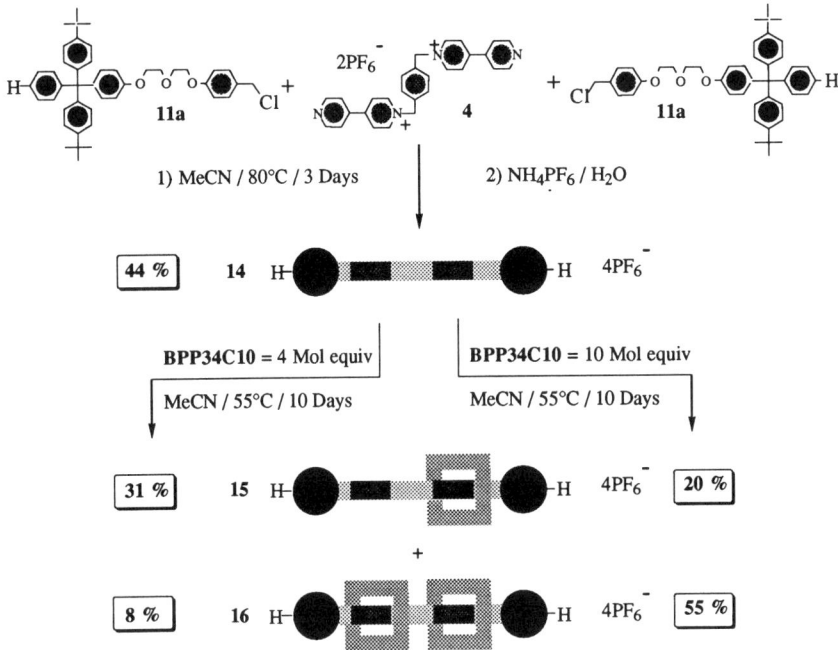

Scheme 5: Synthesis of a [2]- and a [3]-rotaxane by slipping.

After heating the dumbbell **14** with 4 molar equivalents of **BPP34C10** in acetonitrile at 55°C for 10 days we have been able to isolate (Scheme 5) both the [2]rotaxane **15** (31 %) and the [3]rotaxane **16** (8 %). When 10 molar equivalents of the crown ether were employed under the same reaction conditions, the yields of **15** and **16** were 20 % and 55 %, respectively. Similarly, on heating the dumbbell **17** with 20 molar equivalents of **BPP34C10**, we have been able to isolate (Scheme 6) the [2]rotaxane **18** (2 %), the [3]rotaxane **19** (12 %), and [4]rotaxane **20** (19 %).

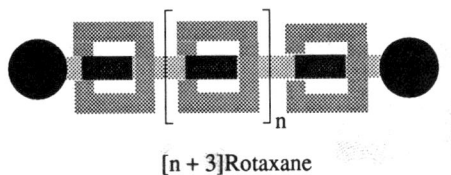

[n + 3]Rotaxane

Figure 5: Schematic representation of a polyrotaxane.

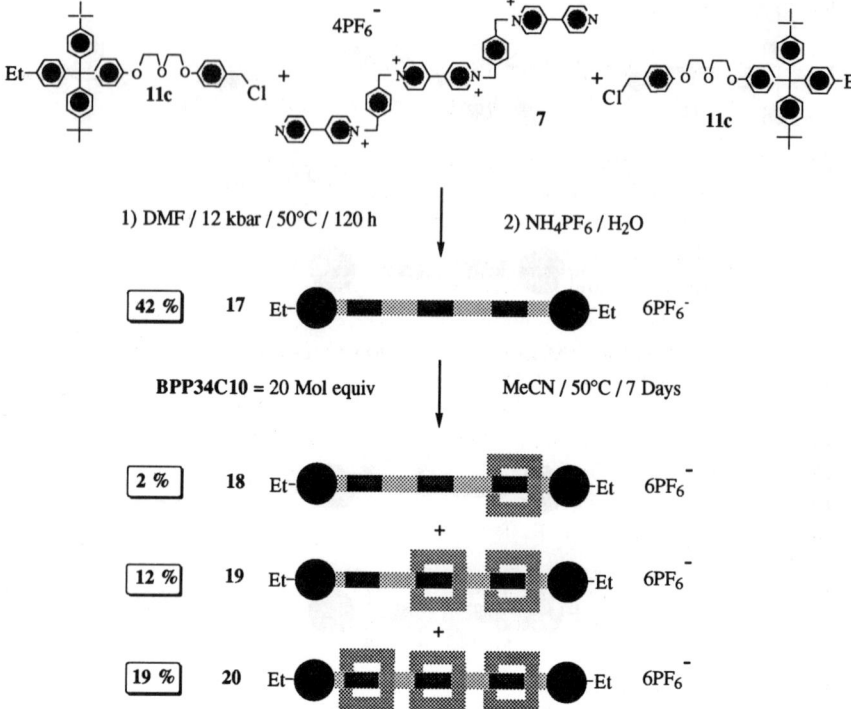

Scheme 6: Synthesis of a [2]-, a [3]-, and a [4]-rotaxane by slipping.

5 CONCLUSIONS

The self-assembly of [2]-, [3]-, and [4]-rotaxanes demonstrates conclusively that it is possible to construct wholly synthetic systems[31-41] with potential device-like properties on the nanometre-scale by relying upon the molecular recognition associated with weak non-covalent interactions between complementary components. In addition to the efficient *threading* procedure which is operative under ultrahigh pressures conditions, we have developed an alternative *slipping* approach based upon the size complementarity between the macrocyclic polyether (**BPP34C10**) and the stoppers on the dumbbell-shaped component. The synthetic simplicity of the *slipping* procedure is a strong recommendation for this methodology as a viable one for the subsequent construction (Figure 5) of oligorotaxanes and polyrotaxanes.[42-51]

REFERENCES

1. J S Lindsey, *New J Chem*, 1991, **15**, 153.
2. D B Amabilino and J F Stoddart, *Pure Appl Chem*, 1993, **65**, 2351.
3. J F Stoddart, *Nature*, 1988, **334**, 10.
4. J M Lehn, *Angew Chem, Int Ed Engl*, 1990, **29**, 1304.
5. P R Ashton, T T Goodnow, A E Kaifer, M V Reddington, A M Z Slawin, N Spencer, J F Stoddart, C Vicent, and D J Williams, *Angew Chem, Int Ed Engl*, 1989, **28**, 1396.
6. D B Amabilino and J F Stoddart, *New Scientist*, 1994, **1913**, 25.
7. C Fouquey, J M Lehn, and A M Levelut, *Adv Mater*, 1990, **2**, 254.
8. R C Merkle, *J Br Interplanet Soc*, 1992, **45**, 407.
9. R P Feymann, *Sat Rev*, 1960, **432**, 45.
10. K E Drexler, *Annu Rev Biophys Biomol Struct*, 1994, **23**, 377.
11. K E Drexler, *J Vac Sci Technol*, 1991, **B9(2)**, 1394.
12. K E Drexler, *J Am Chem Soc*, 1993, **115**, 11657.
13. R C Merkle, *Nanotechnology*, 1991, **2**, 134.
14. C B Musgrave, J K Perry, R C Merkle, and W A Goddard III, *Nanotechnology*, 1991, **2**, 187.
15. R C Merkle, *Nanotechnology*, 1993, **4**, 86.
16. G Schill, 'Catenanes, Rotaxanes and Knots', Academic Press, New York, 1971.
17. D M Walba, *Tetrahedron*, 1985, **41**, 3161.
18. C O Dietrich-Bichecker and J P Sauvage, *Chem Rev*, 1987, **87**, 795.
19. J C Chambron, C O Dietrich-Buchecker, and J P Sauvage, *Top Curr Chem*, 1993, **165**, 131.
20. L A Summers, 'The Bipyridinium Herbicides', Academic Press, London, 1980.
21. P L Anelli, P R Ashton, R Ballardini, V Balzani, M Delgado, M T Gandolfi, T T Goodnow, A E Kaifer, D Philp, M Pietraskiewicz, L Prodi, M V Reddington, A M Z Slawin, N Spencer, J F Stoddart, C Vicent, and D J Williams, *J Am Chem Soc*, 1992, **114**, 193.
22. P R Ashton, D Philp, N Spencer, and J F Stoddart, *J Chem Soc, Chem Commun*, 1992, 1124.
23. D Philp and J F Stoddart, *Synlett*, 1991, 445.
24. P L Anelli, N Spencer, and J F Stoddart, *J Am Chem Soc*, 1991, **113**, 5131.
25. J F Stoddart, *Angew Chem, Int Ed Engl*, 1991, **31**, 846.
26. J F Stoddart, *Chem Br*, 1991, **27**, 714.
27. P R Ashton, M Bĕlohradský, D Philp, and J F Stoddart, *J Chem Soc, Chem Commun*, 1993, 1269.
28. P R Ashton, M Bĕlohradský, D Philp, N Spencer, and J F Stoddart, *J Chem Soc, Chem Commun*, 1993, 1274.
29. I T Harrison, *J Chem Soc, Perkin Trans 1*, 1974, 301.

30   I T Harrison and S Harrison, *J Am Chem Soc*, 1967, **89**, 5723.
31   J F Stoddart, *Chem Aust*, 1992, **59**, 576.
32   P R Ashton, R A Bissell, N Spencer, J F Stoddart, and M S Tolley, *Synlett*, 1992, 914.
33   P R Ashton, R A Bissell, R Gorski, D Philp, N Spencer, J F Stoddart, and M S Tolley, *Synlett*, 1992, 919.
34   P R Ashton, R A Bissell, N Spencer, J F Stoddart, and M S Tolley, *Synlett*, 1992, 923.
35   R A Bissell and J F Stoddart, 'Computation for the Nano-Scale', Eds P E Blöchl, A J Fisher, and C Joachim, Kluwer, Dordrecht, 1993, 141.
36   E Córdova, R A Bissell, N Spencer, P R Ashton, J F Stoddart, and A E Kaifer, *J Org Chem*, 1993, **58**, 6550.
37   R A Bissell, E Córdova, A E Kaifer, and J F Stoddart, *Nature*, 1994, **369**, 133.
38   J C Chambron, S Chardon-Noblat, A Harriman, V Heitz, and J P Sauvage, *Pure and Appl Chem*, 1993, **65**, 2343.
39   R Ballardini, V Balzani, M T Gandolfi, L Prodi, M Ventura, D Philp, H G Ricketts, and J F Stoddart, *Angew Chem, Int Ed Engl*, 1993, **32**, 1301.
40   A C Benniston and A Harriman, *Angew Chem, Int Ed Engl*, 1993, **32**, 1459.
41   A C Benniston, A Harriman, V M Lynch, *Tetrahedron Lett*, 1994, **35**, 1473.
42   M Born and H Ritter, *Makromol Chem, Rapid Commun*, 1991, **12**, 471.
43   A Harada, J Li, and M Kamachi, *Nature*, 1992, **356**, 325.
44   A Harada, J Li, T Nakamitsu, and M Kamachi, *J Org Chem*, 1993, **58**, 7524.
45   G Wenz and B Keller, *Angew Chem, Int Ed Engl*, 1992, **31**, 197.
46   G Wenz, *Angew Chem, Int Chem Engl*, 1994, **33**, 803.
47   C Wu, M C Bheda, C Lim, Y X Shen, J Sze, and H W Gibson, *Polym Commun*, 1991, **32**, 204.
48   Y X Shen and H W Gibson, *Macromolecules*, 1992, **25**, 2058.
49   X Y Shen, D Xie, and H W Gibson, *J Am Chem Soc*, 1994, **116**, 537.
50   X Sun, D B Amabilino, P R Ashton, I W Parson, J F Stoddart, and M S Tolley, *Macromol Symp*, 1994, **77**, 191.
51   D B Amabilino, I W Parson, and J F Stoddart, *Trends Polym Sci*, 1994, **2**, 146.

THE SUPERELECTROPHILIC CHARACTER OF THE 4,6-DINITROBENZOFUROXAN STRUCTURE

François Terrier

URA CNRS 403, Department of Chemistry, University of Versailles, 45, Avenue des Etats-Unis, F-78000 Versailles, France.

1   INTRODUCTION

Although 4,6-dinitro-2,1,3-benzoxadiazole N-oxide, *ie* 4,6-dinitrobenzofuroxan (DNBF), is a long known compound,[1] it was only in the 1960's that its structure was fully characterized[2-4] and in the late 1970's that clear information was obtained that this neutral 10π-electron heteroaromatic substrate exhibits a powerful electrophilic character.[5,6] The finding that DNBF undergoes a facile σ-complexation in the absence of any added base in aqueous or methanolic solution was the first illustration of this behaviour.[7,8] The pK_a's for the formation of the hydroxy and methoxy adducts (**1a**) and (**1b**) according to equation 1 are equal to 3.75 (25°C) and 6.45 (20°C), respectively, in the corresponding solvents.[8] This makes these σ-complexes about 10^{10} times thermodynamically more stable than the analogous adducts (**2a**) and (**2b**) of 1,3,5-trinitrobenzene (TNB), which form exclusively through equation 3 and are the common references in anionic σ-complex chemistry.[5,9]

Interestingly, the pK_a value for water addition to DNBF is also lower than that for pseudobase formation from the positively charged aromatic tropylium cation ($pK_a = 4.7$) in aqueous solution.[10]

The rate constants for the various pathways of equations 1-3 are given in Table 1. As can be seen the greater thermodynamic stability of the DNBF adducts relative to the TNB adducts is the result of much higher rates of nucleophilic attack and of much lower rates of decomposition of these adducts. These data support the idea that two main factors contribute to the exceptional ease of formation and stability of (1a) and (1b).[5] The first is the relatively low aromaticity of the benzofuroxan system which favors the covalent addition of the nucleophile compared to most electron-deficient aromatic and heteroaromatic systems. The second factor is the combination of the strong electron-withdrawing effects of the two nitro groups and the annelated furoxan ring. The result is not only a high electron-deficiency at C-7 of DNBF, and hence a high susceptibility of this position to nucleophilic attack, but also a high electron-delocalizing capability of the negative charge which enhances the stability of the resulting σ-adducts.

Table 1: Rate and equilibrium constants for the formation and decomposition of DNBF and TNB adducts (1) and (2) in aqueous or methanolic solution.[a]

Adduct	$k_2^{RO} dm^3$ $mol^{-1}s^{-1}$	$k_{-2}^{RO} s^{-1}$	$K_2^{RO} dm^3$ mol^{-1}	$k_1^{ROH} s^{-1}$	$k_{-1}^{H} dm^3$ $mol^{-1}s^{-1}$	pK_a^{ROH}
(1a)[a]	33500	2.5×10^{-6}	1.78×10^{10}	0.035	146	3.75
(1b)[a]	1.87×10^6	8.9×10^{-5}	2.1×10^{10}	0.03	4.68×10^4	6.46
(2a)[b]	33.9	8	4.21	-	-	13.37
(2b)[b]	7050	305	23.1	-	-	15.40

(a) 25°C; F Terrier in *Nucleophilic Aromatic Displacement*, ed H Feuer, VCH, New York, 1991; (b) ref 9.

Following the discovery of its unique reactivity towards weak oxygen bases like water and methanol and the demonstration that neither of the two types of tautomerism shown in equations 4 and 5 is important in its reactions,[2,11] the use of DNBF to assess the reactivity of carbon nucleophiles has received considerable attention. It is the purpose of this paper to discuss some of the most remarkable systems studied in this context, focusing in particular on the evidence recently obtained that DNBF is a more powerful electrophile than the *p*-nitrobenzenediazonium cation and a much stronger electrophile than the proton.

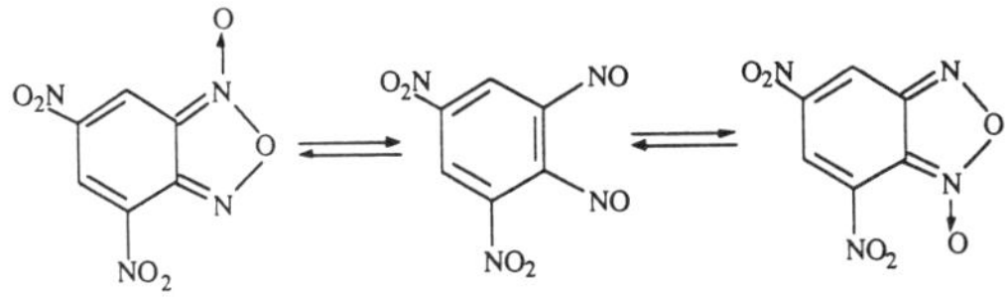

2 σ-COMPLEXATION OF DNBF BY VERY WEAK CARBON BASES

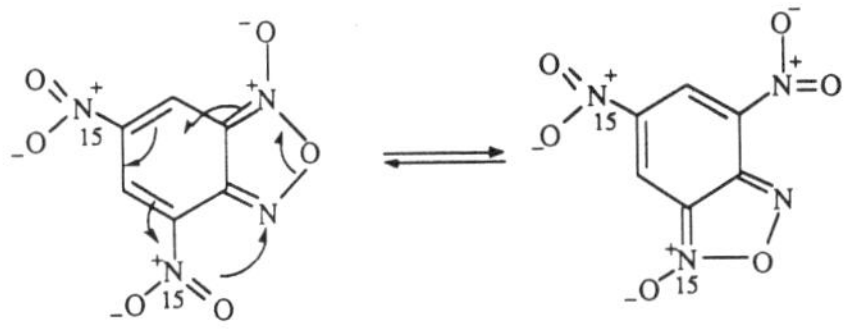

Coupling of DNBF to weakly activated enolic double bonds (pK_a values in the range -2 to +2 have been reported for C-protonation of a number of simple enols)[12] is readily achieved in the absence of any added base in dimethylsulfoxide or acetonitrile solution.[13] A number of ketones and β-diketones thus react according to Scheme 1 to afford quantitatively the keto adducts (3) in their acidic form. With β-diketones, subsequent enolization of (3) to (4) occurs. In none of the reactions, could initial attack via the enolic oxygen atoms to form O-adducts under kinetic control be detected. Nitroalkanes, the conjugate acids of (5a-d), also react in the absence of an external base in Me$_2$SO to give C-adducts of type (6) which are presumably the result of a direct attack of DNBF by equilibrium concentrations of the aci-forms of these carbon acids.[14a] Interestingly, the kinetics of formation of (6a-d) according to the simple equilibrium of equation 6 could be studied by using buffer solutions made up

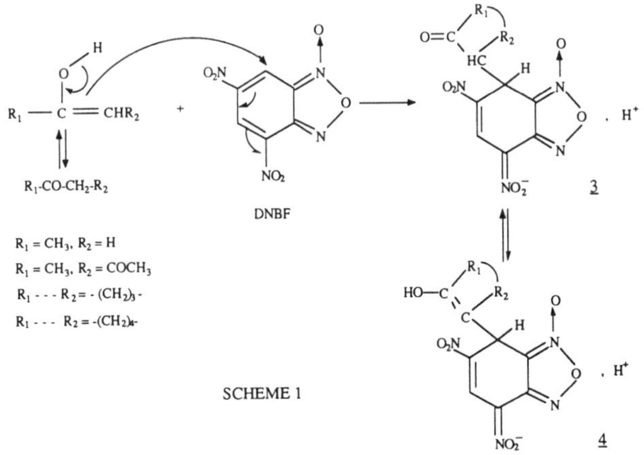

SCHEME 1

DNBF + R₁R₂C = NO₂⁻
5a R₁ = R₂ = H
5b R₁ = H, R₂ = CH₃
5c R₁ = R₂ = CH₃
5d R₁ = H, R₂ = CH₂CH₃

$\xrightleftharpoons[k_{-1}]{k_1}$

(6)

6a-d

↓ Base

7a-d

from the nitroalkanes themselves in aqueous solution. As shown in Table 2, the rate constants k_1 for formation of the adducts (**6a-d**) are very high. For example, the k_1 value for attack of DNBF by nitromethane anion ($pK_a^{H_2O} = 10.25$) is ten times greater that that for the addition of OH⁻, this latter base being 3×10^5 times stronger ($pK_a^{H_2O} = 15.74$) than the carbanion. These results suggests a much higher affinity of DNBF for carbon bases than for oxygen bases of similar pK_a's.[14a]

Table 2: Rate and equilibrium constants for the formation and decomposition of the nitroalkene adducts (**6a-d**) of DNBF in aqueous solution.[1,b]

Nitroalkane	$pK_a^{H_2O}$	Adduct	k_1/dm^3 $mol^{-1}s^{-1}$	k_{-1}/s^{-1}	K_1/dm^3 mol^{-1}
CH_3NO_2	10.25	(**6a**)	2.6×10^5	$ca\ 4 \times 10^{-5}$	$ca\ 6.5 \times 10^9$
$CH_3CH_2NO_2$	8.55	(**6b**)	26700	$< 10^{-6}$	$> 2.67 \times 10^{10}$
$CH_3CH(NO_2)CH_3$	7.85	(**6c**)	2200	1.4×10^{-5}	1.57×10^8
$CH_3CH_2CH_2NO_2$	8.98	(**6d**)	33400	$< 10^{-6}$	$> 3.34 \times 10^{10}$

(a) 25°C; I = 0.1 mol dm³ (KCl); (b) Unpublished results.

A noteworthy feature of the adducts (**6a-d**) is their susceptibility to undergo a base-catalyzed elimination of nitrous acid to afford the olefinic carbanions (**7a-d**) in DMSO.[14b] In general, the nitro group is not a good leaving group in base induced β-eliminations, being only capable of departing when there is a strong electron-withdrawing group in the β position.[15] The formation of (**7a-d**) thus implies that even the negatively charged DNBF moiety of the adducts acts as a powerful electron-withdrawing functionality. Consistent with this idea is the observation that the OH group of the hydroxy adduct (**1a**) undergoes ionization in dilute aqueous hydroxide solutions (pK$_a$ = 11.30).[8a] It is the ease of addition of a number of aromatic and heteroaromatic derivatives which provides the best illustration for the high affinity of DNBF for compounds of low carbon basicity. Treatment of DNBF by aniline in DMSO or methanol results in a rapid formation of the C-adduct (**9**) as a thermodynamically stable species and it is only under some peculiar experimental conditions that the initial and reversible formation of the N-adduct (**8**) under kinetic control can be observed.[16,17] Scheme 2 depicts the interaction in which the carbon attack followed by proton loss from (**9,H**) gives rise to the rearomatized product (**9**) in an effectively irreversible process. In fact, most primary anilines react according to Scheme 2 but, depending upon the structure of the parent amine, the S$_E$Ar substitution of the aniline ring by DNBF takes place at the position para, *eg* 1,2-diaminobenzene, 2,6-dimethylaniline, 2-chloroaniline, *o*-toluidine, or ortho, *eg* *p*-toluidine, 4-chloroaniline, to the amino group.[16-18] Several of the resulting carbon-bonded arylamine complexes were isolated as the zwitterions or the potassium salts of the corresponding anionic species.

Prior to the study of these aniline-DNBF systems, there was no report that aromatic amines could act as carbon nucleophiles in σ-complex formation and related nucleophilic aromatic substitution reactions. The complex (**10**), in which aniline is attached to two DNBF moieties *via* the carbon and nitrogen functionalities, is a nice illustration of the ambident character of arylamines.[17b]

$$\underset{\textbf{11}}{\text{[structure: 1,8-bis(NMe}_2\text{)naphthalene]}} \xrightarrow{\text{DNBF}} \underset{\textbf{12}}{\text{[zwitterionic adduct with H-bridge, DNBF}^-\text{]}} \xrightarrow{\text{excess 11}} \underset{\textbf{12}^-}{\text{[anionic adduct, DNBF}^-\text{]}} \quad (7)$$

The facile σ-complexation of DNBF by 1,8-bis(dimethyl-amino)naphthalene is a spectacular example of the ability of this heterocycle to suffer C-addition by aromatic amines.[19] As is well known, compound (11) is an unusually strong nitrogen base, commonly named PROTON SPONGE, but a very poor nitrogen nucleophile. Also, because of the steric effects related to the proximity of the two dimethylamino groups, there is little possibility of bringing even one of theseamino groups into the plane of the ring. As a result, the conjugation between thenitrogen lone pairs and the aromatic system is very low and the carbon basicity of (11) must be negligible compared to that of other aromatic amines.

However, despite this unfavorable situation, the zwitterionic C-adduct (12) is formed quantitatively on treatment of DNBF with (11) in DMSO, equation 7. Conversion of (12) into its anionic counterpart (12$^-$) occurs in the presence of excess (11), implying a lower basicity of the NMe$_2$ groups of the adduct than of the parent amine. This further confirms that a negatively charged DNBF moiety still exerts a strong acidifying effect.

The pK$_a$ values for the C-protonation equilibria of the trihydroxy and dihydroxy benzenes (13a) and (13d), as well as of the related methyl ethers (13b), (13c), and (13e) have been measured by Kresge et al in concentrated aqueous perchloric and sulfuric acids.[20] These pK$_a$ values are all very negative,

$$\underset{\textbf{13}}{\text{[benzene with R}_1\text{, R}_3\text{, R}_5\text{]}} + \text{DNBF} \underset{k_{-1}}{\overset{k_1}{\rightleftharpoons}} \underset{\textbf{14,H}}{\text{[adduct with H, DNBF}^-\text{, R}_3^+\text{]}} \xrightarrow{k_2} \underset{\textbf{14}}{\text{[adduct DNBF}^-\text{, R}_3\text{]}} + \text{H}^+ \quad (8)$$

(a) R$_1$=R$_3$=R$_5$=OH (b) R$_1$=R$_5$=OCH$_3$, R$_3$=OH (c) R$_1$=R$_3$=R$_5$=OCH$_3$
(d) R$_1$=R$_3$=OH, R$_5$=H (e) R$_1$=R$_3$=OCH$_3$, R$_5$=H (f) R$_1$=R$_5$=H, R$_3$=O$^-$

ranging from -9 for 1,3-dimethoxybenzene (13e) to -3.13 for 1,3,5-trihydroxybenzene (13a). This emphasizes the weak basic character of these compounds which all react very readily with DNBF in acidic media in various solvents to afford exclusively the C-adducts (14a-e) are according to equation 8.[21] No formation of isomeric complexes was observed in the case of the three potentially ambident carbon bases, ie 3,5-dimethoxyphenol (13c), 1,3-dihydroxybenzene (13d), and 1,3-dimethoxybenzene (13e). Interestingly, weaker carbon bases like simple phenol (pK_a = -14), anisole (pK_a = -15) or 1,3,5-trimethylbenzene (pK_a = -13.2) are totally unreactive toward DNBF under the same experimental conditions as those used for the coupling of (13a-e). The C-adduct of phenol does form, however, in basic media (NEt_3). In this case, the phenoxide anion is undoubtedly the reactive species involved in the formation of (14f).[17b]

Table 3: Second-order rate constants k for formation of the adducts (14a-e) according to equation 8 in 50-50 (v/v) H_2O-DMSO.[a]

Starting Aromatic	pK_a[b]	$k/dm^3 mol^{-1} s^{-1}$
1,3,5-Trihydroxybenzene	-3.13	790
3,5-Dimethoxyphenol	-4.35	123
1,3,5-Trimethoxybenzene	-5.72	20.4
1,3,5-Trimethoxybenzene-d_3		5.5
1,3-Dihydroxybenzene	-7.83	1.6
1,3-Dimethoxybenzene	-9.0	0.52

[a] 25 °C; [b] Ref 20.

$$k = \frac{k_1^{DNBF} k_2}{k_{-1} + k_2} \quad (9)$$

A significant feature of the reactions of equation 8 is that they could be kinetically studied in aqueous DMSO mixtures. Equation 9 gives the general expression for the second-order rate constant k associated to the formation of (14), as derived under the assumption that the zwitterions (14,H) are low concentration intermediates. Table 3 summarizes the k values measured for (14a-e) in 50-50 (v/v) H_2O-DMSO as well as some additional data for the DNBF-1,3,5-trimethoxybenzene system. As expected, the measured k values increase with increasing the basicity of the parent aromatic but the observation of a substantial kinetic isotope effect in the reaction of (13c) (k^H/k^D = 3.75) indicates that both the addition and rearomatization steps of equation 8 are of primary kinetic importance in the coupling. Accordingly, the rate constants k_1^{DNBF} for

DNBF addition to (13a-e) cannot be derived and no direct comparison with the corresponding rate constants k_1^H for protonation of these aromatics can be made.

It is from a thorough structural and kinetic study of the reactions of DNBF with a series of 5-X-substituted indoles that the electrophilic reactivity of DNBF could be quantitatively assessed not only with respect to that of the proton but also to that of a common powerful electrophile such as the p-nitrobenzenediazonium cation.[21] The indoles (15a-k) have pK_a values which range from +0.26 for the most basic derivatives, i.e. 2,5-dimethylindole (15j), to -6 for the less basic compound, 5-cyanoindole (15a),[22] but they all add rapidly to DNBF, affording the σ-adducts (16a-k in different solvents, equation 10. These adducts have a high thermodynamic stability, being in particular totally insensitive to decomposition by strong acids (2M H_2SO_4).

| 15 | 16,H | 16 | (10) |

Obviously, the S_EAr mechanism of equation 10 resembles that of equation 8 but no significant dependence of the rates of complexation on the hydrogen or deuterium labeling at C-3 of the indole ring was found in the various systems studied.[21] This shows that proton elimination from the zwitterions (16,H) is rapid ($k_2 \gg k_{-1}$) and that electrophilic attack by DNBF is the rate-determining step of the substitutions of equation 10.[23-26] Accordingly, the measured second-order rate constants k for formation of the adducts (16), as given by equation 9, becomes identical to the rate constant k_1^{DNBF}. The data obtained are summarized in Table 4 and Figures 1 and 2; they call for some interesting comments prior to a comparison with the related information available for protonation and diazocoupling by ArN_2^+ cations of indoles.

Inspection of Table 4 reveals that the rate constants k_1^{DNBF} for the 5-X-substituted indoles (15a-f) increase regularly with increasing the basicity of the indole. This behaviour is also evidenced by Figure 1 which shows that four f) different but parallel straight lines are obtained on plotting the log k_1^{DNBF} values measured in the four solvents studied versus the known $pK_a^{H_2O}$ values for (15a-f).

Table 4: Second-order rate constants k_1^{DNBF} and $k_1^{H_3O^+}$ for addition of DNBF and H_3O^+ to indoles in aqueous solution.[a]

Indole		$pK_a^{H_2O}$	k_1^{DNBF} dm^3 mol^{-1} s^{-1} [b]	$k_1^{H_3O^+}$ dm^3 mol^{-1} s^{-1} [c]
5-Cyanoindole	(15a)	-6.00	2.8 (2.79)[d]	0.095
5-Bromoindole	(15b)	-4.57	125	-
5-Chloroindole	(15c)	-4.53	125.5	-
Indole	(15d)	-3.46	1110 (940)[d]	8.17
5-Methylindole	(15e)	-3.30	5000	17.67
5-Methoxyindole	(15f)	-2.90	5260 (4760)[d]	20.5
5-Chloro-2-methylindole	(15g)	-1.30	1586	-
2-Methylindole	(15h)	-0.28	21000 (19900)[d]	646
5-Methoxy-2-methylindole	(15i)	+0.13	81000	1178
2,5-Dimethylindole	(15j)	+0.26	94000	969
N-Methylindole	(15k)	-2.32	6980	19.38

(a) 25 °C ; (b) k_1^{DNBF} values measured in 70-30 (v/v) H_2O-DMSO ; (c) $k_1^{H_3O^+}$ values calculated from available protiodetritiation k_{exch} rate constants by assuming a k_{-1}^H/k_{-1}^T ratio of 18, see ref 21 for discussion; (d) k_1^{DNBF} values for DNBF addition to indoles-3d.

Thus, while the effect of the X substituent is essentially solvent independent, the rate of complexation of DNBF by a given indole decreases markedly with decreasing the polarity of the solvent: the observed reactivity pattern is 70-30 (v/v) H_2O-DMSO > 50-50 (v/v) H_2O-DMSO > methanol > acetonitrile. This sequence agrees with the idea that the reactions must proceed through a strongly dipolar transition state, namely (17), which is expected to be especially stabilized in aqueous solvents.[27] Also consistent with the development of a significant positive charge on the indole nitrogen in (17), is the fact that relatively large negative ρ values are deduced from the Hammett relationships (not shown) obtained on plotting the log k_1^{DNBF} values versus the substituent constants σ_p (ρ ca. -3.85± 0.1).[28] More importantly, Figure 1 suggests that greater k_1^{DNBF} values would have been obtained for the formation of the adducts (16a-f) in water if the solubilities of (15a-f) made possible measurements in this solvent.

Another interesting point emerges from Figure 2 which shows that linear Brønsted plots can also be drawn from the rate data for DNBF addition to 5-X-2-methylindoles (15g-j) in the various solvents studied. These plots are parallel to, but located 1.5 log k unit below, the corresponding Brønsted lines for the indoles (15a-f). Thus, the 5-X-substituent exerts a similar electronic influence in the two series, but the presence of the methyl group in the position adjacent to the site of

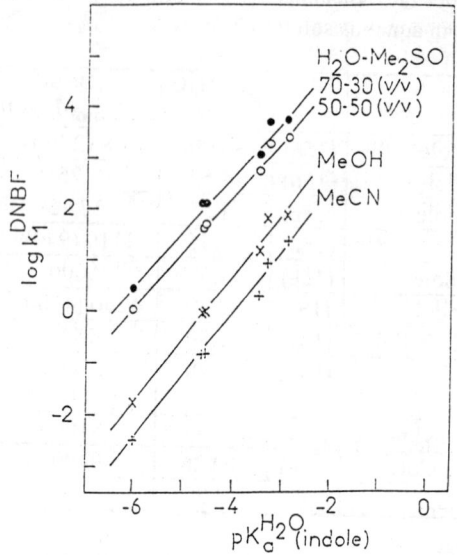

Figure 1: Effect of the indole basicity on the rates of complexation of DNBF by 5-X-substituted indoles at 25 °C in different solvents.

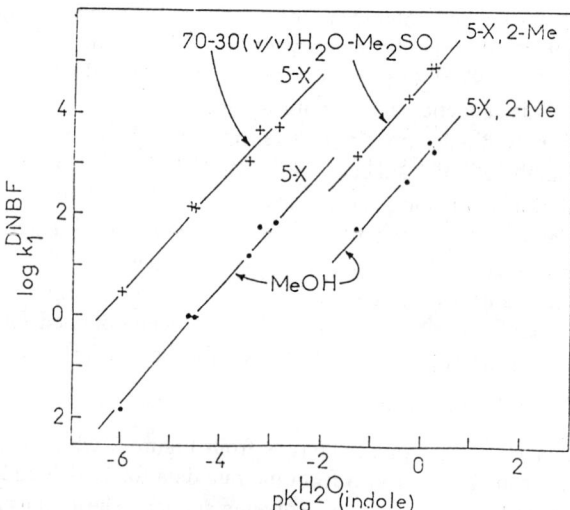

Figure 2: Effect of the presence of a 2-methyl group on the addition of DNBF to 5-X-substituted indoles at 25 °C in different solvents.

electrophilic attack reduces considerably the efficiency of the DNBF addition. This decrease can reasonably be attributed to steric hindrance to the approach of DNBF from the C-3 position of 2-methylindoles.

Keeping in mind that all the k_1^{DNBF} values obtained in 70-30 (v/v) H_2O-DMSO (Table 4) are minimized relative to aqueous solutions and that those for 5-X-2-methylindoles are further minimized for steric reasons, an interesting comparison can be made with available $k_1^{H_3O+}$ values for protonation of indoles, as derived from kinetic studies of protiodetritiation or protiodedeuteration exchange experiments in aqueous solutions.[21b,24] As discussed recently, protonation at C-3 of the indole ring is not subject to steric effects from an adjacent 2-methyl group.[21b] From Table 4, it is then clear that a neutral DNBF molecule behaves as a much stronger electrophile than the positively charged hydronium ion. The difference in electrophilic reactivity corresponds to a $k_1^{DNBF}/k_1^{H_3O+}$ ratio of 30 for 5-cyanoindole but it increases with increasing pK_a(indole), e.g. $k_1^{DNBF}/k_1^{H_3O+} = 283$ for 5-methylindole, as a result of the greater sensitivity of the rates of DNBF addition to that of the protonation rates to increased basicity of the indole reagent ($\beta_{indole}^{H_3O+} = 0.65$; $\beta_{indole}^{DNBF} = 1.15$). On the other hand, it can be seen that, for a given basicity, the $k_1^{DNBF}/k_1^{H_3O+}$ ratio decreases markedly on going grom the 5-X series to the 5-X-2-methyl series. This obviously reflects the occurrence of steric hindrance in the 2-methylindole-DNBF systems.

The data also allow a comparison to be made of the electrophilic reactivity of DNBF with that of p-nitrobenzenediazonium cation (4-$NO_2ArN_2^+$) - a stronger electrophile than the proton.[26] As shown in Table 5, the rate constants for DNBF complexation by indole and N-methylindole (at 25°C) are 1.43 and 3 times greater that the rate constants $k_1^{4-NO_2ArN_2+}$ for coupling of these two heterocycles by p-nitrobenzenediazonium cation (at 30°C) in acetonitrile. This suggests some slightly higher electrophilic character of the neutral molecule in this solvent. Different trends are observed, however, in comparing the behaviour of the two reagents toward 2-methylindole. Then, DNBF becomes the less efficient electrophile in acetonitrile ($k_1^{DNBF}/k_1^{4-NO_2ArN_2+} = 0.23$) but is the more reactive species in aqueous solution ($k_1^{DNBF}/k_1^{4-NO_2ArN_2+} = 7.7$). This reversal is a reflection of the solvation and steric effects which govern the 2-methylindole reactions but the better stabilization of the transition state (**17**) in polar solvents apparently plays a major role in determining the $k_1^{DNBF}/k_1^{4-NO_2ArN_2+}$ ratio in aqueous solution.

Table 5: Relative electrophilic reactivities of DNBF and 4-nitrobenzenediazonium cation towards indoles.

Indole Reagent	$pK_a^{H_2O}$	Solvent	k_1DNBF $dm^3mol^{-1}s^{-1}$	$k_1$4-$NO_2ArN_2^+$ $dm^3mol^{-1}s^{-1}$
Indole (15d)	-3.46	CH_3CN	1.96[a]	1.37[c]
N-Methylindole (15f)	-2.32	CH_3CN	19.6[a]	6.50[c]
2-Methylindole (15h)	-0.28	CH_3CN	99[a]	435[c]
		H_2O	21000[b]	2700[d]

[a] 25 °C; [b] 25 °C in 70-30 (v/v) H_2O-DMSO; [c] 30 °C, ref 26; [d] 25 °C, ref 24.

a) $R_1=R_2=R_3=R_4=R_5=H$
b) $R_1=R_3=R_4=H$, $R_2=R_5=CH_3$
c) $R_3=R_4=H$, $R_1=R_2=R_5=CH_3$
d) $R_1=R_5=H$, $R_2=R_4=CH_3$, $R_3=C_2H_5$

Additional quantitative information regarding the relative reactivities of DNBF, H_3O^+ and 4-$NO_2ArN_2^+$ is provided by the results pertaining to a kinetic study of the reactions of DNBF with the pyrroles (**18a-d**) to give the C_α or the C_β adducts (**19a-d**). Taking into account the unfavorable effect of the DMSO cosolvent on the k_1^{DNBF} rate constants, Table 6 clearly emphasizes the tendency of DNBF to be more electrophilic than the 4-$NO_2ArN_2^+$ cation. On the other hand, the much greater reactivity of DNBF relative to H_3O^+ is confirmed.

Table 6: Relative electrophilic reactivities of DNBF and the H_3O^+ and 4-nitrobenzenediazonium cations towards pyrroles in aqueous solution.

Pyrrole Reagent	$pK_a^{H_2O}$	k_1DNBF $dm^3\ mol^{-1}s^{-1}$	$k_1H_3O^+$ $dm^3\ mol^{-1}s^{-1}$	$k_1$4-$NO_2ArN_2^+$ $dm^3\ mol^{-1}s^{-1}$
Pyrrole (18a)	-3.79	650[a]	1.03[c]	220
2,5-Dimethylpyrrole (18b)	-1.07	11600[a]	760[c]	2000
1,2,5-Trimethylpyrrole (18c)	-0.50	48800[b]	950[c]	1600
Kryptopyrrole (18d)	+3.75	≈ 2x10^6 [a]	5690[c]	6x10^6

[a] 25 °C in 70-30 (v/v) H_2O-DMSO; F Terrier et al, unpublished results; [b] ref 21a; [c] $k_1H_3O^+$ values at 25 °C calculated from protiodetritiation data as discussed in ref 29; [d] 25 °C; ref 25.

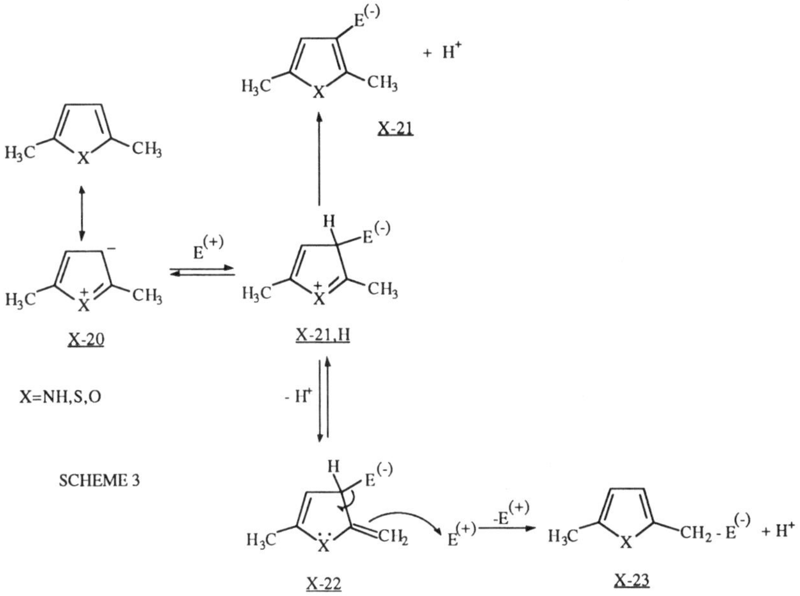

SCHEME 3

X=NH,S,O

An important observation which is qualitatively consistent with our proposal that DNBF is somewhat more electrophilic than the 4-nitrobenzenediazonium cation is the occurrence of a side-chain substitution of the methyl group to give the C-adduct **(O-23)** on treatment of 2,5-dimethylfuran **(O-20)** with DNBF.[30] A similar side-chain coupling occurs on treatment of **(O-20)** with 2,4-dinitrobenzenediazonium cation but not with 4-nitrobenzenediazonium cation.[31]

As discussed previously, the most reasonable mechanism for these reactions is the one shown in Scheme 3 where the first step is the reversible formation of the low concentration intermediate adduct **(O-21,H)**. Due to the relatively low aromaticity of the furan ring, the rearomatization of **(O-21,H)** to the product of C-β substitution **(O-21)** is much less favored than in pyrrole or thiophene series, where the adducts **(NH-21)** and **(S-21)** are rapidly formed. This allows the formation of the very reactive methylene intermediate **(O-22)** to occur if the acidity of the methyl group is sufficiently strong. This latter process actually takes place when the electrophilic moiety bonded to C-3 is very electron-withdrawing, which is the case with E=2,4-NO$_2$ ArN$_2$ and E=DNBF⁻, but not with E=4-NO$_2$ArN$_2$. Then, the formation of the side-chain substitution product **(O-23)** can proceed in a concerted process involving addition of a second electrophile moiety to the methylene carbon of **(O-22)** with concomitant expulsion of the first E structure initially attached at C-β.

3   CONCLUDING REMARKS

The various systems discussed above emphasize the conclusion that, despite its neutral character, DNBF ranks among the most powerful electrophiles known to date, a finding which calls for exploration of this unique property in other systems. In this respect, there have been recently two reports which are worth mentioning. The first is directly connected to the ease of σ-complex formation from a DNBF structure since it pertains to the observation that the S_NAr hydrolysis of 7-chloro-4,6-dinitrobenzofuroxan occurs 10^4 times more rapidly than that of picryl chloride in aqueous solution.[32] The second is more illustrative and deals with the finding that 7-methyl-4,6-dinitrobenzofuroxan (**24**) is a strong carbon acid ($pK_a^{H_2O} = 2.5$) which exhibits one of the lowest intrinsic reactivities so far measured for this type of compounds ($\log k_o = -2.15$).[33] Such a behaviour implies an extensive sp^2 rehybridization of the exocyclic carbon in the resulting carbanion (**C-24**) and therefore a high capability of the DNBF structure to absorb the negative charge, as shown in (**C-24'**).[34] As a matter of fact, nmr data support the olefinic structure of the carbanion.[33]

24   **C-24**   **C-24'**

REFERENCES

1   P Drost, *Justus Liebigs Ann Chem*, 1899, **307**, 49.
2   R K Harris, A R Katritzky, S Oksne, A S Bailey, and W G Paterson, *J Chem Soc*, 1963, 197.
3   A J Boulton and P B Ghosh, *Adv Heterocyl Chem*, 1962, **10**, 1.
4   C K Prout, O J R Hodder, and D Viterbo, *Acta Crystallogr, Sect B*, 1972, **28**, 1523.
5   F Terrier in *Nucleophilic Aromatic Displacement*, ed H Feuer, VCH Publishers, New York, 1991, p.18 and 138 and references therein.
6   F Terrier, *Chem Rev*, 1982, **82**, 77.
7   P B Ghosh and M W Whitehouse, *J Med Chem*, 1968, **11**, 305.
8   Terrier, A P Chatrousse, Y Soudais, and M Hlaibi, *J Org Chem*, 1984, **49**, 4176.
9   C F Bernasconi, *J Am Chem Soc*, 1970, **92**, 4682.
10 (a) C D Ritchie and H Fleischhauer, *J Am Chem Soc*, 1972, **94**, 3481; (b) J W Bunting and M M Conn, *Can J Chem*, 1990, **68**, 537.

11 F Terrier, J C Hallé, P MacCormack, and M J Pouet, *Can J Chem*, 1989, **67**, 503.
12 (a) J Toullec, *Tetrahedron Lett*, 1988, **29**, 5541; (b) J Toullec, *J Chem Soc, Perkin Trans 2*, 1989, 167.
13 F Terrier, M P Simonnin, M J Pouet, and M J Strauss, *J Org Chem*, 1981, **46**, 3537.
14 (a) F Terrier, T Boubaker, and A P Chatrousse, unpublished results; (b) F Terrier, J Lelièvre, A P Chatrousse, T Boubaker, A Bachet, and A Cousson, *J Chem Soc, Perkin Trans 2*, 1992, 361.
15 Feuer and A T Nielsen, VCH Publishers, 1990, Ch.1, p 86.
16 (a) R Spear, W P Norris, and R W Read, *Tetrahedron Lett*, 1983, **24**, 1555; (b) R W Read, R J Spear, and W P Norris, *Aust J Chem*, 1984, **37**, 985.
17 (a) M J Strauss, R A Renfrow, and E Buncel, *J Am Chem Soc*, 1983, **105**, 2473; (b) E Buncel, R A Renfrow, and M J Strauss, *J Org Chem*, 1987, **52**, 488.
18 F Terrier, M J Pouet, E Kizilian, J C Hallé, F Outurquin, and C Paulmier, *J Org Chem*, 1993, **58**, 4696.
19 F Terrier, J C Hallé, MJ Pouet, and M P Simonnin, *J Org Chem*, 1986, **51**, 409.
20 A J Kresge, H J Chen, L E Hakka, and J E Kouba, *J Am Chem Soc*, 1971, **93**, 6174.
21 F Terrier, G Moutiers, and J Morel, *J Chem Soc, Perkin Trans 2*, 1993, 1665.
21 (a) F Terrier, E Kizilian, J C Hallé, and E Buncel, *J Am Chem Soc*, 1992, **111**, 1740; (b) F Terrier, M J Pouet, J C Hallé, S Hunt, J R Jones, and E Buncel, *J Chem Soc, Perkin Trans 2*, 1993, in press.
22 (a) B C Challis and E M Millar, *J Chem Soc, Perkin Trans 2*, 1972, 1111, 1116, & 1618; (b) R L Hinman, and J Lang, *J Am Chem Soc*, 1964, **86**, 3796.
23 R Taylor, *Electrophilic Aromatic Substitutions*, Wiley, New York, 1990.
24 B C Challis and H S Rzepa, *J Chem Soc, Perkin Trans 2*, 1975, 1209.
25 A R Butler, P Pogorzelec, and P T Shepherd, *J Chem Soc, Perkin Trans 2*, 1977, 1452.
26 A H Jackson and P P Lynch, *J Chem Soc, Perkin Trans 2*, 1987, 1483.
27 C Reichardt, *Solvents and Solvent Effects in Organic Chemistry*, VCH Publishers, Weinheim, 2nd Edition, 1988, p.147.
28 T H Lowry and K S Richardson, *Mechanism and Theory in Organic Chemistry*, Harper and Row, New York, 3rd Edition, 1987, p143 and references therein.
29 (a) F Terrier, F Debleds, J F Verchère, and A P Chatrousse, *J Am Chem Soc*, 1985, **107**, 307; (b) F Terrier, A P Chatrousse, J R Jones, S Hunt, and E Buncel, *J Phys Org Chem*, 1990, **3**, 684.
30 F Terrier, J C Hallé, M P Simonnin, and M J Pouet, *J Org Chem*, 1984, **49**, 4363.

31   (a) S T Gore, R K Mackie, and J M Tedder, *J Chem Soc, Perkin Trans 1*, 1976, 1639; (b) M G Bartle, S T Gore, R K Mackie, S Mhatre, and J M Tedder, *ibid*, 1978, 401.
32   F Terrier, L Xiao, M Hlaibi, and J C Hallé, *J Chem Soc, Perkin Trans 2*, 1993, 337.
33   F Terrier, D Croisat, A P Chatrousse, M J Pouet, J C Hallé, and G Jacob, *J Org Chem*, 1992, **57**, 3684.
34   C F Bernasconi, *Adv Phys Org Chem*, 1992, **27**, 119.

ION PAIRS AND ION-MOLECULE PAIRS IN SOLVOLYTIC SUBSTITUTION, ELIMINATION AND REARRANGEMENT REACTIONS

Alf Thibblin

Institute of Chemistry, University of Uppsala, PO Box 531, Uppsala, S-751 21 Uppsala, Sweden.

1 INTRODUCTION

This paper summarizes some of our work on short-lived intermediates in solvolytic elimination, substitution, and rearrangement reactions. Short-lived ion-pair intermediates are well established intermediates in solvolysis reactions in nonpolar and partially polar solvents but the knowledge of such intermediates in highly aqueous solvents, especially in elimination reactions, is not very extensive.

The knowledge of the role of the analogous type of intermediate with an uncharged leaving group, *ie* ion-molecule pairs, is even smaller. Very few reactions have been discussed in terms of such intermediates. The reason is presumably the difficulty in probing these elusive short-lived species.

The equilibrium constant ($K_{as} = k_d/k_{-d}$) for formation of ion pairs from singly charged ions in water is generally < 1 M^{-1}.[1] Thus, it has been pointed out that a substantial reaction of the contact or solvent-separated ion pair with a dilute reactant (< 1 M) in highly aqueous solvents is not very likely since the lifetime of the ion pair should be too short to allow it to encounter and react with reactants other than those that are already close to the ion pair when it is formed.[1] Accordingly, a substantial k_p/k_{-d} ratio (equation 1) seems to require either a preassociation mechanism whereby the dilute reactant C gets into reaction position before the R-X bond is ruptured, or reaction with the solvent. The same reasoning may be applied to reactions via ion-molecule pairs (equation 2).

$$\text{RX} \longrightarrow \text{ion pair(s)} \underset{k_d}{\overset{k_{-d}}{\rightleftarrows}} \text{R}^+ \longrightarrow \text{products} \quad \begin{array}{c} \uparrow k_p' \\ \text{products} \end{array} \quad C \downarrow k_p \quad \text{products} \tag{1}$$

$$\text{RX} \longrightarrow \text{ion - molecule pair} \underset{k_d}{\overset{k_{-d}}{\rightleftarrows}} \text{R}^+ \longrightarrow \text{products} \tag{2}$$

Monomolecular reactions are of course not restricted in this way and a large fraction of products may arise directly from the ion pair or ion-molecule pair (k_p'). Such reactions are, for example, rearrangement reactions and intramolecular leaving-group promoted elimination reactions in which the leaving group is the hydron-abstracting base. The situation when reaction from ion pair or ion-molecule pair is slow, *ie* $k_{-d} \gg k_p[C]$, k_p', corresponds to the classical E1 and S_N1 reactions.

2 ELIMINATION *VIA* AN IRREVERSIBLY FORMED ION PAIR OR ION-MOLECULE PAIR ($D_N^{\#}*A_{xh}D_H$)[1]

It was concluded recently that pure E1 reactions, *ie* rate-limiting ionization followed by elimination from the free carbocation, are probably not common for solvolytic reactions of substrates with leaving groups that are negatively charged or are neutral but efficient bases.[1] Thus, there seems to be no conclusive evidence that elimination from such substrates occurs mainly or exclusively from the free, diffusionally-equilibrated carbocation.

The leaving group of an ion pair is often very efficient in promoting elimination by abstraction of a β-hydron, *even in highly aqueous solvents*. An example is the reaction of 1,1-diphenylethyl derivatives, Ph_2CMeX, which react solvolytically to elimination product via the ion pair (equation 3),[1] despite the relatively high stability of the carbocation.

$$Ph_2CX \atop Me \rightleftarrows Ph_2C^+X^- \atop Me \rightarrow Ph_2C^+ \atop Me \xrightarrow{H_2O} Ph_2COH \atop Me \quad (3)$$

$$\searrow k_e \quad \swarrow k_e'$$

$$Ph_2C=CH_2$$

$$k_e > k_e'$$

It was found that the leaving groups acetate and *p*-nitrobenzoate anions give rise to three times as much elimination as the leaving groups MeOH and HOAc in 20 vol% Me_2SO in water.[5] The substitution product, on the other hand, originates mainly from nucleophilic attack on the free carbocation since the measured nucleophilic selectivities are very similar with different leaving groups. Thus, classical S_N1 reaction competes with classical E1 reaction but most of the elimination product comes directly from the contact ion pair by a mechanism in which the leaving group acts as the hydron-abstracting base.

Competing stepwise elimination and substitution reactions have frequently been postulated, for the sake of mechanistic simplicity, to occur through a common carbocation or ion pair. Independent evidence for such a coupling of the elimination and substitution processes *via* a common intermediate has been presented recently for the reaction of Ph_2CMeCl in methanol-acetonitrile mixtures (equation 4).[1] Analysis of the kinetic deuterium isotope effects for the separate reactions, *ie* the elimination (k_E) and the substitution (k_S), strongly

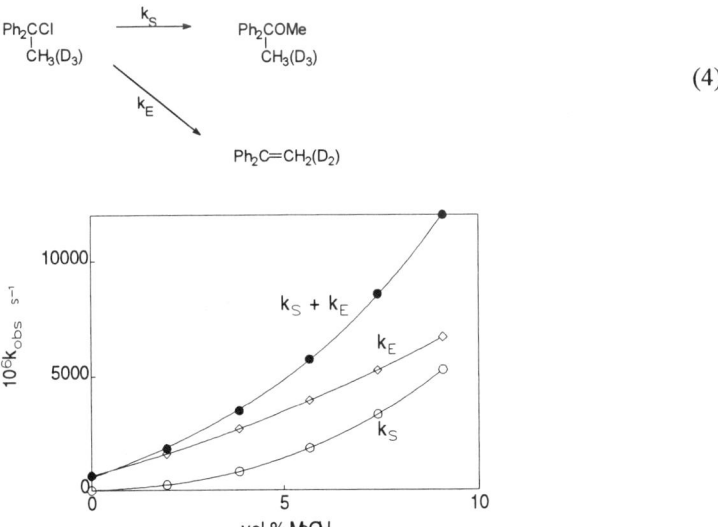

Figure 1: Reaction rate constants for the reactions of Ph_2CMeCl as a function of solvent composition, vol% MeOH in acetonitrile, at 25 °C.

indicates a common intermediate. Very significant base catalysis from the leaving group indicates that this intermediate is the contact ion pair.

Let us inspect the experimental data in some detail. The reaction rates increase rapidly with increasing concentration of methanol (Figure 1). The measured kinetic isotope effects for different solvent compositions are shown in Table 1. The changes in isotope effects are in accordance with competing reactions through a common intermediate. Thus, the substitution isotope effect k_S^H/k_S^D increases with increasing methanol content from 0.84 to 0.96 corresponding to 13 and 44 % substitution, respectively. The elimination isotope effect k_E^H/k_E^D increases from 1.73 to 3.20 when the fraction of substitution increases from 0 to 44 %. These isotope effects and trends in isotope effects are not consistent with two competing parallel reactions which do not have an intermediate in common. However, branching via a common intermediate (equation 5) may account for the results. This conclusion originates from analysis of the expressions for the kinetic isotope effects on k_E and k_S as described shortly below.

Table 1: Kinetic deuterium isotope effects for the reactions of Ph_2CMeCl in methanol-acetonitrile at 25 °C (equation 4).

Solvent	$(k_S^H + k_E^H)/(k_S^D + k_E^D)^a$	k_S^H/k_S^{Db}	k_E^H/k_E^{Dc}
0% MeOH	1.73		1.73
1.96% MeOH	1.81	0.84	2.18
3.85% MeOH	1.72	0.86	2.48
5.66% MeOH	1.63	0.88	2.73
7.41% MeOH	1.58	0.91	2.97
9.09% MeOH	1.58	0.96	3.20
20% MeOH	1.34		

a Estimated maximum error: $\pm 3\%$. b Estimated maximum error: ± 0.06.
c Estimated maximum error: ± 0.12.

$$RX \underset{k_{-1}}{\overset{k_1}{\rightleftarrows}} R^+X^- \overset{k_2}{\underset{k_3}{\diagup\diagdown}} \begin{array}{l} ROS \\ alkene \end{array} \quad (5)$$

$$k_S = k_1 k_2/(k_{-1} + k_2 + k_3) \quad (6)$$

$$k_E = k_1 k_3/(k_{-1} + k_2 + k_3) \quad (7)$$

$$k_S + k_E = k_1(k_2 + k_3)/(k_{-1} + k_2 + k_3) \tag{8}$$

$$k_S^H/k_S^D = (k_1^H/k_1^D)(k_2^H/k_2^D)(k_{-1}^D + k_2^D + k_3^D)/(k_{-1}^H + k_2^H + k_3^H) \tag{9}$$

$$k_E^H/k_E^D = (k_1^H/k_1^D)(k_3^H/k_3^D)(k_{-1}^D + k_2^D + k_3^D)/(k_{-1}^H + k_2^H + k_3^H) \tag{10}$$

$$(k_S^H + k_E^H)/(k_S^D + k_E^D) = (k_1^H/k_1^D)[(k_2^H + k_3^H)/(k_2^D + k_3^D)]$$

$$[(k_{-1}^D + k_2^D + k_3^D)/(k_{-1}^H + k_2^H + k_3^H)] \tag{11}$$

Let us assume for simplicity that internal return is negligible ($k_{-1} \ll k_2, k_3$) and that $k_2^H/k_2^D = 1$. It can be inferred from equation 10 that the isotope effect on the elimination reaction attains a maximum value of $k_E^H/k_E^D = (k_1^H/k_1^D)(k_3^H/k_3^D)$, ie a secondary kinetic deuterium isotope effect multiplied by a primary one, when elimimination is much slower than substitution ($k_3 \ll k_2$). Under these conditions, a maximum isotope effect on the substitution is attained, $k_S^H/k_S^D = k_1^H/k_1^D$ (equation 9). On the other hand, a minimum elimination isotope effect of $k_E^H/k_E^D = k_1^D/k_1^D$ and a minimum substitution isotope effect of $k_S^H/k_S^D = (k_1^H/k_1^D)(k_3^D/k_3^H)$ are observed if elimination is much faster than substitution. The isotope effect of 1.73 (Table 1) measured in pure acetonitrile indicates that the intermediate undergoes some internal return since this value is somewhat larger than the expected maximum secondary isotope effect for a substrate with three β-deuteriums. Reaction branching as the cause of unusually large and unusually small isotope effects has been discussed previously for carbanion reactions[1] and carbocation reactions[7,1,2] and has been generalized and reviewed.[7]

The large catalytic effect of the leaving chloride anion on the elimination reaction strongly indicates that the intermediate is of the contact ion-pair type. The ion pair is estimated to eliminate > 3000 times faster than the free carbocation in 0.4 vol% water in acetonitrile.[6]

Another system which has been studied is the competing elimination and substitution of 9-(2-X-2-propyl)fluorene (**1-X**) in water-acetonitrile mixtures (equation 12).[1] The chloride **1-Cl** was found to yield about 64 % of olefin **2** and 36 % of alcohol and only a trace of the thermodynamically more stable olefin **3** was formed in 25 vol% acetonitrile in water at 25 °C. Addition of the common ion Cl⁻ does not substantially depress the disappearance of the substrate (k_{obs}) but catalyses the formation of alkene **2**. This suggests rate-limiting ionization. Also, weak bases as well as the leaving group catalyse elimination from the ion pair. The catalysis from substituted acetate anions was found to be small, β = 0.05. As shown in Figure 2, the catalysis from halide anions is described fairly well by the same Brønsted line.

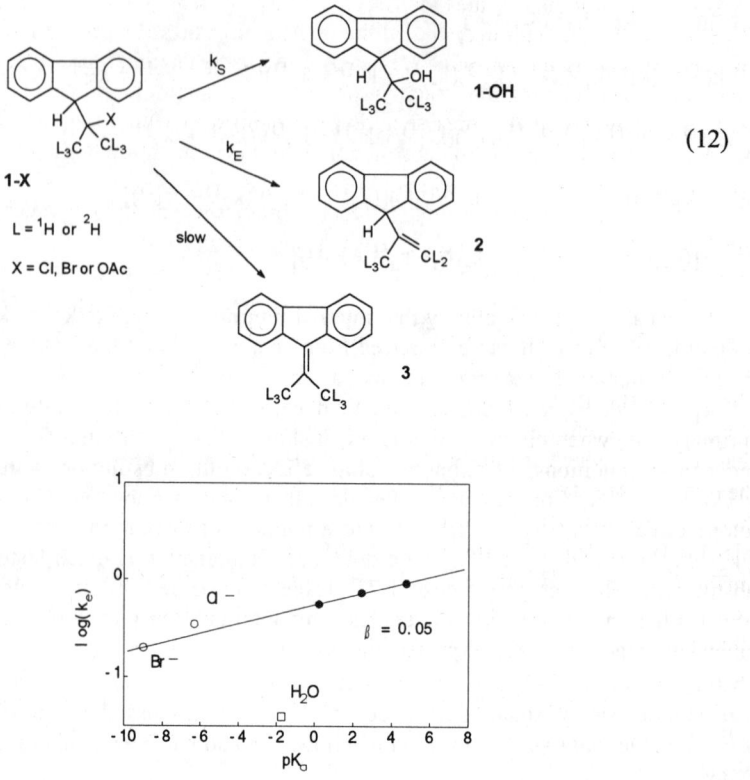

Figure 2: Brønsted plot for the dehydronation of the ion-pair intermediate formed from **1-X** (X = Cl) with substituted acetate anions (•) in 25 vol% acetonitrile in water; ionic strength 0.75 M maintained with sodium perchlorate. The pK_a values refer to water.

The presence of strong base, on the other hand, opens up a parallel bimolecular concerted elimination route (E2).[10,1] This route provides exclusively the more stable olefin **3** and exhibits a large kinetic isotope effect of $k^H/k^D = 8.1$ (substrate deuteriated at the 9-position of the fluorene moiety). The intermediate shows very small discrimination between the nucleophiles azide anion, methanol, and water. Thus, an azide anion is about five times more reactive than a solvent molecule toward the carbocation intermediate, ie $k_{N3}/k_{HOH} \sim 5$. The selectivity is so small that it may represent reaction within a pool of solvent molecules that are present at the time of ionization to the ion pair. The rate constant for the reaction of the intermediate with water to form the alcohol was estimated at $\sim 4 \times 10^{10}$ s^{-1} based upon a diffusion-controlled reaction with azide anion with $k_d = 5 \times 10^9$ M^{-1}s^{-1}.[10,1,2] Accordingly, the rate constant for deprotonation of the intermediate by solvent water is $\sim 7 \times 10^{10}$ s^{-1}. These rate constants are larger, or at least comparable to, the estimated rate of diffusional separation of the ion pair. Thus, it was concluded that the dehydronation of the intermediate and the nucleophilic substitution are processes that occur mainly at the ion-pair stage before the ion pair undergoes diffusional separation. The elimination reaction promoted by addition of acetate anion should occur by a stepwise preassociation mechanism in which the base comes into reaction position for hydron abstraction before the ionization to the ion pair.

The measured kinetic isotope effects support the equation 5 mechanism with $k_{-1} \ll k_2 + k_3$. The isotope effect on the disappearance of the substrate having the methyl groups fully deuteriated was measured as $k^H/k^{D6} = 2.2$ at 25 °C. This large secondary kinetic β-deuterium isotope effect, which corresponds to a value of 1.14 per deuterium, shows that the bonds to the hydrons are weakened considerably in the ionization step. The kinetic isotope effect on substitution and elimination for the solvolysis of the chloride were measured as $k_S^H/k_S^{D6} = 1.4$ and $k_E^H/k_E^{D6} = 3.7$. These isotope effects are in accord with a mechanism in which a rate-limiting ionization step is followed by branching. The competing paths show differences in sensitivity to isotopic substitution. Owing to this competition, the isotope effects on k_S and k_E are attenuated and enlarged, respectively, compared with the isotope effects on the rate-limiting ionization of the substrate. The experimental data for the chloride in 25 vol% acetonitrile in water and the mechanistic model are consistent with $k_2^H/k_2^{D6} = 1.0$ and $k_3^H/k_3^{D6} = 2.8$ (equation 5).

The consistency of the measured isotope effects with branching through a common intermediate both at low and high water concentration indicates that internal return from the ion pairs is not significant. The large ionization isotope effect, $k^H/k^{D6} = 2.2$, suggests that the ionization is accompanied by considerable reorganisation of the carbocation structure and the solvent. These processes slow down the collapse of the ion pair back to covalent material.[1]

Other systems which have been analyzed by using the kinetic deuterium isotope effect tool are solvolysis of PhCMe$_2$X in 25 vol% acetonitrile in water (equation 13)[9] and the solvolyses of the isomeric indene derivatives **4-OAc** and **5-OAc** in the same solvent (equation 14).[1] The proposed mechanisms involve common ion-pair intermediates for elimination, substitution, and rearrangement.

$$\begin{array}{c} \text{CL}_3 \\ | \\ \text{PhCX} \\ | \\ \text{CL}_3 \\ \text{L = }^1\text{H or }^2\text{H} \end{array} \quad \begin{array}{c} k_S \nearrow \\ \\ k_E \searrow \end{array} \quad \begin{array}{c} \text{CL}_3 \\ | \\ \text{PhCOH} \\ | \\ \text{CL}_3 \\ \\ \text{PhC=CL}_2 \\ | \\ \text{CL}_3 \end{array} \quad (13)$$

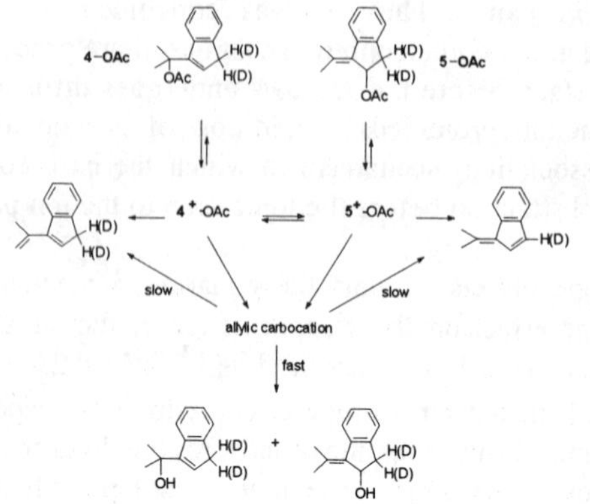

$$\text{(14)}$$

A common ion-molecule pair intermediate was quite recently suggested for the reaction of PhCMe$_2$P$^+$ (P = pyridine) to alkene and alcohol in aqueous acetonitrile.[1] Indications supporting this suggestion were observation of catalysis of the leaving group on the elimination process. For example, eight times more alkene was obtained from this substrate than from the protonated ether PhCMe$_2$OMeH$^+$.

3 ELIMINATION AND SUBSTITUTION *VIA* A REVERSIBLY FORMED ION PAIR ($D_N*A_{xh}D_H^\ddagger$)[3]

Recently it was concluded that PhCH$_2$CMe$_2$Cl (**6**, equation 15) reacts by an E2 mechanism with methoxide anion in methanol to give alkene **7**.[8] This elimination product is also formed by a stepwise route via a reversibly formed ion pair by dehydronation with solvent and added bases, and probably also by the leaving group. The other alkene **8** is only formed by the route through reaction of the ion pair (equation 15). Solvolysis without any base present provides all three products. Let us look briefly at the experimental results on which these mechanistic assignments are based.

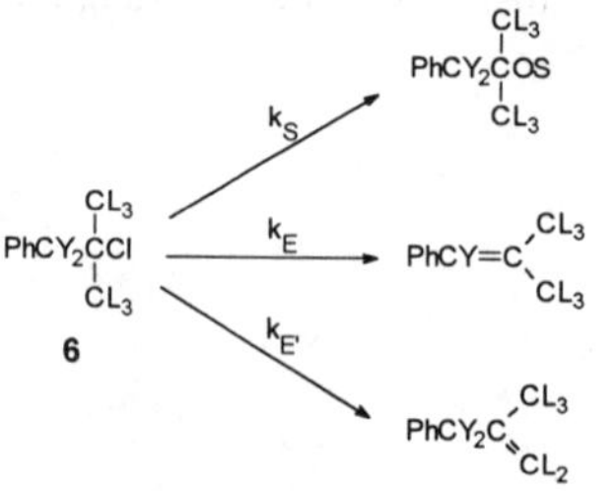

$$\text{(15)}$$

(a) Y = ^{1}H and L = ^{1}H
(b) Y = ^{2}H and L = ^{1}H
(c) Y = ^{1}H and L = ^{2}H

There are several indications for reversible ionization in methanol as well as in highly aqueous solvent. Thus, the solvolysis in 25 vol% acetonitrile in water is somewhat faster in the presence of azide anion or bromide anion than perchlorate anion which suggests nucleophilic attack on a reversibly formed ion-pair intermediate giving rise to a bimolecular contribution to the observed rate. The isotope effect on the total reaction rate also suggests reversible ionization since $k_{obs}{}^H/k_{obs}{}^{D2} = 1.41$ and 1.42 for reaction at 25 °C in the aqueous medium and methanol, respectively, corresponding to an isotope effect of 1.19 per deuterium that is too large for a secondary β-deuterium isotope effect.[1]

Addition of methoxide anion to methanol increases the overall rate of disappearance of the substrate. This increase in total rate is caused by a large increase in the rate of elimination to give **7** but also by an increase in the rate of formation of **8** (equation 15). However, the rate of formation of the ether decreases. There is also a large increase in $k_{obs}{}^H/k_{obs}{}^{D2}$ and $k_E{}^H/k_E{}^{D2}$ but $k_{E'}{}^H/k_{E'}{}^{D2}$ and $k_{E'}{}^H/k_{E'}{}^{D6}$ are not changed (Table 2). These results strongly indicate a parallel, competing methoxide-promoted concerted E2 reaction (equation 16). Further indications are the observed decreases in $k_{obs}{}^H/k_{obs}{}^{D6}$ and $k_E{}^H/k_E{}^{D6}$.

(16)

Table 2: Isotope effects for the reactions of **6** in methanol at 25 °C

Base	$k_{obs}{}^H/k_{obs}{}^i$	$k_S{}^H/k_S{}^i$	$k_E{}^H/k_E{}^i$	$k_{E'}{}^H/k_{E'}{}^i$
		$i = D2$		
none	1.42	1.33	2.53	1.15
NaOAc[b]	1.46	1.36	2.56	1.18
NaOMe[c]	2.39	1.46	3.89	1.15
		$i = D6$		
none	1.81	1.70	1.50	3.4
NaOAc[b]	1.81	1.72	1.47	3.4
NaOMe[c]	1.35	1.47	1.16	3.4

[a] 3.74 vol% water. [b] 0.98 M. [c] 2.00 M.

The results do not indicate a parallel E2 reaction for formation of the other alkene **8** but are completely in accord with base-promoted and solvent-promoted elimination via the reversibly formed ion pair.

The experimentally measured isotope effect $k_E^H/k_E^{D2} = 3.89$ (Table 2) is the isotope effect for formation of alkene **7** both through the E2 route and the ion-pair path. The assumption that the rate constant ratio k_3/k_4 (equation 16, pseudo-first order rate constants for reaction with solvent and base) is approximately the same with and without added base makes it possible to calculate the isotope effect for the E2 reaction with MeO⁻ as $k^H/k^{D2} = 4.9$. The isotope effect for the carbocationic route to **7** is 2.5 and 1.5 for the dideuteriated and the hexadeuteriated substrates, respectively. The values are similar to those obtained without base.

The uncatalyzed elimination to give **7** cannot be a one-step solvent-promoted concerted E2 reaction. Such a mechanism is not consistent with the measured isotope effect of $k_E^H/k_E^{D6} = 1.50$ in methanol which decreases to 1.16 in the presence of 2 M sodium methoxide. The expected value for this secondary β-deuterium isotope effect on a one-step reaction should be very close to unity for reaction both with and without added base. The values strongly indicate a stepwise mechanism for the reaction with pure solvent and elimination mainly through an E2 mechanism in the presence of a substantial amount of lyate anion.

Thus, two concurrent mechanisms having different transition-state structures for base-promoted formation of **7** seem to prevail. The E2 reaction should be about four times faster than the methoxide-promoted reaction via the ion pair. Apparently, the methoxide-promoted reactions are very close to the borderline where both mechanisms have the same activation energy. This borderline does not correspond to merging of transition-state structures.[1]

4 REARRANGEMENT AND SUBSTITUTION REACTIONS *VIA* ION-MOLECULE PAIRS

One example of elimination through an ion-molecule pair is discussed above for the elimination of pyridinium ion from the cumyl derivative. In this section, examples of such intermediates in rearrangement and substitution reactions will be discussed.

An acid-catalyzed intramolecular allylic rearrangement has been employed as an ion-molecule pair probe (equation 17).[1] Accordingly, in the solvolysis of **ROMe** the formation of the allylic isomer **R'OMe** along with the allylic alcohols **ROH** and **R'OH** was taken as evidence that the ion-molecule pair **R⁺** OMeH is an intermediate with a significant lifetime. The studied allylic system was that shown in equation 18.

It was found that acid-catalyzed intramolecular isomerization of the ethers 3-(1-methoxy-1-methylethyl)indene (**4-OMe**) and 2-methoxyisopropylidene-indan (**5-OMe**) accompanies formation of the allylic alcohols **4-OH** and **5-OH** in acetonitrile-water mixtures at 35 °C. For example, in reactions with **5-OMe** the rate constant ratio k_{2S}/k_{21} was measured as 9.2 and 11 in 25 vol% acetonitrile and 9.1 vol% acetonitrile ion water, respectively. This shows that a carbocation-molecule pair is an intermediate since a separate pericyclic reaction for the rearrangement reaction is not reasonable.[19] Analysis of the kinetic data, in combination with results obtained from solvolysis of the corresponding chloride 3-(1-chloro-1-methylethyl)indene (**4-Cl**) suggests that the intramolecular rearrangement of the ethers may in fact have two discrete ion-molecule pairs in common with the hydrolysis to alcohols (equation 19). A similar mechanism, but with common ion pairs, has also been concluded for the solvolysis of **4-OAc** and **5-OAc**.[14] The ion-molecule pairs should be of the contact type with the methanol molecule associated with C-1 and C-3, respectively, of the allylic system. One intermediate or two rapidly equilibrating ion-molecule pair intermediates are not consistent with the experimental data.

$$\text{(19)}$$

An ion-molecule pair has also been concluded to be involved in the acid-catalyzed solvolysis of 9-benzylidene-10-methoxy-10-methyl-9,10-dihydroanthracene (**9-OMe**).[1] The heterolysis is expected to yield the resonance-stabilized free, diffusionally-equilibrated, carbocation with the anthracene structure as the dominant resonance form. Trapping of this intermediate with solvent water yields exclusively the alcohol **10-OH**. However, the dominant product under kinetic control is the anthranyl alcohol **9-OH** which is formed nine times faster than **10-OH** in 25 vol% acetonitrile in water (equation 20). This was concluded to be strong evidence for an ion-molecule pair which is formed in the heterolysis of the protonated ether (equation 21).

Nucleophilic substitution by solvent water on this species gives **9-OH**. This process is obviously 9 times faster than diffusional separation to give the free carbocation. Kinetic experiments with formic acid buffers indicate that the reaction of **9-OMe** is specific acid catalyzed rather than general acid catalyzed. As indicated in equation 20, the acid-catalyzed reaction of **10-OMe** yielded only **10-OH**; *no* trace of **9-OH** was formed.

An alternative mechanism for the formation of **9-OH** may be a concerted S_N2 reaction of the hydronated **9-OMe** with water. However, it is unreasonable that the tertiary species **9-OMeH$^+$** undergoes an S_N2 reaction with solvent water at a rate that is nine times faster than ionization.

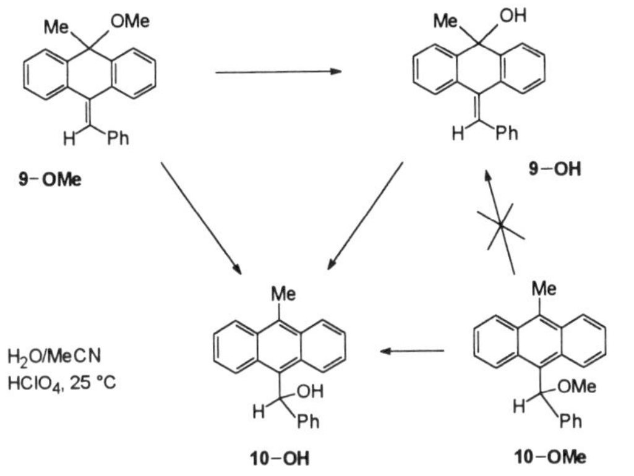

[Scheme (20): 9-OMe → 9-OH and 10-OH / 10-OMe interconversions in H₂O/MeCN, HClO₄, 25 °C]

(20)

[Scheme (21): 9-OMe ⇌ 9-ȮMeH ⇌ cationic intermediate → 9-OH (H₂O) and → 10-OH (H₂O), with rate constant k_{-d}]

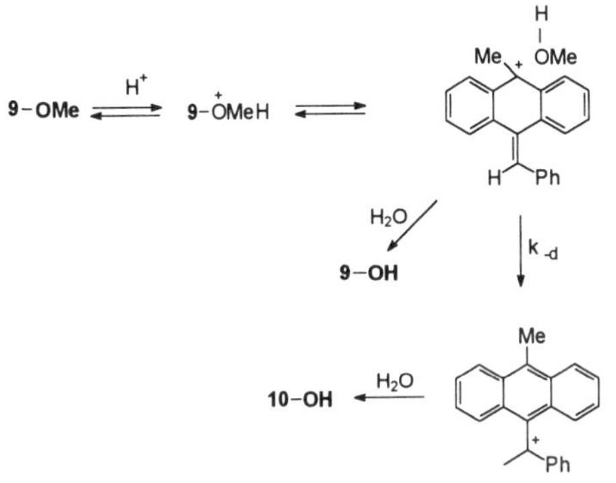

(21)

REFERENCES

1. C W Davies, *Ion Association*, Butterworths, London, 1962, p 77 and 168.
2. W P Jencks, *Chem Soc Rev*, 1981, **10**, 345.
3. Commission on Physical Organic Chemistry, IUPAC, *Pure App Chem*, 1989, **61**, 23. R D Guthrie and W P Jencks, *Acc Chem Res*, 1989, **22**, 343.
4. A Thibblin, *Chem Soc Rev*, 1993, **22**, 427.
5. A Thibblin, *J Phys Org Chem*, 1992, **5**, 367.
6. A Thibblin and H Sidhu, *J Am Chem Soc*, 1992, **114**, 7403.
7. A Thibblin and P Ahlberg, *Chem Soc Rev*, 1989, **18**, 209, and references therein.
8. A Thibblin, *J Am Chem Soc*, 1989, **111**, 5412.
9. A Thibblin, *J Phys Org Chem*, 1989, **2**, 15.
10. A Thibblin, *J Am Chem Soc*, 1987, **109**, 2071.

11   A Thibblin, *J Am Chem Soc*, 1988, **110**, 4582.
12   R A McClelland, V M Kanagasabapathy, N S Banait, and S Steenken, *J Am Chem Soc*, 1991, **113**, 1009.
13   J P Richard and W P Jencks, *J Am Chem Soc*, 1984, **106**, 1373.
14   C Paradisi and J F Bunnett, *J Am Chem Soc*, 1985, **107**, 8223.
15   A Thibblin, *J Chem Soc, Perkin Trans 2*, 1986, 321.
16   A Thibblin and H Sidhu, *J Phys Org Chem*, 1993, **6**, 374.
17   Values of $1.10 \pm 0.05/\beta$-D have been reported for secondary isotope effects: K C Westaway, *Isotopes in Organic Chemistry*, eds E Buncel and C C Lee, Elsevier, Amsterdam 1987, Ch 5.
18   R A More O'Ferrall, P J Warren, and P M Ward, *Acta Univ Ups Symp Univ Ups*, 1978, **12**, 209. W P Jencks, *Chem Rev*, 1985, **85**, 511.
19   A Thibblin, *J Chem Soc, Perkin Trans 2*, 1987, 1629.
20   A Thibblin, *J Chem Soc, Chem Commun*, 1990, 697.

THE COURSE OF OXIDATION PROCESSES OF ORGANIC SUBSTRATES MEDIATED BY Mo(VI) AND W(VI) POLYOXOPEROXO COMPLEXES

Francesco P Ballistreri, Gaetano A Tomaselli,* and Rosa Maria Toscano

*Dipartimento Scienze Chimiche, Università di Catania,
viale A Doria 6, Catania, 95125 Italy.*

SUMMARY

Results concerning oxidation processes of thioethers, alkenes, sulfoxides and amines by Mo(VI) and W(VI) polyperoxo complexes are reported. The data, compared with those obtained for the corresponding mononuclear complexes, seem to rule out a nucleophilic oxygen transfer mechanism. Rather they indicate that polyperoxo complexes as well as mononuclear peroxo complexes behave as electrophilic oxidants toward nucleophilic substrates such as thioethers and alkenes. The bulkiness of the oxidant countercation affects the reactivity of the peroxo complex itself. With sulfoxides the mechanistic picture is less conclusive, whereas oxidation of secondary amines to nitrones appears a complex reaction in which radicals can play a relevant role.

1   INTRODUCTION

In the last decade there has been a rapid growth in oxidation processes both as regards their application in the selective synthesis of oxygenated organic compounds as well as for their role in the elucidation of oxygen transfer mechanisms in biological systems.[1-3]

Among the potential oxygen donors, dilute hydrogen peroxide in the presence of Ti(IV), V(V), Mo(VI) and W(VI) is becoming the most favoured oxidant, since it is relatively cheap, has a high content of active oxygen and yields water as a by-product.

The catalytic activity of these transition metals is ascribable to the formation in solution of peroxometal complexes, which are able to transfer oxygen to organic substrates with a rate several orders of magnitude higher than that of hydrogen peroxide itself.[4] On the other hand, the use of dilute hydrogen peroxide to oxidize organic substrates suffers from the drawback of the unavoidable presence of water. This problem, in most cases, can be resolved by employing phase transfer techniques.[5-6]

Basically the procedure involves the formation in the aqueous solution of a peroxometal complex by 'in situ' reaction of hydrogen peroxide and a metal precursor, generally a molybdate or a tungstate, the transfer of the peroxometal species into the organic phase by a suitable phase transfer agent, and the subsequent reaction with the organic substrate to yield the oxidized products. In most of the reported studies the metal precursor is a sodium molybdate or tungstate, dissolved in fairly acidic aqueous solutions.[6-7]

The efficiency of the catalytic cycle depends obviously on the features of the peroxometal species formed in solution and, therefore, there is a close link between the chemistry of metal catalyzed oxidation and that of peroxometal complexes. These are well defined species which in many instances may be isolated and fully characterized.[8-9]

Recently a new class of metal precursors, *ie* polyoxometalates, has rapidly emerged as one of the most promising group of catalysts of oxidation chemistry.[10] They can be considered a family of anionic inorganic clusters, formed by transition metals, usually in their d^0 electronic configurations, and oxide ions. Mo(VI) and W(VI) polyoxometalates, possessing the so-called Keggin structure, have been mostly used together with hydrogen peroxide. It has been suggested that their catalytic activity is due to the generation in solution of polyoxoperoxo complexes.[11] A series of such peroxometal complexes of general formula $Q_3^+\{PO_4[MO(O_2)_2]_4\}^{3-}$ (Q^+= onium ion) [M= Mo(VI) and W(VI)] has been isolated and the oxidative ability has been tested toward many organic substrates, such as alkenes,[11-12] alkynes,[7,13-14] alcohols,[11,15] diols,[15-16] amines,[17-18] thioethers and sulfoxides.[19]

All these reactions proceed smoothly providing fairly high yields of the oxidized products.

On this basis one might suggest that the versatility of polyperoxo complexes is due to their ability to take part in mechanistically different reactions. On the other hand, mononuclear Mo(VI) and W(VI) peroxo complexes are less versatile.

Aiming at establishing the reactive behaviour of some Mo(VI) and W(VI) polyperoxo complexes and in order to make a comparison with mononuclear complexes, we have performed a mechanistic study concerning oxidation reactions of thioethers, alkenes, sulfoxides and secondary amines.

2 RESULTS AND DISCUSSION

The reaction of oxygen transfer to organic compounds by peroxometal complexes as well as by organic peroxides can occur through both the homolytic and heterolytic oxygen-oxygen bond cleavage. In the latter case, two different mechanistic pathways may operate.[20]

Thus, the transferred oxygen may have either electrophilic or nucleophilic character, although a distinction between the two alternatives is not always straightforward.

2.1. Oxidation of Thioethers

We have carried out a kinetic study related to the oxidation of thioethers to sulfoxides by PCMP and TEAM (Q^+ = Cetylpyridinium and Tetrahexylammonium respectively):

$$\text{X-C}_6\text{H}_4\text{-SCH}_3 + \text{Ln-M}(\text{O})_2 \longrightarrow \text{X-C}_6\text{H}_4\text{-SOCH}_3 + \text{Ln-M=O} \quad (1)$$

Under pseudo-first order conditions with the thioether in large excess (10-80 times) over the oxidant, sulfoxides are quantitatively formed according to equation 1. We observed that the process follows a second order rate law, first order with respect to each reactant. Pertinent second order rate constants are shown in Table 1.

Table 1: Second order rate constants for the stoichiometric oxidation of p-XC$_6$H$_4$SCH$_3$ to p-XC$_6$H$_4$SOCH$_3$ with Mo(VI) peroxocomplexes in CHCl$_3$.

X	$10^2 \times k_2$ TEAM $(M^{-1} s^{-1})^a$	$10^2 \times k_2$ PCMP $(M^{-1} s^{-1})^b$	$10^2 \times k_2$ MoO$_5$HMPT $(M^{-1} s^{-1})^b$	$10^2 \times k_2$ PICO $(M^{-1} s^{-1})^c$	$10^2 \times k_2$ PIC $(M^{-1} s^{-1})^c$
OCH$_3$	1.85	-	-	-	-
CH$_3$	-	5.0	2.1	$(0.33)^d$	-
H	1.13	4.0	1.1	$(0.24)^d$	$(0.29)^d$
NO$_2$	0.25	0.16	0.10	$(0.08)^d$	-

a Rate constants measured at 20 °C. b Rate constants measured at -40 °C.
c Rate constants measured at 40 °C. d Values in DCE from ref 21.

The results indicate that the oxidation rate increases on increasing the nucleophilicity of the substrate, thus establishing that the polyperoxo complex behaves as the electrophilic partner in the reaction. Therefore the kinetic data do not disqualify the hypothesis that thioethers are oxidized to sulfoxides through a simple bimolecular process, which presumably would involve an external nucleophilic attack of the thioether on the peroxidic oxygens.

In order to make a comparison with mononuclear complexes, Table 1 also reports the second order rate constants for the oxidation reaction of thioethers by MoO_5HMPT and MoO_5PICO (PICO = picolinic acid anion N-oxide). The observed reactive behaviour appears similar regardless of the nature of the oxidant, ie poly- or mono-nuclear, neutral or anionic. From the data of Table 1 it is possible to establish the reactivity sequence MoO_5HMPT > PCMP > TEAM > > MoO_5PICO ≈ MoO_5PIC (PIC = picolinic acid anion). The higher reactivity of the neutral oxidant than that of anionic peroxo complexes seems in agreement with the suggested electrophilic oxygen transfer mechanism. In fact, in contrast to MoO_5HMPT, it is the anionic part of the oxidant to bear the transferable oxygen in PCMP, TEAM, and PICO. Therefore the external nucleophilic approach of the substrate to the peroxidic oxygens of such species is less favoured for coulombic reasons. On this basis, the lowest reactivity displayed by PICO or PIC in the anionic peroxo complex series might be attributed to the lower electrophilicity of their peroxidic oxygens. The minor reactivity exhibited by TEAM with respect to PCMP is due to a countercation effect. The larger bulkiness of tetrahexylammonium with respect to cetylpyridinium causes a higher interionic distance in TEAM and therefore the consequent reduced cation-anion interaction lowers the electrophilic character of peroxidic oxygens.[22]

2.2. Epoxidation of Alkenes

Oxidation of alkenes by Mo(VI) peroxo complexes yields epoxides quantitatively. Also in the case of alkenes a second order kinetic process was observed. Second order rate constants are displayed in Table 2.

Table 2: Second order rate constants for the stoichiometric oxidation of alkenes to epoxides in $CHCl_3$ at 40 °C.

ALKENE	10^4 x k_2 PCMP (M^{-1} s^{-1})	10^4 x k_2 TEAM (M^{-1} s^{-1})	10^4 x k_2 MoO_5HMPT (M^{-1} s^{-1})	MoO_5PICO
1-Octene	2.1	-	9.2	nr
Trans-2-octene	4.7	0.09	14	nr
cyclohexene	4.5	-	37	nr
cyclo-octene	6.4	0.66	68	nr

As can be observed, the data find a rationale within the framework of the electrophilic oxygen transfer mechanism. In fact since alkenes, even if weaker than thioethers, are still nucleophilic substrates, the more nucleophilic olefin is also the more reactive. Furthermore, we found again the same reactivity sequence MoO_5HMPT > PCMP > TEAM >> MoO_5PICO already observed in the case of thioether oxidation. MoO$_5$PICO, due to the low electrophilicity of peroxidic oxygens as well as to the low nucleophilicity of the alkenes, is completely unreactive.

2.3. Oxidation of Sulfoxides

Oxidation of sulfoxides yields sulfones quantitatively (equation 2). The kinetics follow a second order process, first order with respect to each reactant.

$$X\text{-}C_6H_4\text{-SOCH}_3 + Ln\text{-}M(O_2) \longrightarrow X\text{-}C_6H_4\text{-SO}_2CH_3 + Ln\text{-M=O} \quad (2)$$

Second order rate constants are reported in Table 3.

Table 3: Second order rate constants for the stoichiometric oxidation of p-XC$_6$H$_4$SOCH$_3$ to p-XC$_6$H$_4$SO$_2$CH$_3$.

X	$10^2 \times k_2$ PCMP (M^{-1} s^{-1})^a	$10^2 \times k_2$ TEAM (M^{-1} s^{-1})^b	$10^2 \times k_2$ PICO (M^{-1} s^{-1})^{b,c}	$10^2 \times k_2$ PIC (M^{-1} s^{-1})^{b,c}
OCH$_3$	2.9	1.6	-	-
CH$_3$	2.1	1.8	0.081	-
H	2.1	1.9	0.074	0.078
Cl	2.4	2.5	0.091	-
NO$_2$	2.2	4.3	0.083	-
BuS(O)Bu	2.6	1.9	0.110	-
PhS(O)Ph	0.79	0.20	0.022	-

^{a} At 20 °C. ^{b} At 40 °C. ^{c} Values in DCE from ref 21.

The data reveal a reactivity pattern very different from that observed for sulfides. The reaction rates, both for PCMP and TEAM, show little dependence on the nature of the substituent. An identical pattern had been already observed for MoO$_5$PICO and MoO$_5$PIC. Indeed it is not easy to rationalize this behaviour within the framework of the electrophilic oxygen transfer mechanism.

On the other hand a different rationale would call for a nucleophilic oxygen transfer mechanism. In such a case a negatively charged oxygen of the peroxo complex would attack the SO functionality, leading to the formation of a peroxide intermediate, which decomposes to products:

$$Ln\text{-MoOO}^- + RSOR \rightleftharpoons \left[Ln\text{-MoOO-S(R)(R)-O}^- \right] \longrightarrow Ln\text{-Mo=O} + RSO_2R \quad (3)$$

Small substituent effects might arise because of the opposite electronic demands of the substituents in the two steps.

However some experimental facts do not support this hypothesis. Particularly, ^{17}O-NMR measurements revealed no presence[23] in solution of species such as Ln-MoOO$^-$. Even the possibility that the negative charge is located mainly on the Mo=O moiety is questioned by ^{17}O-NMR and X-ray data.[23] On the other hand, we were not able to find examples of nucleophilic behaviour by these peroxo complexes. Indeed, both MoO$_5$HMPT and MoO$_5$PICO are completely unreactive toward olefins bearing electron-withdrawing groups, which usually react with nucleophilic oxidants. Also both observed reactivity sequences PCMP > TEAM > MoO$_5$PICO ≈ MoO$_5$PIC and, within a single oxidant, nBuSOnBu ≥ PhSOCH$_3$ > PhSOPh are hardly reconcilable with the nucleophilic mechanism.

A third explanation might involve the occurrence of a single electron transfer (SET) from the sulfoxide to the peroxo complex. The SET process would generate a radical ion pair, ie the radical anion of the oxidant and the radical cation of the sulfoxide, which collapses to products. The sulfoxide would behave as an electrophile in the SET step and as a nucleophile in the collapse step. Balance between these two behaviours would be responsible for the experimental observations.

In conclusion, as far as sulfoxides are concerned, the overall picture is less conclusive than in the case of sulfides and alkenes and further work is in progress on this topic.

2.4. Oxidation of Amines

Oxidation of secondary amines by hydrogen peroxide in the presence of Na$_2$WO$_4$ or PCWP (Q$^+$ = Cetylpiridinium) yields nitrones. In order to collect mechanistic information on this reaction we performed a preliminary kinetic investigation on the stoichiometric oxidation of benzylisopropylamine by WO$_5$PIC, WO$_5$HMPT and WO$_5$PICO. The oxidant disappearance, determined by iodometric titration, follows second order kinetics. Pertinent data are reported in Table 4.

Table 4: Rate constants for the oxidant disappearance in the oxidation of N,N-benzylisopropylamine with some W(VI) peroxo complexes in CHCl$_3$ at 10 °C.

Peroxocomplex	$10^3 \times k_2$ (M^{-1} s^{-1})
WO$_5$PIC	3.12
PCWP	0.43
WO$_5$HMPT	0.30
WO$_5$PICO	0.10

Although the data would seem to indicate a nucleophilic attack of the amine on the oxidant, the whole picture is more complicated. In fact, the observed reactivity sequence $WO_5PIC > PCWP > WO_5HMPT > WO_5PICO$ does not follow the electrophilicity of peroxidic oxygens; the very different reactivity $WO_5PIC > WO_5PICO$ implies that the coordination sphere of the oxidant is somewhere involved in the oxidative process. Experiments performed in the cavity of an ESR spectrometer reveal the presence of the corresponding nitroxide radical.

Further work is in progress in our laboratory to elucidate the course of amine oxidation.

3 CONCLUSIONS

On the basis of the results collected for the oxidation process of organic sulfides and alkenes, we can conclude that Mo(VI) and W(VI) polyperoxo and mononuclear peroxo complexes behave as electrophilic oxidants, ruling out the possibility of a nucleophilic behaviour. The difference in the oxidative behaviour of the two families seems attributable mainly to the larger electrophilic character of polyperoxo species, because of a more effective delocalisation of the negative charges and a larger ion pairing.

As regards sulfoxides, the mechanistic picture is less conclusive, although incursion of SET processes might be likely. Amine oxidation appears to be a complex reaction involving free radicals.

REFERENCES

1. J K Kochi, *'Organometallic mechanism and catalysis'*, Academic Press, New York, 1978.
2. R A Sheldon and J K Kochi, *'Metal-catalyzed oxidations of organic compounds'*, Academic Press, New York, 1981.
3. W Ando, and Y Moro-Oka, *'The role of oxygen in chemistry and biochemistry'*, Elsevier, Amsterdam, 1988.
4. F Di Furia and G Modena, *Rev Chem Intermed*, 1985, **6**, 51.
5. O Bortolini, F Di Furia, G Modena, and R Seraglia, *J Org Chem*, 1985, **50**, 2688.
6. O Bortolini, L Bragante, F Di Furia, and G Modena, *Can J Chem*, 1986, **64**, 1189.
7. F P Ballistreri, S Failla, and G Tomaselli, *J Org Chem*, 1988, **53**, 830.
8. J M Le Carpentier, R Schlupp, and R Weiss, *Acta Cryst*, 1978, **B28**, 1278.
9. G Amato, A Arcoria, F P Ballistreri, G A Tomaselli, O Bortolini, V Conte, F Di Furia, and G Modena, *J Mol Cat*, 1986, **37**, 165.

10 C L Hill in *'Catalytic Oxidations with Hydrogen Peroxide as Oxidant'*, G Strukul, Ed, Kluwer Academic Publishers, 1992, ch 8., p 253.
11 Y Ishii, K Yamawaki, T Ura, H Yamada, T Yoshida, and M Ogawa, *J Org Chem*, 1988, **53**, 1868.
12 A Arcoria, F P Ballistreri, E Spina, G A Tomaselli, and R M Toscano, *Gazz Chim It*, 1990, **120**, 309.
13 F P Ballistreri, S Failla, E Spina, and G A Tomaselli, *J Org Chem*, 1989, **54**, 947.
14 Y Ishii and Y Sakata, *J Org Chem*, 1990, **55**, 5545.
15 C Venturello and M Ricci, *J Org Chem*, 1986, **51**, 1599.
16 Y Sakata and Y Ishii, *J Org Chem*, 1991, **56**, 6233.
17 F P Ballistreri, U Chiacchio, A Rescifina, G A Tomaselli, and R M Toscano, *Tetrahedron*, 1992, **48**, 8677.
18 S Sakaue, Y Sakata, Y Nishiyama, and Y Ishii, *Chem Letters*, 1992, 289.
19 F P Ballistreri, A Bazzo, G A Tomaselli, and R M Toscano, *J Org Chem*, 1992, **57**, 7074.
20 F Di Furia and G Modena, *Pure Appl Chem*, 1982, **54**, 1853.
21 S Campestrini, V Conte, F Di Furia, and G Modena, *J Org Chem*, 1988, **53**, 5721.
22 F P Ballistreri, G A Tomaselli, R M Toscano, V Conte, and F Di Furia, *J Mol Cat*, 1994, **89**, 235.
23 (a) V Conte, F Di Furia, G Modena, and O Bortolini, *J Org Chem*, 1988, **53**, 4581; (b) M Postel, C Brevard, H Arzoumanian, and J G Riess, *J Am Chem Soc*, 1983, **105**, 251; (c) R Curci, G Fusco, O Sciacovelli, and L Troisi, *J Mol Cat*, 1985, **32**, 251.

THEORETICAL MODELLING OF MECHANISMS FOR GLYCOSIDE HYDROLYSIS

John A Barnes and Ian H Williams*

School of Chemistry, University of Bath, Bath BA2 7AY, UK.

1 INTRODUCTION

Glycoside hydrolyses are very important bioorganic processes whose mechanisms pose questions of considerable interest to physical organic chemists. A simple glycoside may undergo reaction with a nucleophile by means of: (i) a stepwise $D_N + A_N$ mechanism (often referred to as S_N1), involving rate-determining heterolysis of the bond to the aglycone leaving group to form a glycosyl cation, followed by nucleophilic attack; (ii) a concerted A_ND_N mechanism (often referred to as S_N2), in which the bond to the nucleophile is made as that to the leaving group is broken; or (iii) a stepwise D_N*A_N pre-association mechanism in which the nucleophile is present in the rate-determining step merely as a spectator. The lifetime of a typical glycosyl oxocarbenium ion in water is only about 10^{-12} s, and the experimental evidence suggests that glycoside hydrolysis occurs right at the S_N1/S_N2 mechanistic borderline.[1,2] Do these processes occur with retention or inversion of configuration at the anomeric centre? What is the role of stereoelectronic effects? Questions such as these demand a knowledge of transition-state (TS) structure which, from an experimental point of view is best probed by means of kinetic isotope effects (KIEs). Primary (1°) KIEs indicate which atoms are directly involved in bond making or breaking in the rate-determining TS. Secondary (2°) KIEs, involving isotopic substitution at positions not directly involved in bond making or breaking, are often used to provide a measure of the location of the TS along the reaction coordinate between reactants and products. To do this requires knowledge of the corresponding isotope effect (EIE) upon the equilibrium; the position of the TS along the reaction coordinate may then be estimated as the fraction $x \approx \ln(\mathrm{KIE})/\ln(\mathrm{EIE})$. The most convincing examples of KIEs as probes of TS structure are those in which the effects of multiple isotopic substitutions are examined. From the viewpoint of theoretical modelling, questions of reactivity and TS structure require the use of a method capable of describing the idiosyncratic behaviour of electrons in making and breaking bonds. Molecular mechanics (MM) is inherently unsuited for this purpose, being an interpolation scheme for known properties of stable species. A very simple way of treating TSs utilizes the bond-energy – bond-order (BEBO) method, but this may involve unsatisfactory guesswork concerning the TS structure.

Quantum-mechanical (QM) approaches are best able to provide the unknown information regarding TSs by means of semiempirical or *ab initio* molecular orbital (MO) theory. For large systems hybrid QM/MM methods may be applied, combining the merits of a QM treatment of bond making/breaking with a MM description of the environment within which the chemical events occur, be it solvent or protein.

2 MODELLING OF AMP HYDROLYSIS

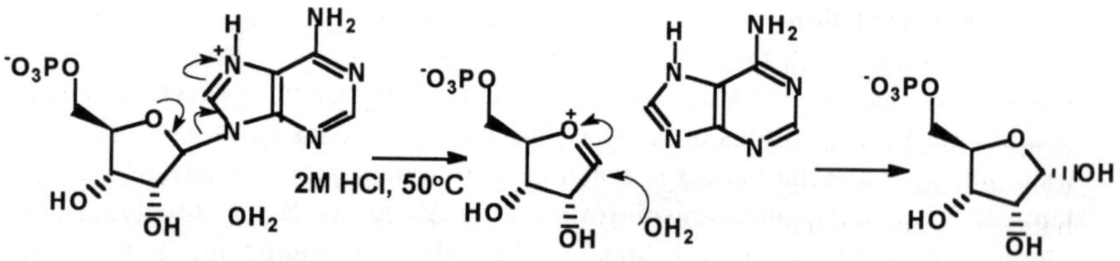

Schramm *et al* measured KIEs for acid-catalyzed hydrolysis of adenosine monophosphate (AMP) with isotopic substitutions in four positions,[3] and performed BEBOVIB calculations[4] upon a cut-off model system to obtain estimates for the TS bond orders.[5] Our intention was to study a model (Scheme 1) for this reaction by means of the AM1 semiempirical MO method[6] in order to characterize the TS structure directly for comparison with the BEBOVIB results. TS structures were determined *in vacuo* for S_N1 and S_N2 mechanisms without and with a water nucleophile respectively.

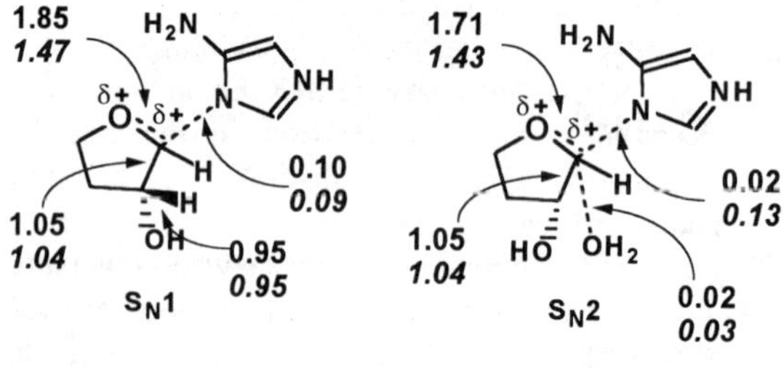

Scheme 1: BEBOVIB and AM1 (*italics*) calculated Pauling bond orders.

Table 1: Experimental and AM1 calculated KIEs (50 °C).

	expt	calc S_N1	calc S_N2
α-^{3}H	1.216	1.200	1.082
β-^{2}H	1.077	1.064	1.037
leaving grp ^{15}N	1.030	1.035	1.027
α-^{13}C	1.044	1.072	1.081

The AM1-calculated Pauling bond orders (*italicized*) are in good agreement with the BEBOVIB estimates (unitalicized) except for the degree of double-bonding to the ring oxygen in both TSs and the extent of leaving-group cleavage in the S_N2 TS. The AM1 calculated KIEs at 50 °C for the S_N1 mechanism are in better accord with the experimental values than are those calculated for the S_N2 mechanism; both AM1 TSs appear to have a 'loose' anomeric carbon too, causing too-large α-^{13}C KIEs, and insufficient conjugation between the anomeric carbon and the ring oxygen. Our conclusion is that the BEBOVIB derived TS structures are certainly no worse than those directly calculated by AM1.

The AM1 *in vacuo* activation enthalpies for the S_N1 and S_N2 mechanisms are very similar, with $DH^\ddagger$ = 32.1 and 33.3 kcal mol^{-1} (132 and 139 kJ mol^{-1}), respectively; the 'S_N2' TS is really a solvated S_N1 TS. This becomes apparent from inspection of the AM1 *in vacuo* energy surface (Figure 1) for the system with the nucleophile present. There is a broad plateau in the saddle region between reactants and products, which contains a shallow well corresponding to an S_N1 intermediate. Use of the AM1-SM1 method, available in the AMSOL program,[7] for treating the effects of aqueous solvation, leads to a surface on which the plateau is pockmarked by numerous small hollows; it is not clear what the significance of these is. The AM1-SM1 *in aquo* activation enthalpies are 38.7 and 48.2 kcal mol^{-1} (162 and 201 kJ mol^{-1}), respectively.

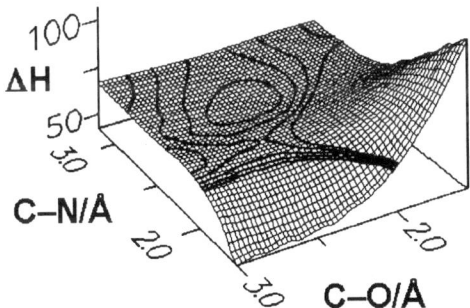

Figure 1: AM1 *in vacuo* energy surface (kcal mol^{-1}) for hydrolysis of AMP model as function of distance from anomeric centre to nucleophile (C–O) and to leaving group (C–N).

3   HYDROLYSIS OF GLUCOPYRANOSIDES

Bennet and Sinnott measured KIEs for acid-catalyzed hydrolyses at 80 °C of methyl α- and β-glucopyranosides with isotopic substitutions in seven positions.[8] These experimental KIEs were interpreted with the aid of calculated EIEs in order to determine the nature of the TS for each anomer. The EIEs (which would not be readily amenable to experimental measurement) were obtained by one of the present authors (IHW) from *ab initio* HF/4-31G MO calculations of harmonic vibrational frequencies for a simple acyclic model (Scheme 2): the glucoside was modelled by a conformer of methanediol and the oxocarbenium ion by protonated formaldehyde. The leaving group ^{18}O KIEs similar to EIEs suggested substantial C···O bond cleavage in the rate-determining TSs for the specific acid-catalyzed reactions, but the anomeric ^{13}C KIEs larger than EIEs suggested that electronic and geometrical reorganization of the glycone lags behind. The α-^{2}H KIE may be interpreted as $x = \ln(KIE)/\ln(EIE) \approx 0.75$, indicating the degree of flattening of the anomeric carbon. The magnitude of the ring ^{18}O KIE depends upon the value of the positive charge at C_α and the dihedral angle ω between the vacant p-orbital at the anomeric centre and the electronic lone pair on the adjacent oxygen.

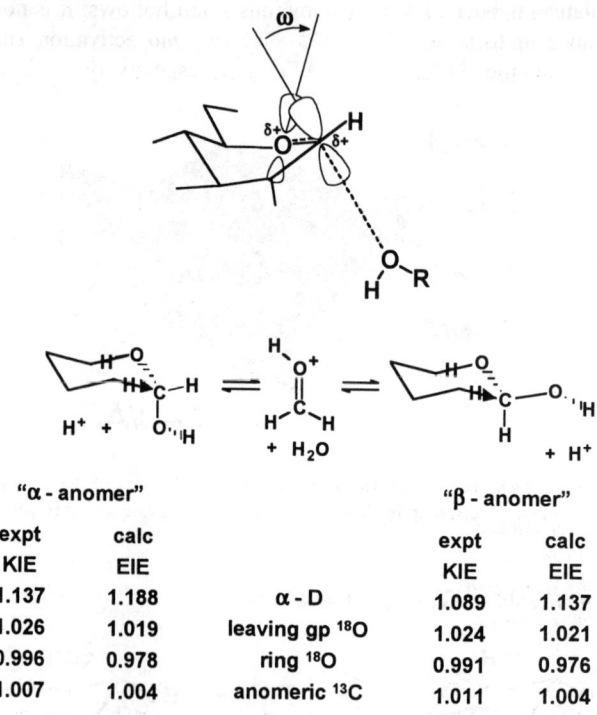

"α - anomer"			"β - anomer"	
expt KIE	calc EIE		expt KIE	calc EIE
1.137	1.188	α - D	1.089	1.137
1.026	1.019	leaving gp ^{18}O	1.024	1.021
0.996	0.978	ring ^{18}O	0.991	0.976
1.007	1.004	anomeric ^{13}C	1.011	1.004

Scheme 2: Experimental KIEs for acid-catalyzed hydrolysis of methyl α- and β-glucopyranosides and HF/4-31G calculated EIEs for model system.

Since the charge is approximately +1, the angle ω may be estimated by substitution of the experimental KIE and calculated EIE (Scheme 2) into the expression $\ln(KIE) \approx \cos^2\omega \times \ln(EIE)$; for the α-anomer this yields $\omega \approx 66°$. In this way – and by means of a similar argument applied to the β-^{2}H KIE – Sinnott proposed that the TS for this anomer adopts an approximately skew-boat conformation (and the β-anomer approximately a chair) 'in contradiction to the theory of stereoelectronic control'. However, it should be appreciated that the validity of this line of reasoning rests upon the reliability of the calculated EIEs.

The methanediol / protonated formaldehyde model system could not be used to provide a calculated β-^{2}H EIE. Results are presented in Scheme 3 for three model systems of increasing complexity. Comparison of the ethanediol / protonated acetaldehyde and 2-methoxytetrahydropyran / tetrahydropyranyl cation systems at the *in vacuo* AM1 level shows that the 'cut-off' approximation employed in the former choice is too severe. Scaling of vibrational frequencies has only a small influence on the calculated EIEs, and the AM1-SM1 *in aquo* treatment reduces the effects by a small amount also. The AM1 method significantly underestimates the EIE relative to HF/4-31G. Consideration of these various contributions allows a 'best' estimate for the isotope effect upon the methyl α-glucopyranoside hydrolysis to be obtained from the AM1 calculated EIE. Utilization of this estimate of 1.177, together with the experimental KIE of 1.073, allows a value of 44° to be calculated for the dihedral angle θ between the vacant p-orbital at C_α and the β-CH bond on the adjacent atom; this compares not too unfavourably with the range of 31° – 43° obtained by Bennet and Sinnott. A similar analysis for the β-anomer produces a 'best' estimate of 1.165 for the EIE which, with the experimental KIE, yields a dihedral angle θ = 52° as compared with the range of 41° – 50° obtained previously.

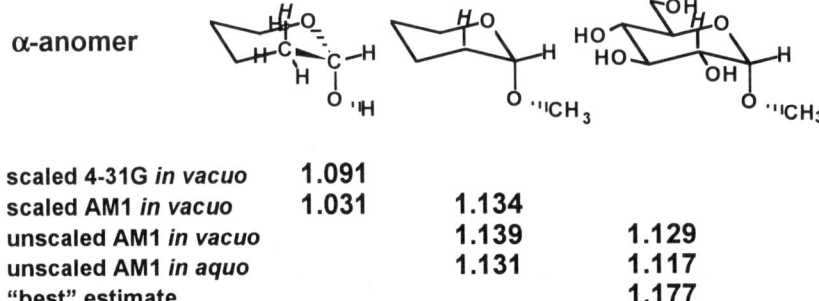

α-anomer			
scaled 4-31G *in vacuo*	1.091		
scaled AM1 *in vacuo*	1.031	1.134	
unscaled AM1 *in vacuo*		1.139	1.129
unscaled AM1 *in aquo*		1.131	1.117
"best" estimate			1.177

Scheme 3: Calculated β-^{2}H EIEs (80 °C) for glucoside hydrolysis models.

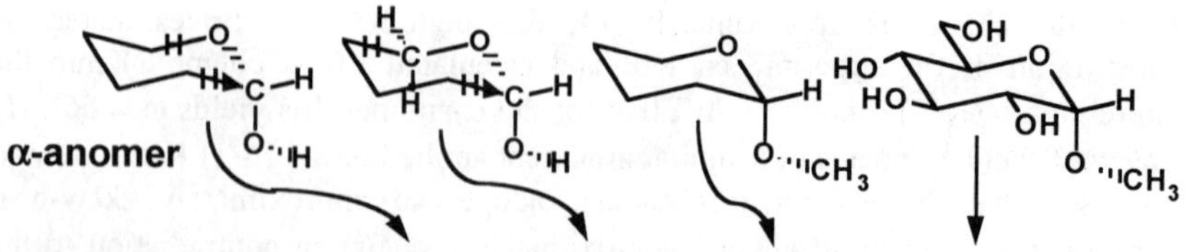

α-anomer

scaled 4-31G *in vacuo*	0.978	0.988		
scaled AM1 *in vacuo*		0.994		
unscaled AM1 *in vacuo*			0.996	0.996
unscaled AM1 *in aquo*			0.995	0.993
"best" estimate				0.987

Scheme 4: Calculated ring ^{18}O EIEs (80 °C) for glucoside hydrolysis models.

A crucial datum for Sinnott's determination of TS structure was the value of the ring ^{18}O EIE. Doubts concerning the reliability of the methanediol/protonated formaldehyde model system (which caused us to defer publication of our earlier calculations) are confirmed by the results presented in Scheme 4. While there is no significant difference between the values calculated for methoxymethanol and 2-methoxytetrahydropyran as models for methyl glucopyranoside, the EIE calculated for the smallest model is appreciably too inverse. The 'best' estimate for the ring ^{18}O EIE for formation of the glucosyl cation from methyl α-glucopyranoside in aqueous solution is 0.987. Combination of this revised value with the experimental KIE, in the manner described above, gives a dihedral angle ω = 56°, as compared with 66° obtained previously. Similarly a revised dihedral angle ω = 30° may now be estimated for the β-anomer, rather less than the original value of 52°. It seems that the details of Sinnott's TS structures should be refined, and his conclusions concerning the role of stereoelectronic effects may need to be moderated somewhat.

Finally, Scheme 5 presents calculated α-^2H EIEs for various models for methyl α-glucopyranoside. The AM1 results are considerably larger (more normal) than the HF/4-31G results, but the most surprising finding is that the AM1-SM1 method leads to very much larger values for these isotope effects. This is puzzling, for it is not obvious to us why this should be so; we are currently investigating other methods for determining the α-^2H EIEs in aqueous solution, in order to assess the validity of these AM1-SM1 results. However, if these results are accepted for the present, then a 'best' estimate of 1.332 is obtained, which would suggest a TS much earlier along the reaction coordinate for re-hybridization at the anomeric centre, with $x = \ln(KIE)/\ln(EIE) \approx 0.45$, as compared with the earlier estimate of 0.75; for the β-anomer the result would be $x \approx 0.39$ as against 0.66.

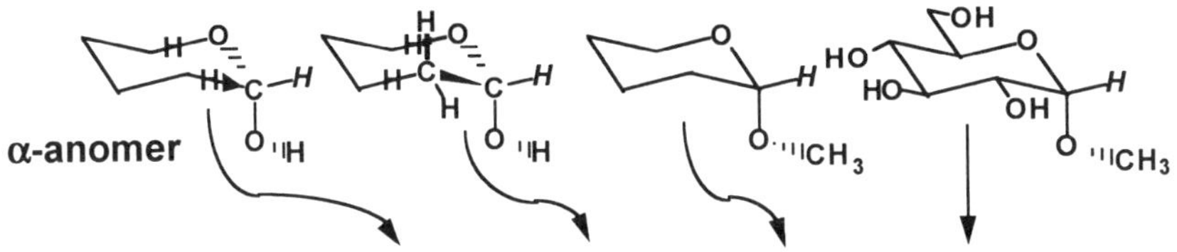

	α-anomer				
scaled 4-31G *in vacuo*	1.188	1.181			
scaled AM1 *in vacuo*		1.274	1.267		
unscaled AM1 *in vacuo*			1.277	1.302	
unscaled AM1 *in aquo*			1.365	1.448	
"best" estimate					1.332

Scheme 5: Calculated α-^{2}H EIEs (80 °C) for glucoside hydrolysis models.

4 CONCLUDING REMARKS

KIEs provide the most subtle and yet the most powerful experimental probe of TS structure, but their interpretation often requires a theoretical framework. QM methods can be employed not only to explore features of energy surfaces governing chemical reactivity – including the direct determination of TSs and intermediates – but also to provide a bridge to experiment by means of calculated KIEs and EIEs. Furthermore, better theoretical methods may be used to validate simpler models. Work is now in progress in our laboratory to apply the AM1/CHARMM hybrid method to aid elucidation of the details of glycoside hydrolysis mechanisms.[9]

We are grateful to the SERC and to Celltech Research for financial support of this work.

REFERENCES

1. T L Amyes and W P Jencks, *J Am Chem Soc*, 1989, **111**, 7888.
2. N S Banait and W P Jencks, *J Am Chem Soc*, 1991, **113**, 7051.
3. D W Parkin and V L Schramm, *Biochemistry*, 1987, **26**, 913.
4. L B Sims, G W Burton, and D E Lewis, QCPE 337 (1977); L B Sims and D E Lewis, *Isotopes in Organic Chemistry*, 1984, **6**, 161.
5. F Mentch, D W Parkin, and V L Schramm, *Biochemistry*, 1987, **26**, 921.
6. M J S Dewar, E G Zoebisch, E F Healy, and J J P Stewart, *J Am Chem Soc*, 1985, **107**, 3902.
7. C J Cramer and D G Truhlar, QCPE 606, *QCPE Bull*, 1991, **11**, 57; *J Am Chem Soc*, 1991, **113**, 8305, 9901.
8. A J Bennet and M L Sinnott, *J Am Chem Soc*, 1986, **108**, 7287.
9. J A Barnes and I H Williams, in *Proc 1st Eur Conf Comput Chem*, Nancy, May 1994.

ROTATIONAL ISOMERISM OF DISUBSTITUTED BENZENES IN THE ALKYLPHENYLDI(1-ADAMANTYL)METHANOL SERIES

John S. LOMAS & V. BRU CAPDEVILLE
ITODYS - UNIVERSITY OF PARIS 7 and CNRS, FRANCE

THEORY

- BENZENE SYMMETRY OUTLAWS THIS SORT OF ISOMERISM :

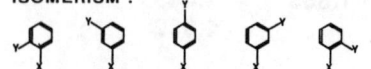

- BUT IF X IS $-CR^1R^2R^3$ AND ROTATION ABOUT THE $sp^2 - sp^3$ BOND IS SLOW, WE HAVE POTENTIALLY :

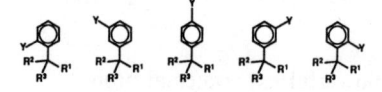

- EVEN IF $R^2 = R^3$

PRACTICE

R^2, R^3 = t-Bu, t-Bu; t-Bu, Ad; Ad, Ad
(Ad = 1-Adamantyl)

- ^{13}C NMR SPECTRA AT 25°C SHOW FIVE AROMATIC C-H SIGNALS

- ROTATION BARRIER FOR R^2, R^3 = t-Bu, t-Bu IS 21 kcal/mol (STERNHELL; BAAS, 1972)

STERIC ENERGIES

- MOLECULAR MECHANICS (MMP2/85; kcal/mol)

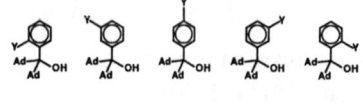

Y = Me	68.1	55.0	55.2	55.2	61.1
Y = t-Bu	105.8	58.6	59.8	59.7	82.4

- NO SIGNIFICANT TEMPERATURE DEPENDENCE OF $\Delta\Delta G°$, I.e. $\Delta\Delta S°$ = ca. 0

- MMP2 APPEARS TO OVERESTIMATE THE STABILITY OF BOTH ANTI-META ROTAMERS BY EXAGGERATING THE CONTRIBUTION OF ATTRACTIVE ALKYL/ADAMANTYL INTERACTIONS

ORTHO ROTAMERS : Y = Me

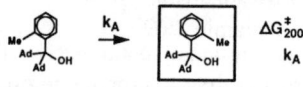

ANTI 11   :   1 SYN

- ROTAMERS EASILY SEPARATED CHROMATO-GRAPHICALLY (Lomas, 1981)
- THERMAL ROTATION IS UNIDIRECTIONAL

	$\Delta G^{\ddagger}_{200}$ kcal/mol
k_A	39.1

META ROTAMERS : Y = Me, t-Bu

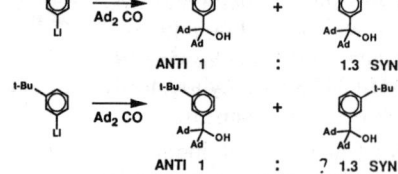

ANTI 1   :   1.3 SYN

ANTI 1   :   ? 1.3 SYN

- ROTAMERS SEPARATED WITH DIFFICULTY BY CRYSTALLISATION
- THERMAL ROTATION IS EQUILIBRATED

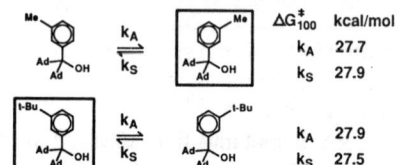

	$\Delta G^{\ddagger}_{100}$ kcal/mol
k_A	27.7
k_S	27.9
k_A	27.9
k_S	27.5

METHANES

- ISOMERICALLY PURE META-Y ALCOHOLS GIVE ISOMERICALLY MIXED METHANES, PROBABLY DUE TO EQUILIBRATION OF THE BROMIDES

SUBJECT INDEX

A

Absorption, multiphoton, 236
Acetohydroxamic acid, 379
Acetone, 324, 335
Acidaminococcus fermentans, 140
Acidity, gas phase, 286
Aconitic acid, 30
N-Acyloxy-2-thiopyridones, 32
3',5"-Adenosyladenosine, 9
Adenosylcobalamin, 209
S-Adenosylmethionine, 94
S-Adenosyl-L-methionine:3'-hydroxy-N-methyl-(S)-coclaurine-4'-O-methyltransferase, 91
S-Adenosyl-L-methionine:(R,S)-norcoclaurine-N-methyltransferase, 91
S-Adenosyl-L-methionine:(R,S)-norcoclaurine-6-O-methyltransferase, 91
S-Adenosyl-L-methionine:(S)-tetrahydroprotoberberine-cis-N-methyltransferase, 100
Alcohols, 430
Aldol condensation, 26
Alignments, multiple sequence, 114
Alkaloid biosynthesis, 89
Alkaloids, 185
  benzophenanthridine, 102
  benzylisoquinoline, 89
Alkane diazoates, 351
Alkenes, 430
Alkylation, 376
  Friedel–Crafts, 305
Alkyl nitrites, 377
Alkynes, 430
Allylmercury compounds, 272
Allyltin compounds, 271
Aluminium, 268
Ambident nucleophiles, 374

Amides, 374
Amines, 430
α-Amino acid esters, 228
Amino acid sequences, 63
  residues, 168, 200
5-Aminolaevulinic acid, 197
Anaerobic ribonucleotide reductase, 66
Anionic surfactants, 231
Anisole, 405
Anona reticulata, 91
Aporphines, 90
Aquocobalamin, 60
L-Arabinose, 33
Arabulose, 33
Arenium ion, 239
L-Arginine, 326
Aromatic nitration, 247
  silylation, 254
Asparagine, 112
Aspartate-84, 201
Aspartic acid, 113
Atomic absorption spectrometry, 329
Atomization, 286
Attack, electrophilic, 323, 406
Attack, nucleophilic, 61, 112, 270, 400, 417, 437
Avidin, 153
Azide 'clock' method, 303

B

Bacteria, anaerobic, 140
Bases, Schiff, 34
Basicity, gas phase, 286
Basis set, 280
Benzaldehyde, 42
1,2-Benzene template, 124
Benzenes, disubstituted, 444

Benzenium ion, 239
Benzocyclobutene, 135
Benzofuran, 134
Benzophenanthridine alkaloids, 102
Benzo[c]phenanthridines, 90
Benzo[b]thiophene, 134
Benzyl alcohol, 42
Benzylamine, 45, 48
Benzylisopropylamine, 434
Benzylisoquinoline alkaloids, 89
Benzylmesitylene, 253
Berberine, 90
Berberis, 90
    beaniana, 94
    stolonifera, 91, 104
    wilsoniae, 95
Bicumene, 309
Biheptaplane, 294
Bilayer membranes, 81
Bioorganic mechanisms, 3
    reactions, 3
Biosynthesis, of alkaloids, 89
    of sterols, 31
    of uroporphyrinogen III, 196
2,2'-Biphenyl template, 124
Bisbenzyliosquinolines, 90
Bisparaphenylene-34-crown-10, 388
Blood pressure, 326
Blowfly larvae, 35
Boron, 268
Botryococcus braunii, 32
Bovine serum amine oxidase, 45, 47
Bowlane, 290
p-Bromobenzyl alcohol, 43
2-Bromoporphobilinogen, 200
Buffer, imidazole, 3, 9
    morpholine, 3, 11
t-Butylbenzene, 257
α-t-Butylbenzyl arenesulfonates, 368
4-t-Butylcatechol, 17

C

Cadmium, 268
Camphor, 27
(S)-Canadine, 94
Cannizzaro reaction, 124
Carbocations, 301, 342
Carbon atom, planar tetracoordinate, 288
Carboxylation, 28
Carboxylic acid, 225
N-Carboxybiotin, 29
Carcinogens, 351, 379
Carnitine, 35
Catalysis, 3, 11, 16, 38, 58, 68, 73, 86, 123, 149, 185, 200, 322, 329, 334, 357, 419, 429, 438
Catalysts, enantioselective, 225
    hydrolytic, 223
Cationic surfactants, 231
Charge-transfer state, 296
(S)-Cheilanthifoline, 99
Chemotherapeutic agents, 351
p-Chlorobenzyl alcohol, 53
3-(1-Chloro-1-methylethyl)indene, 425
Cholesterol, 267
Chromatography, affinity, 152
    gas, 238, 353
    HPLC, 9, 66, 70, 81, 132, 141, 344
    silica gel, 394
Circular dichroism (CD), 22, 97, 98, 132, 212
Claisen condensation, 26, 128
Cleavage, 309, 346
    hydrolytic, 18
    photochemical, 85
    proteolytic, 63
Clostridium aminovalericum, 146
    propionicum, 140
Cobalamin deficiency, 69
Cobalamin-dependent methionine synthase, 58
Cobalt, 74
Cobalt-corrins, 209
Cocculus laurifolius, 97
Codeine, 89

Subject Index 447

Collisionally activated dissociation (CAD), 237
Complex, charge-transfer, 308
Complexation, 331
Computer modelling, 22
Computers, chemistry by, 278
Condensation, aldol, 26
    Claisen, 26, 128
    Pictet–Spengler, 91
Condensed-phase reactions, 235
Conductivity, 302
Constants, dielectric, 26
Copper, 32, 224
Coptis japonica (Ranunculaceae), 92
Corydalis cava, 97
Coulometry, 76
Coupling, phenolic, 103
p-Cresol, 384
Cross-interaction constants, 361
Crown ethers, 388
5-Cyanoindole, 406
Cyclic voltammetry, 74
Cyclization, base catalysed, 124
Cycloalkylamines, 186
β-Cyclodextrin, 16
Cyclodextrin 6-monoimidazole, 21
Cyclohexanol, 258
Cyclohexanone, 258
Cysteine, 112

D

N-Dealkylation reactions, 185
Deamination, 257, 352
Dehydronation, 421
5'-Deoxyadenosine, 67
5'-(2'-Deoxy)-adenosylcobalamin, 214
Deoxyribonucleic acid, cleavage of, 3, 11
Deprotonation, 255, 286, 335
1-Deuterio-1-butylamine, 353
Diazido-4-methoxyphenylmethane, 342
Dichloromethane, 309

5'-(2',3'-Dideoxy)-adenosyl-cobalamin, 215
Dielectric constants, 26
Diels–Alder reaction, 263
Diethylamine, 380
Diethyl azodicarboxylate, 264
Dihydrofolate reductase, 164
1,3-Dihydroxybenzene, 405
cis-1,2-Dihydroxy-1,2-dihydrobenzene, 131
3,4-Dihydroxyphenylacetaldehyde, 91
1,1-Dilithiocyclopropane, 289
3,3-Dilithio-1,2-diboracyclopropane, 289
1,3-Dimethoxybenzene, 310, 405
3,5-Dimethoxyphenol, 405
1,8-bis(Dimethylamino)naphthalene, 404
N,*N*-Dimethylanilines, 186
5,5-Dimethyl-6,7-dimethyl-2-pivaloyl-5,6,7,8-tetrahydro-pteridinium tetrafluoroborate, 62
Dimethyldi(trideuteriomethyl)ethene, 265
2,5-Dimethylfuran, 411
Dimethyl glutarate, 144
Dimethyl 2-hydroxyglutarate *O*-*p*-toluenesulfonate, 144
2,5-Dimethylindole, 406
4,6-Dinitrobenzofuroxan, 399
Dinitrogen trioxide, 322
Diols, 430
α,ω-Diphenylalkanes, 240
1,2-Diphenylethane, 246
Diphenylsulfone, 138
Diphosphomevalonate decarboxylase, 35
Dipyrromethane cofactor, 197
Displacement, nucleophilic, 60
Dissociation, collisionally activated, 237
Dithiothreitol, 60
DNA, 3, 11, 21, 94, 150, 161, 223
Dopamine, 91

E

Eicosanoids, 185
Electrochemical reduction, 74
Electrolysis, 74, 78
Electron, impact, 236
    microscopy, 95
    transfer, 68, 185
Electron spin resonance (ESR), 73, 435
Electrophilic attack, 323, 406
Electrophoresis, gel, 22
Elimination, base-catalysed, 402
    reactions, 334
Enantioselectivity, 81
Energy, vibrational, 280
Enolisation, 324, 401
Enzyme, active sites, 51
    models, 223
    reactions, 38
Enzymes, 3, 16, 25, 58, 89, 149, 161
    mammalian liver, 130
    proteolytic, 110
Enzymic catalysis, 123
Enzymology, low-temperature, 26
Epoxidation, 432
Equilibrium, constants, 51, 325, 334
    structures, 278
Escherichia coli, 58, 63, 70, 141, 150, 197
Eschscholtzia californica, 91, 99
Ethoxybenzene, 138
2,2-bis(Ethoxycarbonyl)-1-bromopropane, 76
2-Ethoxyethyl nitrite, 380
Ethyl thioacetate, 335
Ethyl 3-bromo-2-methoxy-2-phenylpropionate, 81
Ethylenediaminetetra-acetic acid (EDTA), 329

F

[4.4.4.4]Fenestrane, 289
Fischer–Hepp rearrangement, 320, 384
Flash photolysis, 301
Flavodoxin, 68
Fluoronitropyrimethamine, 177
Formaldehyde, 184
Formaldimine, 284
Formation, heats of, 284
Formic acid, 358
Frequency, microwave transition, 282
Friedel–Crafts alkylation, 305
Fumaria capreolata, 106
2-Furaldehyde, 33
Fusobacterium nucleatum, 140

G

Galacturonic acid, 33
Gas chromatography (GC), 238, 353
Gas–liquid chromatography (GLC), 80
Gas-phase reactions, 235
Gel electrophoresis, 22
Glucopyranoside hydrolysis, 439
Glucuronic acid, 33
Glutamate mutase, 73, 84
Glyceryl trinitrate, 327
Glycoside hydrolysis, 437
Gold chloride, 63
Grunwald–Winstein coefficient, 372

H

Haemoglobin, 191
Haloenzymes, 73
Halogenation, 324
Heat of formation, 284
Heptamethyl cobyrinate perchlorate, 74
Heterolytic reactions, 263
Hexafluoroacetylacetone, 324
1,1,1,3,3,3-Hexafluoroisopropyl alcohol, 306
Hexaplane, 294
High performance liquid chromatography (HPLC), 9, 66, 70, 81, 132,

Subject Index

141, 344
Histidine, 112, 168
Hofmann elimination, 36
Homocysteine, 58
Homolytic reactions, 263
Horse liver ADH, 40, 42
Hydration, 124
Hydrogenation, catalytic, 26
Hydrogen peroxide, 429
Hydrogen tunneling, 38
Hydrolysis, alkaline, 124
    base-catalysed, 334
    of glucopyranosides, 439
    of glycosides, 437
Hydrolytic catalysts, 223
Hydron shifts, intraannular, 245
    thermal, 245
β-Hydroxybromoalkanes, 220
4-R-Hydroxychroman, 138
3-Hydroxy-2,3-dihydrobenzofuran, 135
3-Hydroxy-2,3-dihydrobenzo[b]thiophene, 135
1-Hydroxy-1,2-dihydronaphthalene, 136
1-Hydroxy-1,4-dihydronaphthalene, 136
2-Hydroxy-1,2-dihydronaphthalene, 138
4-Hydroxyphenylacetaldehyde, 92
2-Hydroxythiachromene, 138
4-Hydroxythiachromene, 138
Hypsochromic shift, 314

I

Imidazole buffer, 3, 9
1-Indanol, 135
Indan-2-yl arensulfonates, 368
Indoles, 374, 409
Infrared-photon emission, 248
Infrared spectrometry, 212, 317
Intraannular hydron shifts, 245
Ion chemistry, 235
Ionization, 335

    energy, 293
    potential, 380
Ion-molecule pairs, 415
Ion pairs, 415
Ion transfer reaction, 342
Isodurene, 253
Isomerization, 86, 257, 425
Isoparametric phenomenon, 361
Isotope effects, 40, 54, 358, 366, 375, 418, 437
Isotopic shift, 336
    labelling, 162, 238
Itaconic acid, 30

J

Jasmonic acid, 102
Jatrorrhizine, 90

K

Ketonisation, 324
Kinetic solvent isotope effect (KSIE), 369
Knorr reaction, 196

L

Labelling, isotopic, 162
β-Lactamase, 155
Larvae, blowfly, 35
Lead, 33
Ligands, micellar amphiphilic, 224
Lithium, 293
Lithium perchlorate, 273
L-Lyxose, 34

M

Macarpine, 102
Macleaya microcarpa, 92, 100
Magnesium, 268
Mandelic acid, 341
Mass analyzed ion kinetic energy (MIKE) spectrometry, 237

Mass spectrometry, 198, 212, 235, 394
Mechanisms, bioorganic, 3
Membranes, bilayer, 81
β-Mercaptoethanol, 61
Mesitylene, 256
Metalloenzymes, 73
Methanol, 284, 315, 400
Methanolysis, 368
Methionine synthase, 58, 63, 68
Methotrexate, 164
4-Methoxybenzaldehyde, 342
p-Methoxybenzyl alcohol, 53
p-Methoxybenzylamine, 49
2-Methoxyisopropylideneindan, 425
Methoxymethanol, 442
3-(1-Methoxy-1-methylethyl)indene, 425
2-Methoxytetrahydropyran, 442
Methyl 2-acetobenzoate, 127
Methyl 8-acetyl-1-naphthoate, 127
Methylamine, 284
N-Methylaniline, 329
3-Methyl-3-butenyl diphosphate, 30
Methylcobalamin, 59, 210
(R,S)-N-Methylcoclaurine-3'-hydrolase, 91
1-Methylcyclopentanol, 258
2-Methyl-1,3-dimethoxybenzene, 310
Methyl ethyl ketone, 324
Methyl α-glucopyranoside, 442
N-Methylindole, 409
Methylmalonyl-CoA mutase, 63, 73, 83
N-Methyl-N-nitrosoaniline, 384
N-Methyl-N-nitroso-p-toluenesulfonamide, 377
Methylococcus capsulatus, 31
(S)-cis-N-Methylstylopine, 100
Methyltetrahydrofolate, 58
(S)-cis-N-Methyltetrahydroprotoberberine, 100
Mevalonic acid 5-diphosphate, 30
Microscopy, electron, 95
Microwave spectral analysis, 279

Microwave transition frequency, 282
Mitochondria, of rat heart, 35
Modelling, computer, 22
  molecular, 115
Models, enzyme, 223
Molecular modelling, 115
Molecular orbital theory, 112, 278, 438
Molecular structure, 281
Molybdenum(VI) polyoxoperoxo complexes, 429
Monoamine oxidase B, 48
Morinda citrifolia, 90
Morphinans, 90
Morphine, 89
Morpholine buffer, 3, 11
Multiphoton absorption, 236
Mutagenesis, site specific, 43, 56, 111, 206
Mutagens, 351

N

1,8-Naphthalene template, 124
1-Naphthol, 136
Neurine, 35
Nickel, 33
Nitramine rearrangement, 320
Nitration, 321
  aromatic, 247
Nitric oxide, 320
Nitrobenzene, 251
p-Nitrophenyl picolinate, 224
N-Nitrosamide, 352
Nitrosation, 320, 374
Nitrosothiols, 327
Nitrous acid, 321, 375, 402
Nitrous oxide, 69
Nitrovasodilators, 327
Non-interactive phenomenon, 361
(S)-Norcoclaurine synthase, 91
14-Norlanosterol, 31
Norlaudanosoline, 90
Nuclear magnetic resonance (NMR), 92, 61, 103, 131, 161, 196, 212,

Subject Index 451

257, 270, 317, 335, 353, 394, 434
Nucleophiles, ambident, 374
Nucleophilic attack, 61, 112, 270, 400, 417, 437
Nucleophilic displacement, 60

O

Octaplane, 292
N-Octylamine, 352
N-Octyltriazene, 352
Organocobalamins, 209
Oxidation, enzyme catalysed, 130
Oxygen transfer mechanisms, 429
Oxygen, 323

P

[2.2.2.2]Paddlane, 290
Papaveraceae, 91
Papaver somniferum, 99
Paraquat, 388
Pavines, 90
Penicillin, 155
Perchloric acid, 346
Pericyclic reactions, 263
Peroxidases, 187
Phase transfer techniques, 429
4,5-Phenanthrene template, 124
Phenol, 405
Phenolic coupling, 103
Phenylalanine, 168
1-Phenylethyl chloride, 369
N-Phenyltriazolinedione, 265
Phenyltrimethylsilane, 240, 257
Phosphoranes, 8, 9, 10
Phosphoric acid, 225
Phosphorus, 9
Photoionization, 236
Photolysis, 74
Photolytic reactions, 263
Photon emission, 248
Photoprotonation, 306
Phthalideisoquinolines, 90
Picornaviruses, 110

Pictet–Spengler condensation, 91
Platinum chloride, 63
Pneumus boldus, 91
Polar effects, 296
Porphobilinogen deaminase, 196
Preuroporphyrinogen, 196
[1.1.1]Propellane, 281
Propionobacterium shermanii, 63, 212
Proteins, 152, 223
Proteins, P450, 185
Proteolytic enzymes, 110
Protoberberines, 90
Protodesilylation, photochemical, 313
Protonation, 62, 114, 286, 375, 406
    reversible, 14
Proton Sponge, 404
Pseudomonas, 28
    putida, 131
Pyridine, 242, 250
Pyrimethamine, 164
Pyruvate-formate lyase, 66

Q

Quantum mechanics, 438
Quinoline, 32
3-Quinuclidinone, 335

R

Racemisation, 334
Rat heart, mitochondria, 35
Rate constants, 12, 39, 54, 226, 241, 301, 321, 334, 352, 357, 382, 400, 420, 431
Rate–equilibrium correlations, 347
Reaction rate, 375
Reactions, bioorganic, 3
    ring expansion, 85
Rearrangement reactions, 415
Rearrangement, Wagner–Meerwein, 26
Reduction, electrochemical, 74
Relationship, Swain–Schaad, 39

Residues, amino acid, 168, 200
(S)-Reticuline, 90
Rhodobacter sphaeroides, 198
Rhoeadines, 90
Ribonuclease A, 3, 4
Ribonucleic acid (RNA), 110
Ribose, 33
Ribulose, 33
Ring-expansion reactions, 85
Rotational isomerism, 444
[n]Rotaxanes, 387

S

Salutaridine, 103
Sanguinaria canadensis, 99
Sanguinarine, 102
Schenck rearrangement, 266
Schiff bases, 34
(S)-Scoulerine, 92
Self-assembly pathways, 387
Sequences, amino acid, 63
Serine, 112
Shift, hypsochromic, 314
Silica gel chromatography, 394
Silylation, aromatic, 254
(R)-Sinactine, 98
Site directed mutagenesis, 111, 114, 206
Site specific mutagenesis, 43, 56
Slipping approach, 393
Smith rearrangement, 266
S_N1 reactions, 417, 437
S_N2 reactions, 65, 143, 352, 365, 437
Sodium, 293
Sodium cyanoborohydride, 144
Sodium molybdate, 430
Sodium nitrite, 321
Sodium nitroprusside, 327
Sodium tungstate, 430
Solvolysis, 257, 303, 343, 365, 421
Solvolytic elimination reactions, 415
Spectral analysis, microwave, 279
Spectra, vibrational, 282
Spectrometry, atomic absorption, 329
  infrared, 212, 317
  mass, 198, 212, 235, 394
  mass analyzed ion kinetic energy, 237
Spectroscopy, UV/visible, 212, 302
3-Stannylallyl hydroperoxide, 269
Steroids, 185
Sterol biosynthesis, 31
Streptavidin, 153
(S)-Stylopine, 99
Styrene, 307
Substitution reactions, 415
Sulfoxides, 430
Sulfur, 323
Surfactants, 231
Swain–Schaad relationship, 39

T

Tautomerism, ring-chain, 124
Taylor expansion, 361
(S)-Tetrahydrocolumbamine, 94
Tetrahydrofuran, 250
Tetrahydrofurfuryl-(S)-camphor-sulfonates, 217
(S)-Tetrahydropalmatine, 97
(S)-Tetrahydroprotoberberine, 95
Thalictrum bulgaricum, 102
Thermal activation, 52, 56
Thermal hydron shifts, 245
Thermodynamic trap, 393
Thiachromone. 138
Thiacoumarin, 138
Thiamine, 31
Thioethers, 430
Thiomorpholine, 379
Thionitrites, 327
Thioproline, 377
Threading methodology, 390
Titration, EPR, 67
Toluene, 253
Tolyltrimethylsilanes, 241
Transition state, cyclic, 264
  pericyclic, 265
Transition structures, 278

Subject Index

Tributyltin hydride, 76
Triethylamine, 242
Trifluoroacetylacetone, 324
Trifluroethanol, 344
anti-Trifluoroethyldiazoate, 355
1,3,5-Trihydroxybenzene, 405
Trimethoprim, 164
Trimethyl lead, 63
5,6,7-Trimethyltetrahydropterin, 62
1,3,5-Trinitrobenzene, 399
Tropinone, 25
Trypsin, 64
Tungsten(VI) polyoxoperoxo complexes, 429
Tyrosine, 91, 168

U

3',5"-Uridyluridine, 9
Uroporphyrinogen, 30
Uroporphyrinogen III, 196

V

Vanadium trichloride, 86

Vasodilation, 326
Vibrational energy, 280
Vitamin B_{12}, 73, 209
Voltammetry, cyclic, 74

W

Wagner–Meerwein rearrangement, 26

X

X-Ray, analysis, 434
 crystallography, 25, 63, 132
 structures, 40, 117, 200, 204, 209, 212
Xylene, 256

Y

Yeast ADH, 40, 42

Z

Zinc, 43, 268

Also published by The Royal Society of Chemistry...

Ref No 1231

Organic Reactivity: Physical and Biological Aspects
Edited by Bernard T. Golding, *University of Newcastle upon Tyne*
Roger J. Griffin, *University of Newcastle upon Tyne*
Howard Maskill, *University of Newcastle upon Tyne*

Special Publication No. 148	Hardcover	xvi + 454 pages
ISBN 085404 710 7	1995	Price £69.50

Ref No 1230

Seminars in Organic Synthesis Volume 4

Softcover	560 pages	ISBN 88 86208 06 5	1994	Price £49.00

Ref No 1129

Excipients and Delivery Systems for Pharmaceutical Formulations
Edited by D. R. Karsa, *Akcros Chemicals*
R. A. Stephenson, *Chemical Consultant*

Special Publication No. 161	Hardcover	viii + 192 pages
ISBN 0 85404 715 8	1995	Price £39.50

Ref No 1051

Medicinal Chemistry: Principles and Practice
Edited by Frank D. King, *SmithKline Beecham Pharmaceuticals, Harlow, UK*

Softcover	xxiv + 314 pages	ISBN 0 85186 494 5	1994	Price £39.50

Ref No 1101

Current Topics in the Chemistry of Boron
Edited by George W. Kabalka, *The University of Tennessee, USA*

Special Publication No. 143	Hardcover	xiv + 406 pages
ISBN 0 85186 535 6	1994	Price £59.50

Ref No 1011

Advances in the Chemistry of Insect Control III
Edited by G. G. Briggs, *AgrEvo UK Ltd*

Special Publication No. 147	Hardcover	vi + 250 pages
ISBN 0 85186 992 0	1994	Price £47.50

Prices subject to change without notice.

To order please contact:
Turpin Distribution Services Ltd., Blackhorse Road, Letchworth, Herts SG6 1HN, UK.
Tel: +44 (0) 1462 672555. Fax: +44 (0) 1462 480947. Telex: 825372 TURPIN G.

RSC Members should order from Membership Administration at our Cambridge address.

For further information please contact:
Sales and Promotion Department, Royal Society of Chemistry,
Thomas Graham House, Science Park, Milton Road, Cambridge CB4 4WF, UK.
Tel: +44 (0) 1223 420066. Fax: +44 (0) 1223 423623.
E-mail (Internet): RSC1@RSC.ORG.

THE ROYAL SOCIETY OF CHEMISTRY

Information Services

Also published by The Royal Society of Chemistry...

Ref No 848
Molecular Recognition: Chemical and Biochemical Problems II
Edited by S.M. Roberts, *University of Exeter*
Special Publication No. 111   Hardcover   viii + 200 pages
ISBN 0 85186 226 8   1992   Price £42.50

Ref No 245
Chemistry of Biomolecules: An Introduction
By R.J. Simmonds, *The University College of Wales, Aberystwyth*
Softcover   xiv + 276 pages   ISBN 0 85186 883 5   1992   Price £18.50

Ref No 252
100 Modern Reagents
1st Reprint 1992
Edited by N.S. Simpkins (Consulting Editor), *University of Nottingham*
Flexicover   210 pages   ISBN 0 85186 893 2   1989   Price £24.00

Ref No 1034
Further Advances in Chemical Information
Edited by H. Collier, *Infonortics Ltd., Calne, Wiltshire*
Special Publication No. 142   Hardcover   viii + 194 pages
ISBN 0 85186 545 3   1994   Price £45.00

Ref No 1046
The Laboratory Environment
Edited by Rupert Purchase, *Environment Group, The Royal Society of Chemistry*
Special Publication No. 136   Hardcover   x + 258 pages
ISBN 0 85186 605 0   1994   Price £49.50

Ref No 197
Hazards in the Chemical Laboratory - 5th Edition
Edited by S.G. Luxon, *Health and Safety Consultant*
PVC Flexi Cover   xx + 676 pages   ISBN 0 85186 229 2   1992   Price £45.00

Prices subject to change without notice.

To order please contact:
Turpin Distribution Services Ltd., Blackhorse Road, Letchworth, Herts SG6 1HN, UK.
Tel: +44 (0) 1462 672555. Fax: +44 (0) 1462 480947. Telex: 825372 TURPIN G.

RSC Members should order from Membership Administration at our Cambridge address.

For further information please contact:
Sales and Promotion Department, Royal Society of Chemistry,
Thomas Graham House, Science Park, Milton Road, Cambridge CB4 4WF, UK.
Tel: +44 (0) 1223 420066. Fax: +44 (0) 1223 423623.
E-mail (Internet): RSC1@RSC.ORG.

THE ROYAL SOCIETY OF CHEMISTRY

Information Services

Also published by The Royal Society of Chemistry ...

Ref No 243
Carbohydrate Chemistry - Volume 26
Senior Reporter R.J. Ferrier, *Victoria University of Wellington, New Zealand*
Specialist Periodical Reports   Hardcover   xvi + 356 pages
ISBN 0 85186 991 2   1994   Price £122.50
Special package price (Vols. 1-11, 13-19, 21-22, 24-26) £1374.00

Ref No 258
Organophosphorus Chemistry - Volume 25
Senior Reporters D.W. Allen, *Sheffield Hallam University*
B.J. Walker, *The Queen's University of Belfast*
Specialist Periodical Reports   Hardcover   xii + 338 pages   ISBN 0 85186 390 6
1994   Price £15   Special package price (Vols. 1-25) £1701.00

Ref No 259
Photochemistry Volume 25
Senior Reporters D. Bryce-Smith, *University of Reading*
A. Gilbert, *University of Reading*
Specialist Periodical Reports   Hardcover   xxvi + 594 pages   ISBN 0 85186 481 3
1994   Price £160.00   Special package price (Vols. 1-10, 12-25) £1940.00

Ref No 251
General and Synthetic Methods - Volume 16
Senior Reporter G. Pattenden, *FRS, University of Nottingham*
Specialist Periodical Reports   Hardcover   xvi + 600 pages   ISBN 0 85186 834 7
1994   Price £165.00   Special package price (Vols 3, 5-6, 8, 9, 12-16) £1185.00

Ref No 987
Chemistry and Light
By Paul Suppan, *University of Fribourg, Switzerland*
Softcover   xiv + 296 pages   ISBN 0 85186 814 2   1994   Price £19.50

Ref No 792
Current Trends in Sonochemistry
Edited by G.J. Price, *University of Bath*
Special Publication No. 116   Hardcover   viii + 184 pages   ISBN 0 85186 365 5   1992   Price £42.50

Ref No 238
Amino Acids, Peptides and Proteins - Volume 25
Senior Reporter J.S. Davies, *University College of Swansea*
Specialist Periodical Report   Hardcover   xiv + 410 pages   ISBN 0 85186 234 9   1994
Price £110.00   Special package price (Vols 1-9, 11-14, 16-25) £1350.00

Prices subject to change without notice.

To order please contact:
Turpin Distribution Services Ltd., Blackhorse Road, Letchworth, Herts SG6 1HN, UK.
Tel: +44 (0) 1462 672555. Fax: +44 (0) 1462 480947. Telex: 825372 TURPIN G.

RSC Members should order from Membership Administration at our Cambridge address.

For further information please contact:
Sales and Promotion Department, Royal Society of Chemistry,
Thomas Graham House, Science Park, Milton Road, Cambridge CB4 4WF, UK.
Tel: +44 (0) 1223 420066. Fax: +44 (0) 1223 423623.
E-mail (Internet): RSC1@RSC.ORG.

THE ROYAL SOCIETY OF CHEMISTRY

Information Services

Also published by The Royal Society of Chemistry . . .

Monographs in Supramolecular Chemistry

Series Editor: J. Fraser Stoddart, FRS, *University of Birmingham, UK*

Ref No 1130
Membranes and Molecular Assemblies: The Synkinetic Approach
By Jürgen-Hinrich Fuhrhop, *Freie Universität Berlin, Germany*
Jürgen Köning, *Freie Universität Berlin, Germany*
Hardcover   xiv + 228 pages   ISBN 0 85186 732 4   1994   Price £69.50

Ref No 1098
Container Molecules and Their Guests
By Donald J. Cram, *University of California, Los Angeles, USA*
Jane M. Cram, *University of California, Los Angeles, USA*
Hardcover   xiv + 224 pages   ISBN 0 85186 972 6   1994   Price £49.50

Ref No 246
Cyclophanes
By François N. Diederich, *University of California, Los Angeles, USA*
Hardcover   xvi + 314 pages   ISBN 0 85186 966 1   1991   Price £57.50
Softcover   xvi + 314 pages   ISBN 0 85186 405 8   1994   Price £25.00

Ref No 247
Crown Ethers and Cryptands
By George W. Gokel, *University of Miami, Florida, USA*
Hardcover   xiv + 190 pages   ISBN 0 85186 996 3   1991   Price £52.50
Softcover   xiv + 190 pages   ISBN 0 85186 704 9   1994   Price £22.50

Ref No 242
Calixarenes
By C. David Gutsche, *Washington University, St. Louis, USA*
Hardcover   xii + 224 pages   ISBN 0 85186 916 5   1989   Price £39.50
Softcover   xii + 224 pages   ISBN 0 85186 385 X   1992   Price £17.50

Prices subject to change without notice.

To order please contact:
Turpin Distribution Services Ltd., Blackhorse Road, Letchworth, Herts SG6 1HN, UK.
Tel: +44 (0) 1462 672555. Fax: +44 (0) 1462 480947. Telex: 825372 TURPIN G.

RSC Members should order from Membership Administration at our Cambridge address.

For further information please contact:
Sales and Promotion Department, Royal Society of Chemistry,
Thomas Graham House, Science Park, Milton Road, Cambridge CB4 4WF, UK.
Tel: +44 (0) 1223 420066. Fax: +44 (0) 1223 423623.
E-mail (Internet): RSC1@RSC.ORG.

THE ROYAL SOCIETY OF CHEMISTRY

Information Services